AF388928

Buildings and Structures under Extreme Loads

Buildings and Structures under Extreme Loads

Editors

Chiara Bedon
Flavio Stochino
Daniel Honfi

MDPI • Basel • Beijing • Wuhan • Barcelona • Belgrade • Manchester • Tokyo • Cluj • Tianjin

Editors
Chiara Bedon
University of Trieste
Italy

Flavio Stochino
University of Cagliari
Italy

Daniel Honfi
RISE Research Institutes of Sweden
Sweden

Editorial Office
MDPI
St. Alban-Anlage 66
4052 Basel, Switzerland

This is a reprint of articles from the Special Issue published online in the open access journal *Applied Sciences* (ISSN 2076-3417) (available at: https://www.mdpi.com/journal/applsci/special_issues/Buildings_Structures_Extreme_Loads).

For citation purposes, cite each article independently as indicated on the article page online and as indicated below:

LastName, A.A.; LastName, B.B.; LastName, C.C. Article Title. *Journal Name* **Year**, *Article Number*, Page Range.

ISBN 978-3-03943-569-2 (Hbk)
ISBN 978-3-03943-570-8 (PDF)

Cover image courtesy of Joey Banks.

Contents

About the Editors

Chiara Bedon (1983), Assistant Professor. M.Sc. in Civil Engineering and Ph.D. in Structural Engineering (University of Trieste, Italy). Her research activity includes the analysis of structural materials and systems under extreme design loads, with a focus on buckling-related phenomena, blast, fire, earthquakes. She has been involved in numerous European projects and networks (JRC-ERNCIP, NATO-SPS, COST, etc.) since 2009.

Flavio Stochino (1985), Assistant Professor. M.Sc. in Civil Engineering and Ph.D. in Structural Engineering (University of Cagliari, Italy). His scientific research deals with extreme loads on RC structures with a special focus on blast/impulsive loading and fire design of structures. He has extensive experience regarding reliability and seismic analysis of existing structures, isogeometrical analysis, hybrid/mixed finite elements, laminate mechanics, structural optimization, and structure sustainability.

Daniel Honfi (1977), Senior Researcher. M.Sc. in Civil Engineering (Budapest University of Technology and Economics), Ph.D. in Structural Engineering (Lund University). His research topics include structural serviceability and robustness, resilience of infrastructure systems, engineering decision making, and assessment of existing structures.

Editorial

Special Issue on "Buildings and Structures under Extreme Loads"

Chiara Bedon [1,*]**, Flavio Stochino** [2] **and Daniel Honfi** [3]

[1] Department of Engineering and Architecture, University of Trieste, 34127 Trieste, Italy
[2] Department of Civil Environmental Engineering and Architecture, University of Cagliari, 09123 Cagliari, Italy; fstochino@unica.it
[3] RISE Research Institutes of Sweden, 41756 Göteborg, Sweden; daniel.honfi@ri.se
* Correspondence: chiara.bedon@dia.units.it

Received: 17 July 2020; Accepted: 5 August 2020; Published: 15 August 2020

1. Introduction

Exceptional loads on buildings and structures may have different causes, including high-strain dynamic effects due to natural hazards, man-made attacks, and accidents, as well as extreme operational conditions (severe temperature variations, humidity, etc.). All these aspects can be critical for specific structural typologies and/or materials that are particularly sensitive to unfavorable external conditions. In this regard, dedicated and refined methods are required for their design, analysis, and maintenance under the expected lifetime. However, major challenges are usually related to the structural typology and materials object of study, with respect to the key features of the imposed design loads. Further issues can be derived from the need for the mitigation of adverse effects or retrofit of existing structures, as well as from the optimal and safe design of innovative materials/systems. Finally, in some cases, no appropriate design recommendations are currently available in support of practitioners, and thus experimental investigations (both on-site or on laboratory prototypes) can have a key role within the overall structural design and assessment process. This Special Issue presents 19 original research studies and two review papers dealing with the structural performance of buildings and structures under exceptional loads, and can represent a useful answer to the above-mentioned problems.

2. Contents

A first set of papers reports on earthquake structural design of structures and buildings [1–5]. Various kinds of structures have been considered under the effects of seismic loads, including steel frames [1], liquid storage tanks [2], and an experimental prototype of atrium-style underground metro station [3], but also existing masonry structures [4] or new timber buildings [5], presenting a perspective review on their seismic design. Among others, a extreme natural event is certainly represented by windstorms. In this Special Issue, wind load modelling and design is mainly addressed by [6–8], while [9] describes the results of a visual test carried out on a suction caisson that support offshore wind turbines. Finally, the last natural hazard analyzed in the Special Issue is snowdrift. Actually, the effects of snowdrift and snow loads in cold regions have been investigated by [10] and [11], with the proposal of a novel calculation approach and a case-study application, respectively.

The knowledge of material properties and characteristics, as known, represents the first influencing parameter for the load-bearing performance assessment of a given structure. In this regard, the knowledge on the topic has been improved by two interesting research contributions focused on composite concrete-steel shear walls [12] and structural glass members [13], respectively, with the support of laboratory/on-site experiments and numerical analyses.

Another interesting group of papers dealing with soil properties and structures–soil interaction phenomena further extends the research fields covered in this Special Issue. In particular, [14] deals with the determination of Young modulus in bored piles, while [15] presents an investigation on the horizontal axis deviation of a small radius Tunnel Boring Machine (TBM). In this context, it is important to also mention the study in [16], and reporting on the friction resistance for slurry pipe jacking. Finally, an interesting analysis on the effects of derailment and post-derailment of trains is presented in [17], with the support of full-scale testing.

In conclusion, it is known that both man-made attacks and accidents can yield to explosions and fire loads that could push the constructional materials, and thus the structures, to their capacity limits. Blast loads analyses, in this regard, are reported in [18–20], while fire effects on a tunnel structure are analyzed in [21].

Acknowledgments: This Special Issue would not be possible without the contributions of various talented authors, hardworking and professional reviewers, and dedicated editorial team members of the *Applied Sciences* journal. We would like to take this opportunity to record our sincere gratefulness to all the involved scientists, both authors and reviewers, for their valuable contribution to this collection. Finally, we place on record our gratitude to the editorial team of *Applied Sciences*, and special thanks to Felicia Zhang, Assistant Managing Editor for *Applied Sciences*.

Conflicts of Interest: The authors declare no conflict of interest.

References

1. Jeon, S.-H.; Park, J.-H.; Ha, T.-W. Seismic Design of Steel Moment-Resisting Frames with Damping Systems in Accordance with KBC. *Appl. Sci.* **2019**, *9*, 2317. [CrossRef]
2. Jing, W.; Feng, H.; Cheng, X. Dynamic Responses of Liquid Storage Tanks Caused by Wind and Earthquake in Special Environment. *Appl. Sci.* **2019**, *9*, 2376. [CrossRef]
3. Zhang, Z.; Bilotta, E.; Yuan, Y.; Yu, H.-T.; Zhao, H. Experimental Assessment of the Effect of Vertical Earthquake Motion on Underground Metro Station. *Appl. Sci.* **2019**, *9*, 5182. [CrossRef]
4. Stepinac, M.; Kišiček, T.; Renić, T.; Hafner, I.; Bedon, C. Methods for the Assessment of Critical Properties in Existing Masonry Structures under Seismic Loads—The ARES Project. *Appl. Sci.* **2020**, *10*, 1576. [CrossRef]
5. Stepinac, M.; Šušteršič, I.; Gavrić, I.; Rajčić, V. Seismic Design of Timber Buildings: Highlighted Challenges and Future Trends. *Appl. Sci.* **2020**, *10*, 1380. [CrossRef]
6. Li, L.-X.; Zhou, Y.; Wang, H.; Zhou, H.J.; He, X.; Wu, T. An Analytical Framework for the Investigation of Tropical Cyclone Wind Characteristics over Different Measurement Conditions. *Appl. Sci.* **2019**, *9*, 5385. [CrossRef]
7. Chen, X.; Liu, Z.; Wang, X.; Chen, Z.; Xiao, H.; Zhou, J. Experimental and Numerical Investigation of Wind Characteristics over Mountainous Valley Bridge Site Considering Improved Boundary Transition Sections. *Appl. Sci.* **2020**, *10*, 751. [CrossRef]
8. Ke, S.; Zhu, P.; Xu, L.; Ge, Y. Evolution Mechanism of Wind Vibration Coefficient and Stability Performance during the Whole Construction Process for Super Large Cooling Towers. *Appl. Sci.* **2019**, *9*, 4202. [CrossRef]
9. Xie, L.; Ma, S.; Lin, T. The Seepage and Soil Plug Formation in Suction Caissons in Sand Using Visual Tests. *Appl. Sci.* **2020**, *10*, 566. [CrossRef]
10. Liu, M.; Zhang, Q.; Fan, F.; Shen, S. Modeling of the Snowdrift in Cold Regions: Introduction and Evaluation of a New Approach. *Appl. Sci.* **2019**, *9*, 3393. [CrossRef]
11. Zhang, G.; Zhang, Q.; Fan, F.; Shen, S. Research on Snow Load Characteristics on a Complex Long-Span Roof Based on Snow–Wind Tunnel Tests. *Appl. Sci.* **2019**, *9*, 4369. [CrossRef]
12. Jiang, D.; Xiao, C.; Chen, T.; Zhang, Y. Experimental Study of High-Strength Concrete-Steel Plate Composite Shear Walls. *Appl. Sci.* **2019**, *9*, 2820. [CrossRef]
13. Bedon, C. Issues on the Vibration Analysis of In-Service Laminated Glass Structures: Analytical, Experimental and Numerical Investigations on Delaminated Beams. *Appl. Sci.* **2019**, *9*, 3928. [CrossRef]
14. Moayedi, H.; Kalantar, B.; Abdullahi, M.M.; Rashid, A.S.A.; Bin Nazir, R.; Nguyen, H. Determination of Young Elasticity Modulus in Bored Piles Through the Global Strain Extensometer Sensors and Real-Time Monitoring Data. *Appl. Sci.* **2019**, *9*, 3060. [CrossRef]
15. Qiao, S.; Xu, P.; Liu, R.; Wang, G. Study on the Horizontal Axis Deviation of a Small Radius TBM Tunnel Based on Winkler Foundation Model. *Appl. Sci.* **2020**, *10*, 784. [CrossRef]

16. Ye, Y.; Peng, L.; Zhou, Y.; Yang, W.; Shi, C.; Lin, Y. Prediction of Friction Resistance for Slurry Pipe Jacking. *Appl. Sci.* **2019**, *10*, 207. [CrossRef]

17. Bae, H.-U.; Moon, J.; Lim, S.-J.; Park, J.-C.; Lim, N.-H. Full-Scale Train Derailment Testing and Analysis of Post-Derailment Behavior of Casting Bogie. *Appl. Sci.* **2019**, *10*, 59. [CrossRef]

18. Helal, M.; Huang, H.; Fathallah, E.; Wang, D.; Elshafey, M.M.; Ali, M.A.E.M. Numerical Analysis and Dynamic Response of Optimized Composite Cross Elliptical Pressure Hull Subject to Non-Contact Underwater Blast Loading. *Appl. Sci.* **2019**, *9*, 3489. [CrossRef]

19. Stochino, F.; Attoli, A.; Concu, G. Fragility Curves for RC Structure under Blast Load Considering the Influence of Seismic Demand. *Appl. Sci.* **2020**, *10*, 445. [CrossRef]

20. Yussof, M.M.; Silalahi, J.H.; Kamarudin, M.K.; Chen, P.-S.; Parke, G.A.R. Numerical Evaluation of Dynamic Responses of Steel Frame Structures with Different Types of Haunch Connection Under Blast Load. *Appl. Sci.* **2020**, *10*, 1815. [CrossRef]

21. Kim, S.; Shim, J.; Rhee, J.-Y.; Jung, D.; Park, C. Temperature Distribution Characteristics of Concrete during Fire Occurrence in a Tunnel. *Appl. Sci.* **2019**, *9*, 4740. [CrossRef]

 *applied sciences*

Article

Seismic Design of Steel Moment-Resisting Frames with Damping Systems in Accordance with KBC 2016

Seong-Ha JEON [1], Ji-Hun PARK [2,*] and Tae-Woong HA [1]

[1] Graduate Student in Department of Architecture, Incheon National University, Incheon 22012, Korea; wings616@inu.ac.kr (S.-H.J.); 201921129@inu.ac.kr (T.-W.H.)

[2] Professor in Division of Architecture and Urban Design, Incheon National University, Incheon 22012, Korea

* Correspondence: jhpark606@inu.ac.kr; Tel.: +82-32-835-8474

Received: 26 April 2019; Accepted: 31 May 2019; Published: 5 June 2019

Abstract: An efficient design procedure for building structures with damping systems is proposed using nonlinear response history analysis permitted in the revised Korean building code, KBC 2016. The goal of the proposed procedure is to design structures with damping systems complying with design requirements of KBC 2016 that do not specify a detailed design method. The proposed design procedure utilizes response reduction factor obtained by a limited number of nonlinear response history analyses of the seismic-force-resisting system with incremental damping ratio substituting damping devices. Design parameters of damping device are determined taking into account structural period change due to stiffness added by damping devices. Two design examples for three-story and six-story steel moment frames with metallic yielding dampers and viscoelastic dampers, respectively, shows that the proposed design procedure can produce design results complying with KBC 2016 without time-consuming iterative computation, predict seismic response accurately, and save structural material effectively.

Keywords: damping device; seismic design; design base shear; nonlinear response history analysis

---

1. Introduction

There were no seismic design provisions for the application of damping systems in the Korean Building Code (KBC) 2009, therefore, Korean engineers encountered many difficulties in the practical application of damping devices [1]. The KBC was revised in 2016 with the addition of design criteria for structures with damping systems [2]. The design provisions for structures with damping systems in KBC 2016 adopted only nonlinear response history procedure. In the case of ASCE 7, both equivalent lateral force procedure and response spectrum procedure are allowed. However, the use of those two procedures is restricted for strict conditions and nonlinear response history procedure is adopted major design procedure in ASCE 7-16 [3].

In spite of being adopted as a major design procedure, nonlinear response history analysis procedure requires much more computational efforts compared to linear analysis. It is difficult to design damping devices by trial and error using nonlinear response history analysis. Therefore, many design procedures adopt equivalent linearization technique to take into account the nonlinear characteristics of either damping device or building structure [4–10]. Some design procedures adopt nonlinear static analysis to take into account the inelastic behavior of the structure more directly [11,12], which utilizes the equivalent linearization technique to determine the performance point. Many equivalent linearization techniques for different nonlinear damping devices have been proposed [13–15]. Most of them utilize damping correction factors, which represent response reduction for a specified amount of damping ratio developed for linear or nonlinear systems [13,14,16,17]. However, those design procedures based on equivalent linearization technique have limitations in that they assume deformed shape based on elastic analysis and are difficult to identify localized nonlinear behavior of a structure, such as weak story mechanism and force or deformation demands on structural components and damping devices with sufficient accuracy for detailed design. Besides, proposed optimization procedures have been developed. Some of those procedures adopt stochastic analysis [18,19] or linear response history analysis [20]. The others make use of nonlinear response history for evaluation of objective function or boundary conditions [21–23]. However, the latter methodologies repeat nonlinear response history analysis for every iterative step during optimization and, as a result, are computationally demanding in the case of actual building structures with many degrees of freedom

This study is to propose and validate an efficient and systematic design procedure for inelastic multi-degree-of-freedom (MDOF) structures with damping systems complying with KBC 2016, which requires nonlinear response history analysis but does not provide methodology for detailed design. Two design examples for steel moment frames using nonlinear response history analysis are presented. The proposed design procedure makes use of nonlinear response history analysis of an MDOF structure directly to determine design parameters of damping devices and captures complex inelastic behavior of building structures more realistically. Differently from the existing optimization procedure, the design procedure performs only a small number of nonlinear response history analyses, which is usually three to five times. As a result, the proposed design procedure requires only a small amount of computational efforts compared to the existing optimization-based design procedures that perform nonlinear response history analysis repeatedly until convergence. The proposed design procedure can be implemented using commercial structural analysis software and can be applied to design code except for KBC 2016.

2. KBC2016 Seismic Design Provisions for Structures with Damping Systems [2]

2.1. Damping Systems

The damping system is intended to reduce the seismic demand to a structure and refers to a subsystem that includes both damping devices and structural elements that transmit forces from the damping devices to seismic-force-resisting systems or foundations of the structure. The damping device is a structural element that dissipates energy by relative motion between two ends of the device and includes all the elements such as pins, bolts, gusset plates, braces, etc. necessary to install the damping device. The damping device may be installed in a separate structure out of the seismic-force-resisting system or in the seismic-force-resisting system. Figure 1 shows examples of configurations of damping devices and damping systems connected to a seismic-force-resisting system.

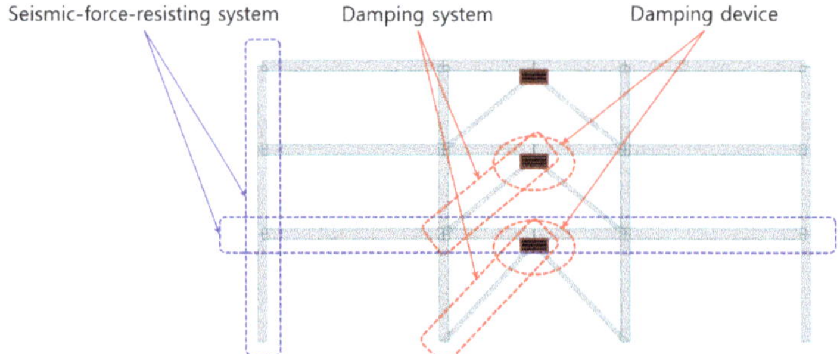

Figure 1. Damping system (DS) and seismic-force-resisting system (SFRS) configurations.

A damping device is classified into a velocity-dependent damping device whose force response depends on the relative velocity between the two ends of the device and a displacement-dependent damping device of which force response is determined by the relative displacement between the two ends of the devices. A mathematical model of the velocity-dependent damping system shall include the velocity coefficient corresponding to the test data. Displacement-response characteristics of the displacement-dependent damping device shall be modeled considering the dependence of seismic force response on the frequency, amplitude, and duration of ground motion clearly.

The components constituting the damping system shall be designed so that the damping device works normally without interruption. Thus, structural elements in the damping system are designed to remain elastic when subjected to design earthquake including forces transmitted from the damping device. The forces from the damping devices shall not be calibrated by the intensity reduction factor or the response correction factor. Moreover, the damping device shall be designed so as not to break when subjected to the maximum considered earthquake.

2.2. Seismic-Force-Resisting System

A building structure to which damping systems are applied shall have a seismic-force-resisting system defined in the KBC in each direction. Table 1 shows design factors of steel moment-resisting frame systems, which are appropriate to install damping devices due to relatively low stiffness and used in design examples of this study. At the initial stage, the seismic-force-resisting system of a building structure with damping systems are designed in order to resist the minimum base shear V_{min} independently. V_{min} is calculated by Equations (1) and (2).

$$V_{min} = \eta V \tag{1}$$

$$\eta \geq 0.75 \tag{2}$$

where V is the design base shear calculated by equivalent lateral force procedure and η is the expected damping correction factor representing the degree of seismic force reduction acting on the structure obtained by damping systems. The expected damping correction factor η shall be validated through nonlinear response history analysis of the seismic-force-resisting systems combined with damping systems as described in the next section.

Table 1. Design factors for steel moment-resisting frame systems.

Steel Moment-Resisting Frame Systems	Design Coefficients		
	Response Modification Factor R	Overstrength Factor Ω_o	Deflection Amplification Factor C_d
Special	8	3	5.5
Intermediate	4.5	3	4
Ordinary	3.5	3	3

2.3. Damping Performance

The expected damping correction factor η applied to the minimum base shear V_{min} of the seismic-force-resisting system shall be higher than or equal to the actual damping correction factor η_h shown in Equation (3).

$$\eta_h = \frac{V_h}{V_{he}} \tag{3}$$

where V_h and V_{he} are the base shear calculated from the analysis of the structure with damping systems and from the analysis of the structure in which velocity-dependent components of damping devices are removed and displacement-dependent components of the damping devices are substituted into the effective stiffness, respectively.

The effective stiffness for the displacement-dependent component of the damping device is calculated based on the peak displacement and corresponding force of the damping device obtained from the analysis to calculate V_h as follows.

$$k_{eff} = \frac{|F^+| + |F^-|}{|\Delta^+| + |\Delta^-|}$$

(4)

where Δ^+ and Δ^- is the peak displacement of each damping device in positive and negative directions, respectively, and F^+ and F^- are corresponding forces, respectively.

3. Damping System Design Procedure

An efficient and systematic seismic design procedure for damping systems using nonlinear response history analysis is proposed in this section. The goal of the proposed procedure is to design structures with damping systems complying with the design requirements of KBC 2016 that does not specify a detailed design method.

3.1. Elastic Design of Seismic-Force-Resisting Systems

Seismic-force-resisting systems are designed to meet strength requirement for the minimum base shear with a target value of η. Then, the story drifts are checked using allowable story drifts corresponding to the seismic risk category. If some story drifts exceed the allowable story drifts, it is necessary to use damping systems in order to reduce the demand for those story drift. The allowable story drifts are 1.0%, 1.5%, and 2.0% for seismic risk category S, I, and II, respectively.

3.2. Target Damping Ratio Calculation

In order to estimate the target damping ratio, nonlinear response history analysis of the seismic-force-resisting system designed in Section 3.1 is performed repeatedly with a damping ratio raised at a constant increment. The peak story drift and peak base shear of the seismic-force-resisting systems are computed for each analysis until those peak responses reach their target values, respectively. Both peak story drift and peak base shear are determined as average responses obtained from nonlinear response history analysis using seven or more ground motion records. The target base shear V^* is determined as the minimum base shear corrected by the demand–capacity ratio (DCR), which is determined in the elastic design of the seismic-force-resisting system and represent redundancy in the design.

$$V^* = \frac{V_{min}}{DCR}$$

(5)

The damping ratio of the seismic-force-resisting system is composed of inherent damping and damping added by damping systems. In this study, the damping ratio of the structure is assumed to be 5%. The target damping ratio β_t is calculated by Equation (6).

$$\beta_t = \beta_{IH} + \beta_v$$

(6)

where, β_{IH} is the inherent damping ratio and β_v is the damping ratio added by damping system. In the nonlinear response history analysis, β_v is increased at a constant increment, and the increment need not be too small because the response can be interpolated numerically. Thus, three to five times of nonlinear response analysis for a given ground motion set is adequate.

Response reduction ratio with respect to effective damping ratio can be plotted in order to determine the target damping ratio and design of damping devices. An illustrative example of such plots is represented in Figure 2, where response reduction ratios for both base shear and maximum story drift are plotted in broken lines.

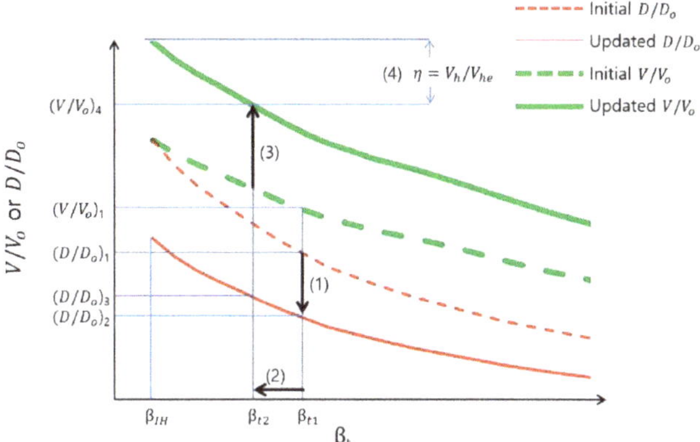

Figure 2. Design process of damping device.

3.3. Damping Device Design

Damping device design is to determine design parameters that can provide damping parameters. Both displacement-dependent damping device and velocity-dependent damping device are addressed for illustrative design examples in this study. Metallic yielding dampers, friction dampers and many other types of dampers belong to displacement-dependent damping devices. The TADAS (triangular-plate added damping and added stiffness) system is adopted as an example of displacement-dependent damping devices in this study. Velocity-dependent damping devices include viscous dampers and viscoelastic dampers of which the latter provides a displacement-dependent force component. The viscoelastic damper is adopted as an example of velocity-dependent damping devices in this study.

3.3.1. Effective Damping of Displacement-Dependent Damping Devices

The force-displacement relationship of the TADAS system is defined as a bi-linear model with a post-yield stiffness ratio of 0.02. The strength and stiffness characteristics of the TADAS system including deformation capacity are calculated by Equations (7) to (9) using metal plate dimensions and material strengths of which details can be found in Ramirez et al. [24].

$$\delta_y = \frac{3}{2}\left(\frac{\varepsilon_y h^2}{t}\right) \tag{7}$$

$$V_y = \frac{F_y b t^2}{4h} \tag{8}$$

$$\delta_{max} = \frac{\varepsilon_{max} h^2}{t} \tag{9}$$

where δ_y, V_y and δ_{max} are yield deformation, yield strength and deformation capacity, respectively, and b, h and t are width, height and thickness of the triangular metal plate, ε_y is the yield strain of the material, ε_{max} is the strain limit of the damping device and can be calculated by the following equation [25–28].

$$\varepsilon_{max} = AN_f^{-B} \tag{10}$$

where N_f is the number of deformation cycles and assumed to be 100, A and B are constant and assumed to be 0.08 and 0.3, respectively [24]. The peak deformation demand on the damping device is calculated as follows.

$$\delta_u = h_i \theta_i \tag{11}$$

where θ_i and h_i are the peak story drift angle (rad) and the story height of the i-th story.

For given peak deformation demands, the effective damping ratio can be calculated by Equations (12) in accordance with ASCE 7-10 based on the energy dissipation due to cyclic deformations illustrated in Figure 3 [29].

$$\beta_v = \frac{\sum E_{dj}}{4\pi E_s} \tag{12}$$

where E_s is the strain energy stored in the structure E_{dj} is the energy dissipated by the j-th damping device. E_s and E_{dj} are calculated by the following equations.

$$E_s = \frac{1}{2}\sum F_i \delta_i \tag{13}$$

$$E_{dj} = 4\left(F_{yj}\delta_{uj} - F_{uj}\delta_{yj}\right) \tag{14}$$

where F_i and δ_i are the peak lateral force and peak lateral displacement at the i-th story, respectively, δ_{yj}, δ_{uj}, F_{yj} and F_{uj} are the yield displacement, peak displacement, yield strength and peak force of the i-th damping device.

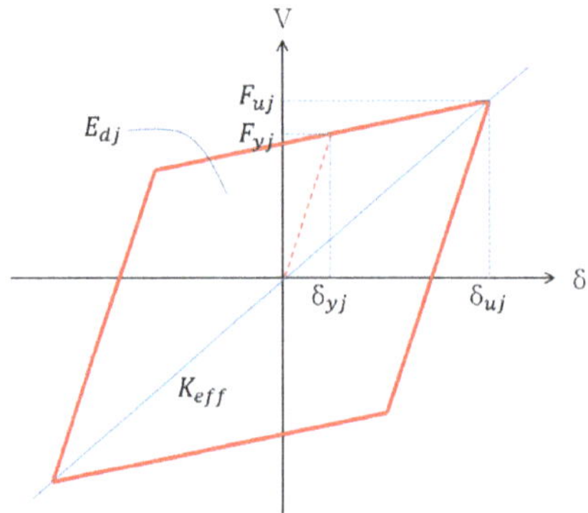

Figure 3. Energy dissipation of metallic damping device.

3.3.2. Effective Damping of Velocity-Dependent Damping Devices

The characteristics of the viscoelastic damping device adopted in this study are based on the experimental results of Soong and Dargush (1997) represented in Table 2 [30]. Kelvin model illustrated in Figure 4 is adopted for numerical modeling of the viscoelastic damping device. Effective stiffness and damping coefficients of the model are given by the following equations

$$K_d = \frac{AG'}{t_d} \tag{15}$$

$$C_d = \frac{\eta_d}{\omega}K_d \tag{16}$$

where G' and η_d are storage modulus and loss factor of the viscoelastic material, respectively, A and t_d are the shear area and thickness of the viscoelastic damper, respectively, and ω is the excitation frequency, which is taken as the fundamental frequency of the structure with effective stiffness of damping devices. Added damping ratio β_v can be calculated by the following equation on the basis of the modal strain energy method [31].

$$\beta_v = \frac{\eta_d}{2}\left(1 - \frac{\omega_n^2}{\overline{\omega}_n{}^2}\right) \tag{17}$$

where ω_n and $\overline{\omega}_n$ are the natural frequency before and after installation of damping systems, respectively. Although the modal strain energy method is applicable to linear elastic structures, it is assumed that nonlinear damping devices is linearized using effective stiffness. Thus, $\overline{\omega}_n$ is calculated by eigenvalue analysis of the structure with the effective stiffness of damping systems added. The thickness of the viscoelastic damper t_{VED} can be calculated as follows.

$$t_{VED} = \frac{\Delta_u}{\gamma_{max}} \tag{18}$$

where Δ_u is the maximum shear deformation of the damping device and γ_{max} is the maximum strain capacity

Table 2. Properties of the viscoelastic damping device [20].

Temperature (°C)	Frequency (Hz)	Strain (%)	Shear Storage modulus, G' (MPa)	Shear loss modulus, G'' (MPa)	Loss Factor η_d
24	1.0	20	0.958	1.151	1.20

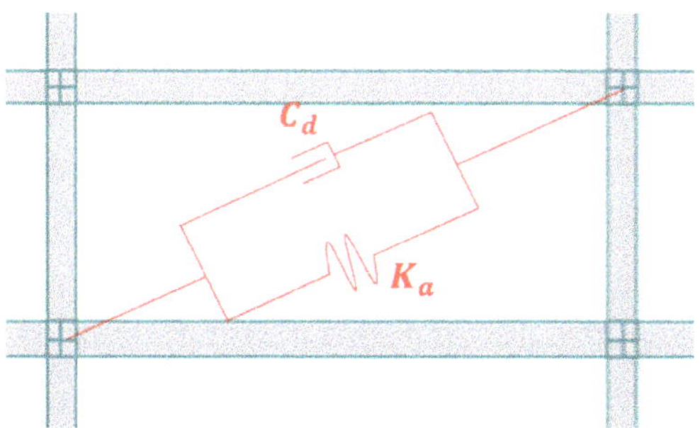

Figure 4. Kelvin model of viscoelastic damping devices.

3.3.3. Design of Damping Devices Considering Change of the Natural Frequency

Once damping device properties are determined, it is necessary to update the effective damping ratio β_v because the effective damping ratio β_v given by Equations (12) or (17) is dependent on the stiffness and/or strength of the damping device. A rational method to update the effective damping ratio considering response reduction and stiffening effects of damping devices is described in this section and Figure 2 illustrates the procedure for updating the effective damping ratio conceptually.

(Step 1) Base shear and maximum story drift reduction factors are plotted with respect to β_t, broken lines in Figure 2.

(Step 2) The target damping ratio β_{t1} corresponding to the initial base shear reduction factor $(V/V_0)_1$ is interpolated from the plot 'Initial V/V_0' (thick broken line) in Figure 2.

(Step 3) The maximum story drift reduction factor $(D/D_0)_1$ is interpolated from the plot 'Initial D/D_0' (thin broken line) in Figure 2.

(Step 4) Damping devices are designed to achieve the target damping ratio Equation (12) or (17) subjected to deformations corresponding to $(D/D_0)_1$. Those deformations can be approximated by multiplying $(D/D_0)_1$ to the initial response for only β_{IH}.

(Step 5) Effective stiffness of damping devices is calculated and added to the structure without damping devices. Then, the fundamental frequencies of the structure are updated.

(Step 6) Response reduction factors are updated corresponding to change of the fundamental frequencies. The peak story drift reduction factor is reduced on the basis of displacement design spectrum, as shown in Figure 5a due to fundamental period shortening. The modified peak story drift reduction factor is plotted as 'Updated D/D_0' (thin solid line) in Figure 2.

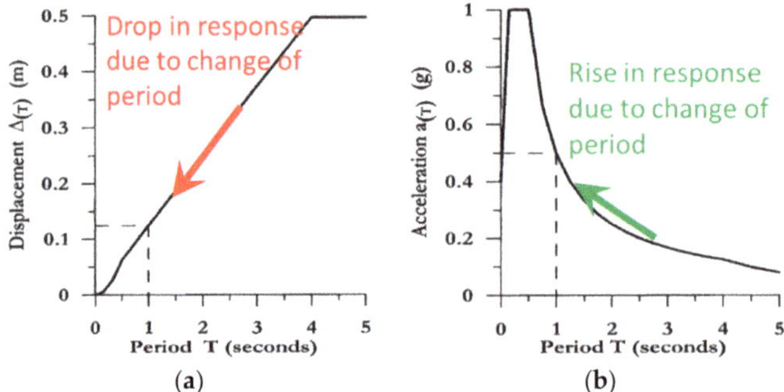

Figure 5. Modification of response due to change of the fundamental period: (**a**) Displacement spectrum and (**b**) pseudo-acceleration spectrum.

(Step 7) The effective damping ratio is updated to β_{t2} considering the decrease of deformations in (Step 6), and corresponding maximum story drift reduction factor is interpolated from the plot 'Updated D/D_0' in Figure 2.

(Step 8) The base shear reduction factor is determined on the basis of pseudo-acceleration design spectrum and increased due to shortening of the fundamental period, as shown in Figure 5b. The modified base shear reduction factor is represented as 'Updated V/V_0' in Figure 2 (thick solid line).

(Step 9) Damping correction factor η for β_{t2} is calculated as a ratio between V/V_0's at β_{t2} and β_{IH} on 'Updated V/V_0' in Figure 2.

(Step 10) Adjust $(V/V_0)_1$ and repeat (Step 2) to (Step 9) until η becomes sufficiently close to the target.

(Step 11) If η converges to the target, nonlinear response history analysis is performed in order to confirm whether the actual response reduction factor satisfies design requirements or not.

The design procedure proposed above utilizes nonlinear response history analysis only at Step 1. Additional response prediction is performed using the elastic design spectrum. Thus, the proposed design procedure is computationally efficient compared to the trial-and-error method based on fully nonlinear response history analysis.

4. Design Example

Two design examples based on the proposed damping system design procedure are presented. The first example is a three-story steel moment-resisting frames with metallic yielding dampers and the second example is a six-story steel moment-resisting frames with viscoelastic dampers.

4.1. Nonlinear Modeling of Structural Elements

Nonlinear modeling of beams and columns of the steel moment frames is performed in accordance with ASCE 41-13 [32]. Common load-deformation relationship for beams and columns subjected to flexure is represented in Figure 6, where the yield rotation angle of beams and columns is calculated by Equations (19) and (20), respectively.

$$\text{Beams}: \theta_y = \frac{ZF_y l_b}{6EI_b} \tag{19}$$

$$\text{Columns}: \theta_y = \frac{ZF_y l_c}{6EI_c}\left(1 - \frac{P}{P_{ye}}\right) \tag{20}$$

where l_b, l_c, Z and l are the moment of inertia for beams and columns, the plastic section modulus and member length, respectively. In addition, E, F_y, P and P_{ye} are the modulus of elasticity, yield strength of steel, the axial force acting on the member and the axial strength of member, respectively.

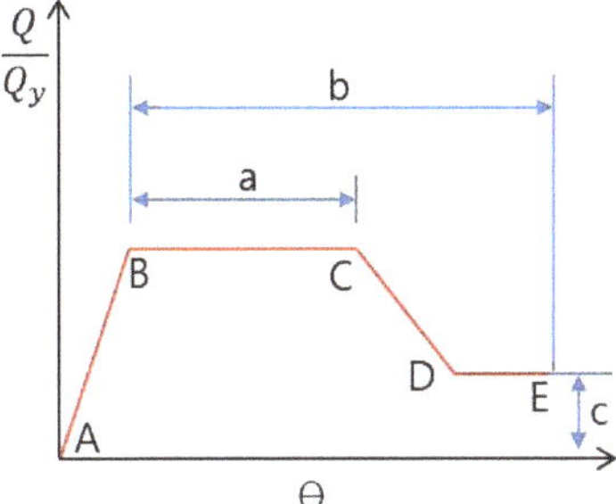

Figure 6. Load-deformation relationship of beams and columns [12].

The panel zone of the steel moment frame was explicitly modeled with Krawinkler's model [33] of which configuration is represented in Figure 7, and the load-deformation relationship is shown in Figure 8. The characteristics of the panel zone model were calculated by the following Equations.

$$K_e = 0.95 d_b d_c t_p G \tag{21}$$

$$K_p = 1.04 b_{fc} t_{fc}^2 G \tag{22}$$

$$M_y = 0.55 F_y t_p 0.95 d_b d_c \tag{23}$$

$$\theta_p = 4\theta_y \tag{24}$$

where d_b, d_c, b_{fc}, t_p, t_{fc}, G and α are beam depth, column depth, column flange width, panel zone thickness, column flange thickness, shear modulus and strain-hardening ratio (0.02), respectively. Perform-3D software was used for modeling and nonlinear response history analysis.

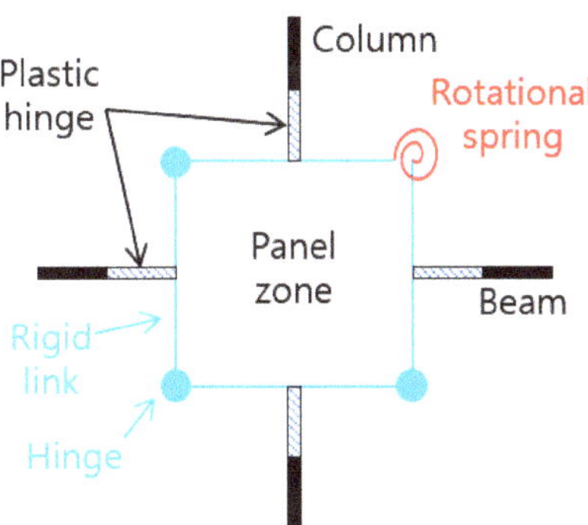

Figure 7. Krawinkler's model for panel zone.

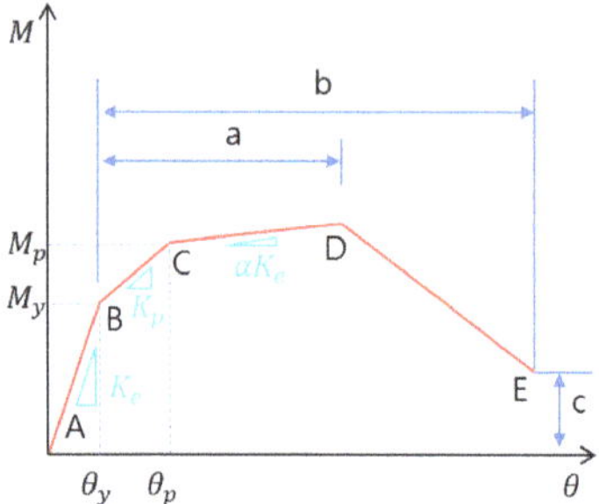

Figure 8. Load-deformation relationship for panel zone [22].

4.2. Three-Story Steel Moment-Resisting Frames with Metallic Yielding Dampers (3F-OMRF-MD)

4.2.1. Initial Design of Seismic-Force-Resisting System

An example building with displacement-dependent damping devices is designed based on the KBC 2016, which is composed of three stories, five spans in X direction, three spans in Y direction. Figure 9a,b shows a three-dimensional view and plan view of the building. All the X-directional internal frames are identical and only one frame is used in this design example and represented in Figure 9c. The dead and live loads of the structure applied to building floors are 5.0 kN/m^2 and 3.5 kN/m^2, respectively, and identical for all stories. The building is assumed to be located in seismic zones I and belong to seismic risk category 'Special'. Site class S_D is assumed for the building. The seismic-force-resisting system of the building is designed as an ordinary moment-resisting frame of which design factors are listed in Table 1. Allowable story drift of 1.0 % for seismic risk category 'Special' is adopted.

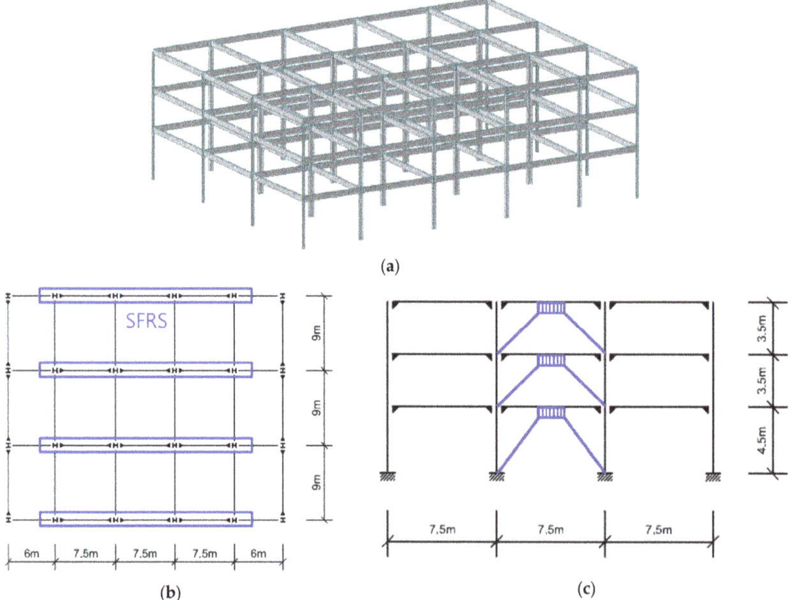

Figure 9. Three-story steel moment resisting frame with metallic yielding dampers: (**a**) Isometric view, (**b**) plan, (**c**) elevation of internal frame.

The initial design of the moment-resisting frame was performed for the minimum base shear with $\eta = 0.75$. The properties of columns and beams of the designed frame are listed in Table 3. The same section is used for each member in all stories. SM490 material was applied to all the members. The DCR of the initial design result is 0.95. Thus, the target base shear reduction factor required for damping systems is moderated to be 0.79, considering the DCR.

Table 3. Properties of moment-resisting frames.

Model	Member	Story	Section	Material
3F-OMRF-MD	Beam	1-3	H-506 × 201 × 11/19	
	Column	1-3	H-394 × 405 × 18/18	
	Brace	1-3	H-200 × 200 × 8/12	SM490
6F-SMRF-VED	Beam	5-6	H-354 × 176 × 8/13	f_y = 315 MPa
		3-4	H-450 × 200 × 9/14	f_u = 490 MPa
		1-2	H-496 × 199 × 9/14	
	Column	4-6	H-414 × 405 × 18/28	
		1-3	H-428 × 407 × 20/35	
	Brace	1-6	H-244 × 252 × 11/11	

From linear dynamic analysis using the response spectrum method, the maximum story drift without damping devices was 1.32%, which occurs at the first story and is higher than the allowable story drift of 1.0%. Therefore, it is necessary to reduce the story drift as well as base shear using damping devices. TADAS damping devices are installed at the center span using brace members listed in Table 3. Thus, the center frame and the damping devices shown in Figure 9c comprises a damping system.

4.2.2. Design of Displacement-Dependent Damping Devices

In order to achieve target reduction factors for base shear and maximum story drift, those two response values are recorded from nonlinear response history analyses repeated with incremental damping of 0.05. Thus, the nonlinear response history analysis was performed only five times. Response reduction factors obtained from the nonlinear response history analysis are plotted in Figure 10.

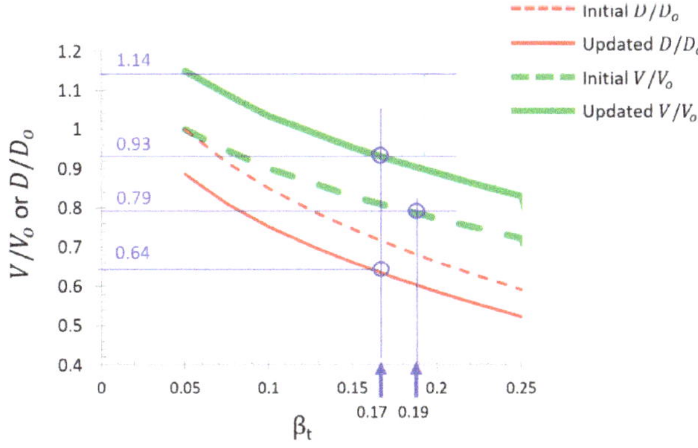

Figure 10. Base shear and peak story drift reduction factors.

Target damping ratio corresponding to the base shear reduction factor 0.79 is interpolated to be 19.1%. The damping devices were designed to achieve an added damping of 0.141 except 0.05 inherent damping of the moment-resisting frame. To design each damping device, the total dissipated energy $\sum E_{dj}$ was calculated from Equation (12) in combination with Equation (13). Then $\sum E_{dj}$ was distributed to each story in proportion to the story shear force. It is taken into account that yield strength or friction force of displacement-dependent damping devices are distributed based on the distribution of story shear force to maximize energy dissipation [34,35]. First estimation of target β_v and corresponding β_t were 0.141 and 0.191. Characteristics of damping devices determined to achieve the target β_v on the basis of Equation (14) are given in Table 4. The post-yield stiffness ratio of the damping device was assumed to be 0.02 in the calculation of dissipated energy.

Table 4. Characteristics of TADAS damping devices to achieve target damping ratio.

β_v	Floor	Story Force Ratio	V_y (kN)	V_u (kN)	δ_y (mm)	δ_u (mm)
	3	0.56	140	147	4.6	17.2
0.141	2	0.82	204	223	4.6	24.8
	1	1.00	249	270	5.9	30.4
	3	0.57	145	152	4.6	15.7
0.120	2	0.81	209	226	4.6	22.6
	1	1.00	255	274	5.9	28.2

The fundamental period of the structure without damping devices was 1.04 sec. The fundamental period of the moment-resisting frame with damping devices substituted by secant stiffness thereof at respective peak deformations of them was reduced to 0.92 sec. Considering the change of the fundamental period, the story drift reduction factor was decreased, as shown in Figure 10. Then, β_v and corresponding β_t were modified into 0.12 and 0.17, respectively, using updated damping device deformations corresponding to the adjusted story drift reduction factor. As a result, the corresponding maximum story drift response is reduced to 64% compared to the structure with 0.05 damping ratio. Using those updated damping device deformations, the fundamental period based on the secant stiffness of damping devices was calculated to be 0.91 second. Then, the base shear reduction factor was elevated corresponding to 0.91 second period as shown in Figure 10. Finally, damping device yield strengths are modified in order to compensate reduced deformations due to period change and to achieve an added damping ratio β_v of 0.12. The final damping device properties are listed in Table 4 for each story. Based on the modified base shear reduction factor represented by the thicker solid line in Figure 10, the expected damping correction factor is 0.93/1.14 = 0.82, which is slightly higher than the target $\eta = 0.79$. However, the damping performance obtained from the results of the nonlinear response history analysis is 0.79, which mean that the expected damping performance goal was achieved with a sufficiently accurate prediction of performance.

The average maximum story drift ratios for seven ground motion records representing design earthquake were 0.66%, 0.65% and 0.48% for the first, second, and third story, respectively, as summarized in Table 5 and all of those values are much smaller than the allowable story drift ratio 1.0%. This is because the base shear reduction factor 0.79 for seismic-force-resisting system governs the design rather than story drift reduction in this design example. The average maximum TADAS damping device deformation for seven ground motion records representing the maximum considered earthquake was maximum at the first story and calculated to be 41.2 mm. The deformation capacity of the example TADAS damping device is 60.3 mm. Therefore, damping devices can maintain the damping performance even under the maximum considered earthquake.

Table 5. Comparison of drift ratio and structural weight.

Seismic Design Model	Story	Story Drift Ratio (%)	Member Section (DCR)	Weight (kN)
3F-OMRF-MD (with damping devices)	3	0.48	Beam: H-506 × 201 × 11/19 (0.79) Column: H-394 × 405 × 18/18 (0.81) Brace: H-200 × 200 × 8/12 (0.29)	149
	2	0.65		
	1	0.66		
3F-OMRF-SD (without damping devices)	3	0.39	Beam: H-692 × 300 × 13/20 (0.57) Column: H-428 × 407 × 20/35 (0.58)	221
	2	0.59		
	1	0.61		

KBC 2016 requires structural elements comprising a damping system to remain elastic subjected to both seismic loads and forces induced by damping devices for design earthquake. DCRs for the frame members and panel zones were computed in terms of rotation angle ductility from nonlinear response history analysis and represented in Figure 11. In the case of columns and braces, higher DCR among bending moment DCR and axial force DCR in a member is given in Figure 11a. All the members that belong to the damping system at the central bay remain elastic since DCRs are lower than 1.0. Therefore, the design result obtained by the proposed procedure satisfies all the requirements of KBC 2016.

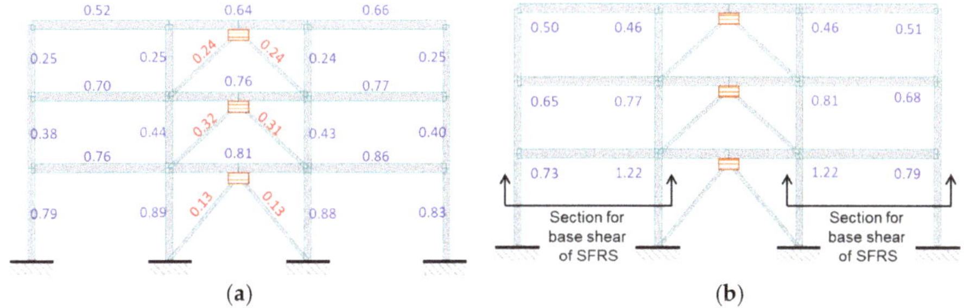

Figure 11. Demand–capacity ratio (DCR) of three-story steel moment-resisting frames with metallic yielding dampers (3F-OMF-MD): (**a**) Frame members, (**b**) panel zones.

To examine the effect of structural steel material reduction by damping devices, a bare ordinary moment resisting frame is designed to achieve story drifts similar to the frame with damping devices. Steel sections and story drifts of two models with and without damping devices are summarized with the respective total weights in Table 5, where 3F-OMRF-SD represents the seismically designed bare frame. The total weight of steel sections for the frame with damping devices is 149 kN, which is 67% of 221 kN for the frame without damping devices. Thus, the proposed procedure can yield efficient structural material-saving design. Ramirez et al. [5] provide similar design example, in which three-story and three-bay frames are designed without and with metallic yielding damping devices using equivalent lateral force procedure although the seismic-force-resisting system is a special moment-resisting frame and target story drift ratio is set to 2% differently from this study. In the comparative design example, the frame with damping devices has 76% of the weight for frames without damping devices. In spite of several different conditions, this comparison supports the ability of the proposed design procedure to reduce seismic demand on the seismic-force-resisting system with supplementary energy dissipation relying on more accurate response prediction by nonlinear response history analysis.

4.3. Six-Story Steel Moment Frames with Viscoelastic Dampers (6F-SMRF-VED)

4.3.1. Initial Design of Seismic-Force-Resisting System

A six-story steel moment-resisting frame is designed in this example. Velocity-dependent damping devices are added for seismic response reduction. The steel moment frame has five spans in the longitudinal direction, and three spans in the transverse direction. The building is assumed to be located in Seismic zone I of KBC 2016 and belong to seismic risk category 'Special' of which importance factor is 1.5. Site class was assumed to be S_D. The overall design was performed under conditions similar to the example building with displacement-dependent damping systems. However, a special moment-resisting frame was adopted for the seismic-force-resisting system. The numerical model of the example building is represented in Figure 12.

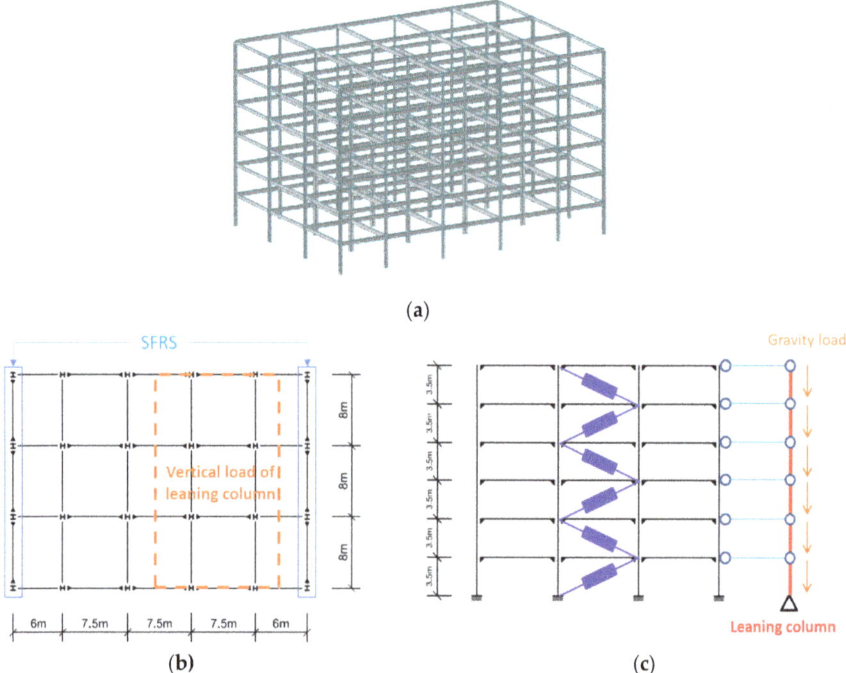

Figure 12. Six-story steel moment-resisting frames with viscoelastic damping devices: (**a**) Isometric view, (**b**) plan, (**c**) elevation of internal frame for design example.

In the transverse direction, only two special moment-resisting frames placed at the outermost part of the building plan play a role of seismic-force-resisting system. Considering geometrical symmetry, only one moment-resisting frame is modeled in the example for simplicity. In addition, the P-Δ effect due to gravity loads at the center of the plan was taken into account using the leaning column as shown in Figure 12c.

The initial design of the special moment-resisting frame was performed for the minimum base shear with $\eta = 1.0$. The DCR of the initial design result is 0.88. Table 3 summarizes sections of members used in the designed frame. Since η assumed in the design equals 1.0, it is unnecessary to confirm whether a target base shear reduction factor is achieved. The damping devices are installed at the center span of the planar frame with braces and illustrated in Figure 12c. Thus, the frame in the central bay and the damping devices shown in Figure 12c comprises a damping system.

From linear dynamic analysis based on response spectrum method with response modification factor and deflection amplification factor defined in KBC 2016, it was observed that the highest peak story drift was 1.75% and observed in the third and fourth stories. It is necessary to reduce the story drift by the damping device because it does not satisfy the allowable story drift of 1.0%.

4.3.2. Design of Velocity-Dependent Damping Devices

The natural frequency of the first mode was 0.5 Hz from the eigenvalue analysis of the structure with only the stiffness component of the viscoelastic damping devices. Viscoelastic damper characteristics corresponding to an excitation frequency of 1.0 Hz, which is the closest one to 0.5 Hz, was adopted among those dependent on excitation frequencies. The stiffness and damping coefficients of each damping device were calculated using Equation (15) and (16).

The average maximum story drift ratios from nonlinear response history analysis for seven ground motion records representing design earthquake are listed in Table 6. The maximum story drift ratio is found to be 1.48% for the third story. Target story drift reduction factor is 1.0%/1.48% = 0.68. The maximum story drift reduction factor was obtained from the nonlinear response history analysis of the moment-resisting frame with an incremental damping ratio of 0.05 that substitute damping devices. Thus, the nonlinear response history analysis was performed only five times and the maximum story drift reduction factor was plotted in Figure 13 and the target damping ratio β_t interpolated from the plot is 0.20. To achieve the target damping ratio, $\beta_v = 0.15$ excluding $\beta_{IH} = 0.05$ is necessary to be added by damping devices.

Table 6. Peak story drifts of six-story steel moment frames with viscoelastic dampers (6F-SMRF-VED).

Seismic Risk Category	Importance Factor	Story	Story-Drift Ratio without Damping Devices (%)		Allowable Story-Drift Ratio (%)
			Equivalent Static Analysis	Nonlinear Response History Analysis	
		6	1.32	1.24	
		5	1.59	1.28	
S	1.5	4	1.75	1.31	1.00
		3	1.75	1.48	
		2	1.54	1.39	
		1	0.93	0.76	

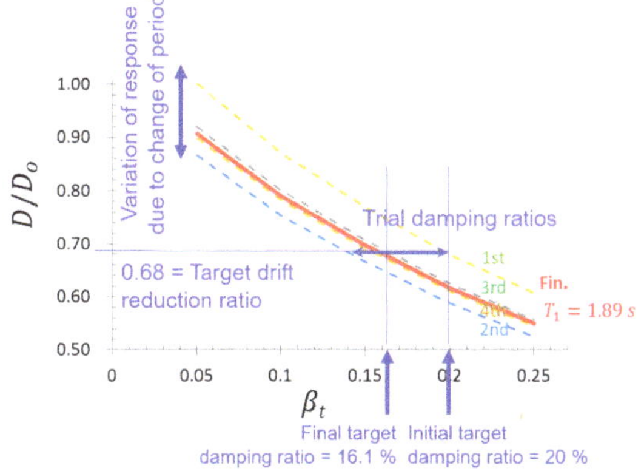

Figure 13. Normalized response vs. incremental damping ratios.

The damping ratio added by viscoelastic damping devices is calculated by Equation (17), in which the added damping ratio is dependent on the fundamental frequency of the structure with effective stiffness of damping devices. In order to design damping devices, the stiffness K_d of damping devices represented by Kelvin model are increased until the fundamental frequency becomes the target value corresponding to the target damping ratio. This work is conducted by eigenvalue analysis of the linear elastic model and does not require additional nonlinear response history analysis. The same stiffness was applied to all the damping devices in this design example, but more efficient distribution may be investigated [18]. When K_d is determined, a corresponding C_d can be calculated using Equation (16).

However, change of the fundamental frequency due to damping devices affects the maximum story drift reduction factor. As a result, the maximum story drift reduction factor in Figure 13 is updated repeatedly. For each update of the maximum story drift reduction factor, the target damping ratio changes correspondingly. Four times of update were performed and updated parameters including target damping ratios and target frequencies are summarized in Table 7. The final fundamental frequency converged to 1.89 sec and K_a and C_d reached 6800 kN/m and 2718 kN·sec/m, respectively.

Table 7. Target damping ratio and frequency.

Structure without Damping Devices		Iteration	Target Damping Ratio β_t	Added Damping Ratio β_v	Target Frequency ω (rad/sec)	Target Period T (sec)
Fundamental Period T (sec)	Fundamental Frequency ω (rad/sec)					
		-	0.050	0.000	3.00	2.09
		1st	0.200	0.150	3.47	1.81
2.09	3.00	2nd	0.142	0.092	3.26	1.93
		3rd	0.166	0.116	3.34	1.88
		4th	0.156	0.106	3.31	1.90
		Fin	0.161	0.111	3.33	1.89

Nonlinear response history analysis was performed using the final stiffness and damping coefficient of the damping devices. It is unnecessary to examine design base shear because the damping correction factor was set to 1.0. The base shear reduction factor was 0.92 which is smaller than 1.0. The peak story drifts of the final design are summarized in Table 8. Compared to Table 6, the maximum peak story drift was reduced to 0.90%, which is slightly lower than the allowable story drift ratio of 1.0%. As a result, the proposed design methodology can design damping systems with a sufficiently accurate prediction of performance.

Table 8. Comparison of peak drift ratio and structural weight.

Seismic Design Model	Story	Story Drift (%)	Member Section (DCR)		Weight (kN)
			Beam	Column or Brace	
6F-SMRF-VED (with damping systems)	6	0.51	H-354 × 176 × 8/13 (0.75)	Column: H-414 × 405 × 18/28 (0.19) Brace: H-244 × 252 × 11/11 (0.43)	297
	5	0.68			
	4	0.83	H-450 × 200 × 9/14 (0.79)		
	3	0.90		Column: H-414 × 405 × 18/28 (0.32) Brace: H-244 × 252 × 11/11 (0.46)	
	2	0.85	H-496 × 199 × 9/14 (0.78)		
	1	0.51			
6F-SMRF-SD (without damping systems)	6	0.62	H-506 × 201 × 11/19 (0.36)	Column: H-498 × 432 × 45/70 (0.07)	590
	5	0.77			
	4	0.90	H-606 × 201 × 12/20 (0.47)		
	3	0.97			
	2	0.88	H-606 × 201 × 12/20 (0.45)	Column: H-498 × 432 × 45/70 (0.15)	
	1	0.48			

As with the displacement-dependent damping system design, the structural elements comprising the damping system must be both elastic against the loads including seismic loads and forces induced by damping devices for design earthquake. Braces to install damping devices transmitting damping device force to the seismic-force-resisting force and the column at the right-hand side of the first story damping device transmitting vertical component of damping device force to the foundation comprises the damping system of the structure. DCRs for the frame members and panel zones were computed in terms of rotation angle ductility from nonlinear response history analysis for design earthquake and represented in Figure 14. In the case of columns and braces, higher DCR between bending moment DCR and axial force DCR is given in Figure 14a. All the members that belong to the damping system in the central bay remain elastic with DCRs lower than 1.0.

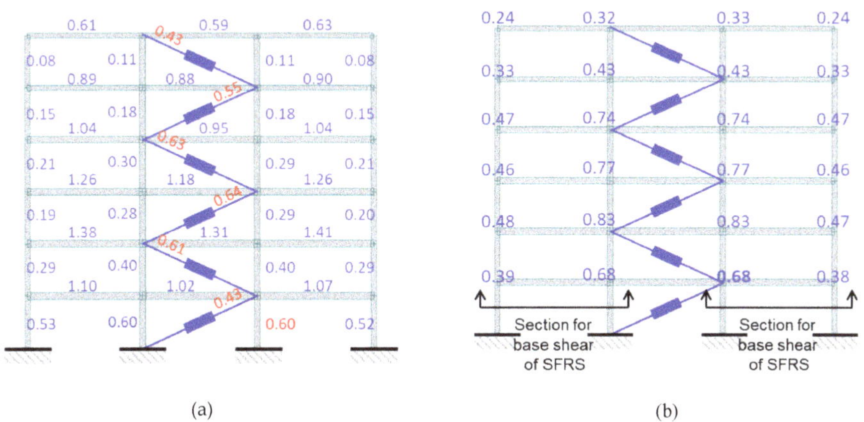

(a)   (b)

Figure 14. DCR of 6F-SMRF-VED. (**a**) Frame members, (**b**) panel zones.

Finally, damping device safety subjected to the maximum considered earthquake was checked. The maximum shear strain of damping devices was 0.455 from the response analysis for the maximum considered earthquake. The experimental data of Soong and Dargush used in the design of the damping device does not provide the deformation capacity of the damping device [20]. Therefore, it is necessary to ensure whether or not the damping device is broken for the strain demand subjected to the maximum considered earthquake. Therefore, the design result satisfies all the requirements of KBC 2016 under the premise that the deformation capacity requirement for the damping device can be met.

Like the three-story frame example, a bare special moment-resisting frame is designed to achieve story drifts similar to the frame with damping devices. Steel sections and story drifts of two models with and without damping devices are summarized with the respective total weights in Table 8 where 6F-SMRF-SD represents the seismically designed bare frame. The total weight of steel sections for the frame with damping devices is 297 kN, which is about 50% of 590 kN for the frame without damping devices. Thus, the proposed procedure can yield efficient structural material-saving design in case of viscoelastic damping devices. Ramirez et al. [5] provide similar design example, in which six-story and three-bay frames are designed without and with viscous damping devices using equivalent lateral force procedure although the damping device does not have stiffness component and target story drift ratio is set to 2% differently from this study. In the comparative design example, the frame with damping devices has 60% of the weight for frames without damping devices. Similar to the preceding design example, the proposed design procedure can design damping devices effectively to reduce seismic demand on the seismic-force-resisting system with better efficiency, which is owing to more accurate response prediction by nonlinear response history analysis.

5. Conclusions

This study proposed an efficient seismic design procedure for building structures with damping systems subjected to requirements of the revised Korean building code, KBC 2016, using nonlinear response history analysis. The proposed design procedure was validated by two design examples of steel moment-resisting frame with metallic yielding dampers and viscoelastic dampers, respectively. The conclusions from this study are summarized as follows.

- The proposed design procedure makes use of nonlinear response history analysis, but does not repeat time-consuming nonlinear response history analysis until convergence of design solution. Instead, design parameters of damping devices are determined using the response reduction curve prescribed by a limited number of response history analyses. Only five times of response history analysis is sufficient for practical application.
- The proposed design procedure can predict seismic response of nonlinear structures with considerable accuracy because basic response reduction factors are obtained through nonlinear response history although equivalent linearization technique is used partially to estimate effects of damping devices with limited computational efforts.
- The proposed design procedure does not require an optimization procedure and can be conducted using commercial structural analysis software. However, the proposed design procedure provides a systematic process to update the design parameters of damping devices and converges to a final design meeting design goals.
- The proposed design procedure for structures with damping systems can reduce structural materials of seismic-force-resisting systems efficiently by 30 to 50% compared to those without damping systems as illustrated by design examples for steel moment-resisting frames.

Author Contributions: Formal analysis, S.-H.J.; writing—original draft, J.-H.P.; writing—review & editing, T.-W.H.

Acknowledgments: This work was supported by Incheon National University Research Grant in 2016.

Conflicts of Interest: The authors declare no conflict of interest.

References

1. Architectural Institute of Korea. *AIK Korean Building Code. KBC 2009*; AIK: Seoul, Korea, 2009.
2. Architectural Institute of Korea. *AIK Korean Building Code. KBC 2016*; AIK: Seoul, Korea, 2016.
3. American Society of Civil Engineers, Structural Engineering Institute. *ASCE/SEI 7-16. Minimum Design Loads for Buildings and Other Structures*; ASCE: Reston, VA, USA, 2016.
4. Lin, Y.Y.; Tsai, M.H.; Whang, J.S.; Chang, K.C. Direct displacement-based design for building with passive energy dissipation systems. *Eng. Struct.* **2003**, *25*, 25–37. [CrossRef]
5. Ramirez, O.M.; Constantinou, M.C.; Whittaker, A.S.; Kircher, C.A.; Johnson, M.W.; Chrysostomou, C.Z. Validation of the 2000 NEHRP provisions' equivalent lateral force and modal analysis procedures for buildings with damping systems. *Earthq. Spectra* **2003**, *19*, 981–999. [CrossRef]
6. Whittaker, A.S.; Constantinou, M.C.; Ramirez, O.M.; Johnson, M.W.; Chrysostomou, C.Z. Equivalent lateral force and modal analysis procedures of the 2000 NEHRP provisions for buildings with damping systems. *Earthq. Spectra* **2003**, *19*, 959–980. [CrossRef]
7. Lin, Y.Y.; Chang, K.C.; Chen, C.Y. Direct displacement-based design for seismic retrofit of existing buildings using nonlinear viscous dampers. *Bull. Earthq. Eng.* **2008**, *6*, 535–552. [CrossRef]
8. Silvestri, S.; Gasparini, G.; Trombetti, T. A five-step procedure for the dimensioning of viscous dampers to be inserted in building structures. *J. Earthq. Eng.* **2010**, *14*, 417–447. [CrossRef]
9. Sullivan, T.J.; Lago, A. Towards a simplified direct DBD procedure for the seismic design of moment resisting frames with viscous dampers. *Eng. Struct.* **2012**, *35*, 140–148. [CrossRef]
10. Palermo, M.; Silvestri, S.; Landi, L.; Gasparini, G.; Trombetti, T. A "direct five-step procedure" for the preliminary seismic design of buildings with added viscous dampers. *Eng. Struct.* **2018**, *173*, 933–950. [CrossRef]

11. Kim, J.K.; Choi, H.H.; Min, K.W. Performance-based design of added viscous dampers using capacity spectrum method. *J. Earthq. Eng.* **2003**, *7*, 1–24. [CrossRef]
12. Lin, Y.Y.; Chang, K.C. An improved capacity spectrum method for ATC-40. *Earthq. Eng. Struct. Dyn.* **2003**, *32*, 2013–2025. [CrossRef]
13. Ramirez, O.M.; Constantinou, M.C.; Gomez, J.D.; Whittaker, A.S.; Chrysostomou, C.Z. Evaluation of simplified methods of analysis of yielding structures with damping systems. *Earthq. Spectra* **2002**, *18*, 501–530. [CrossRef]
14. Park, J.-H. Seismic response of SDOF systems representing masonry-infilled RC frames with damping systems. *Eng. Struct.* **2013**, *56*, 1735–1750. [CrossRef]
15. Chae, Y.; Ricles, J.M.; Sause, R. Development of equivalent linear systems for single-degree-of-freedom structures with magneto-rheological dampers for seismic design application. *J. Intell. Mater. Syst. Struct.* **2017**, *28*, 2675–2687. [CrossRef]
16. Newmark, N.M.; Hall, W.J. *Earthquake Spectra and Design*; Earthquake Engineering Research Institute: Berkeley, CA, USA, 1982; pp. 35–36.
17. Priestley, M.J.N.; Calvi, G.M.; Kowalsky, M.J. *Displacement-Based Seismic Design of Structures*; IUSS Press: Pavia, Italy, 2007.
18. Park, J.-H.; Kim, J.; Min, K.-W. Optimal design of added viscoelastic dampers and supporting braces. *Earthq. Eng. Struct. Dyn.* **2004**, *33*, 465–484. [CrossRef]
19. De Domenico, D.; Ricciardi, G. Earthquake protection of structures with nonlinear viscous dampers optimized through an energy-based stochastic approach. *Eng. Struct.* **2019**, *179*, 523–539. [CrossRef]
20. Ontiveros-Perez, S.P.; Miguel, L.F.F.; Riera, J.D. Reliability-based optimum design of passive friction dampers in buildings in seismic regions. *Eng. Struct.* **2019**, *190*, 276–284. [CrossRef]
21. Lavan, O.; Levy, R. Optimal design of supplemental viscous dampers for irregular shear-frames in the presence of yielding. *Earthq. Eng. Struct. Dyn.* **2005**, *34*, 889–907. [CrossRef]
22. Lavan, O. A methodology for the integrated seismic design of nonlinear buildings with supplemental damping. *Struct. Control Health Monit.* **2015**, *22*, 484–499. [CrossRef]
23. Wang, S.; Mahin, S.A. High-performance computer-aided optimization of viscous dampers for improving the seismic performance of a tall building. *Soil Dyn. Earthq. Eng.* **2018**, *113*, 454–461. [CrossRef]
24. Ramirez, O.M.; Constantinou, M.C.; Kircher, C.A.; Whittaker, A.S.; Johnson, M.W.; Gomez, J.D.; Chrysostomou, C.Z. *Development and Evaluation of Simplified Procedures for the Analysis and Design of Buildings with Passive Energy Dissipation Systems*; Technical Report MCEER-00-0010; Multidisciplinary Center for Earthquake Engineering Research, University at Buffalo: Buffalo, NY, USA, 2000.
25. Manson, S.S. Behavior of Materials under Conditions of Thermal Stress. In *Heat Transfer Symposium*; Engineering Research Institute, University of Michigan: Ann Arbor, MI, USA, 1953; pp. 9–75.
26. Coffin, L.F., Jr. A Study of Effects of Cyclic Thermal Stresses on a Ductile Metal. *Trans. Am. Soc. Mech. Eng. N. Y.* **1954**, *76*, 931–950.
27. Koh, S.K.; Stephens, R.I. Mean Stress Effects on Low Cycle Fatigue for High Strength Steel. *Fatigue Fract. Eng. Mater. Struct.* **1991**, *14*, 413–428. [CrossRef]
28. Mander, J.R.; Panthaki, F.D.; Kasalanati, A. Low-Cycle Fatigue Behavior of Reinforcing Steel. *J. Mater. Civ. Eng.* **1994**, *6*, 453–467. [CrossRef]
29. American Society of Civil Engineers, Structural Engineering Institute. *ASCE/SEI 7-10. Minimum Design Loads for Buildings and Other Structures*; ASCE: Reston, VA, USA, 2010.
30. Soong, T.T.; Dargush, G.F. *Passive Energy Dissipation Systems in Structural Engineering*; John Wiley & Sons Ltd.: London, UK; New York, NY, USA, 1997.
31. Chang, K.C.; Lai, M.L.; Soong, T.T.; Hao, D.S.; Yeh, Y.C. *Seismic Behavior and Design Guidelines for Steel Frame Structures with Added Viscoelastic Dampers*; Technical Report MCEER-93-0009; Multidisciplinary Center for Earthquake Engineering Research, University at Buffalo: Buffalo, NY, USA, 1993.
32. Pekelnicky, R.; Engineers, S.D.; Chris Poland, S.E.; Engineers, N.D. ASCE 41-13: Seismic Evaluation and Retrofit Rehabilitation of Existing Buildings. In Proceedings of the SEAOC 2012 Convention, Santa Fe, NM, USA, 12–15 September 2012.
33. Krawinkler, H. Shear in beam-column joints in seismic design of steel frames. *Eng. J.* **1978**, *15*, 82–91.

34. Lee, S.-H.; Park, J.-H.; Lee, S.-K.; Min, K.-W. Allocation and slip load of friction dampers for a seismically excited building structure based on storey shear force distribution. *Eng. Struct.* **2008**, *30*, 930–940. [CrossRef]
35. Nabid, N.; Hajirasouliha, I.; Petkovski, M. A Practical Method for Optimum Seismic Design of Friction Wall Dampers. *Earthq. Spectra* **2017**, *33*, 1033–1052. [CrossRef]

Article

Dynamic Responses of Liquid Storage Tanks Caused by Wind and Earthquake in Special Environment

Wei Jing [1,*], Huan Feng [2] and Xuansheng Cheng [1]

[1] Western Engineering Research Center of Disaster Mitigation in Civil Engineering of Ministry of Education, Lanzhou University of Technology, Lanzhou 730050, China; cxs702@126.com
[2] School of Civil Engineering & Mechanics, Huazhong University of Science and Technology, Wuhan 430074, China; fengh2528990166@163.com
* Correspondence: jingwei3276@lut.edu.cn

Received: 24 April 2019; Accepted: 3 June 2019; Published: 11 June 2019

Abstract: Based on potential flow theory and arbitrary Lagrangian–Eulerian method, shell–liquid and shell–wind interactions are solved respectively. Considering the nonlinearity of tank material and liquid sloshing, a refined 3-D wind–shell–liquid interaction calculation model for liquid storage tanks is established. A comparative study of dynamic responses of liquid storage tanks under wind, earthquake, and wind and earthquake is carried out, and the influences of wind speed and wind interference effect on dynamic responses of liquid storage tank are discussed. The results show that when the wind is strong, the dynamic responses of the liquid storage tank under wind load alone are likely to be larger than that under earthquake, and the dynamic responses under wind–earthquake interaction are obviously larger than that under wind and earthquake alone. The maximum responses of the tank wall under wind and earthquake are located in the unfilled area at the upper part of the tank and the filled area at the lower part of the tank respectively, while the location of maximum responses of the tank wall under wind–earthquake interaction is related to the relative magnitude of the wind and earthquake. Wind speed has a great influence on the responses of liquid storage tanks, when the wind speed increases to a certain extent, the storage tank is prone to damage. Wind interference effect has a significant effect on liquid storage tanks and wind fields. For liquid storage tanks in special environments, wind and earthquake effects should be considered reasonably, and wind interference effects cannot be ignored.

Keywords: liquid storage tank; earthquake; wind; dynamic response; fluid–solid interaction

1. Introduction

With the development of economy and society, more and more liquid storage tanks are built in seismically active areas, in extreme cases, these areas may also belong to strong wind areas, which leads to the threat of wind and earthquake to large-scale liquid storage tanks in the whole life cycle. Moreover, earthquake and wind-induced damage cases of liquid storage tanks are very common [1–3], two cases corresponding to earthquake and wind are shown in Figure 1. The destruction of the liquid storage tank not only involves the structure itself, but it will also cause huge economic losses, environmental pollution, fire, and so on, and even threaten people's safety.

 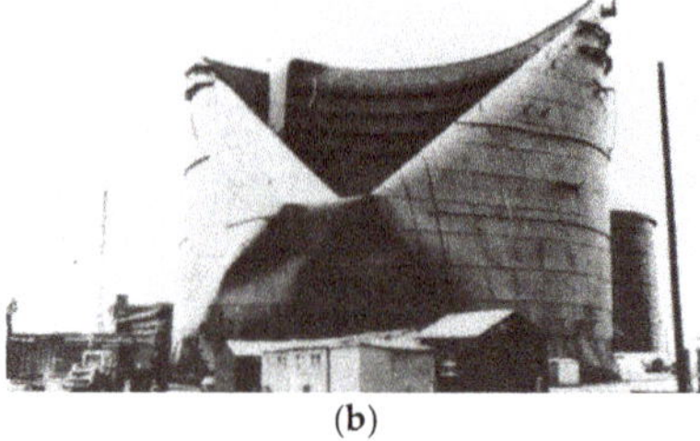

(a)   (b)

Figure 1. Failure cases of liquid storage tank. (**a**) Earthquake. (**b**) Wind [1].

Dynamic responses of liquid storage tanks during earthquakes involves shell–liquid interaction, Rawat et al. [3] used a coupled acoustic–structural (CAS) approach in the FEM for the analysis of the tanks with rigid and flexible walls with varying parameters. Kotrasov et al. [4] simulated the interaction between structure and liquid on the contact surface based on the bidirectional fluid–solid coupling technique and studied the dynamic responses of liquid storage tanks by finite element method. Gilmanov et al. [5] proposed a numerical method to simulate the shell–liquid interaction of elastic thin plate with arbitrary deformation in incompressible fluid. In addition, a large number of studies and post-earthquake investigations show that the failure modes of liquid storage tanks under earthquake basically include liquid overflow, bottom lifting, circumferential tension, and instability. Ishikawa et al. [6] proposed a practical analytical model for shallow excited tank, which exhibited complex behavior because of nonlinearity and dispersion of the liquid. Moslemi et al. [7] conducted nonlinear sloshing analysis of liquid storage tanks and found that the sloshing nonlinearity had a significant effect on the seismic performance of liquid containing structures. Miladi and Razzaghi [8] performed numerical analysis of oil tank by using ABAQUS software, and carried out parametric study to evaluate the effect of amount of stored liquid on seismic behavior and performance of the studied tank. Ormeño et al. [9] performed shake table experiments to investigate the effects of a flexible base on the seismic response of a liquid storage tank, results showed that the axial compressive stresses decreased after a flexible base was considered. Sanapala et al. [10] performed shake table experiments to study the fluid structure interaction effects between the sloshing liquid and the internal structure, and found that when the partially filled storage tank was subjected to seismic excitation, spiky jet-like features were observed over the free surface. Rawat et al. [11] investigated three-dimensional (3-D) ground-supported liquid storage tanks subjected to seismic base excitation by using finite element method based on coupled acoustic–structural and coupled Eulerian–Lagrangian approaches. Generally speaking, dynamic responses of liquid storage tank involves complex fluid–structure interaction, and numerical simulation is an effective means to solve this problem.

Researchers have made certain explorations on the behavior of liquid storage tanks under wind load. Flores and Godoy [12] used numerical methods to study the buckling problem of liquid storage tanks under typhoon, and obtained that bifurcation buckling analysis could better evaluate the critical state of liquid storage tanks. Portela and Godoy [13] used computational model to evaluate the buckling behavior of steel tanks under wind loads. Zhang et al. [14] studied the dynamic responses of flexible liquid storage structure under wind load by multi-material ALE finite element method. Yasunaga et al. [15] used wind tunnel testing and finite element method to study the buckling behavior of thin-walled circular liquid storage tanks, and discussed the effect of wind load distribution on the buckling of liquid storage tanks by comparing it with a static wind load. Chen and Rotter [16] used finite element method to study the buckling of anchored liquid storage tanks with equal wall thickness under wind load. Zhao et al. [17] and Lin et al. [18] used wind tunnel tests to study the distribution of wind pressure and the stability of liquid storage tanks under wind loads.

In view of the structural dynamic response under the combined action of wind and earthquake, Hong and Gu [19] found that for high-flexible structures whose horizontal loads are controlled by wind load, the combined total loads after considering wind and earthquake loads may be more disadvantageous than those when considering wind loads in seismic design. Ke et al. [20] obtained that the structure responses of super-large cooling tower varied significantly along with height under wind load, earthquake, and wind–earthquake. Peng et al. [21] used the method of combining theoretical analysis with numerical simulation to get the position of maximum stress under wind load and earthquake action is different. Sapountzakis et al. [22] studied the nonlinear responses of wind turbine under wind load and earthquake. Mazza [23] synthesized velocity time history of wind based on equivalent spectrum technology, and studied the dynamic responses of steel frame structures under wind load and earthquake action.

To sum up, the dynamic responses of structures under earthquake and wind are obviously different, and the combined action of wind and earthquake will have more adverse effects on the structures, but the research on dynamic responses of liquid storage tanks under wind and earthquake is rare. In this paper, the shell–liquid and the shell–wind interactions are considered, and a refined calculation model of the liquid storage tank is established. The dynamic responses of the liquid storage tank under wind, earthquake, and wind and earthquake are studied in many aspects, which is of great significance to the rationality of the design and the reliability of the operation of the liquid storage tank.

2. Wind Field Control Equations

Large eddy simulation (LES) is used to calculate the wind field, and its control equation is

$$\frac{\partial \bar{u}_i}{\partial t} + \frac{\partial \bar{u}_i \bar{u}_j}{\partial x_j} = -\frac{1}{\rho} \frac{\bar{p}}{\partial x_i} - v \frac{\partial^2 \bar{u}_i}{\partial x_j \partial x_j} + \frac{\partial \bar{\tau}_{ij}}{\partial x_j} \tag{1}$$

$$\frac{\partial \bar{u}_i}{\partial x_i} = 0 \tag{2}$$

where $\bar{\tau}_{ij} = \overline{\bar{u}_i \bar{u}_j} - \overline{u_i u_j}$, $\bar{\tau}_{ij}$ is subgrid-scale stress, namely, SGS stress, which reflects the influence of the motion of small-scale vortices on the motion equation.

If the equations consisting of Equations (1) and (2) are closed, then according to Smagorinsky's basic SGS model, it is assumed that the SGS stress satisfies the following requirements

$$\bar{\tau}_{ij} - \frac{1}{3} \bar{\tau}_{kk} \delta_{ij} = -2\mu_t \bar{S}_{ij} \tag{3}$$

$$\mu_t = (C_s \Delta)^2 |\bar{S}| \tag{4}$$

where $\bar{S}_{ij} = \frac{1}{2} \left(\frac{\partial \bar{u}_i}{\partial x_j} + \frac{\partial \bar{u}_j}{\partial x_i} \right)$, $|\bar{S}| = \sqrt{2\bar{S}_{ij}\bar{S}_{ij}}$, $\Delta = \left(\Delta_x \Delta_y \Delta_z \right)^{1/3}$, μ_t is turbulent viscosity at sublattice scale, Δ is filtration scale of large eddy model, Δ_i represents the grid size along the i-axis, $C_s \Delta$ is equivalent to mixing length, C_s is SGS constant.

3. Structure Control Equations

The structure equation of motion is

$$\mathbf{M}_{ss} \ddot{\mathbf{u}}_s + \mathbf{C}_{ss} \dot{\mathbf{u}}_s + \mathbf{K}_{ss} \mathbf{u}_s = \mathbf{F}_{ss} \tag{5}$$

where $\mathbf{M}_{ss}$, $\mathbf{C}_{ss}$, and $\mathbf{K}_{ss}$ are mass, damping and stiffness matrices of structures, respectively; $\mathbf{F}_{ss}$ is load vector acting on structure, which includes liquid pressure; $\ddot{\mathbf{u}}_s$, $\dot{\mathbf{u}}_s$, and $\mathbf{u}_s$ are vectors of acceleration, velocity, and displacement of structure, respectively.

Newmark method is used to solve the dynamic Equation (5), and the first assumption is

$$\dot{\mathbf{u}}_{s(i+1)} = \dot{\mathbf{u}}_{s(i)} + \left[(1-\beta) \ddot{\mathbf{u}}_{s(i)} + \beta \ddot{\mathbf{u}}_{s(i+1)} \right] \Delta t \tag{6}$$

$$\mathbf{u}_{s(i+1)} = \mathbf{u}_{s(i)} + \dot{\mathbf{u}}_i \Delta t + \left[\left(\frac{1}{2} - \gamma \right) \ddot{\mathbf{u}}_i + \gamma \ddot{\mathbf{u}}_{i+1} \right] \Delta t^2 \tag{7}$$

where β and γ are adjustment coefficients for accuracy and stability.

The incremental forms $\Delta\dot{\mathbf{u}}_s$ and $\Delta\mathbf{u}_s$ of velocity $\dot{\mathbf{u}}_s$ and displacement $\mathbf{u}_s$ can be obtained from Equations (6) and (7), respectively

$$\Delta\dot{\mathbf{u}}_{s(i)} = \dot{\mathbf{u}}_{s(i+1)} - \dot{\mathbf{u}}_{s(i)} = \left(\ddot{\mathbf{u}}_{s(i)} + \beta\Delta\ddot{\mathbf{u}}_{s(i)} \right)\Delta t \tag{8}$$

$$\Delta\mathbf{u}_{s(i)} = \mathbf{u}_{s(i+1)} - \mathbf{u}_{s(i)} = \dot{\mathbf{u}}_{s(i)}\Delta t + \frac{1}{2}\ddot{\mathbf{u}}_{s(i)}\Delta t^2 + \gamma\Delta\ddot{\mathbf{u}}_{s(i)} \tag{9}$$

Acceleration increment $\Delta\ddot{\mathbf{u}}_i$ can be obtained by transforming Equation (9), then taking $\Delta\ddot{\mathbf{u}}_i$ into Equation (8)

$$\Delta\ddot{\mathbf{u}}_i = \frac{1}{\gamma\Delta t^2}\Delta\mathbf{u}_i - \frac{1}{\gamma\Delta t}\dot{\mathbf{u}}_i - \left(\frac{1}{2\gamma} - 1 \right)\ddot{\mathbf{u}}_i \tag{10}$$

$$\Delta\dot{\mathbf{u}}_i = \frac{\beta}{\gamma\Delta t}\Delta\mathbf{u}_i + \left(1 - \frac{\beta}{\gamma} \right)\dot{\mathbf{u}}_i + \left(1 - \frac{\beta}{2\gamma} \right)\Delta t\ddot{\mathbf{u}}_i \tag{11}$$

The incremental form corresponding to Equation (5) is

$$\mathbf{M}_{ss}\Delta\ddot{\mathbf{u}}_{s(i)} + \mathbf{C}_{ss}\Delta\dot{\mathbf{u}}_{s(i)} + \mathbf{K}_{ss}\Delta\mathbf{u}_{s(i)} = \Delta\mathbf{F}_{ss(i)} \tag{12}$$

Taking Equations (9)–(11) into Equation (12)

$$\overline{\mathbf{K}}\Delta\mathbf{u}_{s(i)} = \overline{\mathbf{F}} \tag{13}$$

where $\overline{\mathbf{K}} = \mathbf{K} + \dfrac{1}{\gamma\Delta t^2}\mathbf{M} + \dfrac{\beta}{\gamma\Delta t}\mathbf{C}; \overline{\mathbf{F}} = \Delta\mathbf{F}_{ss(i)} + \mathbf{M}\left[\dfrac{1}{\gamma\Delta t}\dot{\mathbf{u}}_i + \left(\dfrac{1}{2\gamma} - 1 \right)\ddot{\mathbf{u}}_i \right] + \mathbf{C}\left[\left(\dfrac{\beta}{\gamma} - 1 \right)\dot{\mathbf{u}}_i + \left(\dfrac{\beta}{2\gamma} - 1 \right)\Delta t\ddot{\mathbf{u}}_i \right].$

The displacement increment $\Delta\mathbf{u}_{s(i)}$ can be obtained by Equation (13), velocity increment $\Delta\dot{\mathbf{u}}_{s(i)}$ can be obtained by substituting displacement increment $\Delta\mathbf{u}_{s(i)}$ into Equation (11). As a result, the displacement $\mathbf{u}_{s(i+1)}$ and velocity $\dot{\mathbf{u}}_{s(i+1)}$ of $i + 1$ time step can be obtained

$$\mathbf{u}_{s(i+1)} = \mathbf{u}_{s(i)} + \Delta\mathbf{u}_{s(i)} \tag{14}$$

$$\dot{\mathbf{u}}_{s(i+1)} = \dot{\mathbf{u}}_{s(i)} + \Delta\dot{\mathbf{u}}_{s(i)} \tag{15}$$

The acceleration $\ddot{\mathbf{u}}_{s(i+1)}$ of time step $i + 1$ can be obtained by substituting Equations (14) and (15) into Equation (5)

$$\ddot{\mathbf{u}}_{s(i+1)} = \mathbf{M}_{ss}^{-1} \cdot \left[\mathbf{F}_{ss} - \mathbf{C}_{ss} \cdot \dot{\mathbf{u}}_{s(i+1)} - \mathbf{K}_{ss} \cdot \mathbf{u}_{s(i+1)} \right] \tag{16}$$

4. Fluid–Solid Interaction

In order to overcome the defects of large calculation amount and low calculation efficiency, the potential flow theory is used to solve the shell–liquid interaction, and the arbitrary Lagrangian–Eulerian method is used to solve the shell–wind interaction.

4.1. Shell–Liquid Interaction

Because the calculation process involves a large number of nonlinearities, the exact solution of each response can be obtained through multiple equilibrium iterations. $\Delta\phi$ is used to express the increment of the unknown velocity potential ϕ, and $\Delta\mathbf{u}$ is used to express the increment of the unknown displacement $\mathbf{u}$. The shell–liquid interaction dynamic equation based on potential fluid theory is [24]

$$
\begin{bmatrix} \mathbf{M}_{ss} & 0 \\ 0 & \mathbf{M}_{ll} \end{bmatrix} \begin{bmatrix} \Delta\ddot{\mathbf{u}} \\ \Delta\ddot{\boldsymbol{\phi}} \end{bmatrix} + \begin{bmatrix} \mathbf{C}_{uu}+\mathbf{C}_{ss} & \mathbf{C}_{ul} \\ \mathbf{C}_{lu} & -(\mathbf{C}_{ll}+(\mathbf{C}_{ll})_S) \end{bmatrix} \begin{bmatrix} \Delta\dot{\mathbf{u}} \\ \Delta\dot{\boldsymbol{\phi}} \end{bmatrix} + \begin{bmatrix} \mathbf{K}_{uu}+\mathbf{K}_{ss} & \mathbf{K}_{lu} \\ \mathbf{K}_{ul} & -(\mathbf{K}_{ll}+(\mathbf{K}_{ll})_S) \end{bmatrix} \begin{bmatrix} \Delta\mathbf{u} \\ \Delta\boldsymbol{\phi} \end{bmatrix} = \begin{bmatrix} \mathbf{F}_{ss} \\ 0 \end{bmatrix} - \begin{bmatrix} \mathbf{F}_p \\ \mathbf{F}_l+(\mathbf{F}_l)_S \end{bmatrix}
\tag{17}
$$

where $\mathbf{M}_{ll}$ is the liquid mass matrix; $\mathbf{C}_{uu}$, $\mathbf{C}_{lu}$, $\mathbf{C}_{ul}$, and $\mathbf{C}_{ll}$ are the damping matrices of the structure itself, the liquid contributed by the structure, the structure contributed by the liquid and the liquid itself, respectively; and $\mathbf{K}_{uu}$, $\mathbf{K}_{lu}$, $\mathbf{K}_{ul}$, and $\mathbf{K}_{ll}$ are the stiffness matrices of the structure itself, the liquid contributed by the structure, the structure contributed by the liquid and the liquid itself, respectively; $\mathbf{F}_p$, $\mathbf{F}_l$, and $(\mathbf{F}_l)_S$ are the forces acting on the structural boundary caused by the liquid pressure, volume force, and area force, respectively; $\mathbf{F}_l$ is obtained by the volume integral of Equation (18), and $(\mathbf{F}_l)_S$ is obtained by surface integral of Equation (19) [24]

$$
\mathbf{F}_l = \int_V \left(\frac{\partial\rho_l}{\partial h}\dot{h}\delta\phi - \rho_l\nabla\phi \right) dV
\tag{18}
$$

$$
(\mathbf{F}_l)_S = \int_S -\rho_l\mathbf{u}\cdot\mathbf{n}\delta\phi dS
\tag{19}
$$

where ρ_l is the liquid density; V is the liquid domain; S is the liquid domain boundary; $\mathbf{n}$ is the internal normal direction vector of S; and $\dot{\mathbf{u}}$ is the moving speed of the boundary surface S.

The boundary surface adjacent to the structure is represented as S_1, and the force acting on structure boundary $\mathbf{F}_p$ caused by the liquid pressure can be expressed as Equation (20)

$$
-\delta\mathbf{F}_p = -\int_{S_1} pn\cdot\delta\mathbf{u}dS_1
\tag{20}
$$

where $\delta\mathbf{F}_U$ is differentiation of additional forces caused by liquid; $\mathbf{n}$ is normal vector of adjacent interface. Liquid pressure p is calculated by Equation (21)

$$
p = p(h) = p\left[\Omega(\mathbf{x}+\mathbf{u}) - \dot{\phi} - \frac{1}{2}\mathbf{v}_n\cdot\mathbf{v}_n - \frac{1}{2}\mathbf{v}_\tau\cdot\mathbf{v}_\tau \right]
\tag{21}
$$

where Ω is volume acceleration potential energy; $\mathbf{v}_n$ and $\mathbf{v}_\tau$ are liquid normal and tangential velocities on the interaction boundary.

4.2. Shell–Wind Interaction

The wind field equation and the structure equation are expressed by $G_w\left[w,\dot{w}\right] = 0$ and $G_s\left[\mathbf{u},\dot{\mathbf{u}},\ddot{\mathbf{u}}\right] = 0$, respectively, subscript w denotes wind field variables, and subscript s denotes structure variables.

Firstly, the velocity and acceleration of wind field are expressed as [25]

$$
{}^{t+a\Delta t}v = \frac{{}^{t+a\Delta t}u - {}^{t}u}{t} = {}^{t+\Delta t}v\alpha + {}^{t}v(1-\alpha)
$$

$$
{}^{t+a\Delta t}a = \frac{{}^{t+a\Delta t}v - {}^{t}v}{t} = {}^{t+\Delta t}a\alpha + {}^{t}a(1-\alpha)
\tag{22}
$$

where α is stability conditions of compatible time integral.

Velocity and acceleration of Equation (22) at $t + \Delta t$ can be expressed as functions of unknown displacement

$$
\begin{aligned}
{}^{t+\Delta t}v &= \frac{1}{\alpha\Delta t}\left({}^{t+\Delta t}u - {}^{t}u\right) - {}^{t}v\left(\frac{1}{\alpha} - 1\right) = {}^{t+\Delta t}dm + {}^{t}\xi \\
{}^{t+\Delta t}a &= \frac{1}{\alpha^2\Delta t^2}\left({}^{t+\Delta t}u - {}^{t}u\right) - {}^{t}v\frac{1}{\alpha^2\Delta t} - {}^{t}a\left(\frac{1}{\alpha} - 1\right) = {}^{t+\Delta t}dn + {}^{t}\eta
\end{aligned}
$$

(23)

Taking Equations (22) and (23) into wind field equation $G_w\left[w, \dot{w}\right] = 0$ and structure equation $G_s\left[u, \dot{u}, \ddot{u}\right] = 0$

$$
\begin{aligned}
{}^{t+\alpha\Delta t}G_w &\approx G_w\left[{}^{t+\alpha\Delta t}w, \left({}^{t+\alpha\Delta t}w - {}^{t}w\right)/\alpha\Delta t\right] = 0 \\
{}^{t+\Delta t}G_s &\approx G_s\left[{}^{t+\Delta t}u, {}^{t+\Delta t}dm + {}^{t}\xi, {}^{t+\Delta t}dn + {}^{t}\eta\right] = 0
\end{aligned}
$$

(24)

In order to solve the coupled system, Equation (24) is discretized. Assuming that the solution vector of the coupled system is $X = X(X_w, X_s)$, X_w, and X_s represents solution vectors of wind field and structure nodes. Therefore, $u_s = u_s(X_s)$ and $\tau_w = \tau_w(X_w)$, and the shell–wind coupling equation can be expressed as [25]

$$
\begin{aligned}
G_f\left[X_w^k, \lambda_d u_s^k + (1 - \lambda_d)u_s^{k-1}\right] &= 0 \\
G_s\left[X_s^k, \lambda_\tau \tau_w^k + (1 - \lambda_\tau)\tau_w^{k-1}\right] &= 0
\end{aligned}
$$

(25)

where λ_d and λ_τ are displacement and stress relaxation factors.

The above solving process can be illustrated by Figure 2.

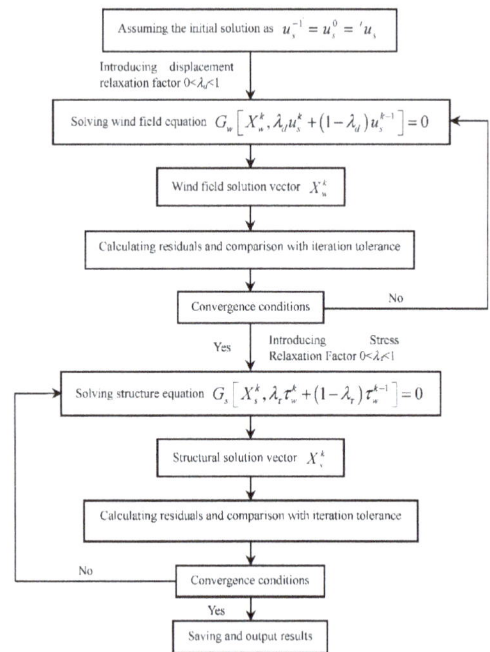

Figure 2. Shell–wind interaction solution.

5. Boundary Conditions

5.1. Wind Field Boundary Conditions

For high Reynolds number incompressible steady flow, velocity-inlet is chosen as the boundary condition at the entrance; pressure-outlet without backflow is chosen as the boundary condition at the outlet, that is, at the exit boundary of the flow field, the diffusion flux of the physical quantity of the flow field along the normal direction of the exit is 0; the non-slip wall boundary is used as boundary condition on the structure surface and ground. Symmetry is chosen as the boundary on both sides and on the top. The boundary conditions for wind field simulation are shown in Figure 3.

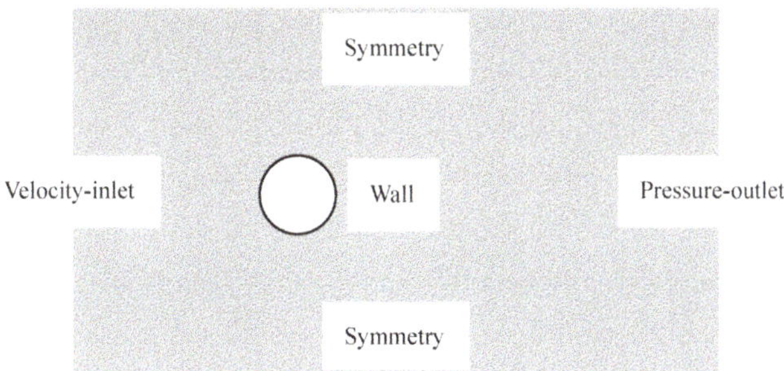

Figure 3. Boundary conditions for wind field simulation.

5.2. Shell–Liquid Interaction Boundary Conditions

The conditions of displacement continuity and force balance need to be satisfied at the shell–liquid interaction interface, namely

$$\mathbf{u}_s = \mathbf{u}_l, \mathbf{F}_s = \mathbf{F}_l \tag{26}$$

where $\mathbf{u}_s$ and $\mathbf{u}_l$ are structure and liquid displacement vectors; $\mathbf{F}_s$ and $\mathbf{F}_l$ are structure and liquid dragging forces.

$$\mathbf{F}_s = \boldsymbol{\sigma}_s \cdot \mathbf{n}_s \tag{27}$$

where $\mathbf{n}_s$ and $\mathbf{n}_l$ are interface normal vector; $\boldsymbol{\sigma}_s$ and $\boldsymbol{\sigma}_l$ are structure and liquid stress vectors.

6. Numerical Example

6.1. Calculation Model

The diameter and height of the tank are 21 m and 16 m, liquid storage height is 8 m. The wall thickness from the bottom to the top is as follows: 0–2 m is 14 mm; 2–4 m is 12 mm; 4–6 m is 10 mm; 6–10 m is 8 mm; and 10–16 m is 6 mm. Bilinear elastic-plastic material and shell elements are used to simulate a liquid storage tank, potential fluid material model and 3D solid element are used to simulated liquid, and liquid free surface is defined to reflect liquid sloshing behavior. El-Centro wave is selected as the ground motion input for time-history analysis.

Since there are a large number of liquid storage tanks in actual oil depots, it is necessary to study the influence of wind interference effect. By comparing the dynamic responses of single tanks and double tanks under wind load, the influence of wind interference effect on liquid storage tanks can be preliminarily discussed. Wind field is simulated by using 8-node 6-hedral FCBI-C element and large-eddy-simulation material. The calculation model material parameters are shown in Table 1, and the calculation model are shown in Figures 4 and 5.

Table 1. Material parameters.

Item	Parameters	Values
Tank	Elastic modulus/Pa	2.06×10^{11}
	Poisson's ratio	0.3
	Yield stress/MPa	235
	Tangent modulus/Pa	2.06×10^{9}
	Density/kg/m^3	7800
Liquid	Bulk modulus/Pa	3×10^{9}
	Density/kg/m^3	1000
	Viscosity coefficient/N'S/m	0.00113
Wind	Density/kg/m^3	1.29
	Viscosity/kg/(m's)	1.74×10^{-5}

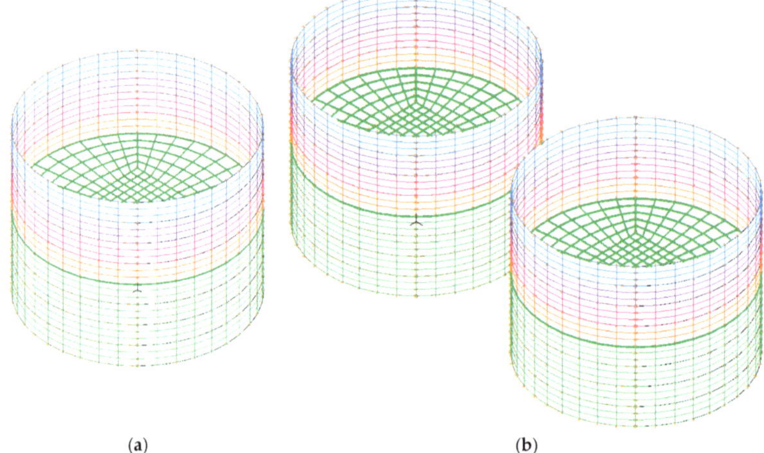

(**a**)　　　　(**b**)

Figure 4. Calculation model of shell–liquid interaction. (**a**) Single tank. (**b**) Double tank.

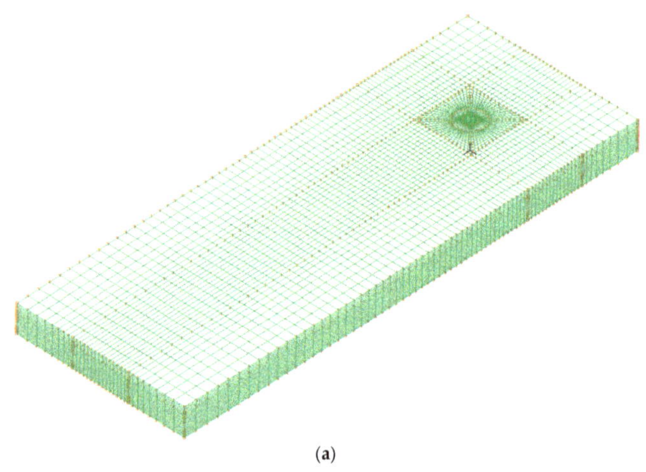

(**a**)

Figure 5. *Cont.*

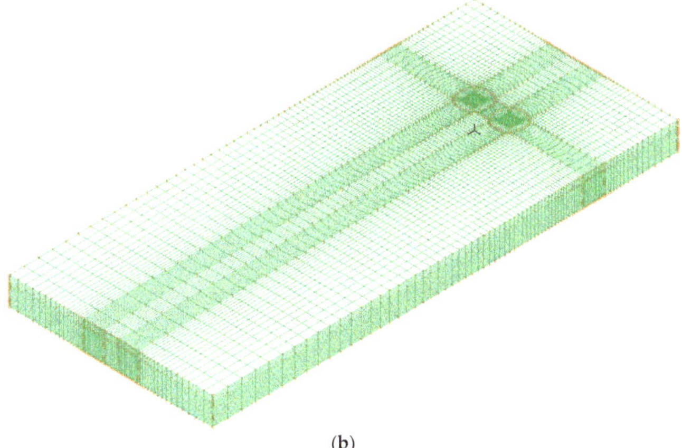

(**b**)

Figure 5. Wind field calculation model. (**a**) Single tank. (**b**) Double tank.

6.2. Comparison of Dynamic Responses under Different Actions

In view of the possibility that the liquid storage tank may be damaged under the action of wind and earthquake, and the combined action of wind and earthquake will have more adverse effects on the structure, a comparative study on the dynamic response of the liquid storage tank under the action of wind, earthquake, and wind and earthquake is carried out, and the specific results are shown in Figures 6–8 and Table 2.

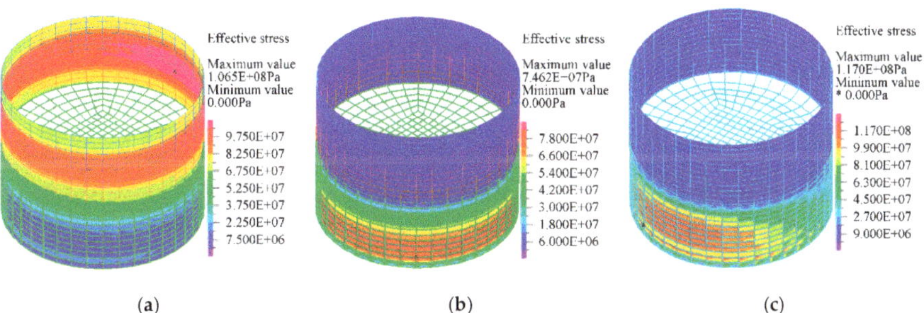

(**a**) (**b**) (**c**)

Figure 6. Comparison of tank effective stress (unit: Pa). (**a**) Wind. (**b**) Earthquake. (**c**) Wind and earthquake.

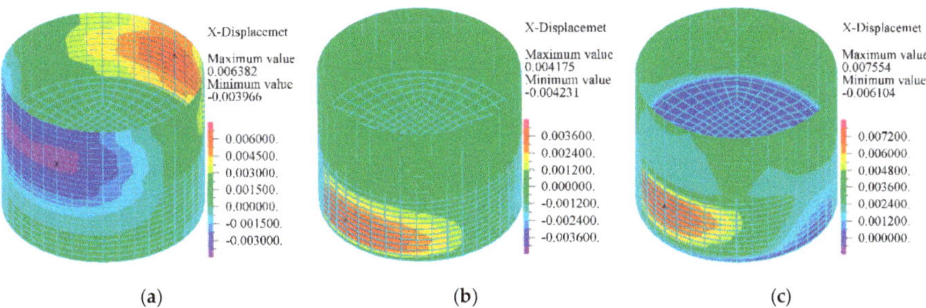

(**a**) (**b**) (**c**)

Figure 7. Comparison of tank displacement (unit: m). (**a**) Wind. (**b**) Earthquake. (**c**) Wind and earthquake.

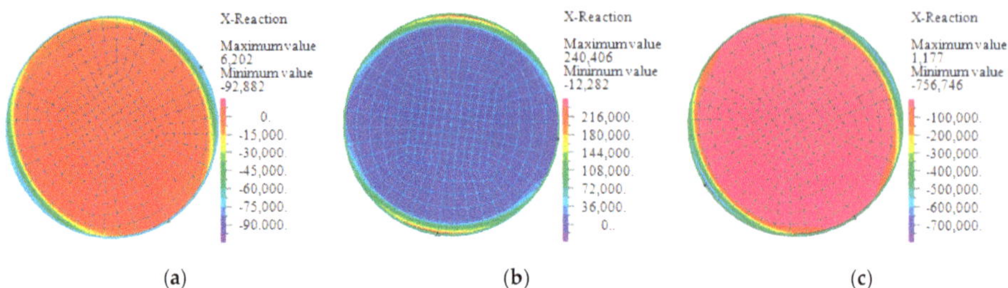

Figure 8. Comparison of tank base shear force (unit: N). (**a**) Wind. (**b**) Earthquake. (**c**) Wind and earthquake.

Table 2. Absolute maximum dynamic response under different actions.

Dynamic Responses	Actions	Maximum Absolute Value
	Wind	106.50
Effective stress/MPa	Earthquake	74.62
	Wind and earthquake	117.03
	Wind	6.38
Displacement/mm	Earthquake	4.23
	Wind and earthquake	7.55
	Wind	92.88
Base shear force/kN	Earthquake	240.41
	Wind and earthquake	756.75

As shown in Figures 6–8, the maximum effective stress and displacement of tank under wind load is located in the unfilled area of the upper part of the liquid storage tank, while the maximum effective stress and displacement of the structure under earthquake is located in the filled area of the lower part of the liquid storage tank, and the maximum of base shear force appears near the contact position between the tank wall and foundation.

As shown in Table 2, it can be seen that when the wind speed is larger, the effective stress and displacement under wind load is greater than that under earthquake. However, a large number of researches on liquid storage tanks have been carried out on the basis of considering only the earthquake action, so there are some defects. Besides, the dynamic responses of tank under the combined action of wind and earthquake are obviously greater than those under the separate action of wind and earthquake. The effective stress, displacement, and base shear force obtained by SRSS are 130.04 Mpa, 7.65 mm and 257.73, respectively, which are very different from considering the interaction of wind and earthquake at the same time.

Therefore, when the wind speed is large, the influence of wind load on the liquid storage tank cannot be ignored. The location of maximum dynamic responses of liquid storage tank under wind and earthquake is different, and the combined effect of wind and earthquake will have a more adverse impact on the liquid storage tank. Therefore, for the liquid storage tank in special areas (such as coastal areas), the combined effect of wind and earthquake should be reasonably considered in its design.

6.3. Influences of Wind Speed on Dynamic Responses

Through the above analysis, it has been found that the influence of wind load on the liquid storage tank cannot be ignored. In order to further discuss the responses of the liquid storage tank under wind load, a comparative study is carried out under the wind speed of 10 m/s and 20 m/s. The nephograms of the effective stress, displacement, and base shear force are shown in Figures 9–11 and Table 3.

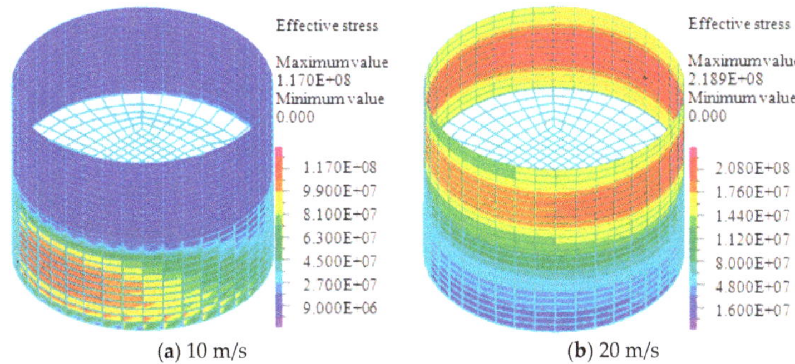

(**a**) 10 m/s   (**b**) 20 m/s

Figure 9. Effect of wind speed on tank effective stress (unit: Pa).

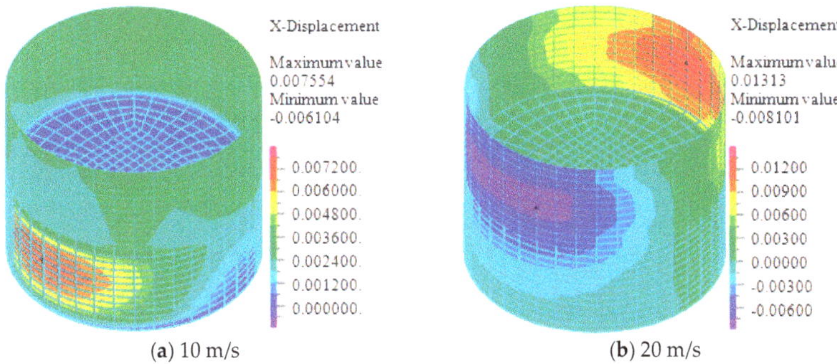

(**a**) 10 m/s   (**b**) 20 m/s

Figure 10. Effect of wind speed on tank displacement (unit: m).

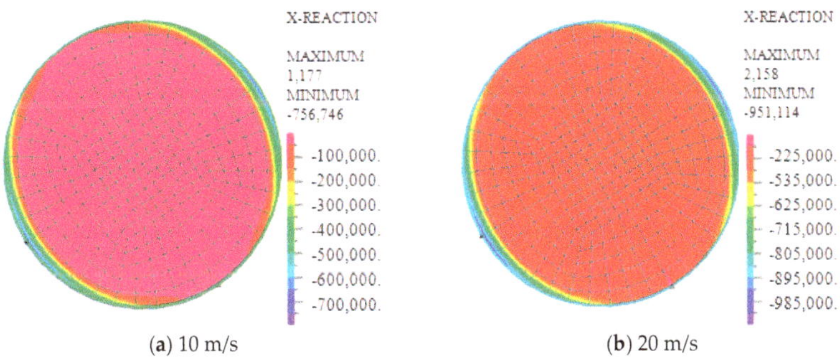

(**a**) 10 m/s   (**b**) 20 m/s

Figure 11. Effect of wind speed on tank base shear force (unit: N).

Table 3. Effect of wind speed on absolute maximum dynamic responses.

Wind Speed	Dynamic Responses		
	Effective Stress/MPa	Displacement/mm	Base Shear Force/kN
10 m/s	117.03	7.55	756.75
20 m/s	218.96	13.13	951.11

As shown in Figures 9–11, under the combined action of wind and earthquake, when the wind speed is 10 m/s, the location of maximum effective stress and displacement of the tank is located in the liquid filled area at the bottom of the liquid storage tank, but when the wind speed increases to 20 m/s, the location of maximum effective stress and displacement of the structure shifts to the unfilled area at the upper part of the liquid storage tank. That is to say, when the wind speed is lower, the responses of liquid storage tanks are dominated by earthquake, on the contrary, when the wind speed is higher, the responses of liquid storage tanks will be dominated by wind.

As shown in Table 3, when the wind speed is increased by 2 times, the effective stress, displacement, and base shear force are significantly increased, especially the effective stress and displacement are approximately increased by 2 times.

Therefore, when the wind speed is high, the probability of damage to the liquid storage tank is relatively high. For liquid storage tanks built in special areas, sufficient attention should be paid to the adverse effects of wind load.

6.4. Wind Interference Effect

Significant wind disturbance effect exists in group structures, Zhao et al. [26] obtained that amplification effect caused by wind disturbance reaches to 20–40% through wind tunnel test. Zhang et al. [27] obtained that the unfavorable influence of double-row arrangement of towers is obviously larger than that of single-row arrangement.

Through the research on the influence of wind speed on the dynamic response of the structure, it is found that the larger the wind speed, the more unfavorable it is to the tank. Therefore, taking the wind speed of 20 m/s as an example, single tanks and double tanks are selected as research objects to study the influence law of wind interference effect on the dynamic responses of liquid storage tanks. The comparisons of effective stress, displacement, base shear force, and velocity field are shown in Figures 12–15 and Table 4.

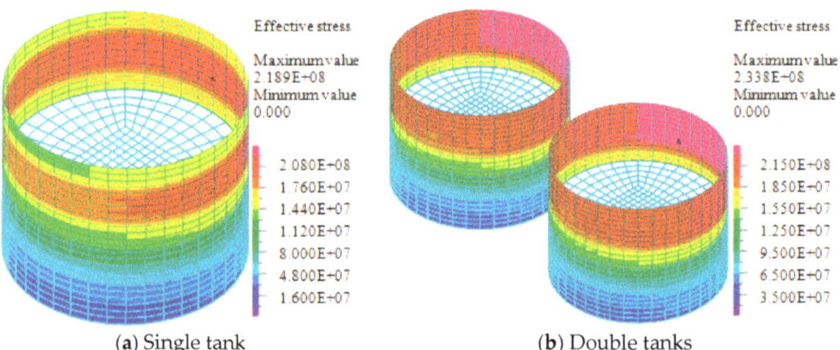

(**a**) Single tank   (**b**) Double tanks

Figure 12. Effect of wind interference on effective stress (unit: Pa).

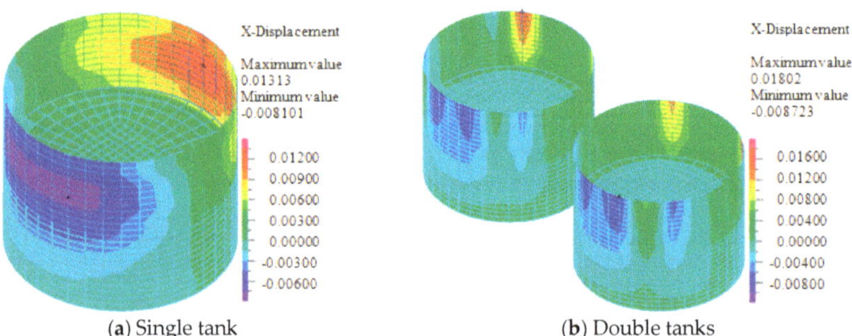

(**a**) Single tank   (**b**) Double tanks

Figure 13. Effect of wind interference on displacement (unit: m).

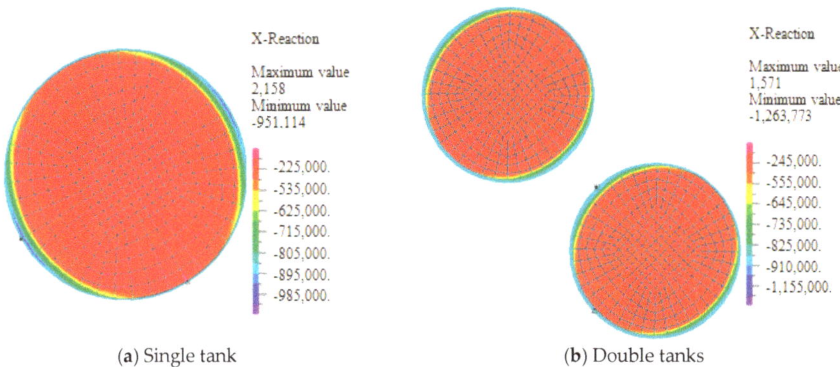

(**a**) Single tank (**b**) Double tanks

Figure 14. Effect of wind interference on base shear force (unit: N).

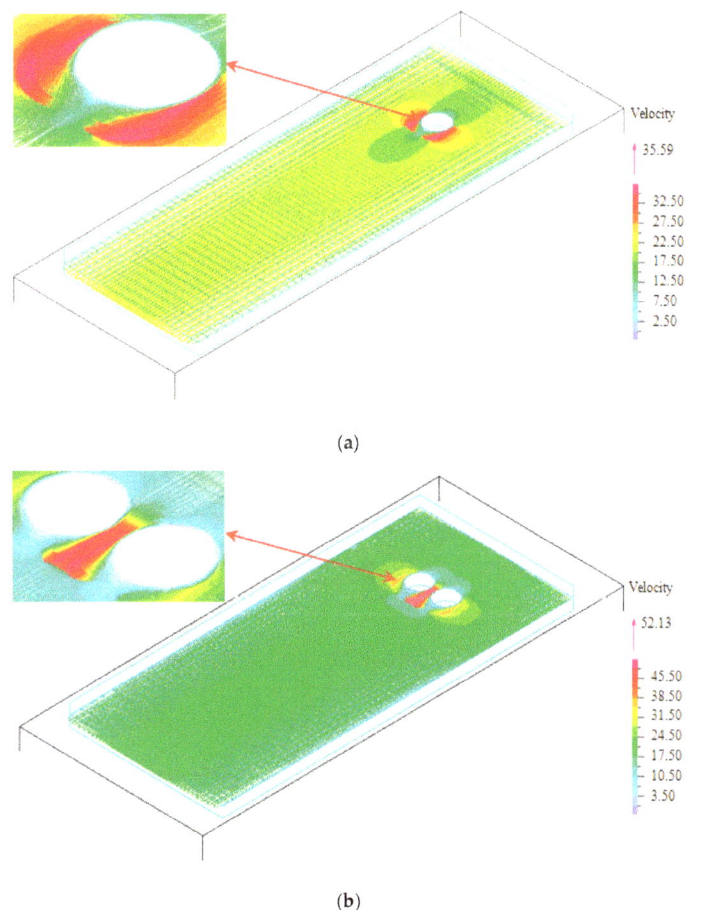

(**a**)

(**b**)

Figure 15. Effect of wind interference on wind velocity field (unit: m/s). (**a**) Single tank. (**b**) Double tank.

Table 4. Effect of wind interference on absolute maximum responses.

Item	Dynamic Responses			
	Effective Stress/MPa	**Displacement/mm**	**Base Shear Force/kN**	**Wind Velocity/m/s**
Single tank	218.96	13.13	951.11	35.59
Double tank	233.81	18.02	1263.77	52.13
Difference ratio	9.07%	37.24%	32.85%	46.47%

As shown in Figures 12–15 and Table 4, without considering and considering the wind interference effect, the maximum absolute values of effective stress of the tank are 218.96 MPa and 233.81 MPa, respectively; the maximum absolute values of the displacement of the tank are 13.13 mm and 18.02 mm, respectively; the maximum absolute values of the base shear force of the tank are 951.11 kN and 1263.77 kN, respectively; and the maximum absolute values of the wind field speed are 35.59 m/s and 52.13 m/s, respectively. The difference ratios corresponding to effective stress, displacement, base shear force and wind velocity are 9.07%, 37.24%, 32.85%, and 46.47%.

For a single tank, the maximum dynamic response is located on the tank axis and the maximum wind speed is located near the tank walls on both sides; while for a double tank, the maximum dynamic response shifts to the side between the two tanks, and the maximum wind speed is located in the area between the two tanks.

It can be seen that the wind interference effect has a great influence on the dynamic response and wind field of liquid storage tanks. Liquid storage tanks in actual oil storage facilities are basically arranged side by side. In order to ensure their safety, it is necessary to consider the wind interference effect.

7. Conclusions

Considering shell–liquid interaction and shell–wind interaction, calculation model of liquid storage tanks is established. The dynamic responses of liquid storage tanks under wind, earthquake, and wind and earthquake are studied comparatively. Besides, the influences of wind speed and wind interference effect on the dynamic responses of the liquid storage tank are discussed. The main conclusions are as follows:

(1) Although it is commonly believed that structure dynamic responses are usually dominated by either wind or earthquake, when wind speed is high, responses of liquid storage tanks under wind are greater than that under earthquake, besides, responses of liquid storage tanks under combination of wind and earthquake are more important. Results indicate that it is necessary to consider the combination of wind and earthquake actions in the design of liquid storage tank.

(2) When wind speed increases from 10 m/s to 20 m/s, under combined action of earthquake and wind, tank dynamic responses are significantly increased; especially, tank wall stress is even close to the yield strength of steel (235 Mpa).

(3) Liquid storage tanks have different performance under earthquake or wind, the maximum responses under wind load are located in the upper region without liquid filling, while the maximum responses under earthquake are located in the lower area of liquid storage tank; while the locations of maximum responses under combination of wind and earthquake are related to wind speed.

(4) Wind disturbance effect has a significant influence on liquid storage tanks and the wind field, and the position of maximum response will be also changed after wind disturbance effect being considered. Besides, dynamic responses corresponding to double tanks are obviously larger than that of single tanks, it can be seen that if the wind disturbance effect is not considered, the responses will be underestimated.

(5)   In order to consider combination of wind and earthquake, the effective stress, displacement, and base shear force obtained by conducting SRSS for wind and earthquake alone conditions are very different from considering wind and earthquake at the same time.

Author Contributions: Methodology, W.J.; software, W.J. and H.F.; formal analysis, W.J. and H.F.; writing—original draft preparation, W.J.; writing—review and editing, X.C.

Funding: This paper is a part of the China Postdoctoral Science Foundation (grant no. 2018M633652XB), a part of the National Natural Science Foundation of China (grant no. 51368039), a part of the Hongliu Outstanding Young Talents Support Program of Lanzhou University of Technology (grant no. 04-061807) and a part of the Open Foundation of International Research Base on Seismic Mitigation and Isolation of Gansu Province (grant no. TM-QK-1904).

Conflicts of Interest: The authors declare no conflict of interest.

References

1.   Holroyd, R.J. On the behaviour of open-topped oil storage tanks in high winds. part II. structural aspects. *J. Wind Eng. Ind. Aerod.* **1983**, *12*, 329–352. [CrossRef]
2.   Vela, R.J.M.; Brunesi, E.; Nascimbene, R. Seismic assessment of an industrial frame-tank system: Development of fragility functions. *Bull. Earthq. Eng.* **2019**, *17*, 2569–2602. [CrossRef]
3.   Rawat, A.; Matsagar, V.A.; Nagpal, A.K. Numerical study of base-isolated cylindrical liquid storage tanks using coupled acoustic-structural approach. *Soil Dyn. Earthq. Eng.* **2019**, *119*, 196–219. [CrossRef]
4.   Kotrasová, K.; Grajciar, I.; Kormaníková, E. Dynamic time-history response of cylindrical tank considering fluid-structure interaction due to earthquake. *Appl. Mech. Mater.* **2014**, *617*, 66–69. [CrossRef]
5.   Gilmanov, A.; Le, T.B.; Sotiropoulos, F. A numerical approach for simulating fluid structure interaction of flexible thin shells undergoing arbitrarily large deformations in complex domains. *J. Comput. Phys.* **2015**, *300*, 814–843. [CrossRef]
6.   Ishikawa, S.; Kondou, T.; Matsuzaki, K.; Yamamura, S. Analysis of nonlinear shallow water waves in a tank by concentrated mass model. *J. Sound Vib.* **2016**, *371*, 171–182. [CrossRef]
7.   Moslemi, M.; Farzin, A.; Kianoush, M.R. Nonlinear sloshing response of liquid-filled rectangular concrete tanks under seismic excitation. *Eng. Struct.* **2019**, *188*, 564–577. [CrossRef]
8.   Miladi, S.; Razzaghi, M.S. Failure analysis of an un-anchored steel oil tank damaged during the Silakhor earthquake of 2006 in Iran. *Eng. Fail. Anal.* **2019**, *96*, 31–43. [CrossRef]
9.   Ormeño, M.; Larkin, T.; Chouw, N. Experimental study of the effect of a flexible base on the seismic response of a liquid storage tank. *Thin-Walled Struct.* **2019**, *139*, 334–346. [CrossRef]
10.   Sanapala, V.S.; Sajish, S.D.; Velusamy, K.; Ravisankar Patnaik, B.S.V. An experimental investigation on the dynamics of liquid sloshing in a rectangular tank and its interaction with an internal vertical pole. *J. Sound Vib.* **2019**, *449*, 43–63. [CrossRef]
11.   Rawat, A.; Mittal, V.; Chakraborty, T.; Matsagar, V. Earthquake induced sloshing and hydrodynamic pressures in rigid liquid storage tanks analyzed by coupled acoustic-structural and Euler-Lagrange methods. *Thin-Walled Struct.* **2019**, *134*, 333–346. [CrossRef]
12.   Flores, F.G.; Godoy, L.A. Buckling of short tanks due to hurricanes. *Eng. Struct.* **1998**, *20*, 752–760. [CrossRef]
13.   Portela, G.; Godoy, L.A. Wind pressures and buckling of cylindrical steel tanks with a dome roof. *J. Constr. Steel Res.* **2005**, *61*, 808–824. [CrossRef]
14.   Zhang, X.; Jin, X.L.; Chen, X.D. Numerical simulation of dynamic characteristics of flexible container under wind load. *J. Vib. Shock* **2009**, *28*, 115–118. [CrossRef]
15.   Yasunaga, J.; Koo, C.; Uematsu, Y. Wind loads for designing cylindrical storage tanks part 2 wind force model with consideration of the buckling behavior under wind loading. *J. Wind Eng.* **2012**, *37*, 79–92. [CrossRef]
16.   Chen, L.; Rotter, J.M. Buckling of anchored cylindrical shells of uniform thickness under wind load. *Eng. Struct.* **2012**, *41*, 199–208. [CrossRef]
17.   Zhao, Y.; Lin, Y.; Yu, S.C. Wind-tunnel test of wind loads on large cylindrical structures with very low aspect ratio. *J. Zhejiang Univ. (Eng. Sci.)* **2014**, *48*, 820–826. [CrossRef]
18.   Lin, Y.; Zhao, Y.; Lin, W. Wind load test and buckling analysis of large steel open-top tanks. *J. Huazhong Univ. Sci. Technol. (Nat. Sci. Ed.)* **2014**, *42*, 66–71. [CrossRef]

19. Hong, X.; Gu, M. Probability model and solution on earthquake effects combination in along wind resistant design of tall-flexible buildings. *Appl. Math. Mech.* **2006**, *27*, 555–563. [CrossRef]

20. Ke, S.T.; Zhao, L.; Ge, Y.J. Comparison of super-large cooling towers under earthquake excitation and wind load. *J. Harbin Inst. Technol.* **2010**, *42*, 1635–1641.

21. Peng, W.S.; Cao, X.W.; Xin, M.D.; Wang, Q.; Liu, W.; Song, C.Y.; Wang, T.T. Analysis of the wind and seismic stability of column-supported conical bottom tank. *Sci. Technol. Eng.* **2014**, *14*, 144–150.

22. Sapountzakis, E.J.; Dikaros, I.C.; Kampitsis, A.E.; Koroneou, A.D. Nonlinear response of wind turbines under wind and seismic excitations with soil-structure interaction. *J. Comput. Nonlinear Dyn.* **2015**, *10*, 1–16. [CrossRef]

23. Mazza, F. Wind and earthquake dynamic responses of fire-exposed steel framed structures. *Soil Dyn. Earthq. Eng.* **2015**, *78*, 218–229. [CrossRef]

24. Sussman, T.; Sundqvist, J. Fluid-structure interaction analysis with a subsonic potential-based fluid formulation. *Comput. Struct.* **2003**, *81*, 949–962. [CrossRef]

25. Zhang, H.; Zhang, X.L.; Ji, S.H.; Guo, Y.H.; Gustavo, L.; Elabbasi, N.; DeCougny, H. Recent development of fluid-structure interaction capabilities in the ADINA system. *Comput. Struct.* **2003**, *81*, 1071–1085. [CrossRef]

26. Zhao, L.; Ge, Y.J.; Xu, L.S. Wind tunnel investigation on wind-induced interference effects for super large cooling towers. *Eng. Mech.* **2009**, *26*, 149–154.

27. Zhang, J.F.; Ge, Y.J.; Zhao, L. Interference effects on global wind loads and wind induced responses for group hyperboloidal cooling towers. *Eng. Mech.* **2016**, *33*, 15–23.

Article

Experimental Study of High-Strength Concrete-Steel Plate Composite Shear Walls

Dongqi Jiang [1],*, Congzhen Xiao [2], Tao Chen [2] and Yuye Zhang [1]

[1]   Department of Civil Engineering, School of Science, Nanjing University of Science and Technology, Nanjing 210094, China
[2]   Structural Institute, China Academy of Building Research, Beijing 100013, China
*   Correspondence: jiangdongqi@njust.edu.cn; Tel.: +86-25-84315773

Received: 17 June 2019; Accepted: 8 July 2019; Published: 15 July 2019

Abstract: Shear walls are effective lateral load resisting elements in high-rise buildings. This paper presents an experimental study of the seismic performance of a composite shear wall system that consists of high-strength concrete walls with the embedded steel plate. Two sets of wall specimens with different aspect ratios (height/width, 1.5 and 2.7) were constructed and tested under quasi-static reversed cyclic loading, including five reinforced concrete shear walls (RCSW) and six reinforced concrete-steel plate shear walls (RCSPSW). The progression of damage, failure modes, and load-displacement responses of test specimens were studied and compared based on experimental observations. The test results indicated that high-strength (HS) RCSPSW system showed superior lateral load strength and acceptable deformation capability. The axial compressive load was found to have an indispensable effect on the ductility of both RCSW and RCSPSW, and an upper limit of axial compression ratio (0.5) is recommended for the application of HS RCSPSW in engineering practices. In addition, the design strength models were suggested for predicting the shear and flexure peak strength values of RCSPSW systems, and their applicability and reliability were verified by comparing with test results.

Keywords: composite shear wall; seismic behavior; quasi-static test; design strength model

1. Introduction

Shear walls serve as an effective structural element for resisting lateral load in tall buildings during seismic events. Conventional reinforced concrete shear walls (RCSW) are considered to be a cost-effective way for preventing structure collapse and ensure life-safety, which has been widely used in the design of low- and medium-rise building structures in seismic regions [1]. With the increase of building height, vertical axial load demands at lower floors exponentially grow, resulting in much thicker RCSW with the consideration of code-specified axial compression ratio limits [2–4]. Under such circumstance, excessively thick walls reduce the usable floor areas and increase the gravity load intensity, causing more seismic forces and severer structural damage. A review of past research work indicated that RCSW were mainly tested under low axial compressive loadings [1,5–7]. The first explanation is that reinforced concrete buildings are usually designed with a low axial compression ratio in high seismicity zones due to code limits. The second possibility is that conventional RCSW undergo significant structural damage in the form of undesirable failure patterns and deteriorated strength and ductility performances when subjected to high axial loadings [8]. Su [9] observed an abrupt out-of-plane compressive failure mode on experimentally tested RCSW under high axial compression ratios, and concluded that significant strength degradation and ductility deterioration would occur with the increase of the axial compression ratio. Moradi [10] conducted a comprehensive research study to set up a library of critical parameters that affect the behavior of shear walls based on a large amount of existing experimental tests, from which a predictive meta-model was developed to forecast the responses of desired shear walls.

Researchers explored the composite steel-concrete wall system by adding structural steel in the wall section, which potentially takes advantages of two materials' mechanical properties in order to improve the seismic performance of concrete shear walls under high axial loading and decrease the wall thickness. Dan [11] tested composite shear walls with I-shape or box-shape steel profiles that were placed at the extremities, and found that the lateral resistance and displacement ductility can be effectively improved by encased steel profiles. Tong and Hajjar [12,13] studied RCSW partially restrained by the steel frame at boundaries. It is concluded from the experimental results that this composite structural system provided adequate strength and stiffness to resist lateral forces and it would be applicable for low- and medium-rise buildings. Some researchers investigated the mixed use of RCSW and circular or rectangular concrete-filled steel tubular (CFST) columns [14–16]. The existing experimental results indicated that the strength and ductility performances were effectively improved by using CFST columns as the boundary elements. Previous research studies show that the structural steel is usually arranged at the extremities of reinforced concrete wall sections to form composite boundary elements; Cho [17], Esaki [18], and Zhou [19] conducted similar research work. For such a structural system, RCSW mainly serve as a shear-resisting component, while the function of composite boundary elements is to resist the overturning moment.

Apart from the use of composite boundary elements, researchers began to make use of the structural steel plate in the wall panel to further improve the lateral load performance in recent years. The concrete-steel plate composite shear wall system can be classified into two types: concrete-filled steel plate (CFSP) shear wall and steel-plate-embedded concrete shear wall. Hu and Nie [20,21] experimentally studied the CFSP composite shear walls and concluded that this wall system has larger lateral strength and deformation capacity under high axial compressive loadings. Hossain [22] and Rafie [23] reported similar findings on the shear wall system that consists of two skins of profiled steel sheeting with an in-fill of concrete. However, corrosion-protection and fire-protection layers are necessary for CFSP walls, because the steel plate is directly exposed to the environment, which increases the difficulty and complexity of construction. Therefore, the steel-plate-embedded concrete shear wall has advantages in engineering practice and has attracted the attention of researchers and practitioners.

Zhao [24] proposed a composite shear wall system comprising an infill steel plate with concrete panels that are attached on one side or both sides, and it is included in AISC Seismic Provisions and denoted as concrete stiffened steel plate shear wall (CSPSW) [25]. Note that a gap exists between concrete panel and the steel frame, thus the concrete panel merely works as a stiffer to prevent the bulking of the infill steel plate. The China Academy of Building Research (CABR) systematically studied the monolithic cast-in-place (CIP) RCSW with embedded steel plate, named as reinforced concrete-steel plate shear walls (RCSPSW). Figure 1 presents configurations of conventional RCSW and RCSPSW. The latter wall system consists of three major components, as depicted in Figure 1b. Concrete and steel reinforcement are identical to RCSW, while the I-shaped steel profiles are encased in boundary elements and the steel plate is embedded in the wall web. Sun [26] and Chen [27] investigated the shear and flexure behavior of normal-strength concrete-SP shear walls and demonstrated their good performances in both strength and deformation capabilities. Xiao [28] studied the effects of aspect ratios on lateral load performances of RCSPSW based on the previous research outcome. Subsequently, RCSPSW was successfully applied in the construction of high-rise buildings in China, such as: Shanghai Tower (632 m in height), Guangzhou East Tower (530 m in height), etc. [29].

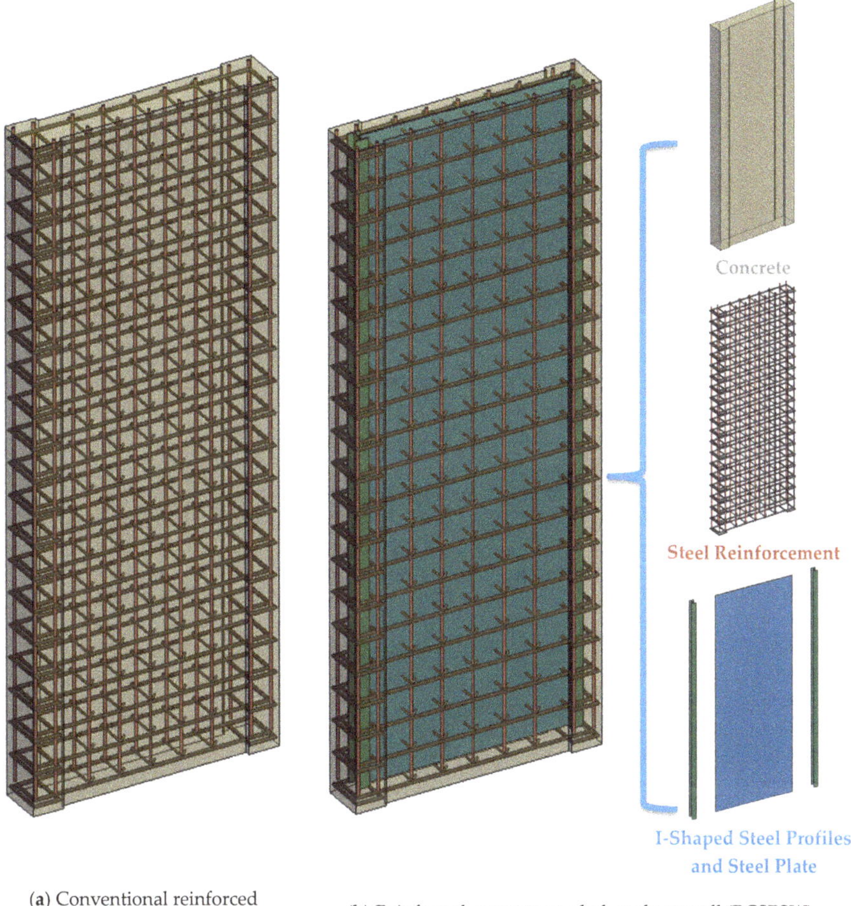

(**a**) Conventional reinforced concrete shear wall (RCSW)

(**b**) Reinforced concrete-steel plate shear wall (RCSPSW)

Figure 1. The Concept of Reinforced Concrete–Steel Plate Composite Shear Wall System.

Not content with the benefits that RCSPSW brought, practitioners claimed that there is still a need to further reduce the wall thickness for super-tall buildings. In response to their expectation, the use of high-strength (HS) concrete in RCSPSW system becomes a potential option, because the high compressive strength of HS concrete can be an advantage when the walls are subjected to high axial compressive loadings. Meanwhile, whether the brittleness of HS concrete affects the ductility of RCSPSW remains in question. Jiang [30] and Xiao [31] investigated the compression-bending behavior of HS concrete shear walls and explore viable structural steel arrangements in the wall section to improve the deformability. The research outcome demonstrated that the flexure strength and ductility were effectively improved by the steel profiles encased in boundary elements. However, the overall seismic performances of HS RCSPSW system remain unknown to researchers. Additionally, the progressions of damage and failure modes of HS RCSPSW with different aspect ratios are not clear. The authors have conducted a comprehensive experimental investigation to study both shear and flexure behavior of HS RCSPSW in order to verify the reliability of HS RCSPSW system in engineering practices. Eleven high-strength concrete shear walls with two different aspect ratios (1.5 and 2.7) were constructed and tested under quasi-static reversed cyclic loading, including five conventional RCSW and six RCSPSW. Lateral load performance and failure modes are compared and discussed based on experimental observations. This paper emphasizes the experimental investigation and the design strength models of HS RCSPSW. Analytical and numerical models, as well as ways to incorporate axial–shear-flexure interaction, will be presented in a follow-up paper.

2. Experimental Investigation

2.1. Specimen Properties

Two batches of specimens were designed, constructed, and tested under quasi-static cyclic loadings, including RCSW and RCSPSW. Table 1 lists key information for test specimen configurations. Specifically, the specimens in Batch No.1 (RCSW 1-X and RCSPSW 1-X) with the aspect ratio (height/width, h/b) of 1.5 were designed to study the shear behavior, while the specimens in Batch No.2 (RCSW 2-X and RCSPSW 2-X) with a higher aspect ratio (2.7) were mainly designed for the flexure behavior investigation. These two aspect ratios were selected based on available references [1,8–10,14,19,32–35] and past engineering experience in high-rise building design to purposely achieve shear failure mode and flexure failure mode, respectively. Apart from the aspect ratio, the vertical axial load is another critical parameter that affects the seismic performance of shear wall specimens. As listed in Table 1, a wide range of vertical load values were applied on shear specimens (900–1500 kN) and flexure specimens (2180–3050 kN) in the quasi-static test to simulate different levels of axial loads that would be expected in the design of high-rise buildings. An axial compression ratio, n, is defined as the ratio of the applied vertical load to the compression capacity of the wall section and it is given by

$$n = \frac{N}{f_c A_c + f_{ay} A_a + f_{py} A_p} \tag{1}$$

where N represents the vertical load that is applied by the hydraulic jack, f_c is the compressive strength of concrete, f_{ay} and f_{py} are yielding strength values of I-shape steel profiles and steel plate, and A_c, A_a, and A_p indicate the area of concrete section, boundary I-shaped steels, and embedded steel plate. For given vertical loads, axial compression ratio ranges from 0.33 to 0.50 for shear specimens and 0.42 to 0.70 for flexure specimens. Note that these values are relatively high for shear walls in building structures, and very few past research work has been performed while using such high axial compression ratios.

Figure 2 shows the cross-sectional dimensions and reinforcing detailing for RCSW and RCSPSW. All of the specimens were designed with the target concrete cubic compressive strength of $f'_{c,cube} = 80$ MPa. The use of high-strength concrete not only effectively increases the lateral capacity, but it also significantly reduces the required wall thickness due to code-specified axial compression ratio limits in the design of tall buildings [3,4]. The geometrical sizes and steel reinforcement were carefully designed, so that desirable failure mechanisms would take place in the test for different batches of specimens. For instance, longitudinal boundary reinforcements with relatively smaller diameters ($\varphi 8$ and $\varphi 10$) were purposely implemented in specimens with the aspect ratio of 2.7, so as to observe a flexure mode of failure prior to achieving a shear mode of failure. The geometric dimensions and reinforcement arrangement were identical for RCSW and RCSPSW specimens with specified aspect ratios. The shear specimens consist of an 800 mm wide by 80 mm thick wall web and two 100 mm wide by 120 mm thick boundary elements. As for flexure specimens, the thickness values for the web and boundary elements are 150 mm and 190 mm, respectively. The distances that were measured from the top surface of the footing to the lateral loading point are 1.2 m and 2.16 m for shear and flexure specimens, as shown in Figure 2. For the RCSPSW specimens, I-shaped steels in boundary elements were welded to the 5 mm thick steel plate that was embedded in the wall web, and steel studs or tie bars were used to ensure the bonding strength between the concrete and steel plate. Longitudinal reinforcement was placed in boundary elements, and steel ratios of 16.9% and 1.35% were used for shear and flexure specimens, respectively. Table 1 lists the steel reinforcement ratios in wall web.

Table 1. Test Specimen Configurations.

Batch Number	Specimen Number	Aspect Ratio	Steel Reinforcement Ratio			Vertical Load (kN)	Axial Compression Ratio
			Web Longitudinal	Web Transverse	Boundary Longitudinal		
1	RCSW 1-1	1.5	0.71%	0.71%	16.9%	900	0.45
	RCSW 1-2					1000	0.50
	RCSPSW 1-1					1000	0.33
	RCSPSW 1-2					1380	0.45
	RCSPSW 1-3					1500	0.50
2	RCSW 2-1	2.7	0.38%	0.67%	1.35%	2180	0.50
	RCSW 2-2					2610	0.60
	RCSW 2-3					3050	0.70
	RCSPSW 2-1					2180	0.42
	RCSPSW 2-2					2610	0.50
	RCSPSW 2-3					3050	0.58

RCSW #-# represents conventional reinforced concrete shear walls and RCSPSW #-# represents new reinforced concrete–steel plate composite shear walls.

Table 2 lists key material properties of test specimens. The 28-day cubic compressive strength, $f'_{ck,cube}$, for shear specimens and flexure specimens are 72.3 MPa and 84.1 MPa, respectively. Two types of steel reinforcements (HPB235 and HRB 335) were used in the wall specimens. Table 2 summarizes the measured yield strength f_{yk} and ultimate strength f_{tk} of steel reinforcements and steel plates.

Table 2. Material Properties.

Material Specimen Number	Concrete		Steel Reinforcement				Steel Plate	
	Compressive Strength $f_{dk,cube}$ (MPa)	Tensile Strength f_{ctk} (MPa)	Yield Strength F_{yk} (MPa)		Ultimate Strength f_{tk} (MPa)		Yield Strength f_{yk} (MPa)	Ultimate Strength f_{tk} (MPa)
			HPB235	HRB335	HPB235	HRB335	Q345	
RCSW 1-X	72.3	3.75	267	454	403	628		
RCSPSW 1-X							367	477
RCSW 2-X	84.1	3.93	295	441	419	595		
RCSPSW 2-X							322	435

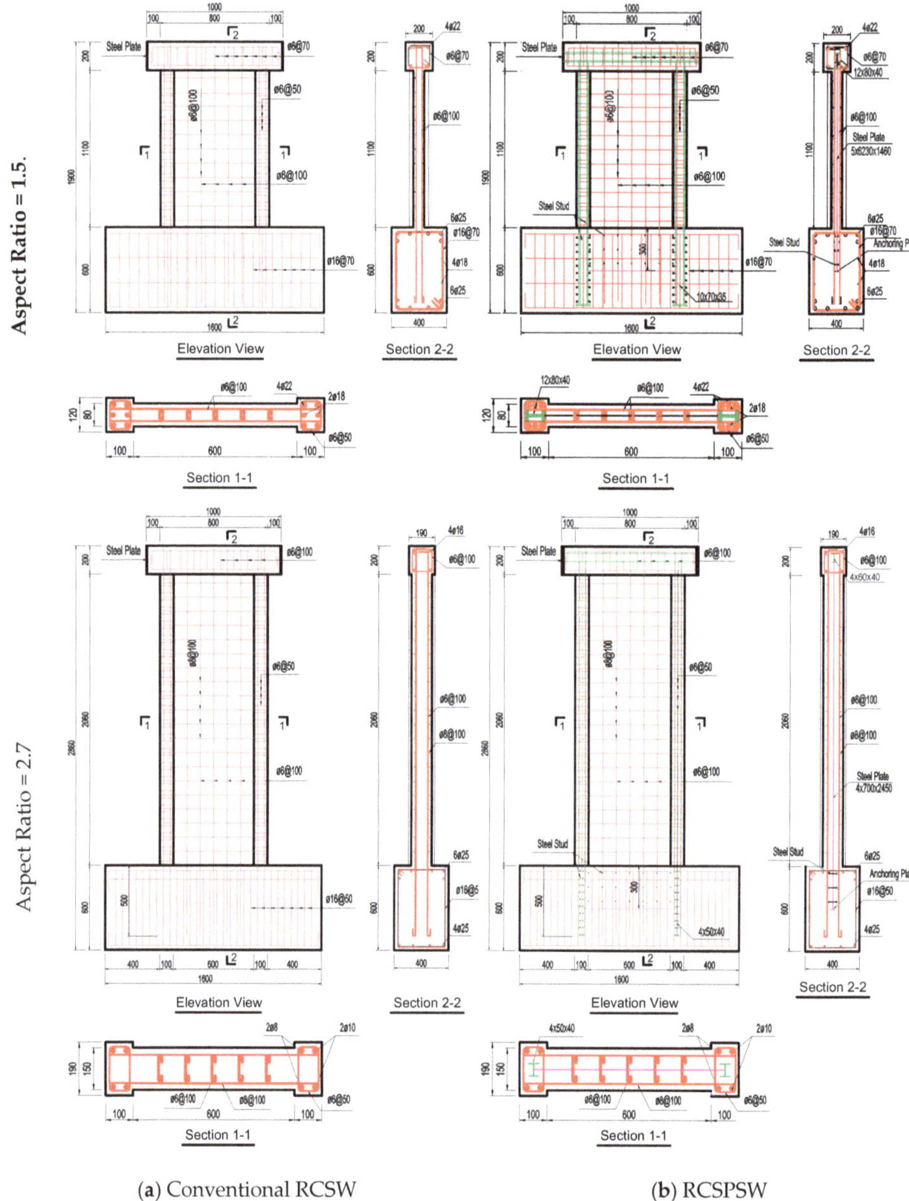

(**a**) Conventional RCSW　　　　　　　　(**b**) RCSPSW

Figure 2. Cross-sectional Dimensions and Reinforcement Details of RCSW and RCSPSW.

2.2. Test Setup, Instrumentation and Load Protocol

The quasi-static load test was conducted on shear and flexure specimens to evaluate the seismic performance of the HS RCSPSW system. Figure 3 presents the experimental setup details. Wall specimen was anchored onto the laboratory rigid floor with fasteners through the footing. One 1500 kN servo-controlled hydraulic actuator along the east-west direction was installed on the reaction wall to provide racking loads on the specimen during the reversed cyclic load test. The vertical load was applied through a 4000 kN capacity hydraulic jack that was installed beneath the rigid steel frame, and the vertical load is uniformly spread along the top surface of the wall specimen by means of two steel distribution beams. Out-of-plan steel bracings were used to keep the specimen movement in-plane and avoid twisting during testing, as can be seen in the setup photograph (Figure 3d).

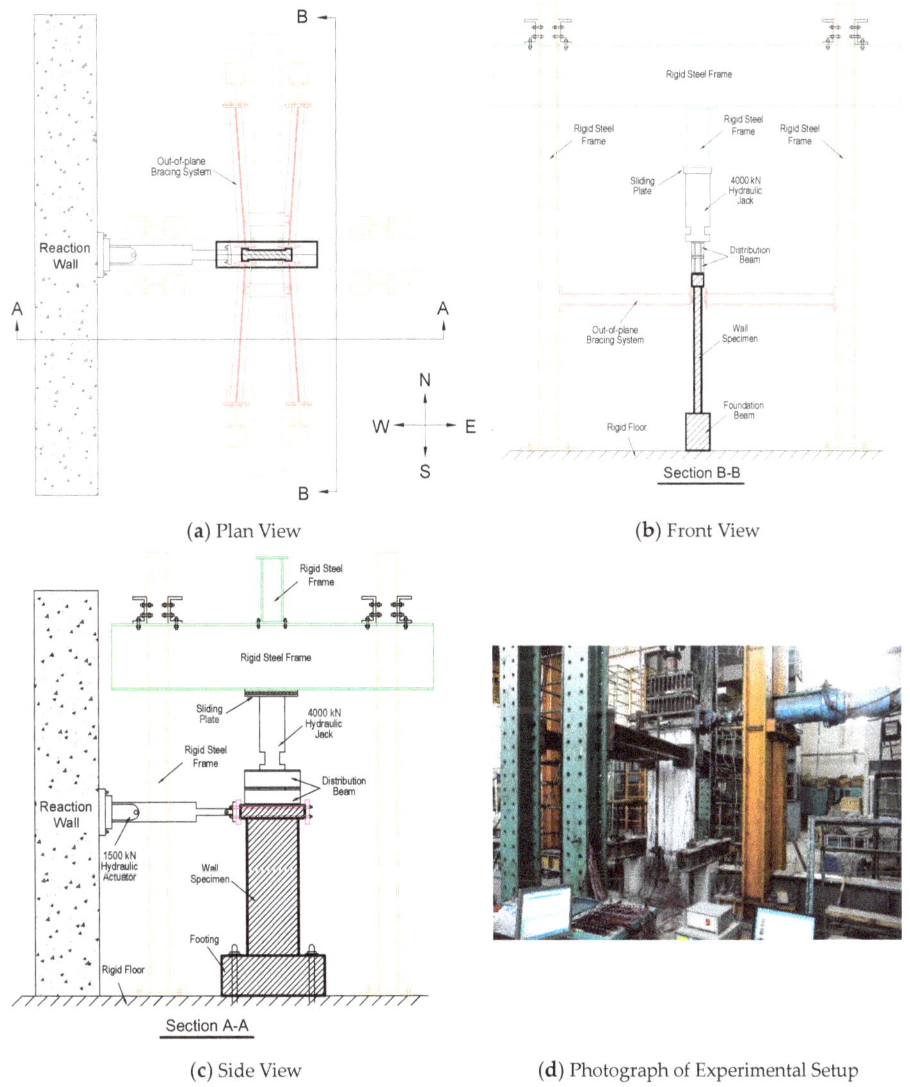

(**a**) Plan View

(**b**) Front View

(**c**) Side View

(**d**) Photograph of Experimental Setup

Figure 3. Details of Experimental Setup (**a**) Plan View; (**b**) Front View (Section B-B); (**c**) Side View (Section A-A); and, (**d**) Photograph of Experimental Setup.

Figure 4 depicts the instrumentation layout of test specimens. The load cells were installed inside the hydraulic jack and hydraulic actuator to measure the vertical and lateral forces during the test. A Linear Variable Differential Transformer (LVDT) was attached to the surface of the loading beam to measure the displacement under racking loads. As shown in Figure 4c, the strain rosettes with 60-mm gauge length were attached in diagonals on the concrete surface of shear specimens to capture the shear cracks. In terms of flexure specimens, individual concrete strain gauges with 60-mm gauge length were installed on the bottom part of wall panel to capture the flexure-tension cracks. Similar strain rosette and strain gauge arrangements were adopted for the embedded steel plate to measure the strain values, but the gauge length is much smaller (5 mm). Additionally, individual strain gauges with 5-mm gauge length were used to monitor the longitudinal strains in the vertical and horizontal reinforcements, as well as the I-shape steel profiles.

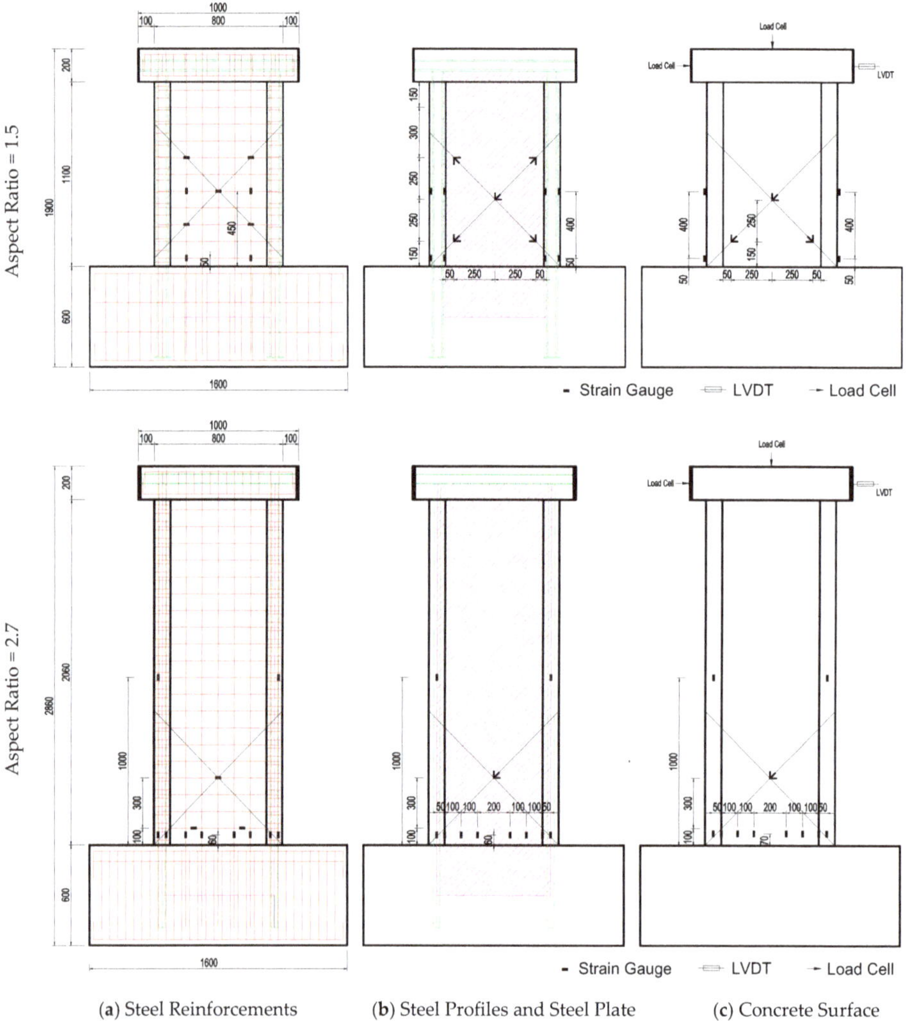

(a) Steel Reinforcements   (b) Steel Profiles and Steel Plate   (c) Concrete Surface

Figure 4. Instrumentation Layout of Test Specimens (**a**) Strain Gauges on Steel Reinforcements; (**b**) Strain Gauges on I-shaped Steel Profiles and Steel Plate; and. (**c**) Strain Gauges, Linear Variable Differential Transformer (LVDT) & Load Cells on Concrete.

Figure 5 shows the cyclic lateral displacement history that was used in the quasi-static test. The specified axial compressive load was constantly applied on the top surface of wall specimen throughout the test. The two phase mixed lateral force/displacement control cyclic loading scheme was used herein, as specified in Chinese Specification for Seismic Test of Buildings [36] and suggested by Zhang [8], Zhou [19], and Nie [21]. Force control cycles were adopted before yielding occurs in shear wall specimens, and only one cycle is applied in each load step. It is possible to determine the accurate lateral forces corresponding to the first visible crack initiation and crack pattern propagation with dividing the first phase into different levels of load cycles. Once the yielding has commenced, displacement-control tests were performed by gradually increasing reversed cyclic displacements in the form of triangular waves and two repeated cycles of the same displacement amplitude were applied for each displacement increment. The magnitude of displacement in each level equals multiple times of the displacement at yield, $y = n\Delta_y$, which also indicates the increasing ductility level at each cycle. The reversed cyclic load test ends until the lateral force drops below 85% of the maximum lateral load capacity. The yield dispalcemnt value, Δ_y, is taken as 2 mm and 6 mm for shear and flexure specimens, respectively, in this series of experimental tests based on a trial test recording.

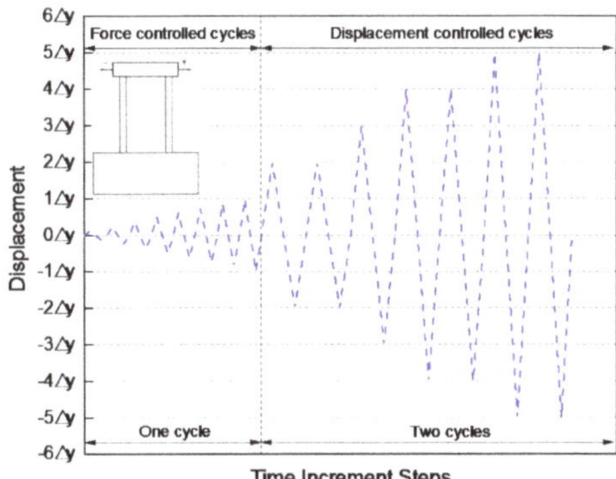

Figure 5. Cyclic Displacement History for Quasi-static Test of Shear Wall Specimens. Δy indicates the lateral displacement when the wall specimen shows yielding behavior.

2.3. Analysis Methods for Experimental Observations

Key information was inferred from the instrumentation recordings in the reversed cyclic load test to evaluate the seismic performance of shear walls, including hysteresis curve, skeleton curve, ductility, energy dissipation ability, etc. Figure 6a presents a typical hysteresis curve that is directly obtained by the lateral force and displacement readings from the load cell and LVDT, and the skeleton curve is determined from the hysteresis curve by joining the peak-load tips of each primary loop.

Ductile structures are preferable in the earthquake-resistant design. Structural members are required to undergo large amplitude lateral deformations without the substantial loss of strength and also dissipate significant amounts of energy in those cyclic deformations. A displacement ductility factor, μ, is defined as the ultimate displacement, Δ_u, to the displacement at yield, Δ_y, and it is given by

$$\mu = \frac{\Delta_u}{\Delta_y} \tag{2}$$

where the ultimate displacement (Δ_u) is defined as displacement value that corresponds to 85% of peak load on descending branch of the skeleton curve as specified in Chinese Specification for Seismic Test of Buildings [36] and suggested by Dan [11], Tong [12], and Liao [37]. The yield displacement (Δ_y) is determined with the method that Priestley recommended [38], which will be described in detail later.

The definition of the yield point in the skeleton curve often causes difficulty in the calculation of ductility factors. Figure 6b illustrates three alternative definitions of the yield displacement and the corresponding yield strength. Priestley [38] suggested that the yield displacement be determined by the equivalent elasto-plastic system with the secant stiffness at 75% of the peak lateral load, P_{max} (Figure 6(b1). Mahin [39] proposed the use of the equivalent elasto-plastic system with the same energy absorption from the origin to the peak lateral load level, as shown in Figure 6(b2)). In ASTM E2126 [40], an equivalent energy elastic-plastic (EEEP) is defined as an elastic-plastic curve that circumscribes an area that is equal to the area enclosed by the skeleton curve between the origin, the displacement axis, and the ultimate displacement (Figure 6(b3)). Park [41] claimed that the first definition is the most realistic option to determine the yield displacement for reinforced concrete structures, which will be used herein to calculate the displacement ductility factors of test shear wall specimens.

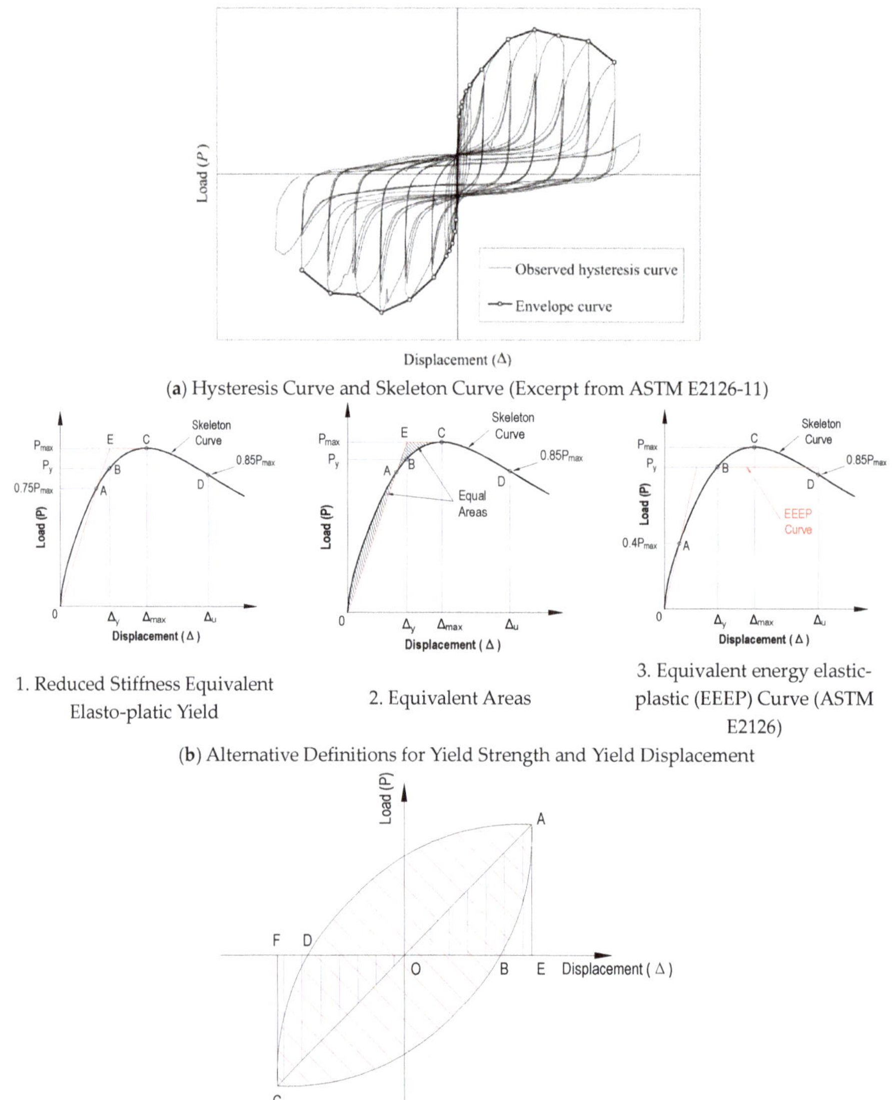

(**a**) Hysteresis Curve and Skeleton Curve (Excerpt from ASTM E2126-11)

<table>
<tr><td>1. Reduced Stiffness Equivalent Elasto-platic Yield</td><td>2. Equivalent Areas</td><td>3. Equivalent energy elastic-plastic (EEEP) Curve (ASTM E2126)</td></tr>
</table>

(**b**) Alternative Definitions for Yield Strength and Yield Displacement

(**c**) Definition for Equivalent Viscous Damping Coefficient

Figure 6. Analysis Methods for Experimental Observations (**a**) Hysteresis Curve and Skeleton Curve; (**b**) Yield Strength and Yield Displacement; and, (**c**) Equivalent Viscous Damping Coefficient.

As shown in Figure 6a, structural members have some deformation capacity beyond the peak lateral load, P_{max}, with undergoing a small reduction of strength. Thus, it is reasonable to recognize part of the post-peak deformation capacity, and the ultimate displacement is defined herein as the point when the load carrying capacity drops below 85% of the peak strength on descending branch of the skeleton curve, or when severe collapse failure takes place in the specimen, whichever occurs first.

The ductility factor is defined in terms of deformation, which provides no indication regarding the energy dissipation capacity of the structural members. The equivalent viscous damping (EVD) coefficient is considered to be a critical parameter in evaluating the energy dissipation ability of given structural members. For shear wall specimens subjected to reversed cyclic loadings, the EVD coefficient can be obtained by a function of the dissipated energy of the structural member, E_{Di}, and the elastic energy stored in an equivalent viscous system, E_{Si}, in cycle i [42]. The energy that is dissipated by the structural member can be expressed as the area enclosed by the hysteresis loop (S_{ABCD}) and the stored energy is measured by areas within two triangles of the first and third quadrants (S_{AOF} and S_{COF}), as depicted in Figure 6c. The formal equation is given by

$$\zeta_{ei} = \frac{1}{2\pi} \frac{E_{Di}}{E_{Si}} \tag{3}$$

where ζ_{ei} is the EVD coefficient for the cycle i.

3. Experimental Results and Discussions

A total of 11 shear wall specimens were tested under reversed cyclic loadings. The progressions of damage and failure modes for all specimens were observed throughout the test procedure. Performance parameters, including lateral load capacity, ultimate displacement, ductility factor, and EVD coefficient, were determined.

3.1. Progression of Damage and Failure Modes

Specimen RCSW 1-1 was conceived to represent a conventional RCSW with low aspect ratio (1.5). The first inclined crack was observed on the web surface during the 200 kN load step, and then horizontal tension cracks formed at the lower part of boundary elements during the 300 kN load step. Afterwards, transverse steel bars in the web and vertical steel bars at the extreme end of the boundary toe yielded at the load step of 400 kN, and the corresponding lateral displacement value is 2 mm. In the following steps, displacement-controlled cycles were adopted and the inclined cracks gradually extended. The quantity of boundary horizontal cracks and the web inclined cracks significantly increased and the crack width reached up to 0.35 mm during the 4 mm load step (0.33% drift ratio). The specimen achieved the maximum lateral load capacity at the 8 mm load step (0.67% drift ratio) and the concrete cover started spalling during that load step. Significant concrete spalling was observed at both wall web and boundary toes, and the lateral load strength substantially reduced at the 16 mm load step (1.33% drift ratio), when the test ended.

In general, specimen RCSW 1-1 showed a typical shear damage pattern as the representative of shear specimens. On the contrary, the flexural damage pattern is primarily observed in specimens with the aspect ratio of 2.7 and the damage progression of RCSW 2-1 is described in detail as an example. The horizontal flexure-tension crack is initiated at the 350 kN load step in the bottom of the boundary elements, and it was then extended to the web surface with the crack width expanding up to 0.35 mm at the 400 kN load step. Longitudinal steel bars in the boundary elements yielded at the 450 kN load step and the lateral displacement reaches 11 mm (0.51% drift ratio). Vertical cracks in boundary toes were observed at the 15 mm load step (0.69% drift ratio) with little concrete cover spalling, and the existing cracks were further diagonally extended to the web surface with the crack widths increasing up to 1.4 mm. Subsequently, the specimen reached the maximum positive and negative lateral loading capacities with more severe concrete spalling at +21 mm (0.97% drift ratio) and −27 mm (1.25% drift ratio) load steps. The test of specimen RCSW 2-1 ended after the cycle of 39 mm (1.81% drift ratio) when the boundary steel rebar fractured and concrete at wall toes also crushed.

Figures 7 and 8 presents crack distributions at different inter-story drifts (ISD) and photographs of failure patterns for all of the specimens. Three ISD values (0.1%, 0.33%, and 1.0%) were selected to evaluate in-service performances of shear walls when subjected to minor, moderate, and major earthquakes. It can be seen from Figures 7a and 8a that no cracks or very few cracks were observed at the ISD of 0.1% and all of the specimens remain in the elastic range at that stage. As the level of ISD increased, more web inclined cracks and boundary horizontal cracks occurred in the shear specimens with the crack width expanding. In terms of flexure specimens, the quantity of boundary horizontal cracks increased, some of which diagonally extended to the web, forming flexure-shear cracks. When the ISD value reached 1%, very dense cracks were distributed on the wall surface. In contrast to crack-riddled shear specimens (Figure 7a), the cracks were mainly distributed in the lower half of flexure specimens, and the upper half almost kept undamaged (Figure 8a). As compared to the RCSW specimens, more densely-distributed cracks were shown in the RCSPSW specimens, but the crack widths are much smaller based on observations. In addition, with the increase of the axial compressive load, the quantity of distributed cracks becomes less and the crack width is smaller due to the compaction of compression forces.

Figure 7b shows the failure modes for shear specimens. Significant spalling of concrete cover along the shear cracks in wall web and concrete crushing in boundary toes were observed in the RCSW shear specimens. As for RCSPSW shear specimens, the embedded steel plate tended to buckle at the failure point, accompanied by severe concrete spalling in the lower part of the wall web. When it comes to flexure specimens, concrete crushing at the toe and longitudinal steel rebar facture usually occurred in the boundary elements, as shown in Figure 8b. Severe spalling of concrete cover in the bottom of wall web is also observed in specimen RCSPSW 2-1. In general, the RCSPSW specimens show a denser crack distribution but the crack width is smaller. Moreover, the damage observed in RCSPSW specimens is not as severe as RCSW.

The failure modes of shear wall specimens were controlled by axial-flexure-shear interactions that were affected by a variety of parameters, including aspect ratio, axial compression ratio, structural steel arrangement, etc. With the variation of key parameters, aspect ratio, in particular, the failure characteristics transit from 'shear failure' to 'flexural failure'.

Figure 9a shows a photograph of shear failure patterns referred to past research studies [1,19,43]. Shear critical walls mainly exhibit a diagonal cracking pattern on the wall surface. The first inclined crack at an angle of approximately 45° is initiated by the shear stress at the corner of the wall and then extends to the mid portion. Upon further loading cycles, new inclined cracks occur and existing cracks extend; the strains of reinforcing bars and the widths of intercrossing main diagonal cracks are developed until the peak lateral load is attained. Afterwards, the spalling and crushing of concrete take place at corners and the diagonal cracking becomes severer until the wall fails. In general, shear-critical walls usually fail by diagonal tension or diagonal compression. Typical shear failure characteristics include the formation of diagonal cracks, yielding of reinforcing bars, and spalling of the concrete cover at the intercrossing diagonal cracks region or wall toes, as illustrated in Figure 9b.

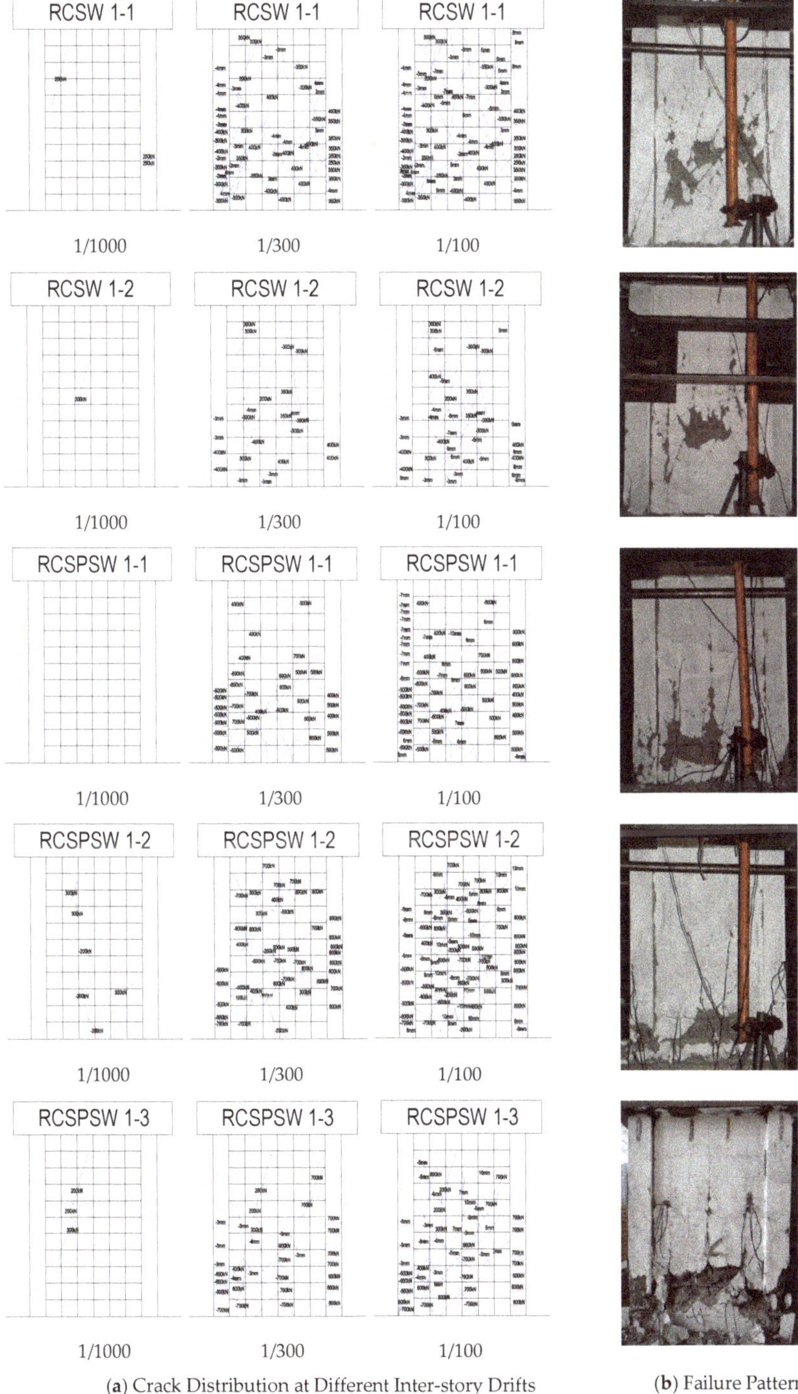

(a) Crack Distribution at Different Inter-story Drifts

(b) Failure Pattern

Figure 7. Crack Distributions and Failure Patterns of Specimens with Aspect Ratio of 1.5 (a) Crack Distributions at Different Drifts and (b) Photograph of Failure Patterns.

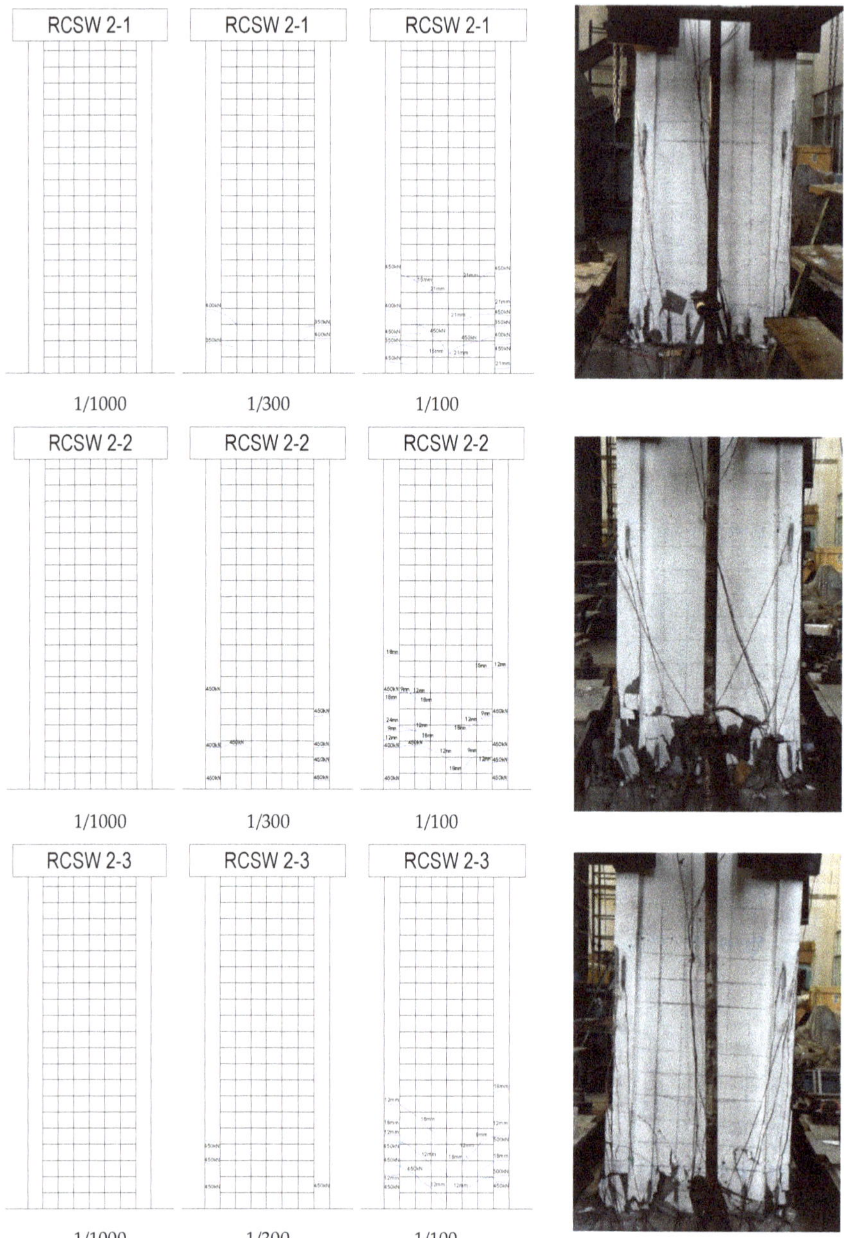

Figure 8. *Cont.*

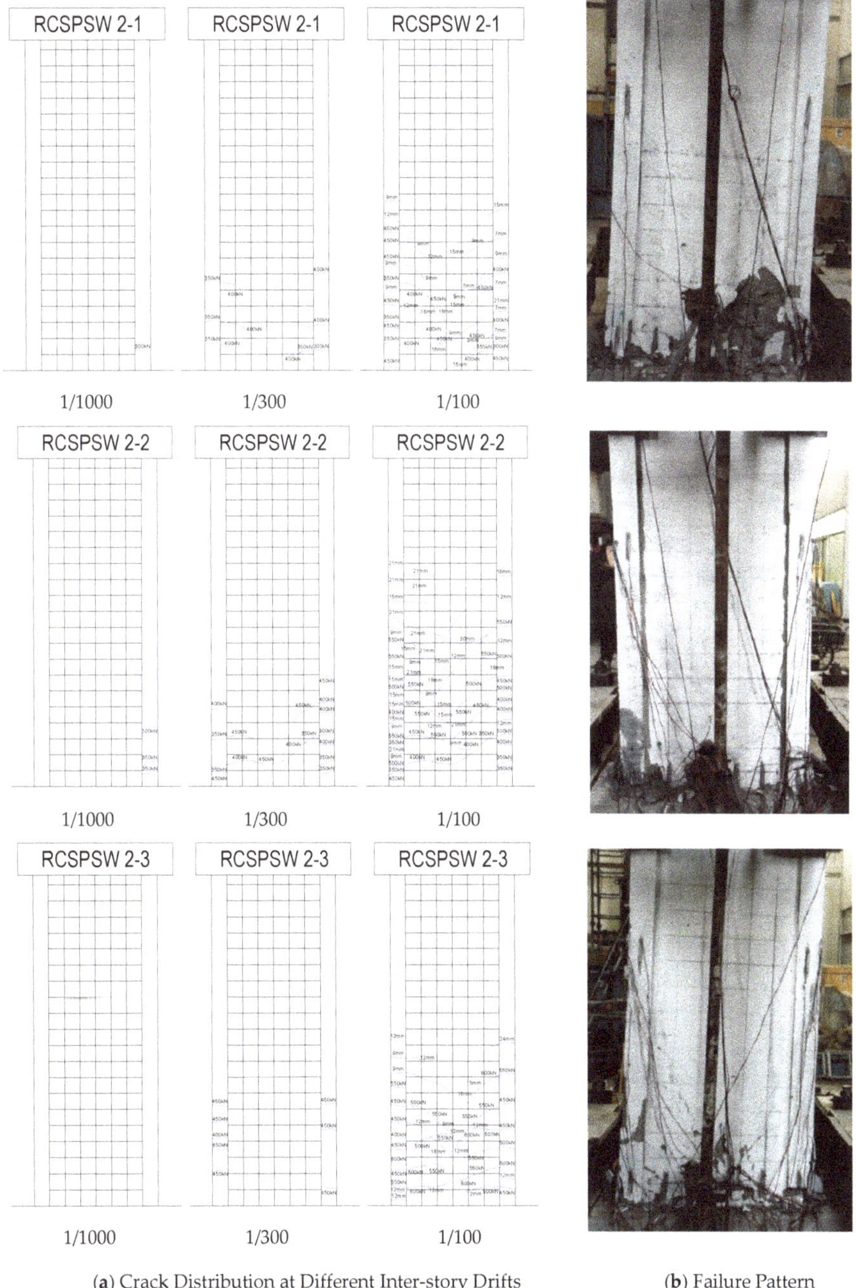

(a) Crack Distribution at Different Inter-story Drifts   (b) Failure Pattern

Figure 8. Crack Distributions and Failure Patterns of Specimens with Aspect Ratio of 2.7 (**a**) Crack Distributions at Different Drifts and (**b**) Photograph of Failure Patterns.

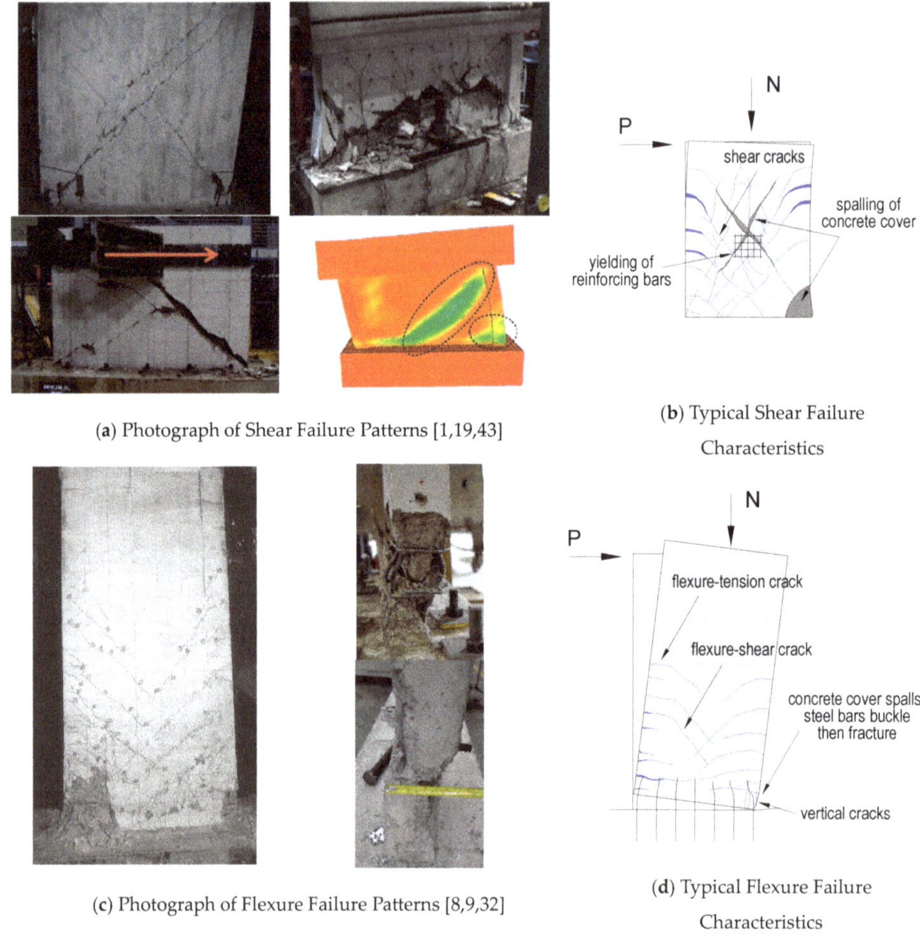

(**a**) Photograph of Shear Failure Patterns [1,19,43]

(**b**) Typical Shear Failure Characteristics

(**c**) Photograph of Flexure Failure Patterns [8,9,32]

(**d**) Typical Flexure Failure Characteristics

Figure 9. Failure Modes of Shear Wall Specimens.

Figure 9c presents the flexure failure patterns and associated crack distribution excerpt from references [8,9,32]. Flexure-dominant failure starts with horizontal cracks that occur at the wall base on the tensile zone. With the load increasing, new horizontal cracks appear along the wall height and existing cracks gradually expand and propagate inwards to the core of the section. These cracks eventually form an inclined cracking pattern in the web. Upon further loading cycles, vertical reinforcing bars at the wall toe yield and significant inclined flexure-shear cracks form on the lower portion of the wall. Vertical cracks appear at the bottom edge of the compression zone with continuing loading. After the lateral load decreases, the concrete cover at the toe in compression spalls off, finally the failure occurs with the crushing of concrete and buckling of reinforcing bars and steel profiles. The typical flexure failure characteristics include horizontal flexure-tension cracks initiated at wall boundaries, inclined flexure-shear cracks on the wall web, longitudinal reinforcement yielding, vertical cracks and concrete cover spalling at the wall toe, concrete crushing in the compression zone, and buckling of reinforcing bars, as shown in Figure 9d.

In the experimental study that is presented herein, specimens with the aspect ratio of 1.5 exhibited a shear failure mode with significant diagonal cracks appearing in the wall web. As for specimens with the aspect ratio of 2.7, the flexure-dominant failure mode is identified with horizontal cracks initiating at the ends, concrete spalling, and tensile fracturing of steel bars. A special characteristic that is observed in RCSPSW specimens is that the embedded steel plate tended to buckle when failure occurred, which was accompanied by severe concrete spalling in the lower portion of the wall. In general, more characteristics of shear failure were observed on the specimens with the decrease of the aspect ratio. A higher axial compression ratio restrained the development of inclined cracks in the web for the reason that the principle tensile stress would be reduced with the increase of the axial load based on experimental observations. Additionally, the "distance" between concrete spalling and wall failure drastically reduces as the axial compression ratio increasing, which results in a more brittle failure mode.

The wall panel in RCSPSW is divided into two halves by the embedded steel plate when compared with conventional RCSW, which results in relatively weaker mechanical collaboration between concrete and steel. It is very important to ensure the effective bonding between two parts to support the engineering application of RCSPSW, although no debonding failure was observed in the quasi-static test. Potential approaches include: (1) using ribbed steel bars with rough surface to increase the mechanical adhesion and friction; (2) specifying sufficient concrete cover to effectively confine steel reinforcement; (3) connecting reinforcement mesh to the embedded steel plate with steel ties, so that the RCSPSW works as a monolithic system; and, (4) adding steel fibers in the concrete to increase the bond strength, as suggested by Dancygier [44] and Harajli [45]. Besides, the use of fiber composite elements is helpful in strengthening the bonding between concrete and steel, and the techniques that were developed by Gattesco [46] using glass fiber reinforced polymer (GFRP) can be deployed.

3.2. Force-Displacement Responses

Figure 10 presents the hysteresis curves (blue line) and skeleton curves (red line) for the shear specimens. Significant "pinching" effects were observed in the hysteresis loops of RCSW shear specimens. With the embedment of the steel plate, the hysteresis curves appear in a plumper shape with higher peak lateral load capacities, which indicated that RCSPSW specimens have better seismic performances. Figure 11 shows comparisons among the skeleton curves of all shear specimens. The blue, red, and green lines in Figure 11a,b represent skeleton curves of specimens under the axial compression ratio of 0.33, 0.45, and 0.50, respectively. The dash and solid lines in Figure 11c show the curves of the RCSW and RCSPSW specimens. It is evident that the use of embedded steel plate significantly increased the lateral load capacity, but a severer and quicker post-peak strength and stiffness degradation was observed in the RCSPSW specimens as compared to RCSW specimens under same axial compression ratios. In terms of the RCSPSW 1-1 specimen, it shows better deformation capability than other RCSPSW specimens, which is attributed to a lower axial compression ratio. Additionally, the positive lateral load re increased after the strength decay in the previous step, which may result from the hardening of the steel plate.

Figure 12 shows the hysteresis and skeleton curves for flexure specimens. Overall, the RCSW specimens show a S-shaped hysteretic behavior with a "pinching" phenomenon. The hysteresis curves of RCSPSW specimens appear in a full bow shape, which indicates that mixed flexure-shear failure mechanisms exist in the damage progression of flexure specimens. From the results that are presented in Figure 12, it is evident that "fatter" hysteresis loops were obtained for specimens when they were subjected to lower axial compressive loads. Figure 13 presents the comparative results of skeleton curves for different flexure specimens. As compared to the RCSW specimens (dash lines), the RCSPSW specimens (solid lines) show higher lateral load capacities. The skeleton curves with the same color represent specimens subjected to the same axial compressive load. With the increase of the axial compressive load, the peak lateral load increased, but the ultimate lateral displacement decreased. A severer and quicker strength and stiffness degradation was observed for RCSPSW specimens with an axial compression ratio higher than 0.50.

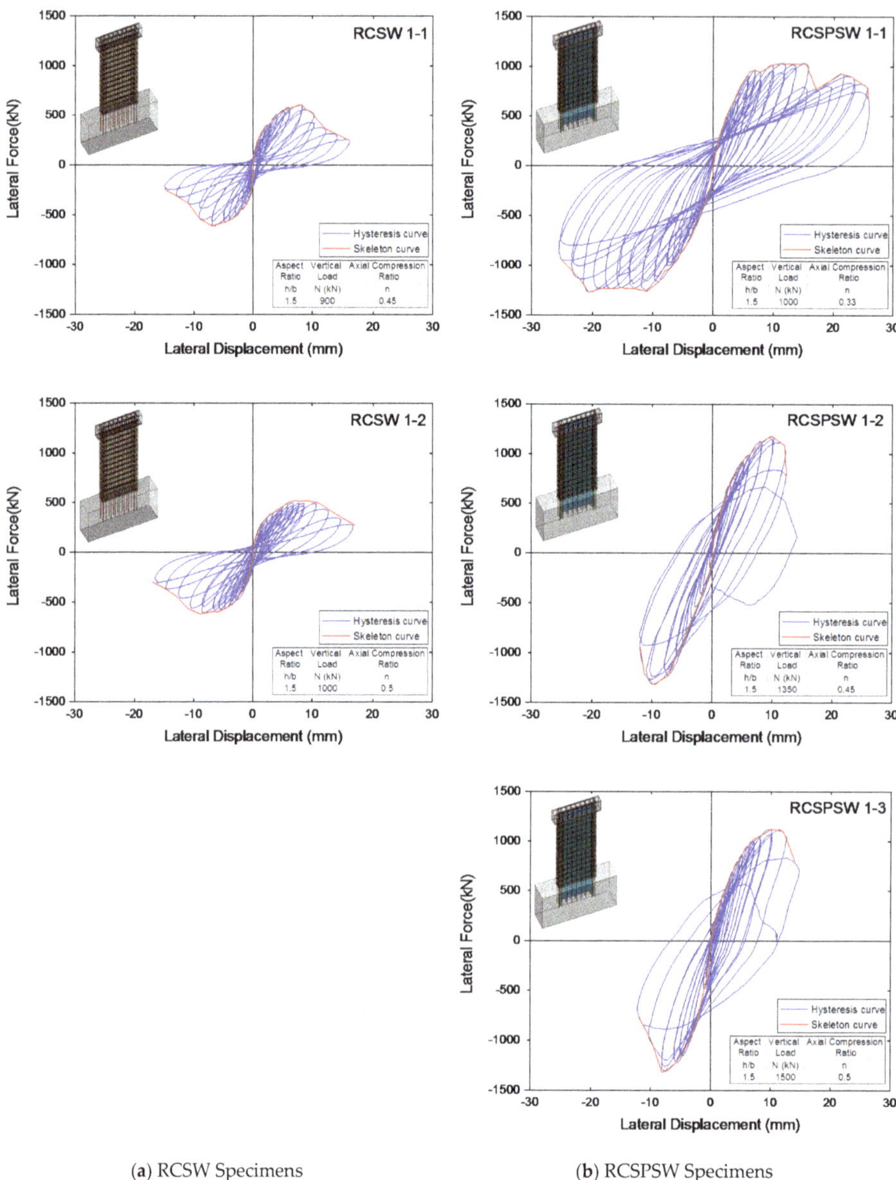

(**a**) RCSW Specimens (**b**) RCSPSW Specimens

Figure 10. Hysteresis and Skeleton Curves for Specimens with Aspect Ratio of 1.5 (**a**) RCSW Specimens and (**b**) RCSPSW Specimens.

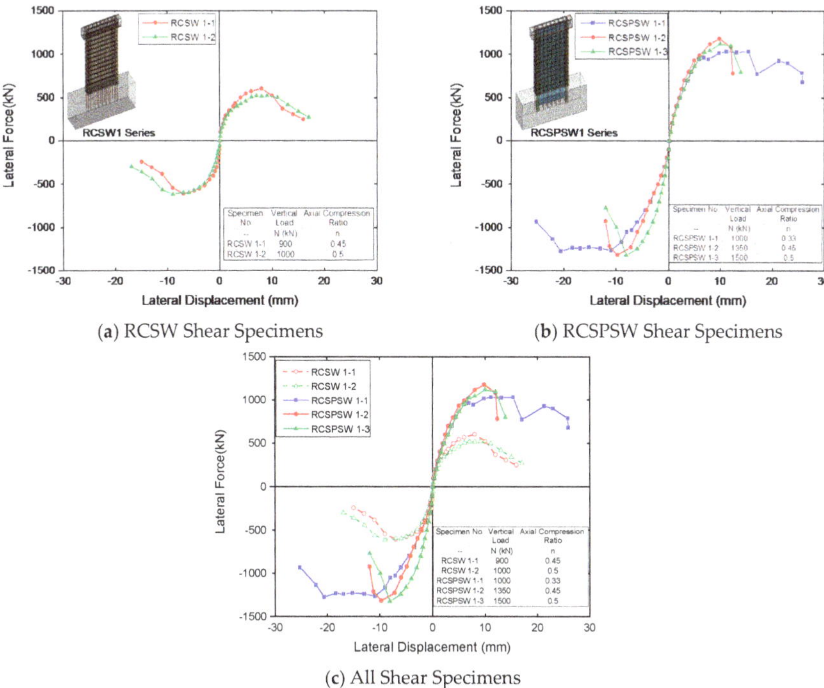

(**a**) RCSW Shear Specimens   (**b**) RCSPSW Shear Specimens

(**c**) All Shear Specimens

Figure 11. Comparison among the Skeleton Curves for Specimens with Aspect Ratio of 1.5 (**a**) RCSW Specimens; (**b**) RCSPSW Specimens; and, (**c**) All Shear Specimens.

The key characteristics were summarized for all shear wall specimens with considerations of the damage progression and force-displacement responses, although differences do exist among their skeleton curves. Figure 14a shows the skeleton curve of RCSW 1-1 specimen as a representative example, where five critical points are defined, including crack point, yield pint, peak point, failure point, and collapse point. Each critical point represents an event when the specimen's behavior is significantly altered. The crack point corresponds to the load step when the first web inclined shear crack or boundary horizontal flexure-tension crack is observed. The determination of yield point, peak point, and failure point has been described in the previous section. The collapse point indicates the tip of last load cycle if the post-ultimate load capacity ($0.85P_{max}$) exists. It is worth mentioning that the yield point and failure point in the load-displacement skeleton curve (Figure 14b) can be determined while using alternative approaches. Smarzewski and Pan [47,48] suggested the equivalent elastoplastic line determine the yield displacement with the secant stiffness at two-thirds of the peak lateral load. As for the failure load, Smarzewski and Lim [47,49] recommended the use of 80% of the peak lateral load, resulting in a relatively larger ultimate displacement as compared to the $0.85P_{max}$ defined herein. Based on trial calculations, the yield displacement determined with the approach that was suggested by Smarzewski is slightly lower than the value that was calculated using the Priestley's method, which is adopted in current research work. Therefore, the ductility factors of shear wall specimens determined with Smarzewski's approach would be larger than the values that are presented in this paper.

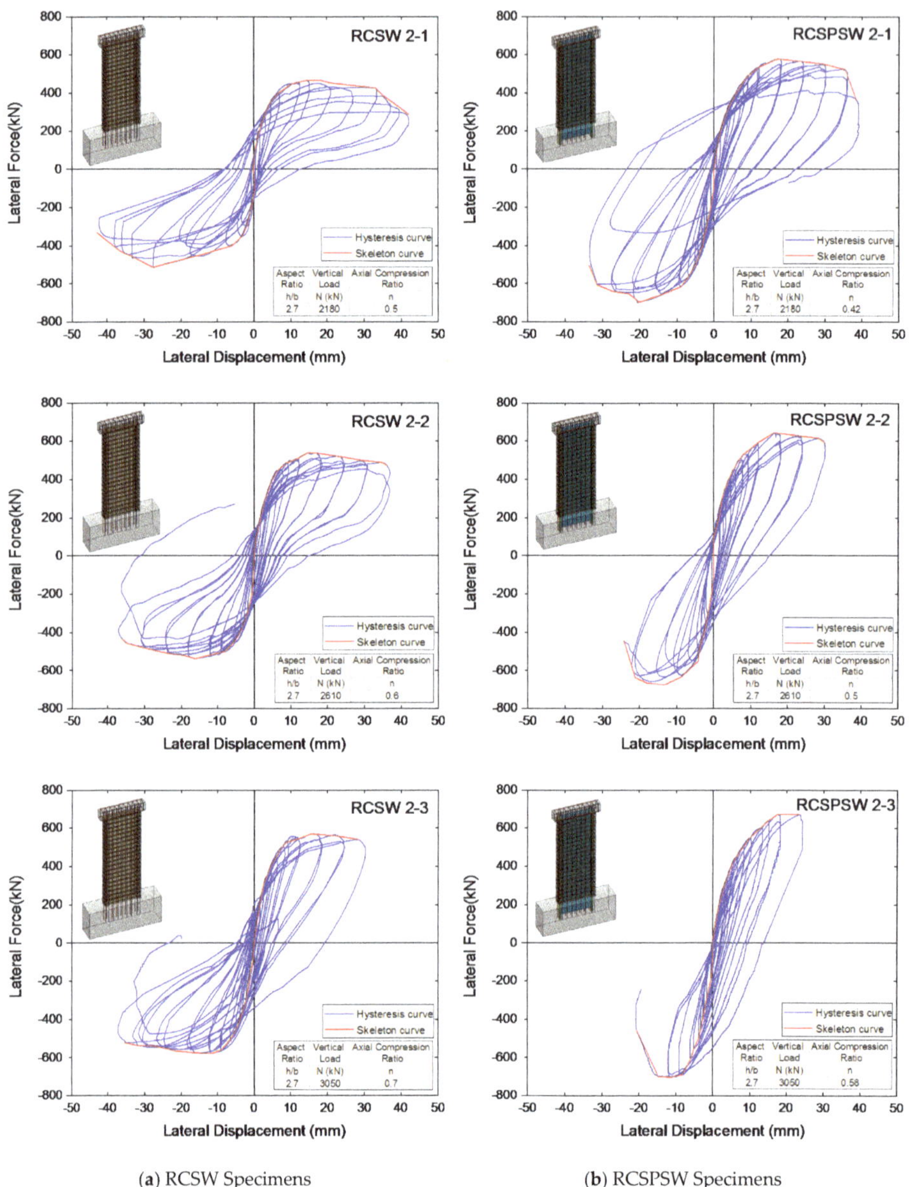

(a) RCSW Specimens   (b) RCSPSW Specimens

Figure 12. Hysteresis and Skeleton Curves for Specimens with Aspect Ratio of 2.7 (**a**) RCSW Specimens and (**b**) RCSPSW Specimens.

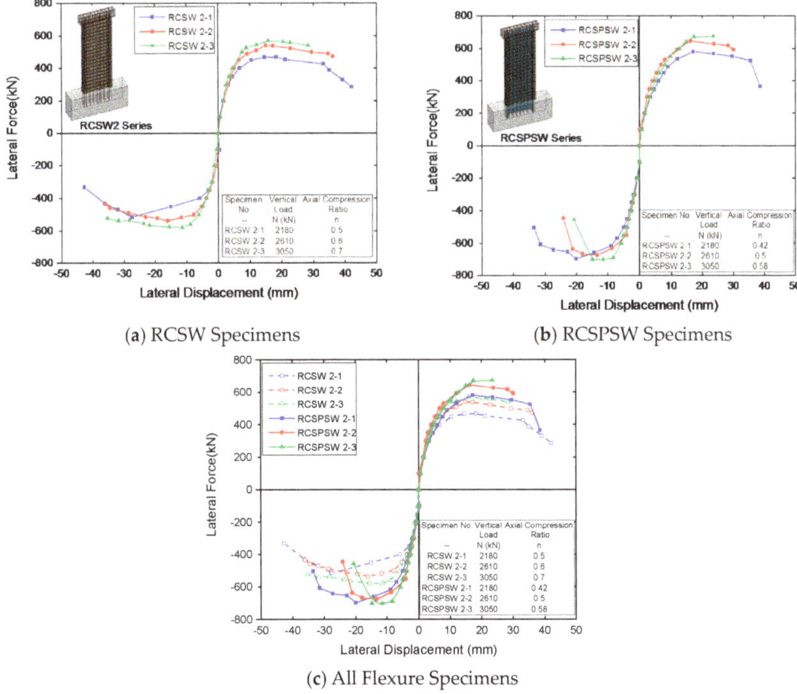

(**a**) RCSW Specimens   (**b**) RCSPSW Specimens

(**c**) All Flexure Specimens

Figure 13. Comparison among the Skeleton Curves for Specimens with Aspect Ratio of 2.7 (**a**) RCSW Specimens; (**b**) RCSPSW Specimens; and, (**c**) All Flexure Specimens.

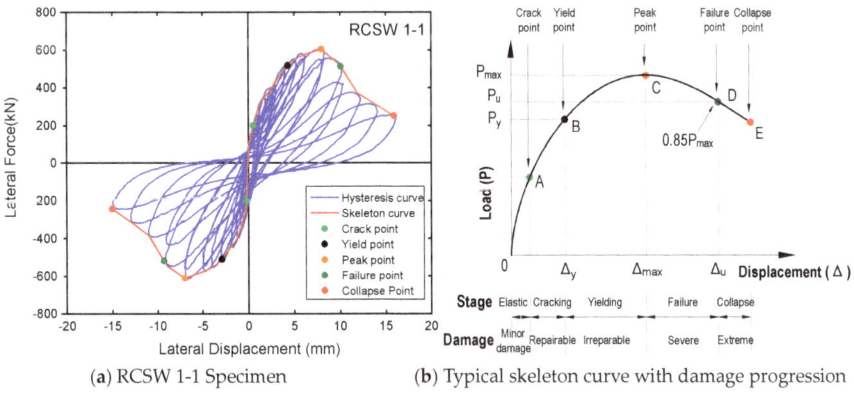

(**a**) RCSW 1-1 Specimen   (**b**) Typical skeleton curve with damage progression

Figure 14. Definition of Critical Points and Working Stages of Shear Wall Specimens.

Based on critical load points, the entire loading procedure was divided into five stages: elastic stage, cracking stage, yielding stage, failure stage, and collapse stage, as shown in Figure 14b. A summary of structural behavior of shear wall specimens in each stage is briefly described, as follows.

(a) Elastic stage (origin to crack point): the specimen keeps intact or minor damage is observed. The structural behavior remains in the elastic range and the load-displacement curve keeps linear.

(b) Cracking stage (crack point to yield point): as the lateral load increasing, more web inclined cracks and boundary horizontal cracks occur in the shear specimens. As for flexure specimens, the quantity of boundary horizontal cracks increases and some of them diagonally extended to the wall web, forming flexure-shear cracks. Crack width expands, but the damage is repairable. The nonlinearity develops in the load-displacement curve.

(c) Yielding stage (yield point to peak point): web transverse steel bars and boundary longitudinal steel bars are gradually yielding in shear and flexure specimens. The crack distribution becomes denser and the crack width further increases. Vertical cracks and concrete cover spalling occasionally take place in the boundary toe. The damage becomes irreparable in this stage. The lateral load continues increasing, but the stiffness value decreases until the peak load is observed.

(d) Failure stage (peak point to failure point): the lateral load capacity starts to reduce. Significant concrete spalling occurs in boundary toes or web panels. Embedded steel plate tends to buckle out-of-plane. The damage is severe and the cycle load test may end in this stage.

(e) Collapse stage (failure point to collapse point): the damage is extremely severe in this stage. The later load capacity continues reducing until the collapse point occurred.

3.3. Lateral Load and Dispalceent Capacity

Table 3 lists lateral load, lateral displacement, and corresponding drift values for the shear specimens at yield point, peak point, and failure point. It is seen that the positive and negative values that were obtained from the cyclic test are asymmetrical, which is also observed in the hysteresis and skeleton curves (Figures 10 and 12). The explanation is when the specimen undergoes damage in one direction; its lateral load capacity is slightly weakened, as it is racked in the opposite direction. The average values are used herein for comparative studies. As compared to RCSW specimens, RCSPSW specimens own approximately 100% higher yield loads, peak loads, and ultimate loads, proving the efficiency of the embedded steel plate in the improvement of shear load capacities. As for the ultimate displacement, the RCSPSW specimens do not show obvious superiority over RCSW specimens, except for RCSPSW 1-1 under lower axial compression ratio (0.33). In general, RCSPSW shear specimens are capable of withstanding an ultimate drift value of 1.0% approximately, showing acceptable deformability for design purpose. The last column lists ductility factors for all shear specimens that were determined by Equation (2). The RCSW specimens own relatively higher ductility factors than RCSPSW specimens under the same axial compression ratio.

Table 4 summarizes the lateral load and displacement capacities of flexure specimens. In general, shear wall specimens with a higher aspect ratio show better ductility performances, because the flexural failure mode governs. Similar with shear specimens, the embedment of steel plate is able to increase the lateral load capacities of the flexure specimens. In particular, the peak load capacities of RCSPSW are 20–30% higher than RCSW under the same axial compressive loads. The RCSW flexure specimens show good deformability with the ultimate drift value of around 1.5% and the ductility factor higher than 4. As for RCSPSW specimens, the deformability is satisfactory for design purposes when the axial compression ratio is lower than 0.50 (ultimate drift is larger than 1.0% and the ductility factor is around 4). As the axial compression ratio increases to 0.58, the ductility factor substantially decreases to 2.61 and the ultimate drift is lower than 1.0%.

Figure 15 presents the effects of the axial compression ratio on peak lateral load capacities, ductility factors, and ultimate drift values of all specimens. Blue lines and red lines represent values for shear specimens and flexure specimens; dash and solid lines are results for RCSW specimens and RCSPSW specimens. It is seen from Figure 15a that all the specimens' lateral load capacities increase with the axial compression ratio, except for conventional RCSW shear specimens (RCSW1 series). For the RCSPSW specimens, the ultimate drift decreases with the axial compression ratio. In particular, the value drops below 1% when it is subjected to the highest axial compressive loads (1500 kN for shear specimens and 3050 for flexure specimens). The ductility factor generally decreases with the axial compression ratio for RCSPSW specimens, but the trend is not as straightforward as the ultimate drift.

The relationship between the deformability and axial compression ratio for RCSW specimens (blue lines) is not clear, as seen in Figure 15b,c, but their ultimate drift and ductility factor values are larger than those of the RCSPSW specimens (red lines) when subjected to the axial compression ratio higher than 0.5.

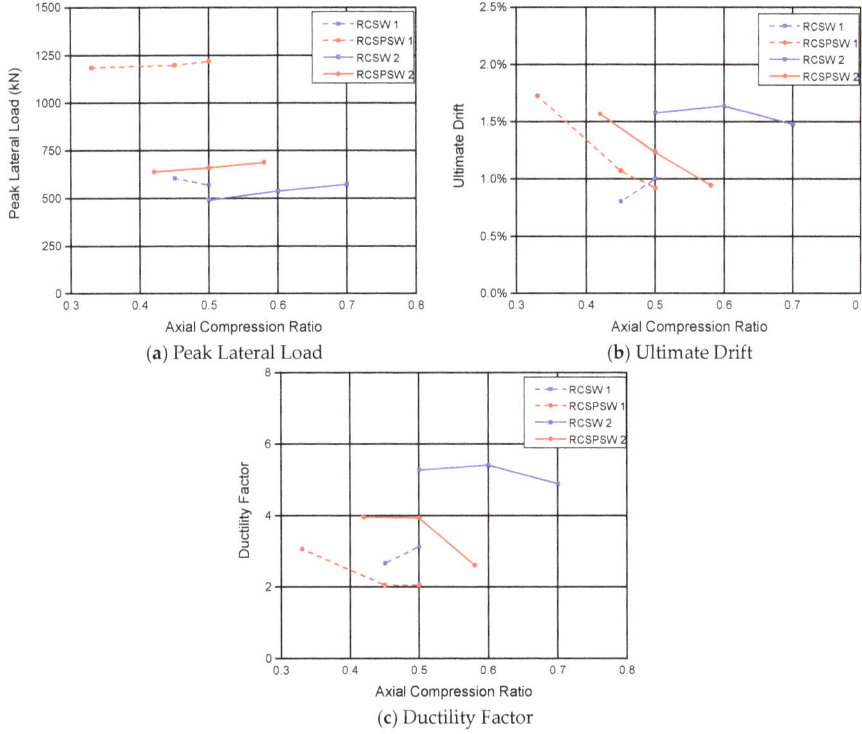

Figure 15. Effects of Axial Compression Ratio on (**a**) Peak Lateral Load; (**b**) Ultimate Drift; and, (**c**) Ductility Factor.

It is observed from the experimental test that the RCSPSW specimens show a relatively lower deformability under high axial compression ratios, which could be attributed to the brittleness of high strength concrete. Another explanation is the weak concrete confinement effect on the embedded steel plate due to the small specimen size. The thickness values of the concrete web on both sides are only 40 mm and 75 mm for shear and flexure specimens. The concrete cover was easily spalled under higher axial compressive loads, and the embedded steel plate tended to buckle when severe concrete spalling took place. Under such circumstance, the strength and stiffness decay quickly and substantially decay. In addition, shear studs were only used in the footing and the connection between the steel plate and concrete wall web is weak, which exacerbate the spalling and strength reduction. Therefore, it is suggested that the shear wall structural component should be designed with caution when subjected to high axial compressive loads and an upper limit of axial compression ratio (0.5) should be set for RCSPSW. Besides, shear studs or steel ties should be used to strengthen the connection between the steel plate and concrete on both sides. Additionally, a higher transverse reinforcement ratio is recommended for further improving the concrete confinement effect. It is worth mentioning that the wall is much thicker in the practical design of high-rise buildings. Thus, the confinement effect and deformability of RCSPSW in future applications could be better than the experimental observations.

Table 3. Lateral Load and Displacement Capacity of Shear Specimens.

Specimen No.	Direction	Yield			Peak			Ultimate			Ductility Factor
		P_y (kN)	Δ_y (mm)	Drift (%)	P_{max} (kN)	Δ_{max} (mm)	Drift (%)	P_u (kN)	Δ_u (mm)	Drift (%)	$\Delta = \Delta_u/\Delta_y$
RCSW 1-1	+	517	4.3	0.36%	604	8.0	0.67%	513	10.1	0.84%	2.34
	−	509	2.9	0.25%	606	7.0	0.58%	515	9.3	0.77%	3.15
	Average	513	3.6	0.30%	605	7.5	0.63%	514	9.7	0.81%	2.67
RCSW 1-2	+	437	4.1	0.35%	522	9.0	0.75%	444	12.4	1.04%	3.00
	−	516	3.5	0.30%	617	8.9	0.74%	524	11.6	0.97%	3.28
	Average	477	3.8	0.32%	570	8.9	0.74%	484	12.0	1.00%	3.13
RCSPSW 1-1	+	941	5.8	0.49%	1070	11.2	0.93%	910	17.0	1.42%	2.92
	−	1043	7.7	0.64%	1301	10.9	0.91%	1106	24.4	2.04%	3.16
	Average	992	6.8	0.57%	1186	11.0	0.92%	1008	20.7	1.73%	3.05
RCSPSW 1-2	+	983	5.9	0.49%	1156	9.7	0.81%	983	13.9	1.16%	2.37
	−	1109	6.7	0.56%	1242	9.8	0.82%	1056	11.8	0.99%	1.77
	Average	1046	6.3	0.52%	1199	9.8	0.81%	1019	12.9	1.07%	2.05
RCSPSW 1-3	+	978	6.4	0.53%	1117	10.0	0.83%	949	12.9	1.07%	2.03
	−	1102	4.5	0.37%	1318	8.1	0.67%	1120	9.2	0.77%	2.06
	Average	1040	5.4	0.45%	1218	9.0	0.75%	1035	11.1	0.92%	2.04

Table 4. Lateral Load and Displacement Capacity of Flexure Specimens.

Specimen No.	Direction	Yield			Peak			Ultimate			Ductility Factor
		P_y (kN)	Δ_y (mm)	Drift (%)	P_{max} (kN)	Δ_{max} (mm)	Drift (%)	P_u (kN)	Δ_u (mm)	Drift (%)	$\Delta = \Delta_u/\Delta_y$
RCSW 2-1	+	386	6.0	0.28%	469	18.2	0.84%	398	32.9	1.52%	5.45
	–	405	6.9	0.32%	514	27.5	1.28%	437	35.4	1.64%	5.12
	Average	396	6.5	0.30%	491	22.9	1.06%	418	34.2	1.58%	5.27
RCSW 2-2	+	455	6.8	0.32%	540	14.9	0.69%	473	36.1	1.67%	5.29
	–	468	6.3	0.29%	536	16.1	0.75%	456	34.8	1.61%	5.54
	Average	461	6.6	0.30%	538	15.5	0.72%	464	35.5	1.64%	5.41
RCSW 2-3	+	487	7.2	0.33%	570	15.8	0.73%	541	28.4	1.31%	3.97
	–	488	5.9	0.27%	578	11.5	0.53%	522	35.4	1.64%	6.01
	Average	488	6.5	0.30%	574	13.6	0.63%	531	31.9	1.48%	4.89
RCSPSW 2-1	+	497	9.6	0.44%	581	17.1	0.79%	493	36.1	1.67%	3.78
	–	581	7.5	0.35%	697	20.1	0.93%	592	31.7	1.47%	4.19
	Average	539	8.5	0.40%	639	18.6	0.86%	543	33.9	1.57%	3.96
RCSPSW 2-2	+	536	8.4	0.39%	646	16.3	0.75%	592	30.0	1.39%	3.59
	–	569	5.2	0.24%	674	13.3	0.62%	573	23.3	1.08%	4.48
	Average	553	6.8	0.31%	660	14.8	0.68%	583	26.7	1.23%	3.93
RCSPSW 2-3	+	560	10.5	0.48%	674	15.0	0.69%	561	23.5	1.09%	2.25
	–	581	5.2	0.24%	701	11.5	0.53%	539	17.3	0.80%	3.33
	Average	571	7.8	0.36%	688	13.2	0.61%	550	20.4	0.94%	2.61

3.4. Energy Dissipation Ability

The EVD coefficients for all specimens were calculated with Equation (3) to compare their energy dissipation abilities. Table 5 lists hysteresis loop areas and EVD coefficients of all the test specimens for test cycles at the peak lateral load (peak point).

Table 5. Equivalent Viscous Damping (EVD) Coefficients of All Specimens.

Specimen Batch	Specimen No.	Hysteresis Loop Area (kN·mm)	EVD Coefficient
Shear Specimens	RCSW 1-1	3518	0.123
	RCSW 1-2	3513	0.109
	RCSPSW 1-1	11600	0.129
	RCSPSW 1-2	11949	0.156
	RCSPSW 1-3	10604	0.157
Flexure Specimens	RCSW 2-1	12064	0.212
	RCSW 2-2	10395	0.184
	RCSW 2-3	11790	0.189
	RCSPSW 2-1	18044	0.202
	RCSPSW 2-2	13442	0.207
	RCSPSW 2-3	11001	0.177

In general, the energy dissipation capacities that were found in shear mode of failure are weaker than those in flexure-controlled failure, and the embedment of steel plate effectively improves the energy dissipation ability of RCSW. For the RCSPSW specimens, the hysteresis loop area decreases with the increase of axial compressive load. Specifically, for the RCSPSW 2-3 specimen that was subjected to the highest axial compressive load (3050 kN), its EVD coefficient is significantly lower than other RCSPSW flexure specimens.

4. Design Models for RCSPSW

4.1. Shear Strength Model

The form of the shear strength equation refers to the design model that was suggested by Chinese Code for Design of Composite Structures [50], given as:

$$V = V_c + V_s + V_a + V_p \tag{4}$$

where the shear strength, V, is contributed by four components: concrete shear-resisting component, V_c, horizontal reinforcement shear-resisting component, V_s, boundary steel profiles shear-resisting component, V_a, and embedded steel plate shear-resisting component, V_p. Equation (5) provides the expressions for different shear strength components.

$$\begin{cases} V_c = 0.67 f_t b_w h_0 + 0.2N\frac{A_w}{A} \\ V_s = f_{yh}\frac{A_{sh}}{s}h \\ V_a = \frac{0.3}{\lambda}f_a A_a \\ V_p = \frac{0.6}{\lambda-0.5}f_p A_p \end{cases} \tag{5}$$

where f_t is concrete tensile strength; b_w is the web width; h_w is the web depth; h and h_0 are the depth and effective depth of the shear wall section; A_w and A represent the area of concrete web and entire section, respectively; N indicates the axial compressive load, $N \leq 0.2 f_c b_w h_w$; f_{yh}, f_a, and f_p mean the yield strength of transverse web reinforcement, boundary I-shape steel profiles, and embedded steel plate; A_{yh}, A_a, and A_p denote the area of those three components; s is the spacing of transverse reinforcement; and, λ is the aspect ratio.

Table 6 summarizes shear strength capacities that were measured from the quasi-static cyclic load test and calculated by shear strength models. Note that the measured characteristic concrete strength values were used in the calculation. It is seen from Table 6 that design models provide 10–15 percent conservative shear strength capacity values for RCSW specimens as compared to the test results. In terms of RCSPSW specimens, shear strength capacities that are estimated by design models are very close to the average experimental data, but 5–10 percent lower shear strength values were observed in the positive direction. A correction factor, k_s, for design models is taken as 0.9 to determine the shear strength capacity in order to achieve a reasonable degree of conservatism. Figure 16a presents the comparison of design strength capacities and experimental results for the shear specimens. Dash red line shows the design values determined by the modified design model with the correction factor. It is evident that the modified design shear capacities are generally larger than the test values in both positive and negative directions, and a reasonable degree of conservatism (approximate 10–20 percent) is obtained by using the shear correction factor.

Table 6. Comparisons of Experimental and Design Strength Capacities for Shear Specimens.

Specimen No.	Experimental Results (kN)			Design Model Results (kN)	Differences		
	Positive	Negative	Average		Positive	Negative	Average
RCSW 1-1	604	606	605	512	15%	16%	15%
RCSW 1-2	522	617	570		2%	17%	9%
RCSPSW 1-1	1070	1301	1186	1199	−12%	8%	−2%
RCSPSW 1-2	1156	1242	1199	1213	−5%	2%	−1%
RCSPSW 1-3	1117	1318	1218		−9%	8%	0%

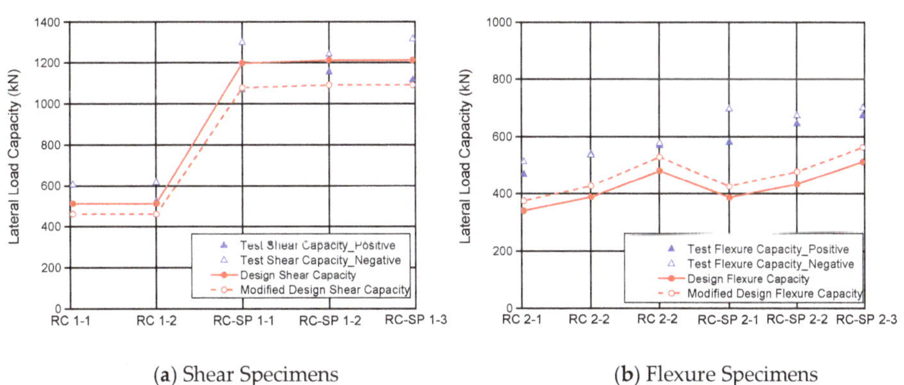

(**a**) Shear Specimens (**b**) Flexure Specimens

Figure 16. Comparisons of Design Strength Capacities and Experimental Results for Shear Specimens and Flexure Specimens.

According to Equation (5), the shear strength capacities contributed by reinforced concrete ($V_c + V_s$), I-shape steel profiles (V_a), and the steel plate (V_p) were quantified and are summarized in Table 7. It is seen that the embedded steel plate takes up approximately 50 percent of the design shear strength capacity. For comparisons, shear strength provided values by each component in the cyclic test were inferred from strain gauge readings. Figure 17a shows the strain gauge arrangement in boundary I-shape steel profiles and the embedded steel plate. Shear strength that was provided by boundary I-shape steel profiles was calculated based on strain values of SG16–SG19 by following Equation (6).

$$V_a = \frac{W_a E_a \left| \frac{\varepsilon_{SG16} + \varepsilon_{SG17}}{2} - \frac{\varepsilon_{SG18} + \varepsilon_{SG19}}{2} \right| / 2}{H} \tag{6}$$

where E_a is measured Young's Modulus of I-shape steels; W_a is the section modulus; and, H is the height of shear specimens.

Table 7. Design Strength Capacities Contributed by Different Components in RCSPSW.

Specimen No.	Shear Strength Capacities (kN)			
	Reinforced Concrete	I-Shape Steels	Steel Plate	Total
RCSPSW 1-1	515 (43%)	116 (10%)	568 (47%)	1199
RCSPSW 1-2	529 (44%)	116 (10%)	568 (46%)	1213
RCSPSW 1-3	529 (44%)	116 (10%)	568 (46%)	1213

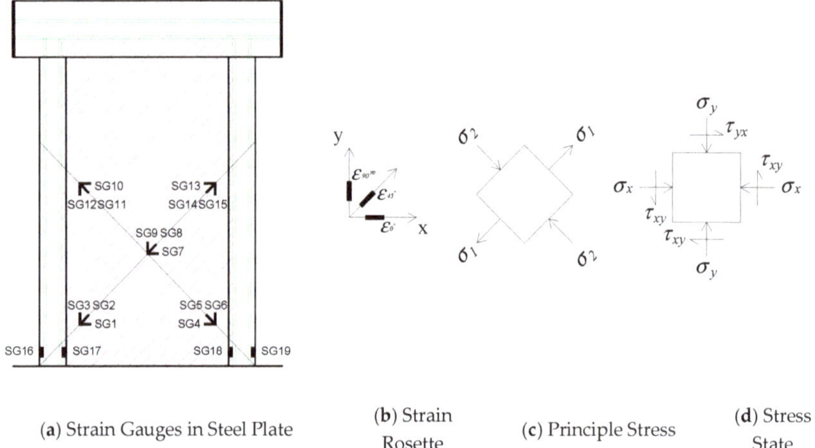

(a) Strain Gauges in Steel Plate   (b) Strain Rosette   (c) Principle Stress   (d) Stress State

Figure 17. Shear Strength of I-shape Steels and Steel Plate Inferred from Strain Gauge Readings (a) Strain Gauge Arrangement in the I-shape Steels and Steel Plate; (b) Strain Rosette; (c) Principle Stress State; and, (d) Required Stress State.

Strain rosettes were attached on the embedded steel plate. Strain values of *SG7*, *SG8*, and *SG9* are recorded and used to infer principle stress state and horizontal shear stress state and further determine the shear strength by using Equation (7).

$$\left\{ \begin{Bmatrix} \sigma_1 \\ \sigma_2 \end{Bmatrix} = \frac{E(\varepsilon_{0^0} + \varepsilon_{90^0})}{2(1-v)} \pm \frac{\sqrt{2}E_p}{2(1+v)} \sqrt{(\varepsilon_{0^0} - \varepsilon_{45^0})^2 + (\varepsilon_{45^0} - \varepsilon_{90^0})^2} \right. \\ V_p = A_s \tau_{yx} \tag{7}$$

where ε_{0^0}, ε_{45^0}, and ε_{90^0} represent strain rosette readings; E_p is the Young's Modulus of steel plate; v is the Poisson's ratio; A_s is the area of steel plate; and, τ_{yx} is the horizontal stress. Note that the shear strength of steel plate is considered to be unchanged after yielding, and shear-resisting component of reinforced concrete is taken as the lateral load strength of RCSW specimens under the same axial compression ratios.

Figure 18 shows the shear-resisting strength that was contributed by different components in the cyclic load test. The blue line, red line, and green line show the capacity trend of reinforced concrete,

I-shape steel profiles, and steel plate. The magenta line presents the experimental skeleton curve for a comparative study. It is observed that the shear strength of all the components increases in the elastic stage. When the specimen goes to the crack stage, the strength increase of reinforced concrete component becomes slower. Moreover, the shear strength of the reinforced concrete component decreases before the specimen reaches the peak lateral load, while the steel plate's strength continues increasing. At the failure point, the summation of shear-resisting strength that is provided by different components is close to the value that is presented in the skeleton curve, showing that the calculation of different components' shear strength is reliable.

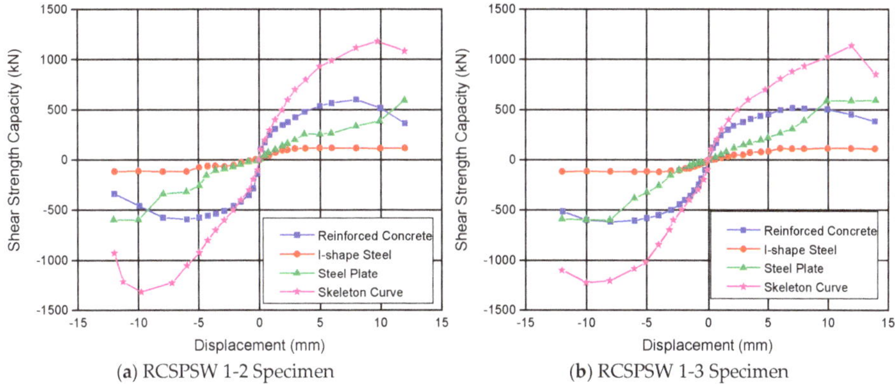

Figure 18. Shear Strength Capacities of RCSPSW Specimens Provided by Different Components.

4.2. Flexural Strength Model

The strain gauges were attached on the bottom part of the embedded steel plate in flexure specimens, as shown in Figure 4b. Figure 19a presents the measured strain distribution along the wall depth at the peak point. It is observed that the strain distribution keeps close to a linear fashion when the specimen reaches the peak lateral load capacity. Under such circumstance, the design flexure strength model is established based on the plain section assumption and the form refers to the bearing capacity equations of eccentrically-compressed members in Chinese Code for Design of Composite Structures [50], as represented in Figure 19b and given in Equation (8):

$$
\begin{cases}
N \leq N_c + f'_a A'_a + f'_y A'_s - \sigma_a A_a - \sigma_s A_s + N_{sw} + N_{pw} \\
N\left(e_0 + \frac{h}{2} - a\right) \leq M_c + f'_a A'_a (h_0 - a'_a) + f'_y A'_s (h_0 - a'_s) + M_{sw} + M_{pw} \\
M = N e_0 \\
V_f = \frac{M}{H}
\end{cases}
\tag{8}
$$

where M and N are the design moment and axial force values; V_f is the design shear strength; e_0 is the eccentricity of the axial compression force; H is the shear wall height; h is the depth of wall section; a is the distance from the extreme tension fiber to centroid of resultant tensile force in tensile I-shape steels and steel reinforcement; h_0 is the effective depth of wall section; f'_a and f'_y are the yield strength of compressive I-shape steels and steel reinforcement; A'_a and A'_y are the areas of compressive I-shape steels and steel reinforcement; σ_a and σ_s are stresses of tensile I-shape steels and steel reinforcement; A_a and A_s are areas of tensile I-shape steels and steel reinforcement; and, a'_a and a'_s are distances from the extreme compression fiber to centroid of resultant compressive force in tensile I-shape steels and steel reinforcement. M_c, M_{sw}, and M_{pw} are the design moment values provided by concrete section, longitudinal steel reinforcement distributed in the wall web, and embedded steel plate, and N_c, N_{sw}, and N_{pw} are the design axial force values that are provided by those three components concrete section, which can be determined by Equations (9)–(12):

when $x \geq h'_\mathrm{f}$

$$\begin{cases} N_\mathrm{c} = \alpha_1 f_\mathrm{c}\left[\xi b_\mathrm{w} h_0 + (b'_\mathrm{f} - b_\mathrm{w})h'_\mathrm{f}\right] \\ M_\mathrm{c} \leq \alpha_1 f_\mathrm{c}\left[\xi(1 - 0.5\xi)b_\mathrm{w} h_0^2 + (b'_\mathrm{f} - b_\mathrm{w})h'_\mathrm{f}(h_0 - 0.5h'_\mathrm{f})\right] \end{cases} \tag{9}$$

when $x < h'_\mathrm{f}$

$$\begin{cases} N_\mathrm{c} = \alpha_1 f_\mathrm{c}\xi b'_\mathrm{f} h_0 \\ M_\mathrm{c} \leq \alpha_1 f_\mathrm{c}\xi(1 - 0.5\xi)b'_\mathrm{f} h_0^2 \end{cases} \tag{10}$$

when $x \leq \beta_1 h_0$

$$\begin{cases} N_{sw} = \left(1 + \frac{x - \beta_1 h_0}{0.5\beta_1 h_{sw}}\right)f_{yw}A_{sw} \\ N_\mathrm{pw} = \left(1 + \frac{x - \beta_1 h_0}{0.5\beta_1 h_\mathrm{pw}}\right)f_\mathrm{pw}A_\mathrm{pw} \\ M_{sw} = \left[0.5 - \left(\frac{x - \beta_1 h_0}{\beta_1 h_{sw}}\right)^2\right]f_{yw}A_{sw}h_{sw} \\ M_\mathrm{pw} = \left[0.5 - \left(\frac{x - \beta_1 h_0}{\beta_1 h_\mathrm{pw}}\right)^2\right]f_\mathrm{pw}A_\mathrm{pw}h_\mathrm{pw} \end{cases} \tag{11}$$

when $x > \beta_1 h_0$

$$\begin{cases} N_\mathrm{sw} = f_{yw}A_\mathrm{sw} \\ N_\mathrm{pw} = f_\mathrm{p}A_\mathrm{p} \\ M_\mathrm{sw} = 0.5 f_{yw}A_\mathrm{sw}h_\mathrm{sw} \\ M_\mathrm{pw} = 0.5 f_\mathrm{pw}A_\mathrm{pw}h_\mathrm{pw} \end{cases} \tag{12}$$

where x is the depth of the compression zone, and $\xi = x/h_0$; α_1 is the concrete stress block factor that is related to equivalent rectangular concrete compressive stress block intensity; f_c is the concrete compressive strength; b'_f and b_w are the width of wall flange and web; h'_f is the depth of wall flange; β_1 is the stress block factor that is related to concrete strength; h_sw and h_pw are the depth of distributed longitudinal reinforcement and embedded steel plate in concrete web; f_{yw} and f_pw are the yield strength of distributed longitudinal reinforcement and embedded steel plate; and, A_sw and A_pw are the areas of distributed longitudinal reinforcement and embedded steel plate.

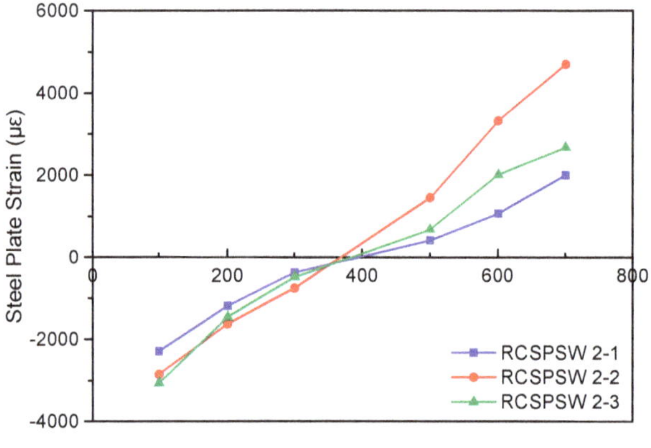

(**a**) Strain Distribution along Wall Depth

Figure 19. *Cont.*

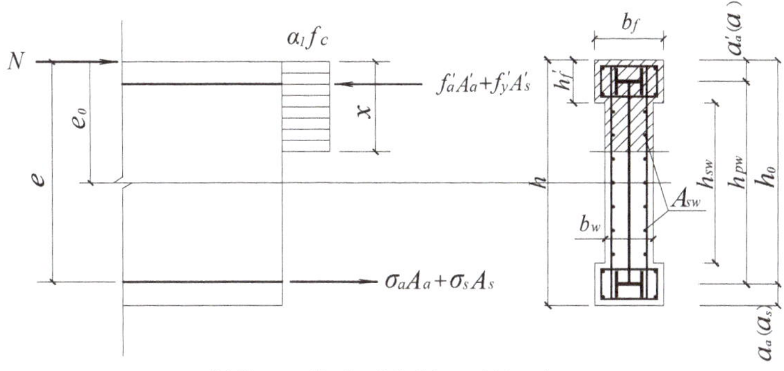

(**b**) Flexure Design Models and Notations

Figure 19. Flexure Design Models Developed based on Plane Section Assumption.

Table 8 summarizes the flexure strength capacities measured from the quasi-static cyclic load test and calculated by flexure strength models. It is seen that the flexure strength capacities predicted by the design models are approximately 20–35 percent lower than the measured values in both positive and negative directions. Similar with the shear model, a correction factor, k_f, of 1.1 is added to achieve a reasonable degree of conservatism. Figure 16b compares the design strength capacities with experimental values for flexure specimens, in which the line and symbol types are the same with Figure 16a. It is observed that the modified design flexure capacities are generally 10–20 percent as compared to the test values, verifying the reliability of the suggested design models.

Table 8. Comparisons of Experimental and Design Strength Capacities for Flexure Specimens.

Specimen No.	Experimental Results (kN)			Design Model Results (kN)	Differences		
	Positive	Negative	Average		Positive	Negative	Average
RCSW 2-1	469	514	492	341	27%	34%	30%
RCSW 2-2	540	536	538	389	28%	27%	28%
RCSW 2-3	570	578	574	481	16%	17%	16%
RCSPSW 2-1	581	697	639	387	33%	44%	39%
RCSPSW 2-2	646	674	660	434	33%	36%	34%
RCSPSW 2-3	674	701	688	512	24%	27%	25%

5. Summary and Conclusions

This paper introduces a high-strength reinforced concrete-steel plate composite shear wall system as lateral load resisting elements the design of high-rise building structures. A total of 11 RCSW and RCSPSW specimens were tested under quasi-static cyclic lateral loading to investigate the effects of critical factors on their seismic performances, including embedment of steel plate, aspect ratio, axial compression ratio, etc. Throughout the test procedure, the progression of damage and failure modes were observed and the key parameters were determined, such as lateral load capacity, ultimate displacement, ductility factor, and EVD coefficient. In addition, the design models were suggested to determine the shear and flexure strength of RCSPSW. By analyzing the data collected from experimental tests, the following conclusions are drawn.

1. Shear and flexure modes of failure dominate in specimens with the aspect ratio of 1.5 and 2.7, respectively. Inclined cracks were observed on the shear specimens, whilst flexure-tension cracks and flexure-shear cracks were mainly distributed in the lower half of boundary elements and wall web in flexure specimens. As compared to the RCSW specimens, more densely-distributed,

but finer, cracks took place in the RCSPSW specimens. With the increase of axial compressive loads, the crack quantity becomes less and the crack width is smaller due to the compaction of compression forces.

2. Obvious pinching" effects were observed in hysteresis loops of RCSW specimens. With the embedment of the steel plate, hysteresis curves appear in a plumper shape with higher peak lateral load capacities, but a severer post-peak strength and stiffness degradation.

3. RCSPSW system demonstrates superior lateral load capacities over RCSW, in general. The RCSPSW shear specimens show acceptable deformability for design purpose with the ultimate drift value of 1.0%. As for RCSPSW flexure specimens, the ultimate drift is larger than 1.0% and the ductility factor is around 4 when the axial compression ratio is lower than 0.50.

4. Axial compression ratio has an indispensable effect on the lateral load performance of wall specimens. In general, the peak lateral load increases, but ultimate displacement decreases as the axial compression ratio increases. A severer strength and stiffness degradation was observed for RCSPSW flexure specimens with the axial compression ratio higher than 0.50. When the axial compression ratio increases to 0.58, the ductility factor substantially decreased to 2.61 and the ultimate drift is lower than 1.0%.

5. The brittleness of high-strength concrete explains the relatively lower deformability of the RCSPSW specimens under high axial compression ratios. Another possibility is the weak concrete confinement effect on the embedded steel plate due to the small wall thickness values of the test specimens. The steel plate tends to buckle when severe concrete spalling occurs under higher axial compressive loads, causing substantial strength and stiffness degradation.

6. Energy dissipation capacities found in shear mode of failure are generally weaker than those in flexure-controlled failure, and the embedment of steel plate effectively improves the energy dissipation ability of RCSW.

7. The suggested design models generally provide conservative design values for shear and flexure strength of RCSPSW. A reasonable degree of conservatism is obtained by using correction factors: k_s, = 0.9 and k_f, = 1.1.

8. Shear walls are suggested to be designed with caution when subjected to high axial compressive loads and an upper limit of axial compression ratio (0.5) should be set for RCSPSW that is based on experimental observations. Steel ties or shear studs should be used to strengthen the connection between the steel plate and concrete on both sides for detailing in high-strength RCSPSW system. Besides, a higher transverse reinforcement ratio is recommended to further improve the concrete confinement effect.

Author Contributions: Conceptualization, D.J. and C.X.; methodology, D.J., C.X. and T.C.; validation, Y.Z.; formal analysis, D.J. and T.C.; investigation, D.J.; C.X. and T.C.; writing—original draft preparation, D.J.; writing-review and editing, C.X. and T.C.; supervision, C.X.; project administration, D.J.; funding acquisition, D.J.

Funding: This project was supported by Natural Science Foundation of Jiangsu Province under Grant No. BK20180487, National Natural Science Foundation of China (NSFC) under Grant Nos. 51808292 and 51508276, and Nanjing University of Science and Technology Start-up Grant AE89958.

Acknowledgments: The authors are grateful to collaborators from China Academy of Building Research (CABR) who provided invaluable assistance in the experimental test of high-strength concrete-steel plate composite shear walls.

References

1. Luna, B.N.; Rivera, J.P.; Whittaker, A. Seismic behavior of low-aspect-ratio reinforced concrete shear walls. *ACI Struct. J.* **2015**, *112*, 593–605. [CrossRef]
2. Taranath, B.S. *Reinforced Concrete Design of Tall Buildings*; CRC Press: Boca Raton, FL, USA, 2009.

3. Chinese Building Standard Committee. *National Standard of the People's Republic of China: Technical Specification for Concrete Structures of Tall Building (JGJ 3-2010)*; China Architecture & Building Press: Beijing, China, 2010.

4. American Concrete Institute (ACI). *Building Code Requirements for Structural Concrete (ACI 318-14) and Commentary*; ACI: Farmington Hills, MI, USA, 2014; p. 520.

5. Salonikios, T.N. Shear strength and deformation patterns of R/C walls with aspect ratio 1.0 and 1.5 designed to eurocode 8 (EC8). *Eng. Struct.* **2002**, *24*, 39–49. [CrossRef]

6. Lopes, M. Experimental shear-dominated response of RC walls Part I: Objectives, methodology and results. *Eng. Struct.* **2001**, *23*, 229–239. [CrossRef]

7. Tasnimi, A. Strength and deformation of mid-rise shear walls under load reversal. *Eng. Struct.* **2000**, *22*, 311–322. [CrossRef]

8. Zhang, Y.; Wang, Z. Seismic Behavior of Reinforced Concrete Shear Walls Subjected to High Axial Loading. *ACI Struct. J.* **2000**, *97*, 739–750.

9. Su, R.K.L.; Wong, S.M. Seismic behaviour of slender reinforced concrete shear walls under high axial load ratio. *Eng. Struct.* **2007**, *29*, 1957–1965. [CrossRef]

10. Moradi, M.J.; Hariri-Ardebili, M.A. Developing a Library of Shear Walls Database and the Neural Network Based Predictive Meta-Model. *Appl. Sci.* **2019**, *9*, 2562. [CrossRef]

11. Dan, D.; Fabian, A.; Stoian, V. Theoretical and experimental study on composite steel–concrete shear walls with vertical steel encased profiles. *J. Constr. Steel Res.* **2011**, *67*, 800–813. [CrossRef]

12. Tong, X.; Hajjar, J.F.; Schultz, A.E.; Shield, C.K. Cyclic behavior of steel frame structures with composite reinforced concrete infill walls and partially-restrained connections. *J. Constr. Steel Res.* **2005**, *61*, 531–552. [CrossRef]

13. Hajjar, J.F. Composite steel and concrete structural systems for seismic engineering. *J. Constr. Steel Res.* **2002**, *58*, 703–723. [CrossRef]

14. Liao, F.-Y.; Han, L.-H.; Tao, Z. Seismic behaviour of circular CFST columns and RC shear wall mixed structures: Experiments. *J. Constr. Steel Res.* **2009**, *65*, 1582–1596. [CrossRef]

15. Qian, J.; JIang, Z.; Ji, X. Experimental study on seismic behavior of steel tube-reinforced concrete composite shear walls with high axial compressive load ratio. *J. Build. Struct.* **2010**, *31*, 40–48.

16. Cao, W.; Wang, M.; Wang, S.; Zhang, J.; Zeng, B. A seismic research of composite shear wall and core walls with rectangular concrete filled steel tube columns. *Eng. Mech.* **2008**, *25*, 58–70.

17. Cho, S.H.; Tupper, B.; Cook, W.D.; Mitchell, D. Structural Steel Boundary Elements for Ductile Concrete Walls. *ASCE J. Struct. Eng.* **2004**, *130*, 762–768. [CrossRef]

18. Esaki, F.; Ono, M. Effect of loading rate on mechanical behavior of SRC shearwalls. *Steel Compos. Struct.* **2001**, *1*, 201–212. [CrossRef]

19. Zhou, Y.; Lu, X.; Dongq, Y. Seismic behavior of composite shear walls with multi-embedded steel sections. Part I: Experiment. *Struct. Des. Tall Spec. Build.* **2010**, *19*, 637–655. [CrossRef]

20. Hu, H.-S.; Nie, J.-G.; Eatherton, M.R. Deformation capacity of concrete-filled steel plate composite shear walls. *J. Constr. Steel Res.* **2014**, *103*, 148–158. [CrossRef]

21. Nie, J.-G.; Ma, X.-W.; Tao, M.-X.; Fan, J.-S.; Bu, F.-M. Effective stiffness of composite shear wall with double plates and filled concrete. *J. Constr. Steel Res.* **2014**, *99*, 140–148. [CrossRef]

22. Hossain, K.M.A.; Wright, H.D. Experimental and theoretical behaviour of composite walling under in-plane shear. *J. Constr. Steel Res.* **2004**, *60*, 59–83. [CrossRef]

23. Rafiei, S.; Hossain, K.M.A.; Lachemi, M.; Behdinan, K.; Anwar, M.S. Finite element modeling of double skin profiled composite shear wall system under in-plane loadings. *Eng. Struct.* **2013**, *56*, 46–57. [CrossRef]

24. Zhao, Q.; Astaneh-Asl, A. Cyclic Behavior of Traditional and Innovative Composite Shear Walls. *ASCE J. Struct. Eng.* **2004**, *130*, 271–284. [CrossRef]

25. American Institute of Steel Construction (AISC). *Seismic Provisions for Structural Steel Buildings (ANSI/AISC 341-10)*; AISC: Chicage, IL, USA, 2010.

26. Sun, J.; Xu, P.; Xiao, C.; Sun, H.; Wang, C. Experimental study on shear behavior of steel plate-concrete composite wall. *Build. Struct.* **2008**, *38*, 6–10.

27. Chen, T.; Xiao, C.; Tian, C.; Xu, P. Experimental study of the compression-bending behavior of composite shear walls of high axial compression ratio. *China Civ. Eng. J.* **2011**, *44*, 9–15.

28. Xiao, C.; Jiang, D.; Xu, Z.; Chen, T. Seismic behavior of steel plate reinforced concrete shear walls. In Proceedings of the 9th CTBUH World Congress, Shanghai, China, 19–21 September 2012; pp. 671–678.

29. Lu, X.; Jiang, H. Recent progress of seismic research on tall buildings in China Mainland. *Earthq. Eng. Eng. Vib.* **2014**, *13*, 47–61. [CrossRef]

30. Jiang, D.; Xiao, C.; Chen, T.; Tian, C.; Xu, P. Experimental study of the compression-bending behavior of high-strength concrete steel composite shear walls. *China Civ. Eng. J.* **2012**, *45*, 17–25.

31. Xiao, C.; Tian, C.; Chen, T.; Jiang, D. Compression-bending behavior of steel plate reinforced concrete shear walls with high axial compression ratio. In Proceedings of the 15th World Conference on Earthquake Engineering, Lisbon, Portugal, 24–28 September 2012.

32. Olivari, C.N.A. Influence of Axial load in the Seismic Behavior of Reinforced Concrete Walls with Nonseismic Detailing. Master's Thesis, Pontificia Universidad Católica de Chile, Santiago, Chile, 2013; p. 108.

33. Greifenhagen, C.; Lestuzzi, P. Static cyclic tests on lightly reinforced concrete shear walls. *Eng. Struct.* **2005**, *27*, 1703–1712. [CrossRef]

34. Hidalgo, P.A.; Ledezma, C.A.; Jordan, R.M. Seismic behavior of squat reinforced concrete shear walls. *Earthq. Spectra* **2002**, *18*, 287–308. [CrossRef]

35. Takahashi, S.; Yoshida, K.; Ichinose, T.; Sanada, Y.; Matsumoto, K.; Fukuyama, H.; Suwada, H. Flexural Drift Capacity of Reinforced Concrete Wall with Limited Confinement. *ACI Struct. J.* **2013**, *110*, 95–104.

36. Chinese Building Standard Committee. *National Standard of the People's Republic of China: Specification for Seismic Test of Buildings (JGJ/T 101-2015)*; China Architecture & Building Press: Beijing, China, 2015.

37. Liao, F.Y.; Han, L.H.; Tao, Z. Performance of reinforced concrete shear walls with steel reinforced concrete boundary columns. *Engineering Structures.* **2012**, *44*, 186–209. [CrossRef]

38. Priestley, M.J.N.; Park, R. Strength and ductility of concrete bridge columns under seismic loading. *ACI Struct. J.* **1987**, *84*, 61–76.

39. Mahin, S.A.; Bertero, V.V. Problems in establishing and predicting ductility in aseismic design. In Proceedings of the International Symposium on Earthquake Structural Engineering, University of Missouri, Rolla, MO, USA, 19–21 August 1976; pp. 613–628.

40. American Society for Testing and Materials (ASTM). Standard test methods for cyclic (reversed) load test for shear resistance of vertical elements of the lateral force resisting systems for buildings. In *ASTM E2126*; ASTM: West Conshohocken, PA, USA, 2011.

41. Park, R. Ductility evaluation from laboratory and analytical testing. In Proceedings of the 9th World Conference on Earthquake Engineering, Tokyo-Kyoto, Japan, 2–9 August 1988; pp. 605–616.

42. Chopra, A.K. *Dynamics of Structures: Theory and Applications to Earthquake Engineering*; Prentice-Hall: Upper Saddle River, NJ, USA, 2011.

43. Hariri-Ardebili, M.A.; Saouma, V.E. Sensitivity and uncertainty analysis of AAR affected reinforced concrete shear walls. *Eng. Struct.* **2018**, *172*, 334–345. [CrossRef]

44. Dancygier, A.N.; Katz, A. Bond over direct support of deformed rebars in normal and high strength concrete with and without fibers. *Mater. Struct.* **2012**, *45*, 265–275. [CrossRef]

45. Harajli, M.H.; Gharzeddine, O. Effect of steel fibers on bond performance of steel bars in NSC and HSC under load reversals. *J. Mater. Civ. Eng.* **2007**, *19*, 864–873. [CrossRef]

46. Gattesco, N.; Amadio, C.; Bedon, C. Experimental and numerical study on the shear behavior of stone masonry walls strengthened with GFRP reinforced mortar coating and steel-cord reinforced repointing. *Eng. Struct.* **2015**, *90*, 143–157. [CrossRef]

47. Smarzewski, P. Analysis of Failure Mechanics in Hybrid Fibre-Reinforced High-Performance Concrete Deep Beams with and without Openings. *Materials* **2018**, *12*, 101. [CrossRef]

48. Pan, A.; Moehle, J.P. Lateral displacement ductility of reinforced concrete flat plates. *ACI Struct. J.* **1989**, *86*, 250–258.

49. Lim, W.-Y.; Hong, S.-G. Shear tests for ultra-high performance fiber reinforced concrete (UHPFRC) beams with shear reinforcement. *Int. J. Concr. Struct. Mater.* **2016**, *10*, 177–188. [CrossRef]

50. Chinese Building Standard Committee. *National Standard of the People's Republic of China: Code for Design of Composite Structures (JGJ 138-2016)*; China Architecture & Building Press: Beijing, China, 2016.

Article

Determination of Young Elasticity Modulus in Bored Piles Through the Global Strain Extensometer Sensors and Real-Time Monitoring Data

Hossein Moayedi [1,2,*], Bahareh Kalantar [3], Mu'azu Mohammed Abdullahi [4], Ahmad Safuan A. Rashid [5], Ramli Nazir [5] and Hoang Nguyen [6]

[1]   Department for Management of Science and Technology Development, Ton Duc Thang University, Ho Chi Minh City 700000, Vietnam
[2]   Faculty of Civil Engineering, Ton Duc Thang University, Ho Chi Minh City 700000, Vietnam
[3]   RIKEN Center for Advanced Intelligence Project, Goal-Oriented Technology Research Group, Disaster Resilience Science Team, Tokyo 103-0027, Japan
[4]   Civil Engineering Department, University of Hafr Al-Batin, Al-Jamiah, Hafr Al-Batin 39524, Eastern Province, Saudi Arabia
[5]   Center of Tropical Geoengineering (Geotropik), School of Civil Engineering, Faculty of Engineering, Universiti Teknologi Malaysia, Johor Bahru 81310, Johor, Malaysia
[6]   Institute of Research and Development, Duy Tan University, Da Nang 550000, Vietnam
*   Correspondence: hossein.moayedi@tdtu.edu.vn

Received: 16 June 2019; Accepted: 12 July 2019; Published: 29 July 2019

Abstract: For friction piles depending on the friction resistance, accurate prediction of unit skin friction around the pile shaft is the dominant resistance to measure the final bearing capacity of a bored-pile. The present study measures the stress–strain transferring in two instrumented bored-piles (BP #1 & BP# 2) embedded within the soil layer in Kuala Lumpur by real-time monitoring global strain extensometer (GSE) sensors. Two bored-piles (i.e., having 1.80 m and 1.0 m diameters, as well as 36 m and 32 m lengths) have been loaded with two times to their design working loads. Extensive data are analyzed to measure the changes in stress–strain in the bored-pile. The effect of loading and unloading stages on the pile's head and base settlement has been monitored, indicating that Young modulus of elasticity in concrete bored-pile (E_c), average strain, and unit skin friction changed along the bored-pile based on the ground site conditions and stress registered. One example of two case studies with great real-time monitoring data has been provided to further design.

Keywords: bored-pile; global strain extensometer; pile friction resistance; real-time monitoring

1. Introduction

In recent years, calculation of pile bearing capacity data in-situ test has been broadly applied by geotechnical engineers and building foundations, because these data are more accurate and reliable than small-scale laboratory tests. In fact, bored-piles have been considered empirically more as an art-work than science [1], and are formed using appropriate machines (capacity-type) to fill the holes with applicable concrete and reinforcement. Their usual sizes are 400 mm to 3,000 mm diameter, with a capacity that reaches up 45,000 kN based on the pile size and geological profile close to the pile, so an excellent pile capacity has reduced the pile cap size and pile numbers in the group [2,3]. Therefore, bored-pile designing in most countries has relied on the results of the conventional standard penetration test (SPT). According to the literature, the procedure of bored-pile design has consisted of stages such as: (1) the calculation of the end bearing capacity (f_b) in bored-pile; (2) calculation of the shaft bearing capacity in bored-pile (f_s), note that, the sum of both values is the ultimate bearing capacity of an individual pile; and (3) pile working load which has been presumed from the ultimate bearing capacity by using a safety factor that permits the piles' interaction within the group [4,5].

Regarding the empirical approach of f_s in relation to $K_s \times$ SPT, f_b has been related to $K_b \times$ SPT widely applied in pile designing. To evaluate K_s (i.e., skin friction ratio, f_s/SPT) and K_b (i.e., end bearing ratio, f_b/SPT), the value including local soil condition has required vibrating wire strain gauges (VWSG) and a mechanical tell-tale rod installed within the pile to measure the axial stress–strain relation and movement in different levels down to the pile's toe and shaft. Bored-pile of length (L) and diameter (D_s) supporting a vertical head load (P_h) by the mobilized shaft and base resistance $(P_s - P_b)$ is illustrated in Figure 1a. The vertical displacement of the pile's base and head are defined as $\Delta_b - \Delta_h$, followed by the pile compressing (shortening) as e_p ignoring the pile weight as it is in Equation (1) [6]:

$$P_h = P_s + P_b \tag{1}$$

According to the cross-sectional area of concrete A, the stress along with the pile σ_h is as Equation (2) [6]:

$$\sigma_h = \frac{P_h}{A} \tag{2}$$

$$\Delta_h = e_p + \Delta_b \tag{3}$$

The length δ_z has been located at depth z below the pile head level as shown in Figure 1a–c. Shaft resistance δP_s mobilized along the segment δ_z equals the axial force changing in axial force δP_z. The unit shaft resistance $f_s(z)$ mobilized in δ_z has been related to $P(z)$ as shown in Equation (4):

$$\frac{dP(z)}{dz} = -\pi D_s f_s(z) \tag{4}$$

The negative sign shows that P_z decreased as z increased (Figure 1c). Omer et al. [7] mention that if the pile's elastic feature has been denoted (E_p), ignoring all vertical soil moving and pile displacement $w(z)$ (e.g., pile movement at depth z below the pile head level) at depth z have been offered as Equation (5):

$$\frac{dw(z)}{dz} = \frac{-4P(z)}{\pi D_s^2 E_p} \tag{5}$$

Equation (5) is different with z and $dP(z)/dz$, replaced by using Equations (1)–(4) as Equation (6):

$$\frac{d^2 w(z)}{dz^2} = \frac{4}{D_s E_p} f_s(z) \tag{6}$$

All the mentioned equations according to force equilibrium and displacement correspondents are valid irrespective of pile type and soil grouping [7].

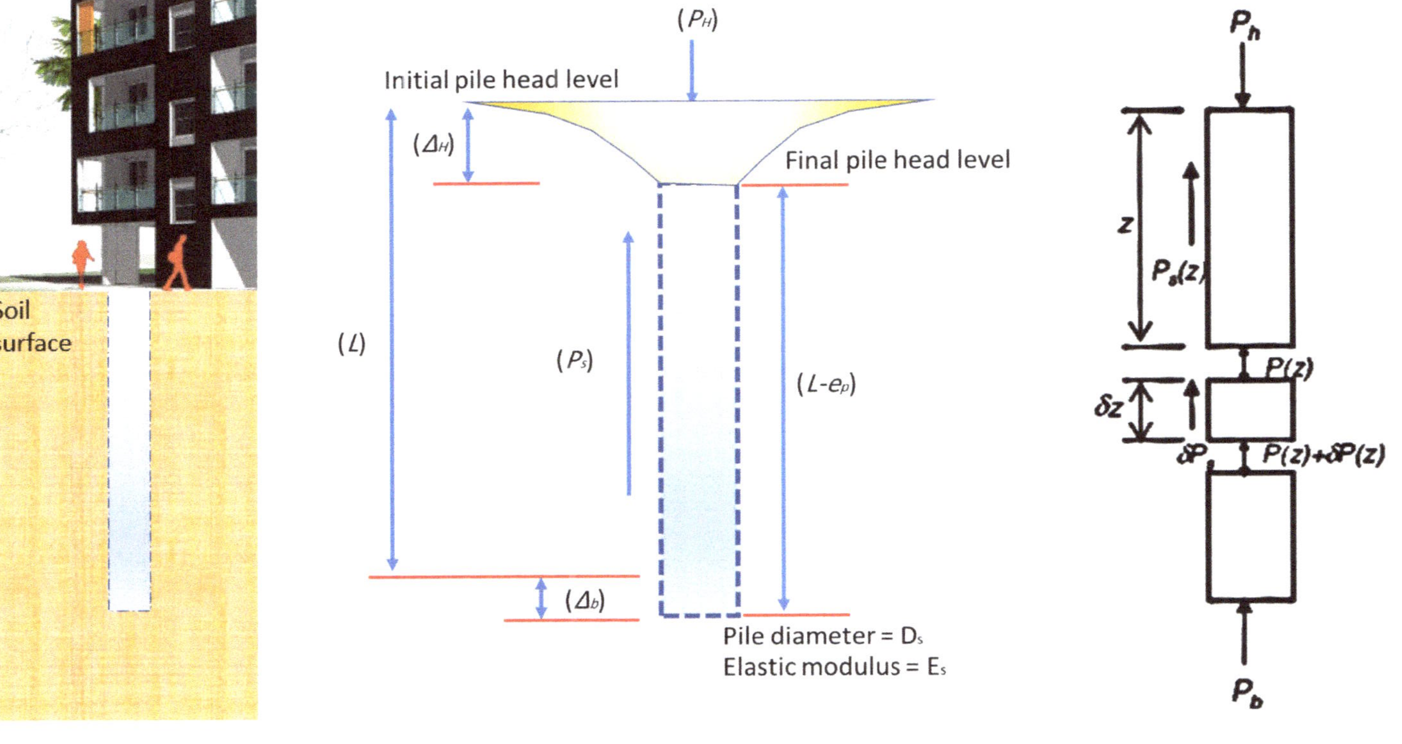

(**a**) schematic representation   (**b**) forces on a pile element in "equilibrium".   (**c**) force details in element with length of δ_z

Figure 1. Displacement and force of loaded bored pile–soil; (**a**) schematic representation, (**b**) forces on a pile element in "equilibrium., (**c**) force details in element with length of δ_z.

Few more studies have been performed on longitude strain, which have measured the instrumented piles [8–13], transferring of load in rapid pile axial loading [14], static, dynamic, seismic, and cyclic lateral load of pile classifications [15–17], rigid and flexible pile behaviors in diverse soft soils [18–20], and skin friction resistance measurement in piles [21]. Some studies have been conducted based on numerical simulations, however, others have been performed on field monitoring (Ng and Sritharan [22]; Hung et al. [23]; Tafreshi et al. [24]; Mascarucci et al. [25]; and Lee et al. [26]). Ochiai et al. [27] propose a reliable designing model for bored-piles following in-situ tests by SPT. Poulos [28] has also introduced an appropriate design for piled rafts, comprising three stages: (1) assessing the feasibility of piled raft application accompanied by the required pile number, (2) evaluating where piles are needed and its general features, and (3) obtaining the optimum number, location, and configuration of a pile and computing the settlement distributions, bending moment and shear in the raft, and the pile loads and moments.

Sego et al. [29] studied the effect of an enlarged base on the total and end bearing resistance of a pile for use in ice-rich permafrost. Therefore, the total capacity of a bored-pile has mainly inclined by belled-pile usage. Shariatmadari et al. [30] studied the bearing capacity of driven piles in sands following SPT by applying 60 previous cases. Data included 43 full-scale, 17 dynamic tests, and static pile load testing analyzed by control and provisioning of wireless access points (CAPWAP). Note that, SPT data have been used close to the pile locations. Another model, as standard penetration test N(SPT-N), has been conducted and proved to have less scatter with higher accuracy. Zhang et al. [31] applied an elastic–plastic model showing the load–settlement relationship, and provided a simple method to analyze the behavior of a pile group and/or a single pile embedded in multilayered soils using two methods like an approach to enable a quick estimation of pile group settlement and/or a single pile embedded in multilayered soils providing time and cost saving. Ruan and Zuo [32] explained the relations between the ultimate vertical bearing capacity and SPT for the jacked pile. These case studies confirm the accuracy and reliability of the formula, mainly in silent pipe pile bearing capacity calculation. Assessing the friction capacity of driven piles in clay has been performed through the use of multivariate adaptive regression spline (MARS) by Samui [33], which led to D (diameter), L (length), undrained shear strength, and effective vertical stress as MARS input and friction capacity as output compared using artificial neural network (ANN). MARS, as a robust model, has been applied to predict the friction capacity in driven piles installed in clay.

Meanwhile, another research carrying the same approach has been conducted on cohesionless soil by Samui [34]. Chae et al. [35] investigated the uplift capacity of belled-piles in weathered sandstones of the Persian Gulf coast. Accordingly, few full-scale pullout load tests on belled tension piles in Abu Dhabi have also been performed. Comparing the results from the field, 3D finite element (FE), and theoretical methods have overestimated the ultimate pullout resistance of belled-pile without bell shape considerations [35]. Sakr [36] compared the results of high strain dynamic and static load tests of single helical-screw piles in cohesive soils. High strain dynamic pile load experiments have been performed on both driven steel open-pipe and helical piles. Case Pile Wave Analysis Program (CAPWAP) and full-scale static load test have confirmed high strain dynamic testing (HSDT) as a reliable tool for assessing static helical pile capacity. Zhang et al. [37] stated that in a typical design, the skin friction and the base resistance are majorly independent. Furthermore, the ultimate bearing capacity of a single pile has been composed in ultimate end resistance and confining skin friction.

Yao and Chen [38] promoted a flexible plastic solution for the uplift belled-pile. In the comparison of the provided solution outcome to the theoretical calculation one, the theoretical method results revealed that the elastic–plastic analytical solutions is a good method. Aksoy et al. [39] developed a new chart to estimate the friction angle between the pile and soil materials. Accordingly, in the current research, soil including different internal friction angles (ϕ) has been initially provided, then the skin friction angles (delta) of the mentioned soils with fiber-reinforced plastic (FRP) like a composite material, wood (pine), and steel (st37) have been defined by undertaking the interface shear test to provide a pile design diagram to determine skin friction angles of the soils and pile materials.

Wang et al. [40] investigated the controlling effectiveness and settlement behavior of two types of rigid pile structure embankments (PSE) constructed on collapsible loess soils beneath a high-speed railway. The results have shown that this type of PSE has mainly reduced embankment settlement so that embankments have to be maintained on collapsible loess. Meanwhile, pile spacing has significant efficiency in settlement reduction. Therefore, the current study has focused on stress–strain transferring throughout the instrumented bored-pile within the layered soil to measure the parameters of soil-pile interaction comprising Young elasticity module in concrete (E_c), average strain, and unit skin friction changed along with the pile.

2. Material and Methods

2.1. Testing

Maintained load test (MLT), is known as static loading test (i.e., the load will remain constant until the settlement ended to small values), has been followed by ASTM Standard (D1143/D1143M-07). An apparent distance between the reaction pile and test pile should not be less than five times to diameter (D) of the immense pile. Considering setup, the piles are loaded by applying hydraulic jacks toward the main beam operated by an electric pump. Then, the applied load is accurately calibrated by vibrating wire load cells (VWLC) (Table 1). Both instrumental piles are located in Jalan Ampang and Kuchai Lama, Kuala Lumpur, Malaysia.

Table 1. Instrumental bored pile load test summary.

Pile No	Diameter (mm)	Working Load (kN)	Pile Length (m)	Pile Area (m^2)	Test Load (kN)	Type of Instrument
BP# 1	1800	22,200	36.95	2.5447	44,400	GSE
BP# 2	1000	6750	32.56	0.7854	13,500	Conventional

2.2. Loading Procedure

Poulos [28] states that applied loading is crucial to the bending moment and differential settlements, but less critical to load-sharing or maximum settlement between the piles and raft. In bored-piles, time for loading test has been determined by the piles' concrete strength (Zhang et al. [37]). In addition, Tomlinson and Woodward [1] suggest that at the testing time, concrete should be in its seven days (at least) with at least doubled strength of applied stress, moreover, in the current pile testing, the load cycles have started 28 days after pile construction. A schematic view of the MLT is shown in Figure 2. Instrumented piles are tested by MLT per two loading-cycle calculated by calibrated VWLCs in every 10 min for one hour (Figure 3).

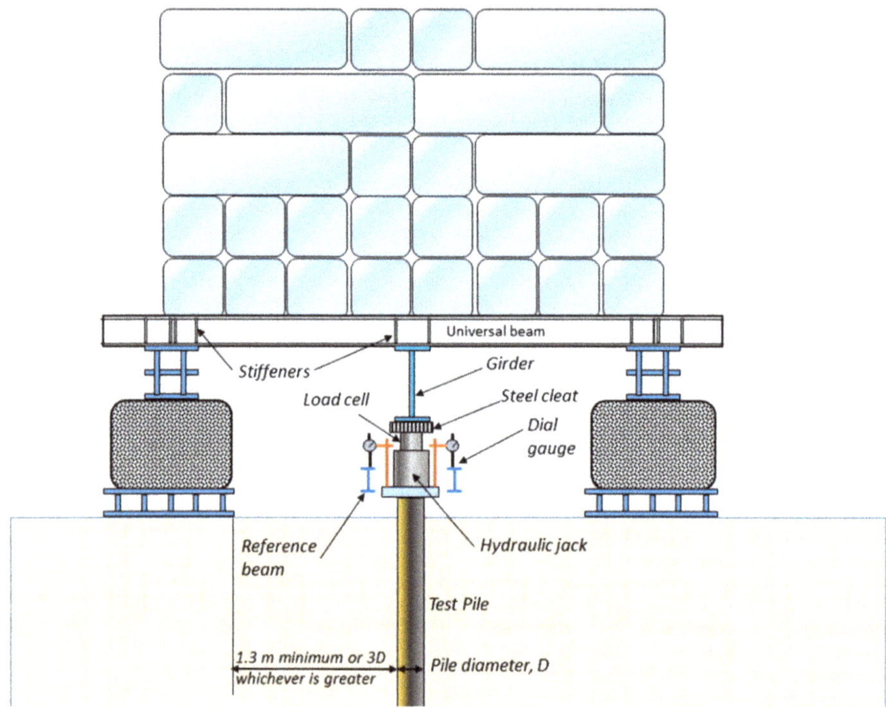

Figure 2. Schematic view of the maintained load test (MLT)

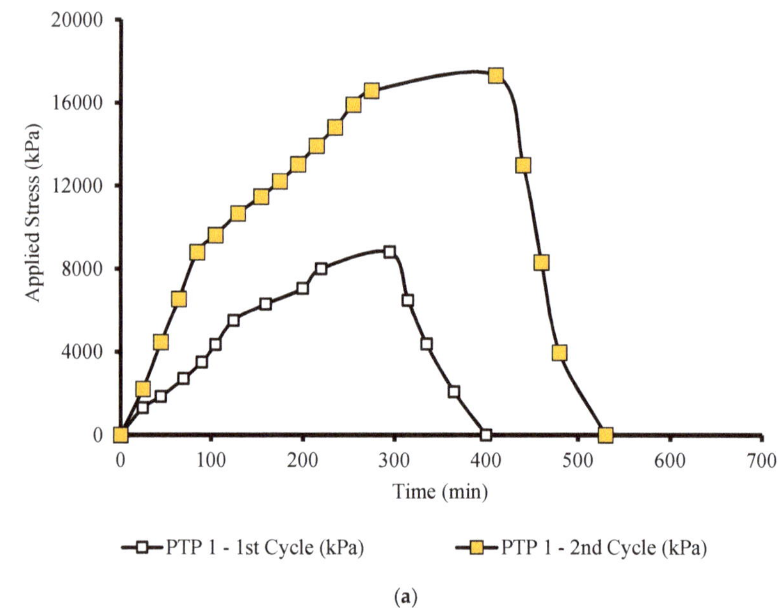

(**a**)

Figure 3. *Cont.*

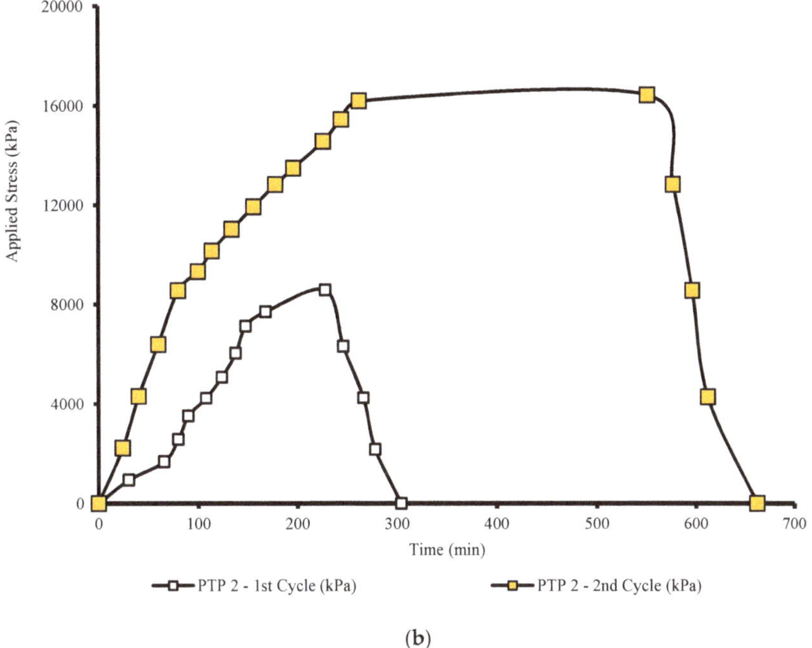

(**b**)

Figure 3. Variation of the total applied static stress on pile versus time, (**a**) Bored pill (BP)# 1 and (**b**) BP# 2.

2.3. Instruments Monitoring System

The influence of geologically weak zones through multilayer site conditions has dramatically changed the designing parameters when the majority of these parameters change by depth. In other words, new global strain extensometer (GSE) sensors have recently measured the change of design parameters through the depth, so the pile top settlement has been monitored using: (1) four linear variation displacement transducers (LVDTs) mounted on the reference beams with their plungers vertically placed against glass plates fixed on the pile top, and (2) vertical scale rules attached to the pile top and sighted by a precise level instrument. Vertical scales have also been shown on the reference beams to check any movement in the loading test. Indeed, VWLC, global strain extensometer (GSE) sensors, VWSG, and LVDTs have been automatically logged across the use of Micro-10x data logger and Multi logger at 5-minute spaces for precise controlling during loading/unloading stages, adding that only accurate level readings have been taken manually. Figure 4 shows the cross section of BP# 1 (D = 1,800 mm) and BP# 2 (D = 1,000 mm) and the sensors' placement along the main steel rods.

Figure 4. Cross-section of BP# 1 (D = 1,800 mm) and BP# 2 (D = 1,000 mm).

Accordingly, the bored-pile has been tested using MLT through the reaction pile. All instruments using Micro-10 data logger and multilevel software have been automatically logged. The conventional instrumented method used VWSG and mechanical tell-tales. While by being attached to the steel cage of bored-pile, VWSG and mechanical tell-tales have permanently been embedded in the concrete. GSE is the second instrument applied in axial load measurement and settlement distribution in the bored-pile.

On static load testing, loaded pile deformation has produced a related moving between every two anchored intervals changed in strain gauge wire tension, in addition to a corresponding change in its resonant vibration-frequency measured by plucking GSE sensors/transducers through a signal cable to readout box/data logger to measure the frequency and display of the shortening and strain reading. Considering the installation set up, GSE has measured shortening and strains on all test pile sections in every load step of a static pile load test, so it has integrated the strains on a larger and more representative sample. Therefore, using a defined instrumental scheme, data derived from instrumented load testing have provided reliable information. The results of GSE have been compared to the conventional instrumentation bored pile results. Subsequently, regarding the test piles, including BP# 2, the Geokon VWSG and tell-tale extensometers have been installed internally to monitor the strain developing and shortening of pile behavior on the test. BP# 1, with the instrumentation of GSE, has been placed with seven levels depending on the pile length and vertical varieties of sub-soil cases in sonic logging tubes (Figure 5).

Due to calibration of the applied axial load and the measured average strain, a calibrated GSE sensor has been installed near the pile's head, in which there is no interaction between the soil friction and pile shaft. The GSE sensor measured the strain of other levels to determine the axial load transferring in all pile shaft sections, also to measure the loads contributed from the toe or/to end bearing resistance. VW Extensometer is installed on the anchored interval at eight levels (Figure 6). Pile deformation in loading has produced a related motion between every two anchored intervals producing strain gauge wire tension alteration in VW transducers, in addition to a correspondent variation in its frequent resonant vibration measured by plucking GSE sensors to the readout box/data logger. Therefore, the process measured the frequency displayed by the shortening reading and strain reading. VWSG for BP# 2 is also installed at five levels (A to E) in four numbers per level (Figure 6). The connected VWSG to the steel cage and electrical lead wires from the sensors coming to the top, have been illustrated in Figures 7 and 8.

The pile head displacement has also been analyzed using dial gauges and LVDTs, resulting in a reading with 0.01 mm accuracy mounted on stable reference beams and protected from direct sunlight and disturbance of personnel in the whole system. Settlement measuring through the use of proper leveling techniques has also been implemented as a useful backup to check the reference beams' movement. Vibrating wire load cells, strain gauges, retrievable extensometers, and LVDTs have been automatically logged by applying a Micro-10x Datalogger and Multilogger software at the 3-minute spacing for precise control during load/unload stages. Only accurate level readings are manually considered.

Briefly, the pile movement monitoring system, such as pile top and bottom settlements, has been monitored using the vertical scale rule fixed to the pile top observed by tunnel boring machine (TBM) for correction purposes. On the other hand, LVDTs and GSEs mounted to the reference beam accompanied by a plunger pressed vertically against a glass plate fixed to the pile's top. The vertical scales have been provided to monitor any movement in the load test. Indeed, strain gauges are manufactured in Geokon, USA. VWSG principles in strain measurement were presented due to the frequent natural vibration of taut wire restrained at both ends and varied with the square root of wire tension. Indeed, this change of tension in the wire has shown all the strain alterations in the structural steel member where the gauge is mounted. Moreover, all strain gauge has been mounted to two end blocks as arc-welded to the main reinforced bar at a proper level. The signal cables of the picked-up sensors fixed to the strain gauges have been tightly tied in the reinforced bars to the top of the piles terminated in a multiplexer box and observed using a Micro-10 Datalogger.

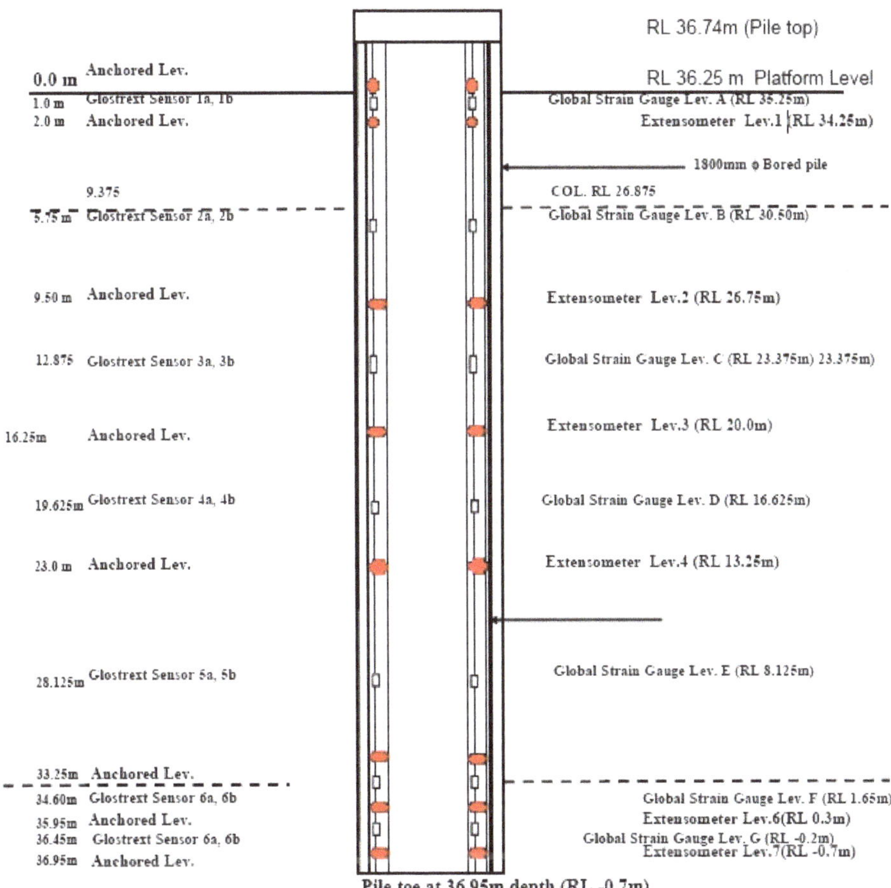

Figure 5. Various instrumental installments at different global strain extensometer sensors levels in BP# 1.

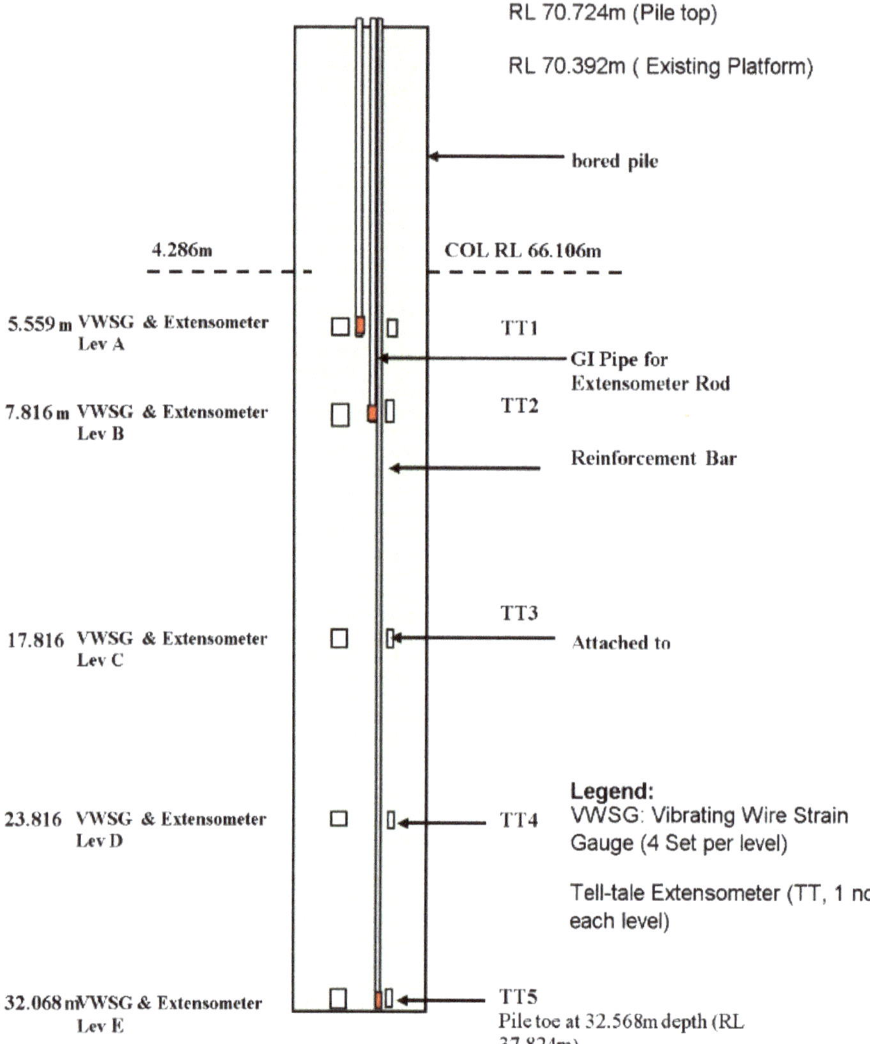

Figure 6. Vibrating wire strain gauges and tell-tales' extensometer arrangement in BP# 2.

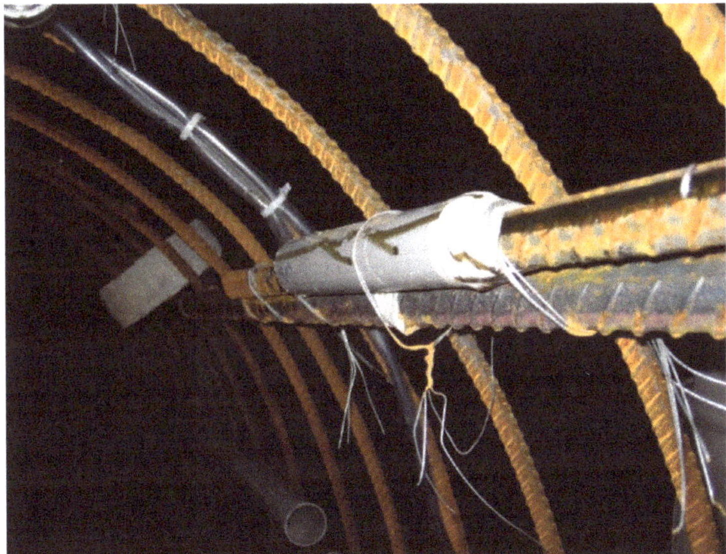

Figure 7. VWSG attached to the steel cage.

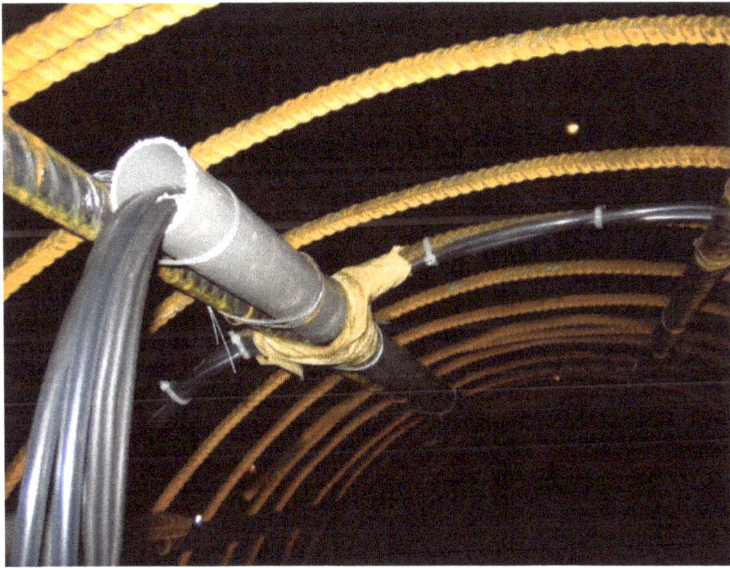

Figure 8. The electrical lead wires from the sensors come to the top.

2.4. Site Condition

In the current paper, two full-scale maintained static load experiments on bored-piles are conducted. The first experiment (BP# 1) was performed at Cadangan Pembangunan, Lorong Stonor, Kuala Lumpur, Malaysia. The test pile was initially loaded up to 2 times to pile structural capacity, therefore, regarding BP# 1 with the structural capacity of 22,200 kN, the nominal diameter of 1,800 mm with a penetration depth of 36.95 m from the current piling platform level is RL 36.25 m. The pile was initially examined by up to 44,400 kN (2 x working load) in two loading cycles. The second test (BP# 2) was applied at Utama Lodge, Jalan Senangria, Kuala Lumpur, Malaysia. The summary of soil types, besides the SPT-N values measured near the pile location, are illustrated in Table 2. Noting that the soil stratum was classified according to the unified soil classification system.

Table 2. Soil profile for tested bored pile after the standard penetration test (SPT).

Test Pile		Soil Stratum	Depth (m)	SPT-N Values *	Average SPT-N
BP# 1	L_1	Stiff Sandy Silt with little gravel	0–8	3–16	15.50
	L_2	Very Stiff Sandy Silt with little gravel	8–10	16–50	27.5
	L_3	Hard Yellowish Sandy Silt with little gravel	10–17	50–111	110
	L_4	Hard Yellowish Sandy Silt with little gravel	17–24	111–150	122
	L_5	Fractured limestone	24–36.95	143–150	150
BP# 2	L_1	Sandy SILT	0–12	4–30	30
	L_2	Sandy CLAY	12–17	19–39	39
	L_3	SILT	18–23	54–125	122
	L_4	Weathered Sandstone	25–31.65	176–200	195

* Note: SPT-N is the number of SPT driven into the ground (e.g., at the bottom of a borehole) by blows from a slide hammer with a mass of 63.5 kg.

The gauges were investigated prior and after installation, after cage placement in the borehole, and after concreting. The strain gauges' signal cables were reserved for testing approximately after 28 days, allowing for the concrete to achieve the design strength. Therefore, on test day, the strain gauges' cable was linked to the switch box connected to the data logger to ensure the functional sequences. Regarding the rod extensometer, galvanized iron (GI) pipes were tied to the main reinforced cage with steel wires at each terminating depth (Figure 9). A mild steel rod (10 mm) was inserted untill it touched the bottom of the pipe. In addition, a steel plate was welded to the rod's end for the plunger to sit on along the experiment (Figure 10).

Figure 9. global strain extensometer pipe for tell-tales extensometer is pre-installed at VWSG.

Figure 10. Existing platform in full-scale maintenance of the pile load test.

3. Results and Discussion

3.1. Stress–Strain Variation in the Piles

Stress distributions along the piles (BP# 1 and BP# 2) in two continuous loading/unloading cycles have been illustrated in Figures 11–14, showing that in a specified load, normal stress used in the pile's surface area is reduced by depth. Therefore, the reduction rate varied along with the pile—it is low in soft soil layers (having low SPT-N) but sharp in stiff layers (with high SPT-N). These results indicate great skin friction capacity of soil where rapid change between two different depths.

Tosini et al. [41] have declared that the forecast of deep foundation settlements in a layered soil profile is not always straightforward due to the problems in value defining of mechanical parameters affecting them. Distribution of average change and their cycles, in both strains (BP# 1 & BP# 2), in addition to the distribution of back-calculated modulus of concrete as E_c (kN/mm^2) in both strains (BP# 1 & BP# 2) are shown in Figures 15 and 16. In BP# 1, E_c for the entire length remained constant at 25 kN/mm^2 except at depths below 5 m within stiff sandy silt with little gravel when the higher load showed slightly lower E_c (Figure 15a). Therefore, the results obtained for E_c (in BP# 2) changed based on the depth and applied load. The lower applied stress through the first cycle provided minor changes in E_c distribution along the pile length. However, higher applied load amounts caused a higher change in E_c with depth (Figure 15). The highest and lowest E_c for the depth = 17.816 m (in BP# 2) were 38.2 and 33.9 kN/mm^2, resepectively, for the applied stress of 1,750 kPa and 12,904 kPa. The reverse relation of E_c with applied stress occurred when higher applied stress on the pile cap showed lower E_c (Figure 16a). Consequently, the highest variation in E_c was observed within the silty and/or sandy silt layers, while the lowest change in E_c was measured at the weathered sandstone zone at depths above 20 m (Figure 16b).

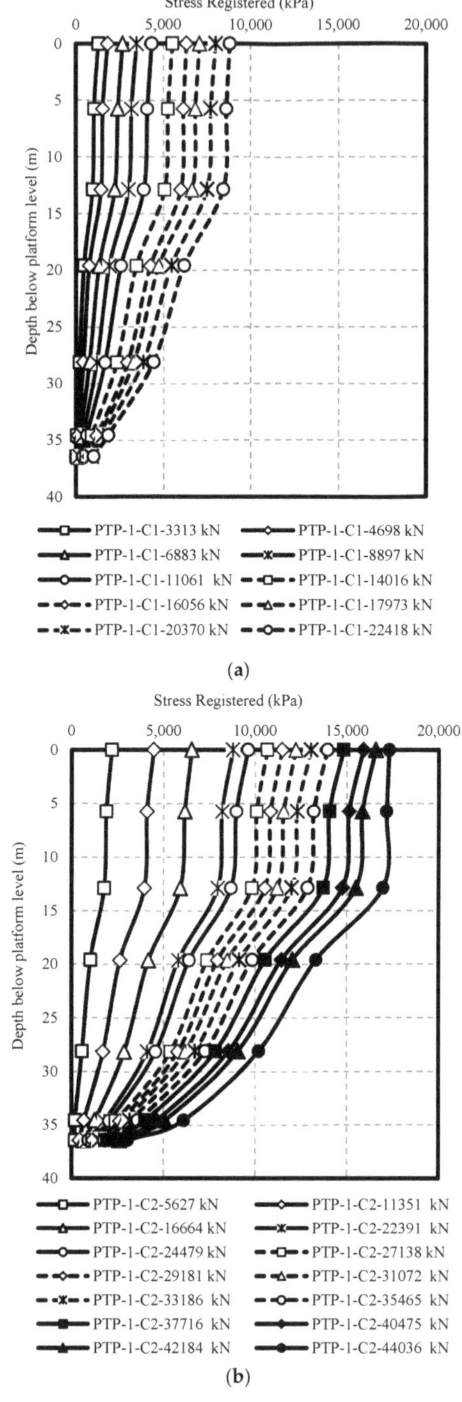

(**a**)

(**b**)

Figure 11. Stress distribution measured along BP# 1 in (**a**) 1st cycle and (**b**) 2nd cycle.

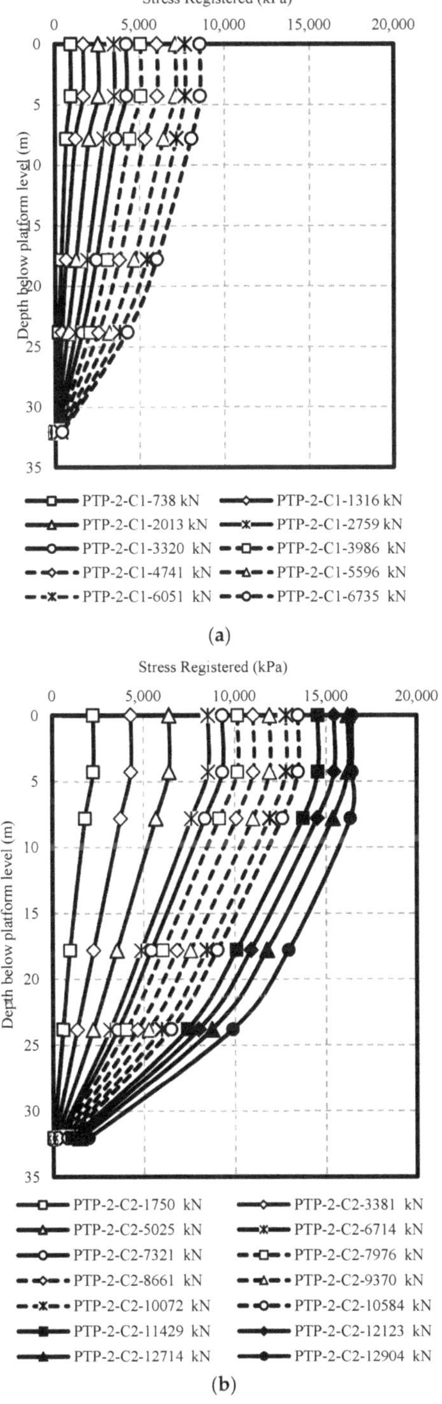

Figure 12. Stress distribution measured along BP# 2 in (**a**) 1st cycle and (**b**) 2nd cycle.

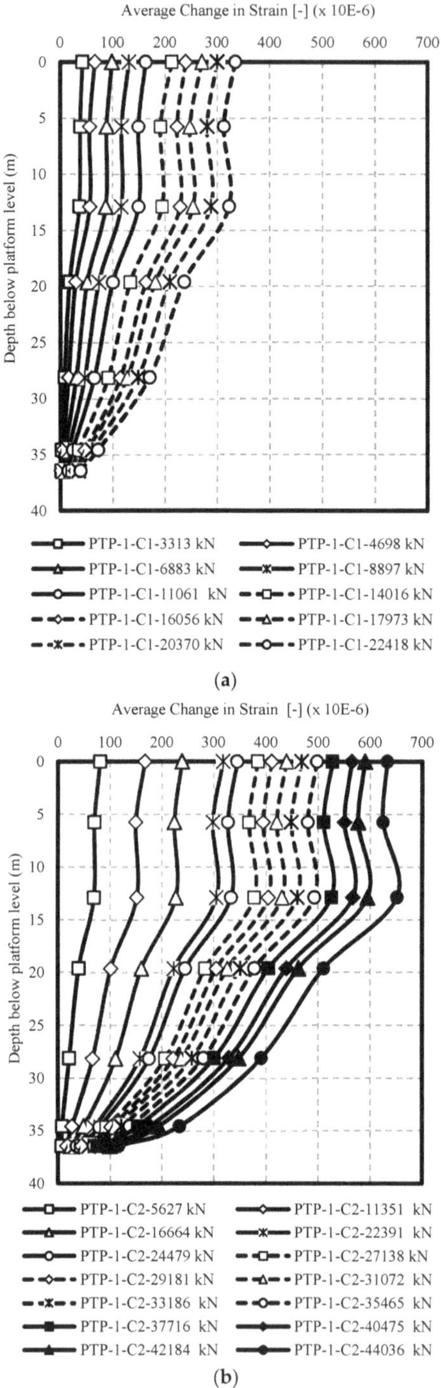

Figure 13. Distribution of average change in strain measured along BP# 1 in (**a**) 1st cycle and (**b**) 2nd cycle.

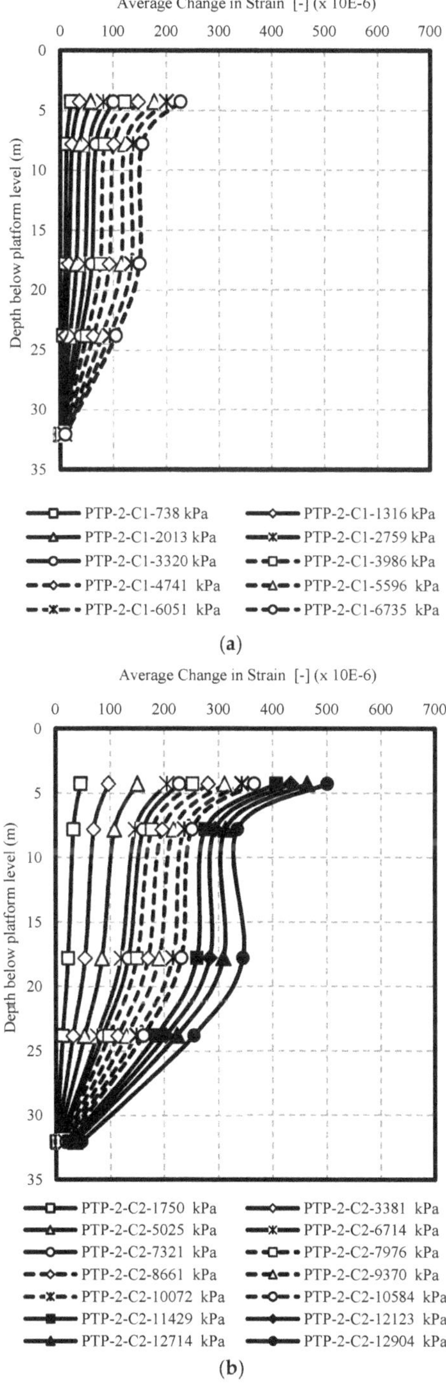

(**a**)

(**b**)

Figure 14. Distribution of average change in strain measured along BP# 2 in (**a**) 1st cycle and (**b**) 2nd cycle.

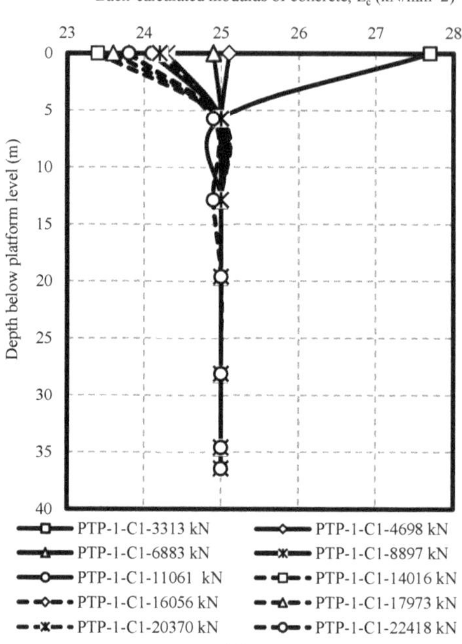

(**a**)

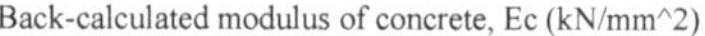

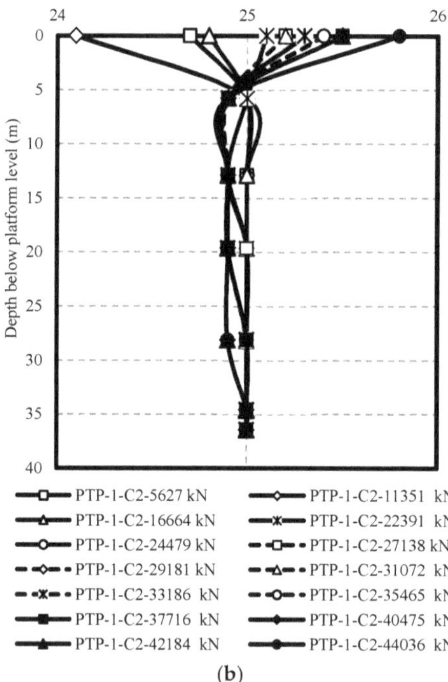

(**b**)

Figure 15. Distribution of back-calculated elastic modulus of concrete, E_c (kN/mm^2) for BP# 1; (**a**) 1st cycle and (**b**) 2nd cycle.

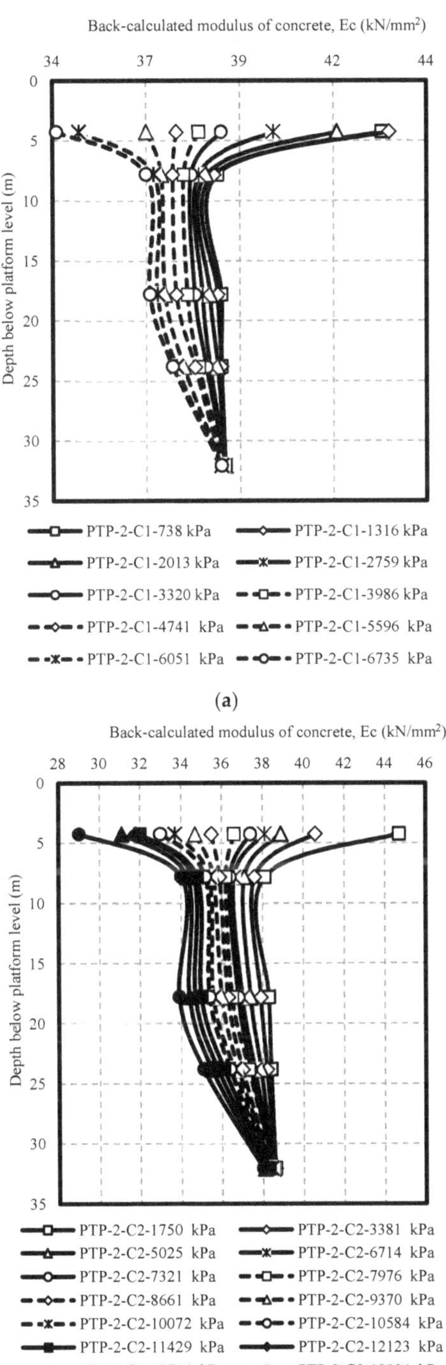

(**a**)

(**b**)

Figure 16. Distribution of back-calculated elastic modulus of concrete, E_c (kN/mm²) for BP# 2; (**a**) 1st cycle and (**b**) 2nd cycle.

Average SPT-N and unit skin friction in BP# 1 & BP# 2 presented in Figure 17, indicate the range of ultimate skin friction with value change of SPT-N. Furthermore, in multilayer site conditions, the least SPT-N sum for a soil layer provided the least unit skin friction and vice versa. The outcome derived from GSE sensors was computed according to the displacement amounts recorded by the sensors. A higher alteration for the recorded axial force within two continuous levels provided a larger unit skin friction for a specified soil layer (Figure 17).

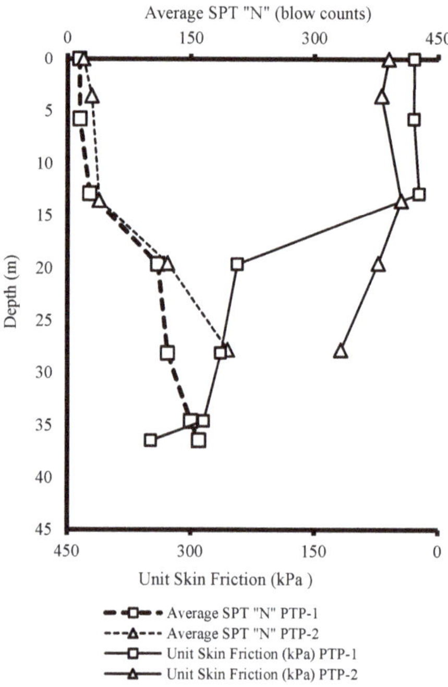

Figure 17. Average number of standard penetration test and the unit skin friction in BP# 1 and BP# 2.

3.2. Pile Movement Monitoring

Zhang et al. [42] state that the stress load-settlement curves reflect (1) the pile–soil interaction law, (2) the load transfer law, and (3) the pile load destruction mode. According to the previous explanations, the pile top and base settlements have been monitored for each load increment by applying the dial and strain gauges. Applied stress alteration versus total pile top and base settlement for two well-instrumented field tests (BP# 1, and BP# 2) in multilayered soils are presented in Figures 18 and 19, respectively. In the 1st cycle, the highest sighted pile top settlement at loading 22,418 kN was 9.60 mm. During unloading to zero, the pile was rebounded to a residual settlement of 0.36 mm. On the contrary, in the 2nd cycle, the maximum sighted pile top settlement at the peak load of 44,036 kN was 24.63 mm, so during unloading to zero, the pile was rebounded to a residual settlement of 5.34 mm. Similarly, residual settlements of 2.4 mm and 6.58 mm were recorded for the BP# 1 and BP# 2, respectively. According to Omer et al. [7], the variation of $f_s(z)$ with depth (z) is affected by different parameters including pile–soil properties, such as(1) pile–soil interface geometry and slip properties, (2) stress performance on the pile–soil interface, (3) pile installment technique, and (4) pile load method and ratio.

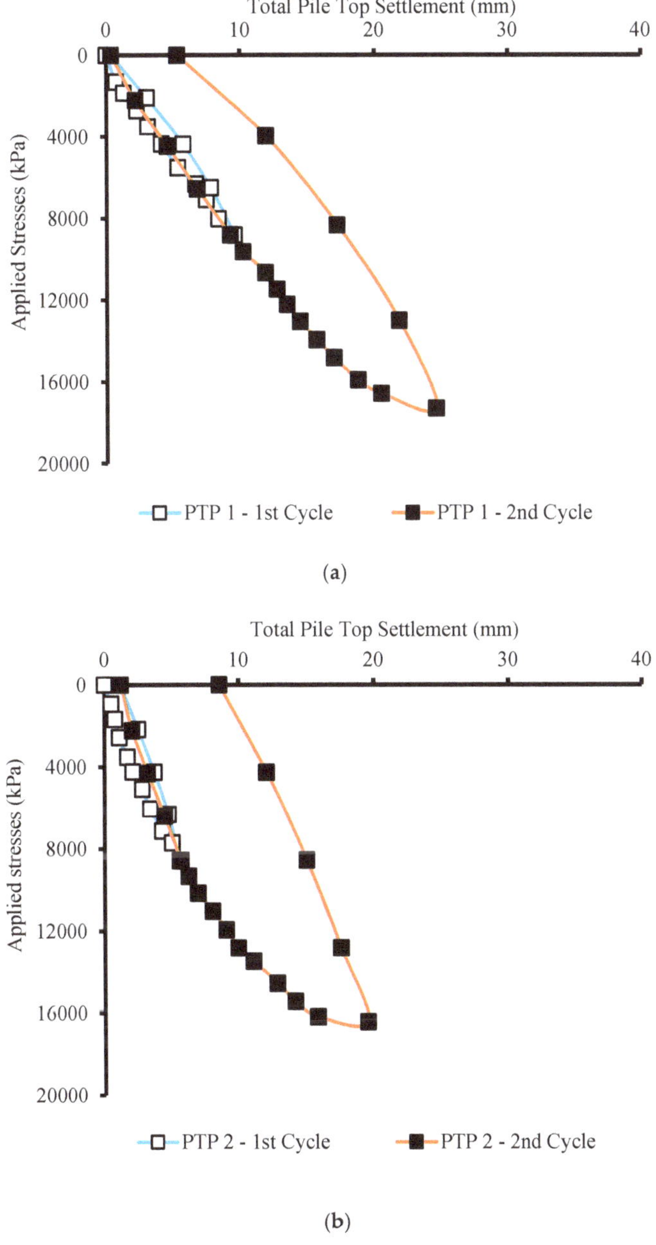

(**a**)

(**b**)

Figure 18. Different applied stresses versus total pile top settlement in (**a**) BP# 1 and (**b**) BP# 2.

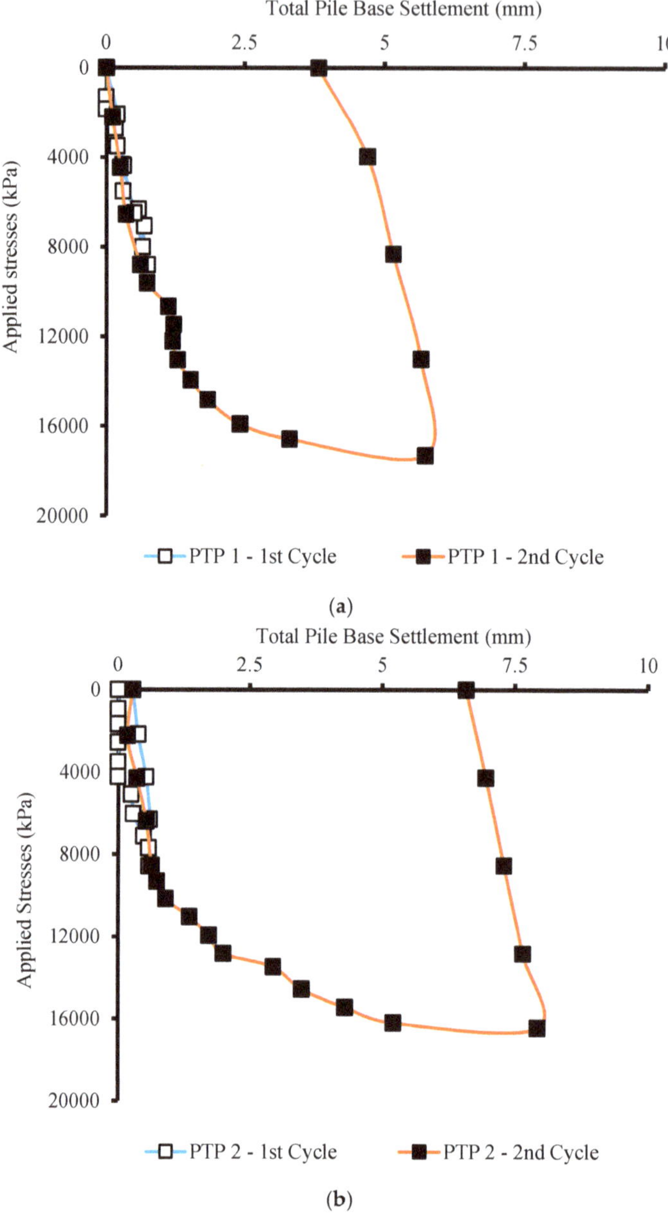

Figure 19. Different applied stress versus total pile base settlement in (**a**) BP# 1 and (**b**) BP# 2.

Different total pile settlement versus time during BP# 1 and BP# 2 is presented in Figure 20. The settlement rate in the pile's head is almost linear in the loading steps. However, when unloading begins, the settlement rate of unloading steps significantly increased depending on the loading/unloading sequences, showing that the loading time for the second cycle of BP# 1 and BP# 2 were 400 and 600 min. The measured settlement on the pile's head was rebounded to permanent vertical deformation of 5.34 mm and 8.55 mm for both tests (BP# 1 and BP# 2) after unloading to zero.

Correspondingly, the highest settlements of 24.63 mm and 19.54 mm were recorded for the pile's head vertical deformation.

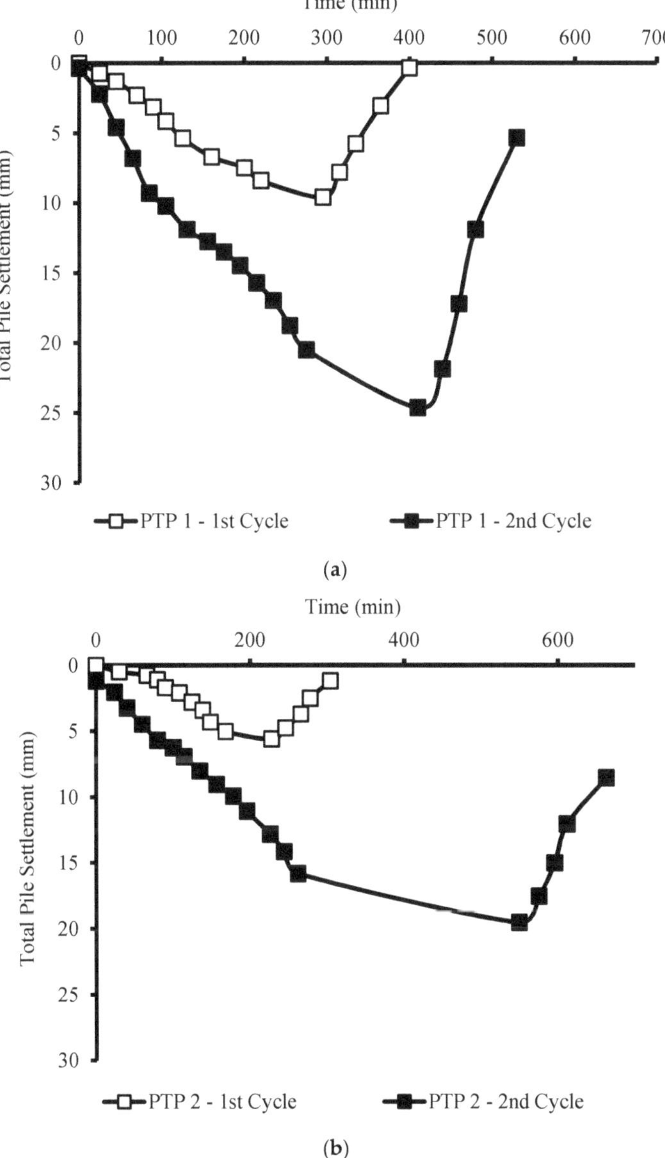

Figure 20. Variation of the total pile settlement versus time during pile test in (**a**) BP# 1 and (**b**) BP# 2.

Some of the critical factors that were considered to be constant during pile load tests (e.g., BP# 1 and BP# 2.) were (i) piling technology, (ii) concrete maturity, (iii) the location of the groundwater, and most importantly, and (iv) the soil parameters (e.g., both mechanical and physical properties changes). Such a problem may affect E_c measurement during the pile servicing period. For instance, it is established that piling technology will influence soil–pile interactions [43–45]. Piling techniques that lead to changes in soil properties can affect the axial force, load-displacement response, and tip resistance of each model pile. In most cases, to assess such influences, a large number of small-scale experimental works, or real-scale numerical and analytical studies are helpful.

On the one hand, the other factor that can influence the results of such investigations is concrete maturity [46,47]. Concrete maturity factor reveals the amount of concrete strength gain during the curing period, which is typically challenging to be taken into consideration as a separate variable on the full-scale experimental programme. It is important to note that the lack of such information may cause a significant change in the load-settlement behaviors of the pile during its working lifetime. Factors such as soil properties (e.g., shear strength parameters such as soil internal friction angle, cohesion, etc.) as well as groundwater levels, are primary terms that can remarkably alter the soil–pile responses to heavy external loadings. As an example, saturating the soil can cause soil shear strength reduction, which will influence the pile settlement as well as reducing the pile bearing capacities [48]. Another critical issue that can increase the complexity of soil–pile reactions, as well as load-settlement responses, as highlighted by Chisari et al. [49]. In the study provided by Chisari et al. [49], the influence of dynamic and static loading conditions (e.g., for identification of the primary material properties of a base-isolated bridge) are investigated. Their results showed that static identification is much less complicated compared to dynamic analysis. Although the current study covers the static load test, future work could evaluate the effect of a dynamic loading test in real-time monitoring of Y_c.

4. Conclusions

The main objective of this study was to find a reliable estimation of the E_c in the installed bored pile. Two full-scale maintained a static loads test on instrumented bored-piles had been conducted in Kuala Lumpur, Malaysia to obtain a reliable range for ultimate skin friction with SPT-N value (i.e., blow counts) alteration. The effects of geologically weak zones through the layered soil ground conditions in crucial parameter-design changing such as elastic concrete modulus and strain–stress along the piles have also been researched. The details of the conclusion are as follows.

- Distribution of concrete modulus has been measured from the stress–strain behavior of tested piles and using back analysis. In BP# 1, along with the pile, the E_c value is almost constant at 25 kN/mm^2, however, this value is between 30 and 45 kN/mm^2 in BP# 2. This has indicated that the applied stress is the dominant factor in E_c alteration because of the various soil reaction systems responding to the stresses released from the pile. The maximum change has been measured at the pile's head where the maximum stresses have been recorded, in contrast, E_c variation at the pile's toe is negligible since the least stresses have been applied in the pile base. The stresses of the pile have been declined by the depth because the skin friction of the pile has carried a large portion of applied load in the pile. Maximum permanent (plastic) deformations of 24.63 mm and 19.54 mm have been measured in the pile head for BP# 1 & BP# 2 correspondent to applied stresses of 17305 kPa ($p = 44306$ kN) and 16430 kPa ($p = 12905$ kN). The pile top settlement has been rebounded to a residual value of 2.4 mm and 8.55 mm in BP# 1 & BP# 2 after unloading to zero.

- The obtained result will be helpful for real-time assessment of bored pile during its service life. However, this should be noted that due to the heterogeneous characteristics of the soil, the measured E_c of the tested soil may vary. Variables such as concrete maturity, piling technology, soil parameters, and groundwater level can have a significant influence on the soil-pile interactions as well as concrete pile characteristics.

Author Contributions: Data collection and experimental works: H.M., B.K., H.N., Writing, discussion, analysis: H.M., A.S.A.R., and R.N., M.M.A.

Funding: This research has received funding from Ton Duc Thang University.

Conflicts of Interest: The authors declare no conflict of interest.

List of Abbreviations

CAPWAP	control and provisioning of wireless access points
D_s	shaft diameter
FE	finite element
FRP	fiber-reinforced plastic
GI	galvanized iron
GSE	global strain extensometer
HSDT	high strain dynamic testing
L	length
LVDTs	linear variation displacement transducers
MARS	multivariate adaptive regression spline
MLT	maintained load test
PSE	pile structure embankments
SPT	standard penetration test
SPT-N	standard penetration test—number of blows
TBM	tunnel boring machine
VWLC	vibrating wire load cells
VWSG	vibrating wire strain gauges

References

1. Tomlinson, M.J.; Woodward, J. *Pile Design and Construction Practice*; Taylor & Francis: Boca Raton, FL, USA, 2003.
2. Badrun, M. Prediction of Ultimate Bored Pile Capacity Using Global Strain Extensometer. Master's Thesis, Faculty of Civil Engineering, Universiti Teknologi Malaysia, Skudai, Malaysia, 2011.
3. Moayedi, H.; Hayati, S. Artificial intelligence design charts for predicting friction capacity of driven pile in clay. *Neural Comput. Appl.* **2018**, 1–17. [CrossRef]
4. Moayedi, H.; Nazir, R.; Mosallanezhad, M. Determination of reliable stress and strain distributions along bored piles. *Soil Mech. Found. Eng.* **2015**, *51*, 285–291. [CrossRef]
5. Barr, L.; Wong, R.C.K. Shaft resistance of bored cast-in-place concrete piles in oil sand—Case study. *Geomech. Eng.* **2013**, *5*, 119–142. [CrossRef]
6. Coduto, R. *Foundation Design Principles And Practices*; Prentice Hall: Englewood Cliffs, NJ, USA, 2001.
7. Omer, J.R.; Delpak, R.; Robinson, R.B. An Empirical Method for Analysis of Load Transfer and Settlement of Single Piles. *Geotech. Geol. Eng.* **2010**, *28*, 483–501. [CrossRef]
8. Fellenius, B.H.; Kim, S.R.; Chung, S.G. Long-term monitoring of strain in instrumented piles. *J. Geotech. Geoenviron. Eng.* **2009**, *135*, 1583–1595. [CrossRef]
9. Moayedi, H.; Nazir, R.; Mosallanezhad, M.; Noor, R.B.M.; Khalilpour, M. Lateral deflection of piles in a multilayer soil medium. Case study: The Terengganu seaside platform. *Measurement* **2018**, *123*, 185–192. [CrossRef]
10. Mosallanezhad, M.; Moayedi, H. Developing hybrid artificial neural network model for predicting uplift resistance of screw piles. *Arab. J. Geosci.* **2017**, *10*, 479. [CrossRef]
11. Nazir, R.; Moayedi, H.; Mosallanezhad, M.; Tourtiz, A. Appraisal of reliable skin friction variation in a bored pile. *Proc. Inst. Civ. Eng.-Geotech. Eng.* **2015**, *168*, 75–86. [CrossRef]
12. Nazir, R.; Moayedi, H.; Subramaniam, P.; Gue, S.-S. Application and Design of Transition Piled Embankment with Surcharged Prefabricated Vertical Drain Intersection over Soft Ground. *Arab. J. Sci. Eng.* **2018**, *43*, 1573–1582. [CrossRef]
13. Niroumand, H.; Kassim, K.A.; Nazir, R.; Faizi, K.; Adhami, B.; Moayedi, H.; Loon, W. Slope stability and sheet pile and contiguous bored pile walls. *Electron. J. Geotech. Eng.* **2012**, *17*, 19–27.
14. Brown, M.J.; Hyde, A.F.L.; Anderson, W.F. Analysis of a rapid load test on an instrumented bored pile in clay. *Geotechnique* **2006**, *56*, 627–638. [CrossRef]
15. Brown, D.A.; O'Neill, M.W.; Hoit, M.; McVay, M.; El Naggar, M.H.; Chakraborty, S. *Static and Dynamic Lateral Loading of Pile Groups*; Transportation Research Board: Washington, DC, USA, 2001.

16. Boulanger, R.W.; Curras, C.J.; Kutter, B.L.; Wilson, D.W.; Abghari, A. Seismic soil-pile-structure interaction experiments and analyses. *J. Geotech. Geoenviron. Eng.* **1999**, *125*, 750–759. [CrossRef]

17. Gao, W.; Dimitrov, D.; Abdo, H. Tight independent set neighborhood union condition for fractional critical deleted graphs and ID deleted graphs. *Discret. Contin. Dyn. Syst. Ser.* **2018**, *12*, 711–721. [CrossRef]

18. Moayedi, H.; Mosallanezhad, M.; Nazir, R. Evaluation of Maintained Load Test (MLT) and Pile Driving Analyzer (PDA) in Measuring Bearing Capacity of Driven Reinforced Concrete Piles. *Soil Mech. Found. Eng.* **2017**, *54*, 150–154. [CrossRef]

19. Moayedi, H.; Nazir, R.; Ghareh, S.; Sobhanmanesh, A.; Tan, Y.C. Performance analysis of piled-raft foundation system of varying pile lengths in controlling angular distortion. *Soil Mech. Found. Eng.* **2018**, *55*, 265–269. [CrossRef]

20. Gao, W.; Guirao, J.L.G.; Abdel-Aty, M.; Xi, W. An independent set degree condition for fractional critical deleted graphs. *Discret. Contin. Dyn. Syst. Ser.* **2019**, *12*, 877–886. [CrossRef]

21. Hutchins, N.; Choi, K.S. Accurate measurements of local skin friction coefficient using hot-wire anemometry. *Prog. Aerosp. Sci.* **2002**, *38*, 421–446. [CrossRef]

22. Ng, K.W.; Sritharan, S. A procedure for incorporating setup into load and resistance factor design of driven piles. *Acta Geotech.* **2016**, *11*, 347–358. [CrossRef]

23. Hung, L.C.; Nguyen, T.D.; Lee, J.H.; Kim, S.R. Applicability of CPT-based methods in predicting toe bearing capacities of driven piles in sand. *Acta Geotech.* **2016**, *11*, 359–372. [CrossRef]

24. Tafreshi, S.N.M.; Javadi, S.; Dawson, A.R. Influence of geocell reinforcement on uplift response of belled piles. *Acta Geotech.* **2014**, *9*, 513–528. [CrossRef]

25. Mascarucci, Y.; Miliziano, S.; Mandolini, A. A numerical approach to estimate shaft friction of bored piles in sands. *Acta Geotech.* **2014**, *9*, 547–560. [CrossRef]

26. Lee, C.; An, S.; Lee, W. Real-time monitoring of SPT donut hammer motion and SPT energy transfer ratio using digital line-scan camera and pile driving analyzer. *Acta Geotech.* **2014**, *9*, 959–968. [CrossRef]

27. Ochiai, H.; Otani, J.; Matsui, K. Performance factor for bearing resistance of bored friction piles. *Struct. Saf.* **1994**, *14*, 103–130. [CrossRef]

28. Poulos, H.G. Piled raft foundations: Design and applications. *Geotechnique* **2001**, *51*, 95–113. [CrossRef]

29. Sego, D.C.; Biggar, K.W.; Wong, G. Enlarged base (belled) piles for use in ice or ice-rich permafrost. *J. Cold Reg. Eng.* **2003**, *17*, 68–88. [CrossRef]

30. Shariatmadari, N.; Eslami, A.; Karimpour-Fard, M. Bearing capacity of driven piles in sands from SPT-applied to 60 case histories. *Iran. J. Sci. Technol. Trans. B-Eng.* **2008**, *32*, 125–140.

31. Zhang, Q.Q.; Zhang, Z.M.; He, J.Y. A simplified approach for settlement analysis of single pile and pile groups considering interaction between identical piles in multilayered soils. *Comput. Geotech.* **2010**, *37*, 969–976. [CrossRef]

32. Ruan, X.; Zuo, R.Y. Research on the Relationship between SPT and the ultimate vertical bearing capacity of jacked pile in Shenyang district. In *Frontiers of Green Building, Materials and Civil Engineering, Pts 1-8*; Sun, D., Sung, W.P., Chen, R., Eds.; Trans Tech Publications Ltd.: Stafa-Zurich, Switzerland, 2011; Volume 71–78, pp. 3308–3311.

33. Samui, P. Multivariate adaptive regression spline applied to friction capacity of driven piles in clay. *Geomech. Eng.* **2011**, *3*, 285–290. [CrossRef]

34. Samui, P. Determination of ultimate capacity of driven piles in cohesionless soil: A Multivariate Adaptive Regression Spline approach. *Int. J. Numer. Anal. Methods Geomech.* **2012**, *36*, 1434–1439. [CrossRef]

35. Chae, D.; Cho, W.; Na, H.Y. Uplift Capacity of Belled Pile in Weathered Sandstones. *Int. J. Offshore Polar Eng.* **2012**, *22*, 297–305.

36. Sakr, M. Comparison between high strain dynamic and static load tests of helical piles in cohesive soils. *Soil Dyn. Earthq. Eng.* **2013**, *54*, 20–30. [CrossRef]

37. Zhang, Q.Q.; Zhang, Z.M.; Li, S.C. Investigation into Skin Friction of Bored Pile Including Influence of Soil Strength at Pile Base. *Mar. Geores. Geotechnol.* **2013**, *31*, 1–16. [CrossRef]

38. Yao, W.J.; Chen, S.P. Elastic-plastic analytical solutions of deformation of uplift belled pile. *Teh. Vjesn.* **2014**, *21*, 1201–1211.

39. Aksoy, H.S.; Gor, M.; Inal, E. A new design chart for estimating friction angle between soil and pile materials. *Geomech. Eng.* **2016**, *10*, 315–324. [CrossRef]

40. Wang, C.D.; Zhou, S.H.; Wang, B.L.; Guo, P.J.; Su, H. Settlement behavior and controlling effectiveness of two types of rigid pile structure embankments in high-speed railways. *Geomech. Eng.* **2016**, *11*, 847–865. [CrossRef]

41. Tosini, L.; Cividini, A.; Gioda, G. A numerical interpretation of load tests on bored piles. *Comput. Geotech.* **2010**, *37*, 425–430. [CrossRef]

42. Zhang, R.K.; Shi, M.L.; Wang, J. Settlement Analysis of Single Large-Diameter and Super-Long Bored Piles in Cohesive Soils. *Adv. Mater. Res.* **2012**, *594*, 320–326. [CrossRef]

43. Xu, M.J.; Ni, P.P.; Mei, G.X.; Zhao, Y.L. Load-settlement behaviour of bored piles with loose sediments at the pile tip: Experimental, numerical and analytical study. *Comput. Geotech.* **2018**, *102*, 92–101. [CrossRef]

44. Zhou, J.J.; Gong, X.N.; Wang, K.H.; Zhang, R.H. Shaft capacity of the pre-bored grouted planted pile in dense sand. *Acta Geotech.* **2018**, *13*, 1227–1239. [CrossRef]

45. Zhou, J.J.; Gong, X.N.; Zhang, R.H. Model tests comparing the behavior of pre-bored grouted planted piles and a wished-in-place concrete pile in dense sand. *Soils Found.* **2019**, *59*, 84–96. [CrossRef]

46. Bojovic, D.; Basic, N.; Jankovic, K.; Senic, A. Assesment of concrete compressive strength using different maturity functions: case study. *Gradevnski Mater. I Konstr. Build. Mater. Struct.* **2018**, *61*, 55–65.

47. Vazquez-Herrero, C.; Vilar, J.; Mendoza, C.J.; Meli, R.; Aire, C. Sustainable production of precast combined pile-cap and column elements via statistical analysis of self-compacting concrete kinetics. *Constr. Build. Mater.* **2018**, *190*, 326–341. [CrossRef]

48. Zhang, W.G.; Goh, A.T.C.; Goh, K.H.; Chew, O.Y.S.; Zhou, D.; Zhang, R.H. Performance of braced excavation in residual soil with groundwater drawdown. *Undergr. Space* **2018**, *3*, 150–165. [CrossRef]

49. Chisari, C.; Bedon, C.; Amadio, C. Dynamic and static identification of base-isolated bridges using Genetic Algorithms. *Eng. Struct.* **2015**, *102*, 80–92. [CrossRef]

Article

Modeling of the Snowdrift in Cold Regions: Introduction and Evaluation of a New Approach

Mengmeng Liu [1,2], Qingwen Zhang [1,2,*], Feng Fan [1,2] and Shizhao Shen [1,2]

[1] Key Lab of Structures Dynamic Behavior and Control of the Ministry of Education, Harbin Institute of Technology, Harbin 150090, China

[2] Key Lab of Smart Prevention and Mitigation of Civil Engineering Disasters of the Ministry of Industry and Information Technology, Harbin Institute of Technology, Harbin 150090, China

* Correspondence: zhangqw@hit.edu.cn; Tel.: +86-150-4581-1536

Received: 17 July 2019; Accepted: 14 August 2019; Published: 17 August 2019

Featured Application: A new approach for investigating snowdrifts around buildings and snow loads on building roofs based on a new facility named the "snow–wind combined experimental facility" is proposed in this paper. First, the method proposed in this paper is convenient and economical, and also the experimental results prove that the method is feasible and reliable. In this paper, only experiments on snowdrift around a cube were chosen for the verification for the sake of generality, but this method could also be adopted for the prediction of snowdrift around complex building environments and snow loads on various shapes of building roofs.

Abstract: Unbalanced, or non-uniform, snow loads caused by snow drifting or sliding in cold regions with heavy snowfalls, can be a serious problem for the building industry. However, the methods for predicting snow distribution still need to be improved. Field observation is the most direct and reliable method to study snow distribution, but because the natural environment is uncontrollable and varies dramatically, sometimes conclusions may be confused under the influence of the many variables in the investigation. This paper proposes a snowing experiment approach using an outdoor snow–wind combined experiment facility for the study of snow distribution. The facility can produce a stable and controllable wind field and snowfall environment. Experiments which focused on snowdrift around a building were conducted during the winter to make an evaluation of the repeatability and reliability of the new approach. Finally, from the analysis of results, it was demonstrated that the experimental facility was stable and that the similarity criterion adopted for the snowing pattern was reliable. Especially, the minimum value of the friction speed ratio was suggested to ensure the test accuracy.

Keywords: snow–wind combined experiment facility; snowdrift; field observation; scale experiments; similarity criterion

---

1. Introduction

Since ancient times, buildings have been constructed to protect people from the natural environment. The environment dictates what type of building should be constructed, and snow loads are one of the dominant live loads that should be considered in the design of buildings in cold regions with a severe winter climate. Investigations show that snow loads are usually unbalanced, due to the action of the wind. And snowdrifts around buildings may cause problems for the vehicular traffic and pedestrians.

Currently, field observations, numerical simulations, and wind tunnel tests are the three main methods to estimate snow distribution, and much work has been undertaken to explore how the wind affects snow distribution.

Field observation is the most direct and reliable approach for obtaining information on snow distributions. Investigations of snowdrifts around an isolated building or group of buildings have been carried out in several previous studies [1,2]. Thiis and Gjessing [3] investigated the snowdrifts around three different model buildings in a valley in Spitsbergen, Norway, and the results showed that a change of roof shape could lead to large differences in snow distribution around the building. Beyers et al. [4] carried out research to investigate the snowdrift around a $2.0 \times 2.0 \times 2.0\ \text{m}^3$ cube at the SANAE IV research station, Antarctica, during the summer of January 2002 and verified the accuracy of a numerical simulation model using the results of field observations. Høibø [5,6] measured snow loads on nearly 200 building roofs under different wind and sun/shade environments from 1966 to 1986 in southern Norway. This method has strict requirements on the environment of the test site, such as wind velocity, temperature, humidity, and snowfall, but these important parameters in the observation were unstable and uncontrollable.

In 1991, Uematsu [7] first used numerical simulation to analyze snowdrifts, and now this method is widely used by researchers. Tominaga et al. [8] analyzed the snowdrift around an actual apartment building using the revised k-ε model. Beyers et al. [4] predicted transient snowdrift around a cubic structure using the standard k-ε model. And a large amount of numerical work has been done recently due to its significant advantage in efficiency and convenience [9–12]. However, only a few attempts have been made to apply CFD (Computational Fluid Dynamics) to roof loading so far [13]. Thiis et al. [14,15] predicted the snow distribution on a curved roof of a sports hall located in Oslo and compared the results with measurements. Some researchers are confused by the selection of the turbulence models, hence there is no approved and unified model.

Wind tunnel testing [16–18] is the most effective method for investigating snow distribution under the action of airflow. Kind [19,20] identified the important similarity criteria by analyzing the saltation process and discussed the necessary compromises in the modeling procedures. Delpech et al. [21] used artificial snow to simulate snow drifting around buildings for the Antarctic Concordia research station. Okaze et al. [22] investigated the development of drifting snow in a boundary layer wind tunnel. LÜ et al. [23] conducted a series of experiments in a wind tunnel to investigate the motion of natural snow and found that the threshold wind speeds for fresh and old snow were 6.2 m/s and 6.8 m/s (at 10 m height). Zhou et al. [24,25] investigated the snow distribution on a stepped flat roof using different granular materials in an open straight-circuit wind tunnel and also conducted a series of experiments on snow loads on flat roofs using high-density silica in the open test section of a boundary-layer wind tunnel. However, the snowdrift simulations in a boundary-layer wind tunnel struggle with scaling issues and similarity criteria. Some previous experiments on snowdrifts using wind tunnels have used bran, or other particles, as a substitute for natural snow, but these cannot completely replicate the distribution of natural snow. Tests have been carried out in advanced climate wind tunnels [26], where the influencing factors that may affect the results, such as the wind speed, temperature, and snowfall rate, are fully controllable. However, they are very expensive and not readily available for most researchers.

Snowdrift loads around buildings or on the roofs, which have a great impact on functionality and security of buildings, require particular attention. However, for the existing research methods, numerical simulation still needs to be improved with data from field observations and wind tunnel tests. Field observations are rarely available when considering security and technical issues, even though they should provide the most reliable data. Wind tunnel tests are the main source of data, but the expense of experiments in climate wind tunnels is so high that few experiments are reported, and it has restricted further studies. What is more, the wind speed and direction of field observations vary dramatically, so some conclusions may be confused when the natural environment is problematic. Hence, this paper proposes a new approach to investigate the snowdrift loads with a facility in which the wind speed and direction are controllable and a snowing environment, just like natural snowfall, is available.

The purpose of this facility is to provide a new approach to study snow distribution under the action of wind. In comparison with traditional wind tunnel tests, there will be a 90% or more decrease in the construction cost because the facility need not be equipped with a refrigerating system. Besides that, the operating expense of the facility is much lower than climate wind tunnels.

Transportation of snow particles can be classified into three processes, i.e., creep, saltation, and suspension. Saltation, which is widely accepted as contributing most to the total transfer volume, is a process in which snow particles move with repeated leaping up, or jumping, on the snow surface.

The height of saltation is usually below 0.1 m. Generally, the length of saltation L [27] is suggested to be

$$L \cong 10\,h. \tag{1}$$

Yet, to ensure that the saltation process could be accurately reproduced on the roofs or around the building, it is suggested that the characteristic length of the prototype model is larger than the length of saltation L.

In this study, the design parameters of the facility, experiment method, and the adopted similarity criteria are introduced and discussed. Finally, a prototype model with external dimensions of $1.0 \times 1.0 \times 1.0$ m^3 was selected, taking both economy and theory into consideration, to study the surrounding snow distribution and evaluate the repeatability and reliability of the new approach.

2. Design of the Snow–Wind Combined Experiment Facility

2.1. Brief Introduction of the Facility

In the study of snow distribution, numerical simulations and wind tunnel tests should be verified by field observations. For this purpose, a snow–wind combined experiment facility as shown in Figure 1a has been constructed at the Harbin Institute of Technology (HIT) to produce stable wind fields to investigate snow distribution [28] The climatic characteristics of the winter months in Harbin provide convenient conditions for the experiments. Harbin has a long winter, which can last for almost six months, with snowy conditions and a temperature range from −30 °C to 0 °C. The location of the facility is shown in Figure 1b. Because the facility was to be sited outdoors, a narrow strip between two buildings was chosen to reduce the interference of the external environment. The dimension of available ground is 12 m × 20 m, and the heights of the surrounding buildings are all greater than 10 m.

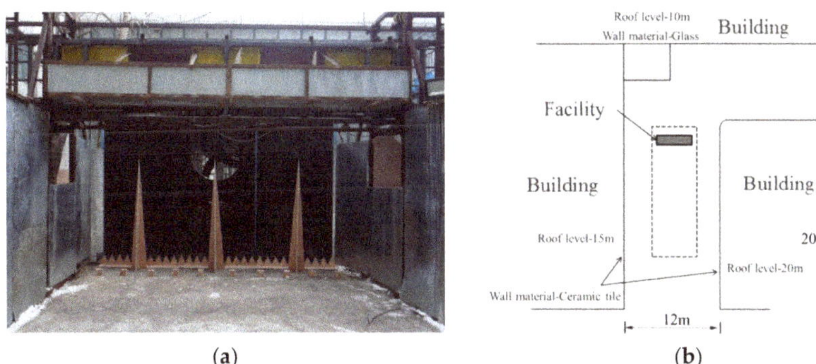

(a) (b)

Figure 1. (a) Photo of the snow–wind combined experiment facility. (b) Location of the facility.

2.2. Introduction of the Main Component

The facility, as shown in Figure 2, consists of a power section, a flow conditioning section, an experimental section, and a snowfall simulator.

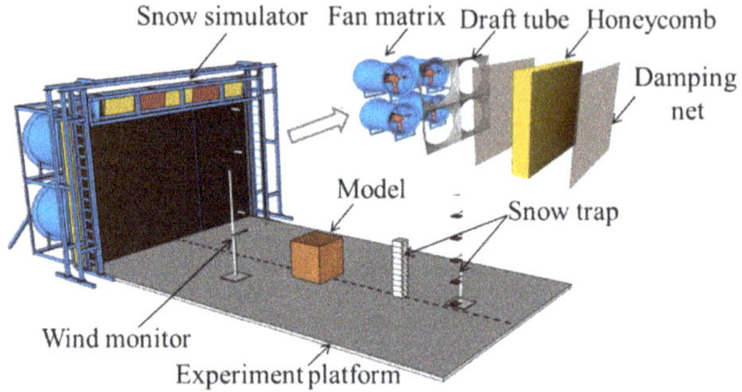

Figure 2. Schematic view of the snow–wind combined experiment facility.

The power station is the 3×2 fan-matrix, and the area affected by the model during a snowdrift is thought to be about three times larger than the characteristic length of the model [27]. Therefore, the cross-section of the facility was designed to be 4.5×3.0 m^2. Considering the fact that there are usually light winds when it is snowing, the facility was designed to be able to produce a stable wind field for a specific duration with a range of speeds from 0.5 to 11.5 m/s. Wind pressure is the main indicator of the output of the fan-matrix and is defined as

$$P = \frac{\rho k U_0^2}{2},\tag{2}$$

where P is the wind pressure; ρ is the density of air, and $\rho = 1.395$ kg/m^3 when the temperature is -20 °C; k is the resistance coefficient of pressure of the flow conditioning section, which is defined as the ratio of the outlet area of the facility to the total outlet area of the six airflow fans, and k = 1.734; U_0 is the wind speed at the outlet or immediate output from the fan matrix, and U_{max}, which is measured at the section 4 m away from the outlet, is regarded as the available maximal wind speed of the facility and $U_{max} = 11.5$ m/s for design purposes.

Because the facility is open to the natural environment, a numerical simulation was undertaken to explore the relationship between U_0 and U, where U is the wind speed at the experiment section. Based on the environment of the experimental site, a calculation model of the facility, with a computational domain of $10 \times 10 \times 20$ m^3, was formulated. The result showed that when the initial velocity of the wind export of the facility was 15 m/s, the uniform wind speed in the range of 4–20 m was about 11.5 m/s, hence, the value of U_0 is approximately equal to 15 m/s.

Accordingly, the minimum wind pressure is 272.1 Pa by calculation, and a fan with 286 Pa wind pressure was chosen. The maximum speed measured in the field was 11.5 m/s, which reasonably meets the requirements of the experiments.

The flow conditioning section, which stabilizes and smooths the airflow produced by the fan, is a combination of the draft tube, the honeycomb, and the damping net. Two pieces of damping net were stuck on both import and export sections of the honeycomb, and the draft tube served as connection between the power section and the honeycomb. The flow conditioning section was designed to be short to reduce the loss of wind power because of the open experimental environment. Considering the square pattern of the cross-section and the design experience from the wind erosion tunnel designed by Saxton [29], a square cross-section with the dimension of 30×30 mm^2 was chosen as the single unit of the honeycomb with a length of 300 mm. The experimental section is a platform with a circular dial, which is used to fix the model at the required angle.

Snowfall suitable for the experiments was not always available, so to accumulate a large amount of data on snow distribution, it was necessary to find a way to simulate snowfall using natural or artificial snow particles.

In this work, a snowfall simulator situated 1 m away from the wind export was designed to simulate snowfall during the experiments. Before the experiments started, snow, which had been collected in heat-insulation boxes, was placed in a metal box with wire gauze on the bottom. Then, shaking the metal box with an eccentric motor to generate friction between the snow and wire gauze, made snow particles fall from the mesh into the stable wind field caused by the fan-matrix so that a snowfall environment was formed in the test section. Different snowfall conditions are available using different kinds of wire gauze with varied vibration frequency. There are three different kinds of the aperture of wire gauze, 2 mm, 3 mm, and 5 mm, and the vibration frequency ranges from 1 to 5 Hz. The metal box, which is 5 m wide, can be raised and lowered freely in the range of 0.5–3 m.

2.3. A Brief Introduction to the Experimental Procedure

Figure 3 shows a sketch of the experimental procedure. The facility can provide a controlled snowfall environment in the test section so that experiments examining snow distribution around the building or on roofs can be undertaken.

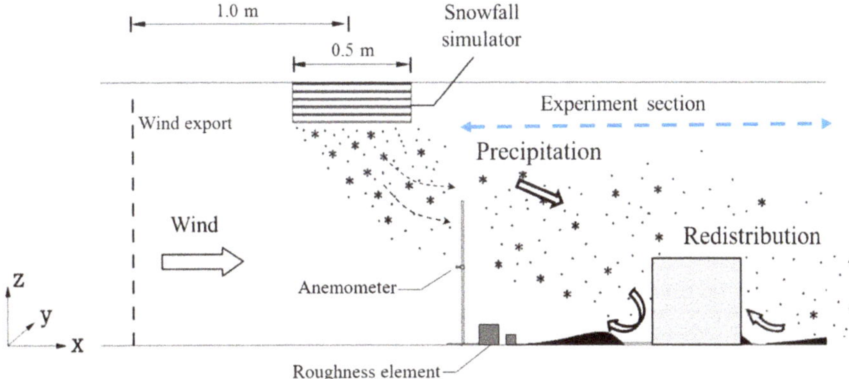

Figure 3. A sketch of the experimental procedure.

When an experiment is going to be conducted, there are four steps: First of all, enough snow should be put into the box of the snow simulator, next the model should be fixed according to the experiment design along with an anemograph for monitoring the wind speed at the reference point; then the fan-matrix is started and adjusted until the wind speed reaches a given value, and then the snow simulator can be turned on for the required duration; finally, the snow load is measured after all of the equipment has been turned off.

3. Verification of the New Approach

3.1. Field Measurements of Wind Field Produced by the Facility

Obviously, the wind field has a significant impact on the results of the experiments, so field measurements of the actual wind field were also made to investigate it and determine the best location of the experimental platform.

Figure 4 shows the method for measuring the wind field. Because the cross-section is symmetrical, the measuring points are arranged symmetrically in the full section. The wind speed was measured by a hot-wire Thermo-Anemometer and each measuring point was measured at least 30 s with a sampling frequency of 1 Hz. The vertical and horizontal measurement points in each cross-section formed a matrix of 13 × 9. Before the measurements, a design speed U was defined. If the measured wind speeds at the 1.5 m point on the first line of the five sections were all between 0.9~1.1 U, this would be recognized as the design speed U. Five cross-sections were measured with a design speed (U) of 3.0 m/s and the distance between them and the export was 1 m, 3 m, 4 m, 5 m, and 7 m.

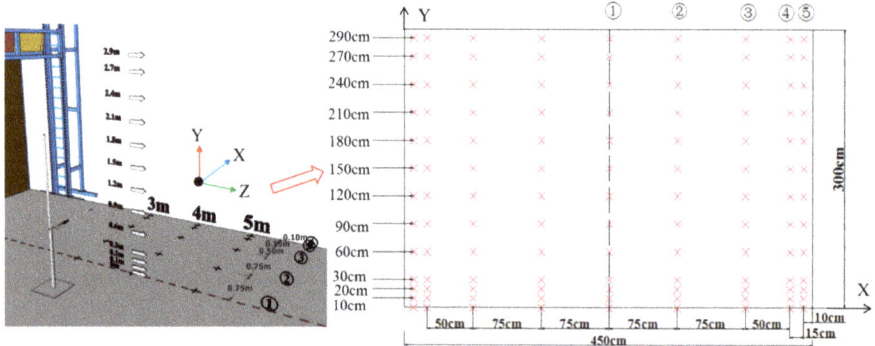

Figure 4. A sketch of the measuring method.

Figure 5a shows that the mean wind speed on the first line of five cross-sections varies with height; the Y-axis represents the height of the cross-section and the X-axis represents the ratio between U^* (the measured wind speed) and U. Figure 5b shows the turbulence intensity of wind on each measuring point. It is apparent that the turbulence intensity is higher in the section before 3 m and the wind is more uniform in the range 3–7 m; this was also shown in the numerical simulation. However, the wind speed decreased faster with increasing distance from the export, which is slightly different from the numerical simulation. To ensure a stable and uniform wind field during the experiments, the effective range of the experimental section was 3–7 m away from the wind export.

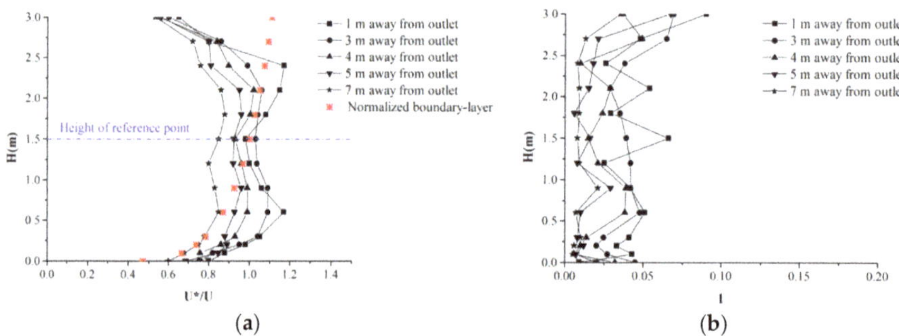

Figure 5. Wind field on the first line of the five sections: (**a**) mean wind and (**b**) turbulence intensity.

Figure 6a shows how the mean wind speed varies with width in the cross-section 4 m away from the export, and Figure 6b shows the turbulence intensity of wind on each measuring point. And from the results, it is apparent that the wind speed decreased rapidly close to the edge of the cross-section. To ensure the veracity of the experiments, the effective width of the experimental section was 4 m.

From a comparison of the measured wind speed with the normalized wind speed in the atmospheric boundary layer, the discrepancy above 1.5 m is non-negligible, but it is relatively acceptable for the given experiment. The turbulence intensity was small except at some points close to the wind export and at the edges of the cross-section, which indicates that the quality of the flow field in the open test section was reasonable and acceptable.

During the experiments, temperature, the natural wind field, and humidity were measured on a PC-4 automatic weather station, which is located on the side of the facility, to rectify the experimental wind field. This has three anemometers, at heights of 0.5 m, 1.5 m, and 2.0 m, and the sampling interval was 1 min.

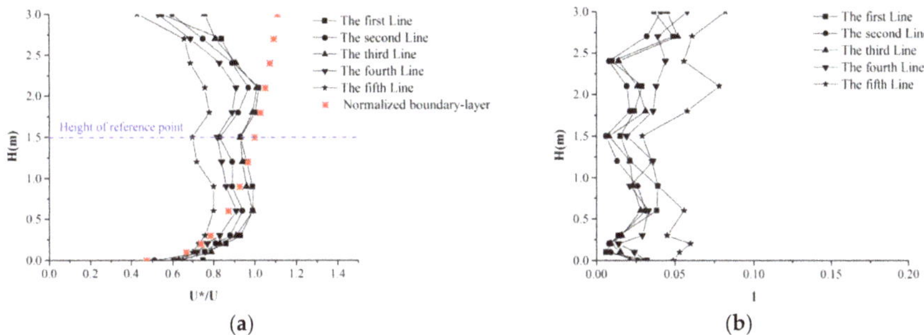

Figure 6. Wind field on the section which is 4 m away from the outlet: (**a**) mean wind and (**b**) turbulence intensity.

3.2. Physical Properties of Snow

Properties of snow particles have a great influence on the result of the test [23] so that the physical properties of the stored natural snow particles were measured before the experiments. The diameter and shape of the snow particles were observed by an optical microscope and the angle of repose was decided from accumulated experimental observations. The terminal velocity of the snow particles was decided by measuring the time that the snow particles freely fell from 2 m to the ground level. A comparison of the physical parameters of the snow particles is given in Table 1.

Table 1. Comparison of the physical parameters of the snow particles.

Reference	Density (kg/m^3)	Terminal Velocity (m/s)	Diameter (mm)	Angle of Repose (°)
Field observation of fresh snow in this paper	100~150	0.30	0.15~1.0	65
Stored snow in this paper	220~310	0.50	0.2~0.5	55
Thiis and Ramberg [14]	50	0.50	0.20	—
Tominaga and Okaze [30]	150	0.20	0.15	—
Oikawa and Tomabechi [27]	50~150	0.20	0.15	60

3.3. Experiments for the Verification of Repeatability

To confirm the repeatability of the new approach, 20 repetitive experiments using stored snow on the snowdrift surrounding a 1 m cube were undertaken. Environmental conditions, namely temperature, humidity, and natural wind field were measured by an automatic weather station that was sited nearby the test section and results are shown in Table 2. The reference point for the experimental wind speed was set at 1 m in front of the model and 0.5 m above the ground. The snowing rate was defined as the accumulated snow depth on the ground in unit time and it was determined by the kind of wire gauze and vibration frequency. For the 20 experiments, the snowing rate was controlled to be the same. Figure 7 shows some photos of the experiment results.

Figure 8 gives the distribution of average snow depth and the standard deviation of each measured point along lateral and streamwise lines crossing the model. It is clear that the distribution of snow depths in the 20 experiments is relatively stable. The Pearson Correlation Coefficient for any two experimental results along the measuring lines shown in Figure 8 was above 0.90, which indicates that there is a strong correlation between the results from the 20 experiments.

Table 2. Environmental conditions of the 20 repetitive experiments.

Experiment Wind Speed	Temperature	Humidity	Max Natural Wind	Bulk Density of Snow	Testing Time
2.5 m/s	0 °C	78%	0.13 m/s	250.6 kg/m^3	1 h
2.5 m/s	0 °C	78%	0.30 m/s	260.2 kg/m^3	1 h
2.5 m/s	−1 °C	76%	0.28 m/s	255.8 kg/m^3	1 h
2.5 m/s	−1 °C	76%	0.19 m/s	236.7 kg/m^3	1 h
2.5 m/s	−1 °C	76%	0.22 m/s	227.5 kg/m^3	1 h
2.5 m/s	−1 °C	76%	0.23 m/s	240.4 kg/m^3	1 h
2.5 m/s	−2 °C	74%	0.18 m/s	266.6 kg/m^3	1 h
2.5 m/s	−2 °C	73%	0.15 m/s	270.1 kg/m^3	1 h
2.5 m/s	−2 °C	73%	0.22 m/s	252.4 kg/m^3	1 h
2.5 m/s	−2 °C	68%	0.14 m/s	261.9 kg/m^3	1 h
2.5 m/s	−2 °C	68%	0.20 m/s	242.3 kg/m^3	1 h
2.5 m/s	−2 °C	67%	0.30 m/s	248.6 kg/m^3	1 h
2.5 m/s	−2 °C	68%	0.17 m/s	233.6 kg/m^3	1 h
2.5 m/s	−2 °C	66%	0.19 m/s	239.5 kg/m^3	1 h
2.5 m/s	−3 °C	64%	0.16 m/s	257.4 kg/m^3	1 h
2.5 m/s	−3 °C	63%	0.11 m/s	244.5 kg/m^3	1 h
2.5 m/s	−3 °C	63%	0.21 m/s	221.0 kg/m^3	1 h
2.5 m/s	−3 °C	61%	0.15 m/s	235.2 kg/m^3	1 h
2.5 m/s	−3 °C	60%	0.12 m/s	243.3 kg/m^3	1 h
2.5 m/s	−3 °C	60%	0.12 m/s	266.4 kg/m^3	1 h

Figure 7. Some photos of the repetitive experiments.

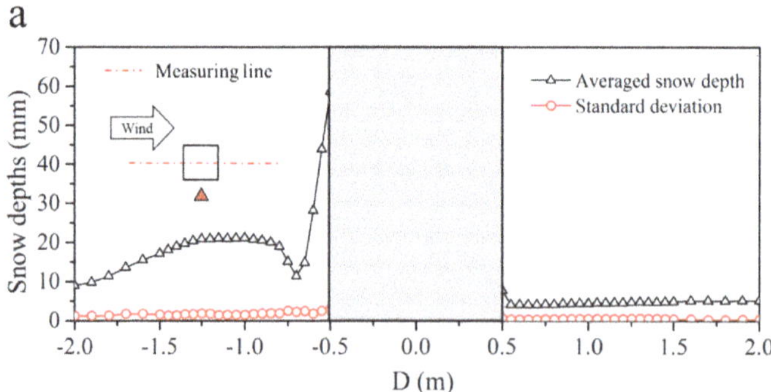

Figure 8. *Cont.*

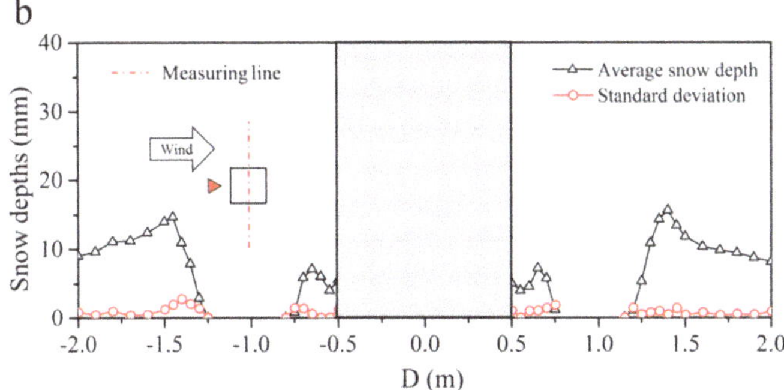

Figure 8. Results of the 20 experiments: (**a**) streamwise direction (center section) and (**b**) lateral section (center section).

From the above analysis, it can be concluded that the results of the 20 experiments, with the same test conditions, show good correspondence with each other. This indicates that the method provides good stability and repeatability.

3.4. Experiments for the Verification of Reliability

3.4.1. A Field Observation on the 1 m Cube Prototype Model

To confirm the reliability of the new approach, the comparison of field observations and experiments should be done. So, the field observation on a 1 m cube prototype model was conducted on the HIT campus on 11 September 2015 from 0:00 to 6:00. To obtain a reasonably stable wind direction, the test site was located on a narrow strip between two buildings, as shown in Figure 1b. Figure 9 shows the method of observation; the wind field and environment conditions were measured by a PC-4 automatic weather station with a sampling interval of 1 min and the box-type snow flux trapper draws on the design experience summarized by Kimura [31].

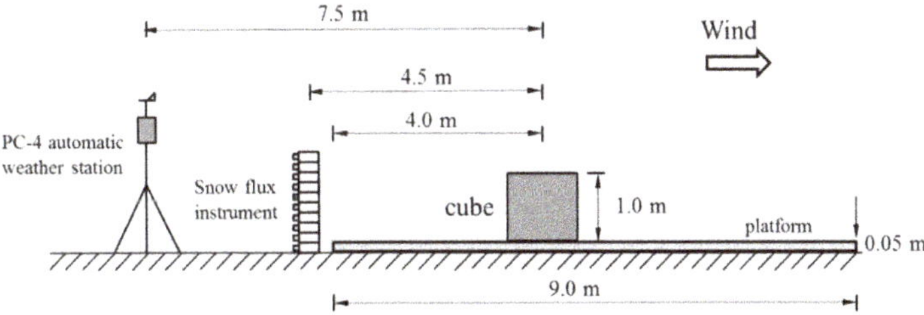

Figure 9. Abridged general view of field observation.

The weather station and snow flux instrument were all mounted on the centerline of the model. The wind field during the observation is given in Figure 10. The average snowfall was 29 mm, which was measured on an open and flat ground 20 m away from the model. Figure 11 shows the photo of the snow distribution of the field observation.

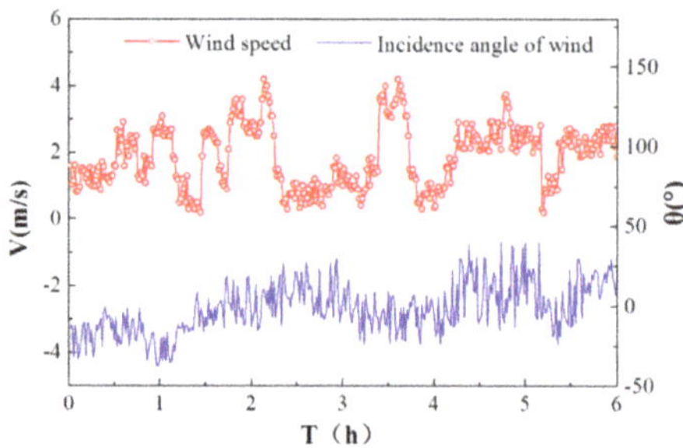

Figure 10. Wind field during the observation.

Figure 11. Photo of snow distribution for the field observation.

3.4.2. Test Similarity Criteria Based on Snowfall Mode

A reliable scale in experimental modeling depends on the right similarity criterion. In order to reproduce the non-uniform snowdrift phenomena caused by structures with scale models, the widely accepted fact is that it is necessary to meet geometric, kinematic, and dynamic similitude requirements.

Since the beginning of snowdrift research,, many different similarity criteria have been proposed, validated, and applied by some famous scholars, such as Strom [32], Odar [33], Calkins [34], Kind [19], Iversen [16], Anno [35], Isyumov [18], Naaim [36], Delpech [21], Beyers [2], and so on.

However, just as it is not possible to satisfy the Froude numbers and Reynolds numbers simultaneously [37], it has been proved that some of the similarity criteria reveal incompatibility, therefore only the most important and widely accepted similarity parameters are selected and discussed.

In the present study, considering the fact that the snow particles falling into the snow bed are provided by the sowing pattern, the experimental modeling is mainly focused on the reliable simulation of the Froude number and drifting volume. The original form of the Froude number is defined as

$$\frac{U^2}{gL},\tag{3}$$

where U is the reference wind velocity, g is gravitational acceleration (m/s^2), and L is the reference length (m). But considering the discrepancy of particle density between the prototype and the test, Odar [33] and Calkins [34] proposed the densimetric Froude number:

$$\frac{\rho}{\rho_p}\frac{U^2}{gL}\tag{4}$$

where ρ is the air density, and ρ_p is the particle density. To confirm the reliable simulation of drifting volume, Anno [35] suggested that the test should satisfy the following time scaling parameter:

$$\frac{TQ\eta}{\rho_p L^2} \tag{5}$$

where T is experiment time (s), Q is the transport rate of snow (kg/m·s), and η is the snow collection efficiency. Except for the above similarity, a number of similarity requirements, like the flow field, ejection process, particle trajectory, and deposition pattern, must be satisfied in the experiments on snow drifting on roofs to ensure a reliable simulation.

First, the near ground turbulent wind velocity profile for stable atmospheric conditions is

$$U = \frac{u_*}{\kappa} \ln(\frac{z}{z_0}), \tag{6}$$

where κ is the Von-Karman constant, u_* is the friction velocity, z is the height above the surface, and z_0 is the roughness height parameter. Geometric similarity should be enforced through the relationship

$$D_p/L. \tag{7}$$

Except for the densimetric Froude number expressed as Equation (4), the following Froude number based on the threshold friction velocity should also be satisfied to simulate the shear stress of particles near the ground [16]:

$$\frac{\rho}{\rho_p - \rho} \frac{u_{*t}^2}{g D_p} \tag{8}$$

where D_P is the particle diameter, u_{*t} is the threshold friction velocity, and v is the kinematic viscosity.

To ensure a similar movement trajectory of a particle, Kind [24] indicated that the following equations must be satisfied:

$$\frac{\rho}{\rho_p - \rho} \frac{u_{*t}^2}{g L}, \tag{9}$$

$$w_f/U, \tag{10}$$

where w_f is the settling velocity of snow particles (m/s). The deposition or erosion mechanisms could be modeled satisfactorily through the similarity of the particle ejection process [19,38]:

$$u_*/u_{*t}. \tag{11}$$

In the present study, because the simulation of the Froude number is considered to be more important, the limit of the Reynold number is released. However, to ensure that the inertial forces are dominating the flow, Kind [19] suggested that the following relationship should be satisfied:

$$\frac{u_{*t}^3}{2gv} > 30. \tag{12}$$

In summary, the parameter (4) and (5) were adopted for determining the wind velocity and experiment time, and several similarity parameters were also chosen for evaluating the results when using a sowing pattern to study snowdrift. In order to verify the reliability of this method, four scale models were selected for the verification test. The experimental conditions are listed in Table 3 and the prototype is the field observation presented above. The reference point for the experimental wind speed was set at 1 m in front of the model along the central line and the height above the ground was equal to the side length of the model. The major similarity parameters for the prototype and scale model are given in Table 4.

Snow flux in the vertical plane was measured by additional experiments with only the snow flux trapper on the testing ground. The snow load coefficient is the ratio of the depth of snow on the roof to the average snowfall. For the experiments, the average snowfall on the ground was determined by an additional experiment conducted for the same duration but without any obstacles in the test section. After the experiments, the snowdrift geometry was measured with the laser total station for about 200~300 points around the model and the measured snow depth data contributed to the contour maps of the snowdrift geometry.

Table 3. Environmental conditions of the scale experiments.

Model Scale	Velocity (m/s)	Snow Mass Flux (kg/m^2·s)	Experiment Time (min)
Prototype	2.475	1.93×10^{-3}	360
1	3.5	2.81×10^{-2}	50
0.5	2.5	1.52×10^{-2}	46
0.25	1.8	8.49×10^{-3}	41
0.1	1.1	3.68×10^{-3}	38

Table 4. Similarity parameters of prototype and scale models.

Similarity Parameters	Prototype	1	0.5	0.25	0.1
D_p/L	0.002	0.003	0.006	0.012	0.03
$\dfrac{\rho}{\rho_p-\rho}\dfrac{u_{*t}^2}{gD_p}$	0.0116	0.0154	0.0154	0.0154	0.0154
$\dfrac{\rho}{\rho_p-\rho}\dfrac{u_{*t}^2}{gL}$	2.32×10^{-5}	4.61×10^{-5}	9.23×10^{-5}	1.85×10^{-4}	4.61×10^{-4}
w_f/U	0.121	0.143	0.200	0.278	0.455
u_*/u_{*t}	0.717	0.507	0.362	0.261	0.159
$\dfrac{u_{*t}^3}{2gv} > 30$	-	99.92	99.92	99.92	99.92

3.4.3. Results and Discussion

There are two key factors to evaluate the accuracy of snowdrift given by the experiments. One is the similarity of the shape of snowdrift; another is the similarity of the snow distribution coefficient. The distribution coefficient is the ratio of measured snow depth and average snowfall at any measuring point.

Figure 12 shows the contour maps of the snowdrift geometry around the model of field observation and scale experiments. The photos of scale experiments are shown in Figure 13. All of the contour maps showed a similar shape which was formed like a horseshoe, in which snow mainly accumulated at the front of the windward side and beginning at the two corners of the windward side, there were two distinct erosion areas near the lateral sides along the downwind direction. From the comparison, if just focusing on the overall distribution shape of the snowdrift, it seemed that the result when the scale was 1/10 was more similar to the field observation and the main erosion areas (near the lateral sides) were larger than field observations for the other scale experiments. The basis for explaining this phenomenon needs an objective and appropriate understanding of the result of the field observation. The main erosion is more distinct in the experiment due to the stable wind speed. However, sometimes there were slight or no wind during the field observation, and snowfall contributes more to decrease the erosion effect in this condition, which would decrease the non-uniform snow distribution to some extent.

However, considering that two distinct snowdrifts were formed in front of the windward side, the results of the 1/1, 1/2, and 1/4 scale experiments had a more similar characteristic of snow distribution. When the scale was 1/10, the snow only accumulated at the feet of the windward side, indicating that wind velocity was excessively reduced; the near to the ground reverse flow in front of the windward side was weak and could not transport snow particles away from the windward side, so that only one distinct snowdrift was formed.

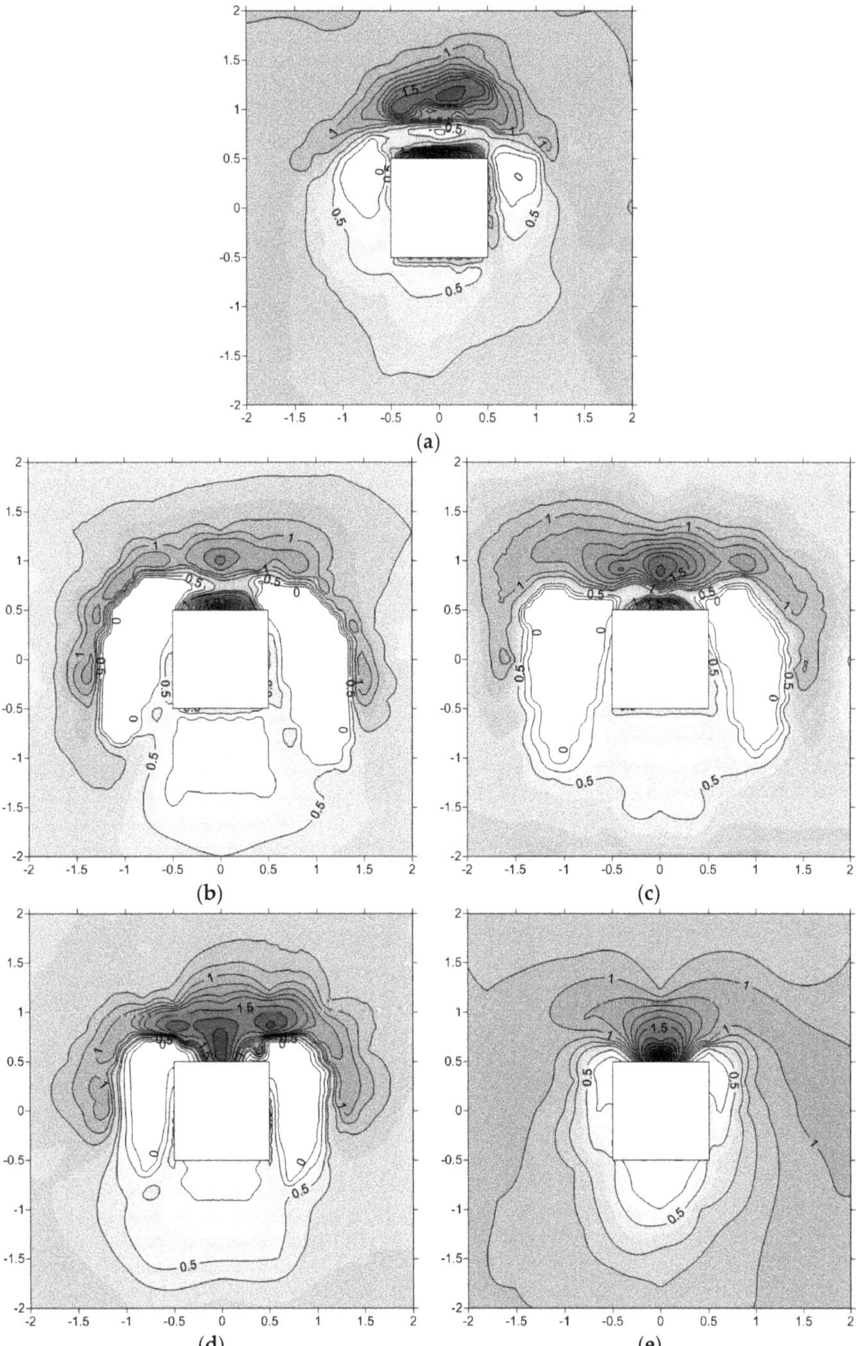

Figure 12. Contour maps of the snowdrift geometry around the model: (**a**) field observation; (**b**) experiment with a 1 m Cube model; (**c**) experiment with a 0.5 m Cube model; (**d**) experiment with a 0.25 m Cube model; and (**e**) experiment with a 0.1 m Cube model.

Considering that the effect of fluctuant wind velocity in the field observation could not be ignored, the 1/1 model may be treated as the self-test prototype. In this situation, according to the comparison of results, the 1/2 and 1/4 scale experiment results would have relatively acceptable accuracy of the reproduction of the snowdrift shape of the field observation, but the 1/10 scale experiment result would have a lower precision reproduction.

In order to compare the snow distribution in a more precise method, Figure 14 compares the snow distributions along lateral and streamwise lines crossing the model.

Figure 13. Photos of scale experiments.

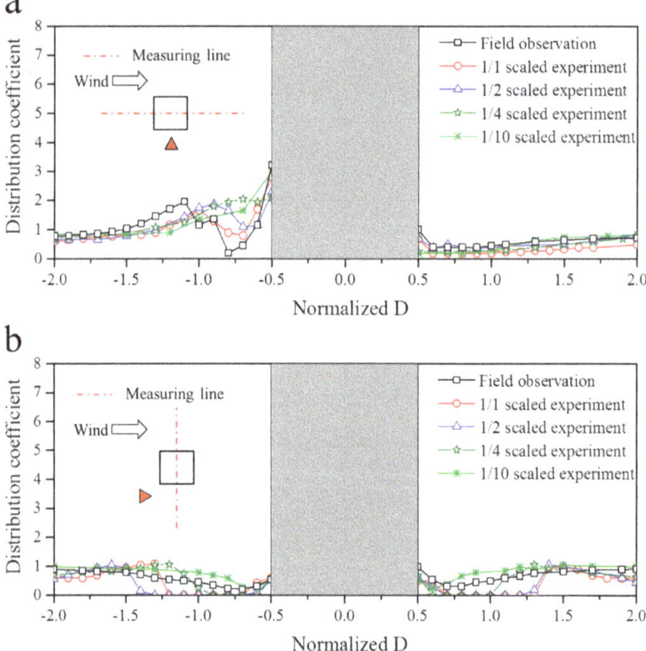

Figure 14. Comparison between normalized snow depths obtained from field observations and scale experiments: (**a**) streamwise direction (center section) and (**b**) lateral direction (center section).

In Figure 14a, it could be seen that the snow distribution of the field observation has obvious characteristics: Near the windward side, the snow formed a compressed "N" shape, and near the leeward side, there formed a compressed "U" shape. When the scale was 1/1 and 1/2, the distribution results showed a good agreement with the field observation, even though the position where the extreme value occurred moved towards to the windward side as the scale ratio decreased. But, as the scale decreased, especially when the scale was 1/10, the snow tended to only accumulate at the foot of the windward side, and the snowdrift away from the windward side disappeared. As for the snow distribution near the leeward side, when the scale was 1/1 and 1/2, the experiments could precisely reproduce the snow distribution, even though the overall value of the coefficient decreased. When the scale was 1/4 and 1/10, the results gave a lower estimate of the snow accumulation, but for the snow distribution away from the leeward side, the results showed better agreement with the field observation both in value and shape.

Snow distribution along the lateral direction is shown in Figure 14b. Except when the scale was 1/10, the snow distribution showed a U-shaped distribution for both field observation and scale experiments. For the 1/10 scale experiment, a similar distribution was found, but the location where the minimum value occurred was much closer to the lateral side. This phenomenon indicates that the erosion effect was excessively decreased.

In summary, considering both the shape of the snowdrift and snow distribution coefficient, except when the scale was 1/10, the results of scale experiments reproduce the field observation accurately, suggesting that the method is reliable.

Especially, it is known that the relationship between friction velocity u_* and threshold friction velocity u_{*t} is of great significance in the study of the snowdrift. From the analysis of the similarity parameters in Table 4, it could be derived that a too-small value of the parameter u_*/u_{*t} may adversely affect the accuracy of reproduction. Therefore, it is recommended to limit the minimum value of the parameter u_*/u_{*t} for scale experiments.

Figure 15 shows the Pearson correlation coefficients of the distribution coefficient between the prototype and scale experiments along the measured line defined as the front, rear, left, and right.

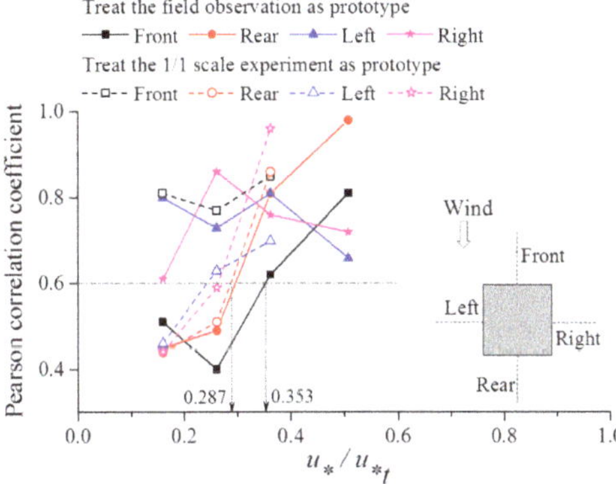

Figure 15. Pearson correlation coefficients of the distribution coefficient between the prototype and scale experiments.

It could be concluded from Figure 15 that the results are of poor quality when the value of the parameter u_*/u_{*t} is lower. Assuming that the experiment results are accepted when the Pearson correlation coefficient is higher than 0.6, the lower limit values of the parameter u_*/u_{*t} are 0.353 and 0.287, respectively, when the field observation or 1/1 scale experiment is chosen as the prototype. At present, the result of the field observation is affected by the adverse effects of the unstable wind field and the quantitative assessment of the effect is still unclear. So, when giving a suggestion, the analysis based on treating the 1/1 scale experiment as a prototype should be more respected and, finally, 0.3 is suggested to be the lower limit value of parameter u_*/u_{*t} in the present study.

4. Conclusions

In the present study, a new approach based on a snow–wind combined experiment facility, for investigating snow distribution, is introduced, including its concept, design, operation, and verification. According to the above work, the following conclusions could be drawn:

(1) According to 20 repetitive experiments, the results show high correspondence with each other and this indicates that the experimental conditions provided by the new experiment facility are stable. From the comparison between field observation and scale experiments, it proves that the proposed experimental approach and adopted similitude criteria are reliable. The two pieces of evidence indicate that the new approach is credible and feasible.

(2) The comparison of results from field observations and scale experiments indicate that when the wind velocity is excessively reduced, the result is not satisfactory. The lower limit value of the parameter u_*/u_{*t} is suggested for ensuring the accuracy of the reproduction. In the present study, taking into consideration the instability of field observations objectively and combining with the results of scale experiments, 0.3 is suggested to be the lower limit value of parameter u_*/u_{*t}.

In this paper, only experiments on snowdrift around a cube were chosen for the verification for the sake of generality, but this method could also be adopted for the prediction of snowdrift around complex building environments and snow loads on various shapes of building roofs. It could provide a basis for site selection planning for buildings and the safety design of roof structures in snowy areas.

Author Contributions: Conceptualization, Q.Z.; data curation, M.L.; formal analysis, M.L. and Q.Z.; funding acquisition, F.F.; investigation, M.L.; methodology, M.L. and Q.Z.; project administration, F.F.; resources, F.F.; software, M.L.; supervision, Q.Z. and S.S.; validation, M.L., Q.Z., F.F. and S.S.; visualization, Q.Z.; writing—original draft, M.L.; writing—review and editing, Q.Z.

Funding: This research was funded by the National Natural Science Foundation of China, grant number 51508133, 51478147, and the National Outstanding Youth Science Fund Project of the National Natural Science Foundation of China, grant number 51525802.

Acknowledgments: This work was financially supported by the National Natural Science Foundation of China (Grant no. 51508133, 51478147), the National Outstanding Youth Science Fund Project of National Natural Science Foundation of China (51525802), and Foundation of Key Laboratory of Structures Dynamic Behavior and Control (Ministry of Education) in Harbin Institute of Technology (HITCE201704). The authors are grateful to the members of the Space Structures Research Center in Harbin Institute of Technology, for providing invaluable information and advice in this study.

Conflicts of Interest: The authors declare no conflict of interest.

References

1. Thomas, K. This Large scale studies of development of snowdrifts around buildings. *J. Wind Eng. Ind. Aerodyn.* **2003**, *91*, 829–839.
2. Beyers, J.H.M.; Harms, T.M. Outdoors modelling of snowdrift at SANAE IV Research Station, Antarctica. *J. Wind Eng. Ind. Aerodyn.* **2003**, *91*, 551–569. [CrossRef]
3. Thomas, K.; Thiis, Y.G. Large-scale measurements of snowdrifts around flat-roofed and single-pitch-roofed buildings. *Cold Reg. Sci. Technol.* **1999**, *30*, 175–181.
4. Beyers, J.H.M.; Sundsb, P.A.; Harms, T.M. Numerical simulation of three-dimensional, transient snow drifting around a cube. *J. Wind Eng. Ind. Aerodyn.* **2004**, *92*, 725–747. [CrossRef]
5. Høibø, H. Snow Load on Gable Roofs—Results from Snow Load Measurements on Farm Buildings in Norway. In Proceedings of the First International Conference on Snow Engineering, Santa Barbara, CA, USA, 1988; pp. 95–104, Special Report 89–6.
6. Høibø, H. Form factors for snow load on gable roofs -Extending use of snow load data from inland districts to wind-exposed areas. In Proceedings of the 11th International Congress on Agricultural Engineering, Dublin, Ireland, 4–8 September 1989.
7. Uematsu, T.; Nakata, T.; Takeuchi, K. Three-dimensional numerical simulation of snowdrift. *Cold Reg. Sci. Technol.* **1991**, *1*, 65–73. [CrossRef]

8. Tominaga, Y.; Mochida, A. CFD prediction of flowfield and snowdrift around a building complex in a snowy region. *J. Wind Eng. Ind. Aerodyn.* **1999**, *81*, 273–282. [CrossRef]

9. Beyers, M.; Waechter, B. Modeling transient snowdrift development around complex three-dimensional structures. *J. Wind Eng. Ind. Aerodyn.* **2008**, *96*, 1603–1615. [CrossRef]

10. Tominaga, Y.; Okaze, T.; Mochida, A. CFD modeling of snowdrift around a building: An overview of models and evaluation of a new approach. *Build. Environ.* **2011**, *46*, 899–910. [CrossRef]

11. Okaze, T.; Takano, Y.; Mochida, A.; Tominaga, Y. Development of a new k–ε model to reproduce the aerodynamic effects of snow particles on a flow field. *J. Wind Eng. Ind. Aerodyn.* **2015**, *144*, 118–124. [CrossRef]

12. Liston, G.E.; Haehnel, R.B.; Sturm, M.; Hiemstra, C.A.; Berezovskaya, S.; Tabler, R.D. Simulating complex snow distributions in windy environments using SnowTran-3D. *Glaciology* **2007**, *53*, 241–256. [CrossRef]

13. Tominaga, Y. Numerical simulation of snowdrift around buildings: Past achievements and future perspectives. In Proceedings of the Snow Engineering VIII, Nante, France, 14–17 June 2016.

14. Thiis, T.K.; Ramberg, J.F. Measurements and numerical simulations of development of snowdrifts of curved roofs. In Proceedings of the Snow Engineering VI, Whistler, CO, Canada, 1–5 June 2008.

15. Thiis, T.K.; Potac, J.; Ramberg, J.F. 3D numerical simulations and full scale measurements of snow depositions on a curved roof. In Proceedings of the 5th European & African Conference on Wind Engineering, Florence, Italy, 19–23 July 2009.

16. Iversen, J.D. Comparison of wind-tunnel model and full-scale snow fence drifts. *J. Wind Eng. Ind. Aerodyn.* **1981**, *8*, 231–249. [CrossRef]

17. Sant'Anna, F.D.M.; Taylor, D.A. Snow drifts on flat roofs: Wind tunnel tests and field measurements. *J. Wind Eng. Ind. Aerodyn.* **1990**, *34*, 223–250.

18. Isyumov, N.; Mikitiuk, M. Wind tunnel model tests of snow drifting on a two level flat roof. *J. Wind Eng. Ind. Aerodyn.* **1990**, *36*, 893–904. [CrossRef]

19. Kind, R.J. A Critical examination of the requirements for model simulation of wind-induced erosion/deposition phenomena such as snow drifting. *Atmos. Environ.* **1976**, *10*, 219–227. [CrossRef]

20. Kind, R.J. A Review of Modelling Methods. *Cold Reg. Sci. Technol.* **1986**, *12*, 217–228. [CrossRef]

21. Delpech, P.H.; Palier, P.; Gandemer, J. Snowdrifting Simulation around Antarctic Buildings. *J. Wind Eng. Ind. Aerodyn.* **1998**, *74–76*, 567–576. [CrossRef]

22. Okaze, T.; Mochida, A.; Tominaga, Y. Wind tunnel investigation of drifting snow development in a boundary layer. *J. Wind Eng. Ind. Aerodyn.* **2012**, *104–106*, 532–539. [CrossRef]

23. Hui, L.; Huang, N.; Tong, D. Wind tunnel experiments on natural snow drift. *Sci. China Technol. Sci.* **2012**, *55*, 927–938.

24. Zhou, X.; Hu, J.; Gu, M. Wind tunnel test of snow loads on a stepped flat roof using different granular materials. *Nat. Hazards* **2014**, *74*, 1629–1648. [CrossRef]

25. Zhou, X.; Kang, L.; Yuan, X.; Gu, M. Wind tunnel test of snow redistribution on flat roofs. *Cold Reg. Sci. Technol.* **2016**, *127*, 49–56. [CrossRef]

26. Delpech, P.H.; Thomas, K.T. Applications of snowind engineering-climatic wind tunnel methods, Technical transaction iss.12. *Civil Eng.* **2015**, *2-B*, 381–403.

27. Oikawa, S.; Tomabechi, T.; Ishihara, T. One-day Observations of Snowdrifts Around a Model Cube. *J. Snow Eng. Jpn.* **1999**, *15*, 283–291. [CrossRef]

28. Liu, M.; Zhang, Q.; Fan, F.; Shen, S. Experiments on natural snow distribution around simplified building models based on open air snow-wind combined experiment facility. *J. Wind Eng. Ind. Aerodyn. Shizhao* **2018**, *1–13*, 173. [CrossRef]

29. Pietersma, D.; Stetler, L.D.; Saxton, K.E. Design and aerodynamies of a Portable wind tunnel for soil erosion and fugitive dust researeh. *Trans. ASAE* **1996**, *39*, 2075–2083. [CrossRef]

30. Tominaga, Y.; Okaze, T.; Mochida, A.; Shida, T.; Yoshino, H. CFD prediction of snowdrift around a cubic building model. In Proceedings of the Snow Engineering VI, Whistler, CO, Canada, 1–5 June 2008.

31. Kimura, T. Measurements of Drifting Snow Particles. *J. Geogr.* **1991**, *100*, 250–263. [CrossRef]

32. Strom, G.H.; Kelly, G.R.; Deitz, E.L.; Weiss, R.F. *Scale Model Studies on Snow Drifting*; Research Report 73; U.S. Army Snow, Ice and Permafrost Research Establishment: Hanover, NH, USA, 1962.

33. Odar, F. *Simulation of Drifting Snow*; Research Report 174; Cold Regions Research and Engineering Laboratory: Hanover, NH, USA, 1965.

34. Calkins, D.J. *Model Studies of Drifting Snow Patterns at Safeguard Facilities in North Dakota, United States Army, Corps of Engineer*; Cold Regions Research & Engineering Laboratory: Hanover, NH, USA, 1974; 20p.

35. Anno, Y. Requirements for modeling of a snowdrift. *Cold Reg. Sci. Technol.* **1984**, *8*, 241–252. [CrossRef]

36. Naaim-Bouvet, F. Comparison of requirements for modeling snowdrift in the case of outdoor and wind tunnel experiments. *Surv. Geophys.* **1995**, *16*, 711–727. [CrossRef]

37. White, F.M. *Viscous Fluid Flow*; McGraw-Hill: New York, NY, USA, 1991.

38. Kind, R.J.; Murray, S.B. Saltation flow measurements relating to modeling of snowdrifting. *J. Wind Eng. Ind. Aerodyn.* **1982**, *10*, 89–102. [CrossRef]

Article

Numerical Analysis and Dynamic Response of Optimized Composite Cross Elliptical Pressure Hull Subject to Non-Contact Underwater Blast Loading

Mahmoud Helal [1,2,3,4]**, Huinan Huang** [1,2]**, Elsayed Fathallah** [1,2,5,*]**, Defu Wang** [1,2,*]**, Mohamed Mokbel ElShafey** [6] **and Mohamed A. E. M. Ali** [5]

[1]   College of Engineering, Northeast Agricultural University, Harbin 150030, Heilongjiang, China
[2]   Key Laboratory of Swine Facilities Engineering, Ministry of Agriculture, Harbin 150030, Heilongjiang, China
[3]   Production and Mechanical Design Dept., Faculty of Engineering, Mansoura University, Mansoura 35516, Egypt
[4]   Department of Mechanical Engineering, Faculty of Engineering, Taif University, Taif 21974, Saudi Arabia
[5]   Department of Civil Engineering, Military Technical College, Cairo 11865, Egypt
[6]   Civil Engineering Department Canadian International College, Cairo 11865, Egypt
*   Correspondence: saidhabib2000@hotmail.com (E.F.); dfwang0203@163.com (D.W.);
    Tel.: +20-1-09900177 or +20-1-12341914 (E.F.)

Received: 1 August 2019; Accepted: 19 August 2019; Published: 23 August 2019

Abstract: Among the most important problems confronted by designers of submarines is to minimize the weight, increase the payload, and enhance the strength of pressure hull in order to sustain the hydrostatic pressure and underwater explosions (UNDEX). In this study, a Multiple Intersecting Cross Elliptical Pressure Hull (MICEPH) subjected to hydrostatic pressure was first optimized to increase the payload according to the design requirements. Thereafter, according to the optimum design results, a numerical analysis for the fluid structure interaction (FSI) phenomena and UNDEX were implemented using nonlinear finite element code ABAQUS/Explicit. The propagation of shock waves through the MICEPH was analyzed and the response modes (breathing, accordion and whipping) were discussed. Furthermore, the acceleration, displacement and failure index time histories at different locations were presented. The results showed that the greatest acceleration occurred in the athwart direction, followed by the vertical and longitudinal directions. Additionally, the first bubble pulse has a major effect on athwart acceleration. Moreover, the analysis can be effectively used to predict and calculate the failure indices of pressure hull. Additionally, it provides an efficient method that reasonably captures the dynamic response of a pressure hull subjected to UNDEX.

Keywords: underwater explosion; composite pressure hull; whipping; breathing; failure index

---

1. Introduction

Significant research work has been presented in order to simulate the behavior of underwater vehicles under severe loading conditions. For instance, Reddy [1] illustrated the effect of shock pressure loading on a ring-stiffened submersible hull using finite element analysis. The failure analysis indicated that fibers failed in tension while matrix failed in shear when the explosion charge exceeded 25 kg TNT. Furthermore, Jen [2] worked on minimizing the weight of pressure hull by enhancing the pressure hull strength taking into consideration both hydrostatical pressure and underwater explosion. Additionally, the results indicated that the dynamic motion of the pressure hull has an accordion mode, a whipping mode and a breathing mode. Also, Cho et al. [3] derived an empirical formula for predicting the collapse strength of composite cylindrical structures under hydrostatic pressure as a function of important design parameters such as the geometric dimensions and the layered angle. In addition, Chen et al. [4] experimentally investigated the dynamic performance of ship body coated by rubber. The results demonstrated that the coating was not effective at reducing the low-frequency whipping motion excited by the bubble pulse. However, it was able to moderate the high-frequency response excited by shock wave. Moreover, Ramajeyathilagam et al. [5] numerically and experimentally investigated the effect of underwater explosions (UNDEX) on rectangular plates. It was revealed that the underwater explosion failures can be predicted using the strain rate effects. Also, Liu et al. [6] illustrated the global responses of ship subjected to UNDEX. The results demonstrated that the UNDEX waves changed the added masses on the ship and effectively affected the global responses of its body. On the other hand, Liang et al. [7] examined the transient dynamic responses of submarine pressure hull exposed to hydrostatic pressure and shock loading. It was concluded that the collapse depth (maximum diving) of the submarine pressure hull was about 700 m. In addition, they observed that the loading condition depends not only on the hydrostatic pressure but also on the shock loading. Likewise, Kwon and Fox [8] studied experimentally and numerically the dynamic response of a cylinder subjected to UNDEX. The results illustrated that the largest strains on a cylinder subjected to a far-field side-on UNDEX occurred near the two ends on the near side of the cylinder to a far field explosive charge. Additionally, the damage occurred at the center on the opposite side of the cylinder. Similarly, McCoy and Sun [9] combined FSI code and finite element modeling techniques to investigate the dynamic response of a thick-section hollow composite cylinder. The results showed that the fluid-structure coupling has a significant effect on stress distributions within the structure.

Furthermore, Shin and Hooker [10] predicted numerically the damage response of submerged imperfect cylindrical structures exposed to UNDEX. Based on these results, the introduction of initial imperfections greatly affected the response of the cylinder when compared with the response of a perfect cylinder. On the other hand, Qiankun and Gangyi [11] demonstrated the shock response of a ship section to non-contact UNDEX using the finite element software package ABAQUS. It was revealed that the fluid thickness and size of fluid mesh effectively affects and improve the modeling accuracy. Additionally, Adamczyk and Cichocki [12,13] performed a numerical study to obtain the shock response of an underwater hybrid structure subjected to UNDEX. Zhao et al. [14] predicted the damage features of RC slabs subjected to air and UNDEX. The shock wave propagation and damage mechanisms from contact explosions in air and water were compared. The dynamic response of the RC slab is highly localized in the air contact explosion. Furthermore, the crater failure is observed at the top surface of the RC slab due to the direct impact of the air contact blast loading and the spalling failure occurs at the bottom of the RC slab. On the other hand, when the RC slab subjected to underwater contact explosion the top surface of the reinforced concrete RC slab almost completely destroyed. Therefore, underwater contact explosion can cause significantly more damage to the RC slab than the same amount of explosive in air. Rajendran and Narasimhan [15] investigate the damage of clamped circular plates subjected to contact UNDEX. The deformation contours were a spherical viewing the maximum absorption of energy for the depth of bulge attained. Also, Jacinto et al. [16] applied a linear dynamic analysis of plate models under explosions. The element size has a great effect on the results. Likewise, Kumar et al. [17] experimentally studied the blast effect on carbon composite panels the results showed that. There were two types of dominant failure mechanisms observed, fiber breakage and inter-layer delamination. Furthermore, Guo et al. [18] presented a new shock factor of twin hull water plane (catamaran) subjected to UNDEX. The shock factor parameter is used to describe the response of ships exposed to this loading condition based on shock wave energy. Similarly, Wang et al. [19] proposed a new method involving an analytical technique connected to elastic dynamic response of laminated plates exposed to UNDEX. The method was validated by comparing its results with those achieved by a semi analytical method and the experiment results. Zhang et al. [20] predicted the dynamic bending moment of a UNDEX bubble acting on a hull. The predicted numerical results showed that UNDEX bubble propagated a longitudinal bending which caused sagging and hogging damage for the ship. Furthermore, Gong and Khoo [21] presented a transient response of an UNDEX bubble on a glass/epoxy composite deep-submersible pressure hull. It was observed that the UNDEX bubble produced a huge deformation around the stand-off point in the pressure hull immediately after the collapse of the bubble, and the minimum volume was observed beneath the composite hull. On the other hand, Wang et al. [22] investigated numerically and experimentally the failure mode and dynamic response of a ship structure subjected to shock wave and bubble pulse. Jun et al. [23] investigated the impact environment characteristics of floating shock platform subject to UNDEX. Young and Leonard [24] modeled and simulate a surface ship shock to UNDEX, the results showed that the cavitation effect must be taken into account in the ship shock simulation, and that cavitation volume must be large enough. Gannon [25] investigated the response of a submerged stiffened cylinder to UNDEX using a coupled Eulerian Lagrangian model and experimental approaches.

This present study developing a procedure and describes a numerical modelling methodology for calculating the dynamic response of optimized Multiple Intersecting Cross Elliptical Pressure Hull (MICEPH) exposed to non-contact UNDEX. First, a submarine with pressure hull in the form of MICEPH is optimized using non-linear finite element analysis software ANSYS. Thereafter, according to optimization, the finite element model is built using non-linear finite element code ABAQUS/Explicit to examine the dynamic response of the pressure hull exposed to non-contact UNDEX.

2. Analysis of Underwater Shock Loading and Bubble Pulse

The most important element of any submersible body is the pressure hull. It contributes about one-fourth to one-half of the total underwater vehicle weight. Figure 1 presents various pressure hull configurations used in submersible bodies [26–28]. Exposing the hull structure to shock wave and bubble pulsation upon UNDEX events leads to great damage to the hull structure [2]. Figure 2 presents the different configurations that occur during UNDEX events [29,30]. The loading mechanisms resulting from UNDEX include incident wave, free surface reflection, shockwave, bottom reflection wave, gas bubble oscillation, bubble-pulse loading and bulk cavitation [31]. The compressive load on the structure of the hull is increased due the reflections from the bottom of the ocean which propagates from the shock wave, while the reflection of the shock wave from the ocean's free surface causes a reduction in the pressure [32,33]. Equation (1) presents the evaluation technique of pressure time history $(P_{in}(t))$ for the pressure profile, as follows [2,34–38]:

$$P_{in}(t) = P_{max}e^{-(\frac{t-t_1}{\theta})} \quad (\text{Mpa})? \quad t \geq t_1 \tag{1}$$

where (t) denotes the time elapsed after detonation of charge (ms), (P_{max}) denotes the peak pressure (Mpa), (t_1) denotes the arrival time of shock wave to the target after the detonation of the charge (ms), and (θ) denotes the time decay constant that describes the exponential decay (ms). The peak pressure and decay constant depends on the size of the explosion and the stand-off distance from the charge at which the pressure is measured. The peak pressure (P_{max}), decay constant of the wave (θ), impulse (I), bubble oscillation period (T), the maximum radius of the first bubble of explosive gas (R_{max}), and the energy flux density/energy per unit volume (E) can be expressed as per Equations (2)–(7) [39–41].

$$P_{max} = K_1 \left(\frac{W^{\frac{1}{3}}}{R} \right)^{A_1} \quad (\text{Mpa}) \tag{2}$$

$$\theta = K_2 W^{1/3} \left(\frac{W^{\frac{1}{3}}}{R} \right)^{A_2} \quad (\text{ms}) \tag{3}$$

$$I = K_3 W^{1/3} \left(\frac{W^{\frac{1}{3}}}{R} \right)^{A_3} \quad (\text{Mpa-sec}) \tag{4}$$

$$T = K_5 \frac{W^{\frac{1}{3}}}{(D + 9.8)^{5/6}} \quad (\text{sec}) \tag{5}$$

$$R_{max} = K_6 \left(\frac{W}{(D + 9.8)} \right)^{1/3} \quad (\text{m}) \tag{6}$$

$$E = K_4 W^{\frac{1}{3}} \left(\frac{W^{\frac{1}{3}}}{(R)} \right)^{A_4} \quad (\text{m} \cdot \text{kPa}) \tag{7}$$

where $(K_1, K_2, K_3, K_4, K_5, K_6)$ and (A_1, A_2, A_3, A_4) are constants that depend on the explosive charge type with values as in [39,42]. These input constants are illustrated in Figure 2. Phenomena of the underwater explosions (UNDEX): shock wave, high pressure and bubble motion.

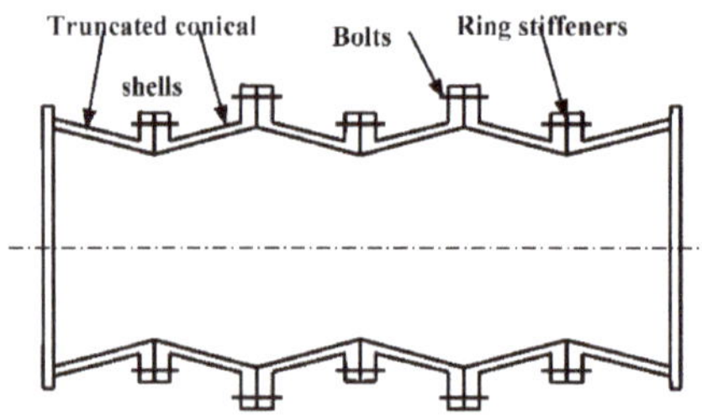

(**a**) Ring-stiffened corrugated pressure hull.

(**b**) Sandwich hull wall construction.

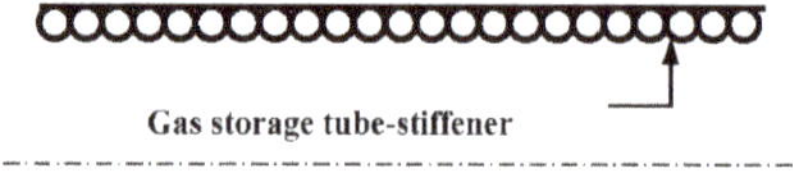

(**c**) Modular composite pressure hull made of truncated spherical section.

(**d**) Gas-storage toroidal tube-stiffener hull wall construction.

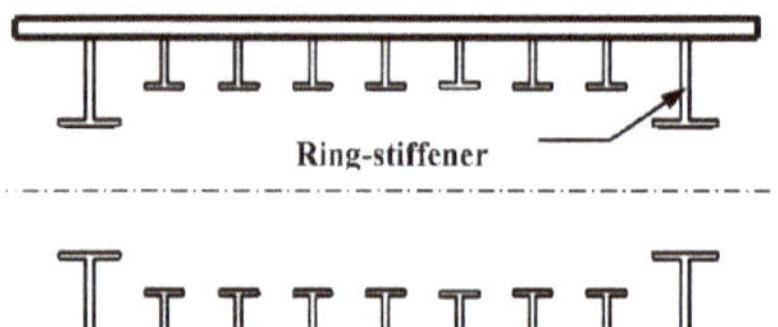

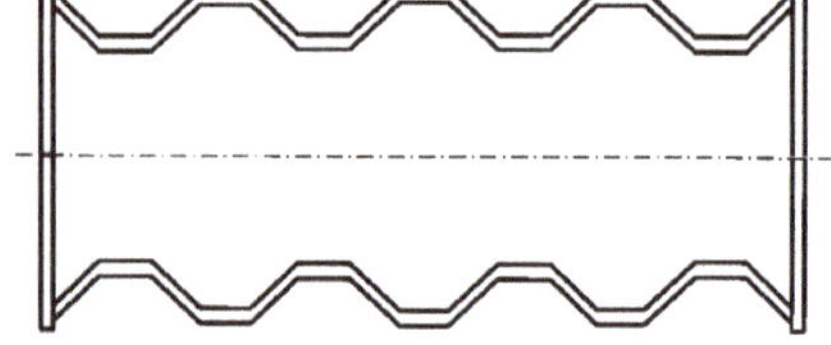

(**e**) Traditional ring-stiffened hull wall construction.

(**f**) Trapesoid corrugated hull wall construction.

Figure 1. Various wall structures utilized for pressure hulls. (**a**) Reproduced with permission from [26], Copyright Elsevier, 2011. (**b,d–f**) reproduced with permission from [27], Copyright Elsevier, 2003. (**c**) reproduced with permission from [28], Copyright Elsevier, 2016.

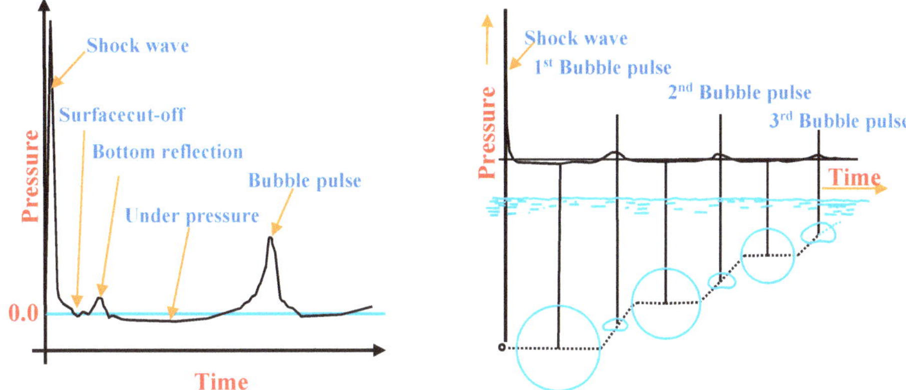

Figure 2. Phenomena of the underwater explosions (UNDEX): shock wave, high pressure and bubble motion.

(*W*) denotes the mass of the explosive charge in (Kg), (*D*) denotes the charge depth in *m*, and (*R*) denotes the distance between the explosive charge and target in (m). In UNDEX events, there are two types of subsequent cavitations that occur: the bulk and the local cavitation. The latter is caused by the reflection of the shock wave at the free surface and must be taken into consideration during the analysis of the surface ship. When the shock wave impinges upon the structure, then the total pressure $P_{tot}(t)$ at the fluid structure interface can be expressed according to Equation (8) [43]:

$$P_{tot}(t) = P_{in}(t) + P_c(t) + P_{st} \tag{8}$$

where $P_{in}(t)$ denotes the incident shock wave, $P_c(t)$ denotes the scattered pressure, and P_{st} denotes the hydrostatic pressure. Accordingly, the local cavitation occurs in the water as the pressure drops to vapor pressure (about 0.3 psi) [31]. Thereafter, the cavitation collapses and reload the structure.

3. Composite Failure Criteria

An appropriate failure criterion is needed in order to find the maximum permissible load before lamina failure. Therefore, it is necessary to develop theories to compare the state of stresses and strains in materials [44–46].

3.1. Maximum Stress Failure Criteria

The maximum stress criteria are single mode failure criteria. Fracture occurs if stresses at the principal material coordinates are higher than their respective strength. The failure index (I_F) is presented in Equation (9) [47]:

$$I_F = \max \begin{cases} \sigma_{11}/X_t & \text{if } \sigma_{11} > 0 \text{ or } -\sigma_{11}/X_c \text{ if } \sigma_{11} < 0 \\ \sigma_{22}/Y_t & \text{if } \sigma_{22} > 0 \text{ or } -\sigma_{22}/Y_c \text{ if } \sigma_{22} < 0 \\ |\tau_{12}|/S \end{cases} \tag{9}$$

where X_t, X_c, Y_t, Y_c and S denote the ultimate longitudinal, transversal and shear strength constants, respectively. Also, (σ_{11}, σ_{22} and τ_{12}) denote the applied longitudinal, transversal and shear stress components, respectively, and can be calculated using Equation (10) [48]:

$$\begin{Bmatrix} \sigma_{11} \\ \sigma_{22} \\ \tau_{12} \end{Bmatrix} = \begin{bmatrix} Q_{11} & Q_{12} & 0 \\ Q_{12} & Q_{22} & 0 \\ 0 & 0 & Q_{66} \end{bmatrix} \begin{Bmatrix} \varepsilon_{11} \\ \varepsilon_{22} \\ \gamma_{12} \end{Bmatrix} \tag{10}$$

where $Q_{11} = (1 - v_{12}v_{21})^{-1}E_1$, $Q_{12} = (1 - v_{12}v_{21})^{-1}E_1v_{21}$, $Q_{22} = (1 - v_{12}v_{21})^{-1}E_2$, $Q_{66} = G_{12}$.

3.2. Tsai-Hill Failure Criteria

Tsai-Hill failure criterion assume that there is an interaction between longitudinal, transversal and shear strength in the damage progress. The Tsai-Hill failure criterion can be expressed as in Equation (11) [49]:

$$\sigma_{11}^2/X^2 + \sigma_{22}^2/Y^2 - \sigma_{11}\sigma_{22}/X^2 + \tau_{12}^2/S^2 = 1 \tag{11}$$

where σ_{11}, σ_{22} and τ_{12} are the applied longitudinal, transversal and shear stress components, respectively, and can be calculated using Equation (10). Meanwhile, X and Y are the longitude and traverse strength, respectively, whether in tension or compression, which depends on the stress status in the laminates. S is the shear strength constant.

3.3. Tsai-Wu Failure Criteria

The Tsai-Wu failure criterion is the most generalized criterion that distinguishes between compressive and tensile strength. The criterion can be expressed as shown in Equation (12) [50,51]:

$$FI = \sigma_{11}\left(\frac{1}{X_t} - \frac{1}{X_c}\right) + \sigma_{22}\left(\frac{1}{Y_t} - \frac{1}{Y_c}\right) - \frac{\sigma_{11}^2}{X_t \times X_c} - \frac{\sigma_{22}^2}{Y_t \times Y_c} - \frac{\tau_{12}^2}{S^2} \tag{12}$$

where FI is the failure index, X_t, Y_t, X_c, Y_c, σ_{11}, σ_{22}, τ_{21} and S are as aforementioned. The failure occurs when the calculated stresses reaches the ultimate stresses and the FI reaches or exceeds the value 1 [52]. In finite element procedures, failure criteria are presented using a defined failure index and can be presented as per Equation (13):

$$FI = \frac{Stress}{Strength} \tag{13}$$

The results presented in this work are based on maximum stress, Tsai-Hill and Tsai-Wu failure criteria.

4. Optimization and Geometrical Configuration of MICEPH

In this work, the proposed form of submarine pressure hull is a multiple intersecting cross-elliptical hull, constructed from Carbon/Epoxy composite (USN-150) with stacking sequence $[(\alpha/-\alpha)_4]_s$. The optimization is performed using the ANSYS V14.5 (ANSYS, Canonsburg, PA, USA) parametric design language (APDL) to maximize buckling load and minimize the buoyancy factor under the constraints of failure strength and deflection (δ_{max}) according to the design requirements. The model was built using SHELL99 for the shell and BEAM189 for the ring and long beams [53,54]. The material properties and the strength parameters are presented in Table 1 [55]. Figure 3 shows the proposed MICEPH utilized in this study. Figure 3 shows the multi-objective optimization procedure flow chart. The random design generation method (RDGM), which is a sub type of the sub problem approximation used in (ANSYS) will be considered in this study. On the other hand, Table 2 illustrates the results of the multi-objective optimization for MICEPH which indicates the optimal design point and objective function. The table contains the failure index (FI) for the maximum stress (F_{MAXF}) and Tsai-Wu (F_{TWSR}) failure criteria. The optimal objective function (MOF) is 0.0487, the buoyancy factor ($B.F$) is 0.197, and the buckling strength factor is 4.056, with a total hull weight of 394.57 kg. The optimal angle-ply orientation (α) is 49°, with a layer thickness (t) equal to 1.294 mm. The major diameter (D_{major}) and minor diameter (D_{minor}) are 2.0 m and 1.6836 m, respectively. The intersecting angel (θ) is 40°, and the radius (R) is equal to 0.50 m.

Table 1. Strengths of unidirectional composites and material properties of the sandwich components.

Material	Material and Strength Properties
Carbon/epoxy composite (USN-150)	E_{11} = 131 GPa, E_{22} = 10.8 GPa, E_{33} = 10.8 GPa, G_{12} = 5.65 GPa, G_{23} = 5.65 GPa, v_{12} = 0.28, v_{23} = 0.0.59, X_t = 2000 MPa, X_c = 1400 Mpa, Y_t = 61 MPa, Y_c = 130 MPa, S = 70 MPa, ρ = 1540 kg/m^3

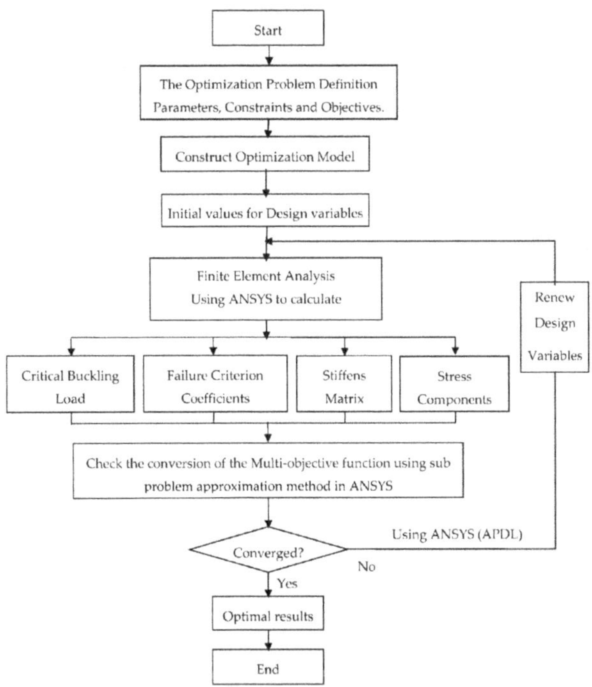

Figure 3. Multi-objective optimization procedure flow chart.

Table 2. The results of optimal design for pressure hull without core (Carbon/epoxy composite (USN)).

	Tsai-Wu Failure (F_{TWSR})	Maximum Stress Failure (F_{MAXF})		Tsai-Wu Failure (F_{TWSR})	Maximum Stress Failure (F_{MAXF})	Buckling Strength Factor (λ)	4.05636769
FI_1	0.741309805	0.678269306	*FI_12*	0.821134469	0.752770265	*Layer* thickness (*t*)	1.2946 mm
FI_2	0.722204168	0.663666634	*FI_13*	0.862284155	0.786597570	*B.F*	0.197617464
FI_3	0.719470819	0.648617196	*FI_14*	0.870768505	0.827885448	**Total weight**	394.574729 kg
FI_4	0.701161177	0.639770414	*FI_15*	0.915660726	0.858129679	θ	40°
FI_5	0.698515548	0.634590148	*FI_16*	0.964664573	0.999689193	α	49°
FI_6	0.689078750	0.625456838	D_{major}	2.0 m		h_1	75.4 mm
FI_7	0.705410043	0.620743340	D_{minor}	1.6836 m		*b1*	56 mm
FI_8	0.707480479	0.671368673	**Operating depth (*H*)**	500 m		h_2	50 mm
FI_9	0.698181393	0.664485365	**Maximum deflection (δ_{MAX})**	4.65067861 (mm)		*b2*	5 mm
FI_10	0.772656298	0.678194589	**R**	0.50 m		h_3	50 mm
FI_11	0.810096248	0.714633754	*MOF*	$4.871783801 \times 10^{-2}$		*b3*	5 mm

5. Modeling and Simulation of MICEPH

The fluid structure interaction (FSI) and UNDEX phenomena are implemented in non-linear finite element code ABAQUS V6.13 (Dassault Sys Simulia Corp, Providence, RI, USA)/Explicit. The MICEPH used in this study consists of five cross-ellipses, as demonstrated in Figure 4. The optimized structural hull and water domain are imported from ANSYS Program as (* iges). Since the water domain does not have internal space for the pressure hull, it therefore needs to be modified to have identical internal shape for the hull. This can be simply done using the merge/cut function in the CAE model part in ABAQUS. The pressure hull was modelled using the shell elements. The stiffeners were added in the radial and longitudinal directions to provide structural integrity. The fourth node shell element (S4R) was used to model the pressure hull and stiffeners. An assemblage of 4-node acoustic tetrahedral elements (AC3D4) was used to represent the external fluid. The total horizontal length of the fluid model with the spherical ends was 11.56 m. The vertical length of the fluid domain was 7 m. The modeled water domain part includes the hull structure. The model and the charge mass were located at depth 100 m below the free surface. In dynamic problems that involve fluid and coupled solid medium, the interface between the two domains must be identical. Also, the water domain must have bulk modulus and density since it is acoustic domain. Figure 5 shows the finite element model and the meshing technique for the MICEPH surrounded by the fluid. The boundaries of the fluid around the MICEPH may cause shock wave reflection or refraction, which may cause a change in its superposition or cancellation by the incident wave [56,57]. To overcome this problem, the boundary condition of the fluid is executed as a non-reflective boundary condition during the analysis. The pressure hull was exposed to UNDEX produced by various amounts of explosives and offset distances (30 kg TNT at 5.5 m, 20 kg TNT at 5.5 m, 15.5 m and 20.5 m). In this study, the stand-off distance is located at the right side of the cross-elliptical submersible pressure hull. The stand-off point represents the location where the incident wave defined and represented by reference point1 (RP_1). The location of the charge (source point) defined as reference point 2 (RP_2), which represents the position of the charge as illustrated in Figure 6. There are three types of input parameters for UNDEX: physical charge, material and bubble model. Defining the UNDEX bubble in the interaction module was applied by defining the source point, stand off point which specifying the wave properties and charge depth. Taking into consideration that the infinite surface shouldn't reflect the shock wave; thus, acoustic impedance (non-reflecting) has to be applied. The initial boundary condition for the infinite surface of water domain (acoustic wave) was set to zero, which means that the domain is calm water, i.e., no interference [58].

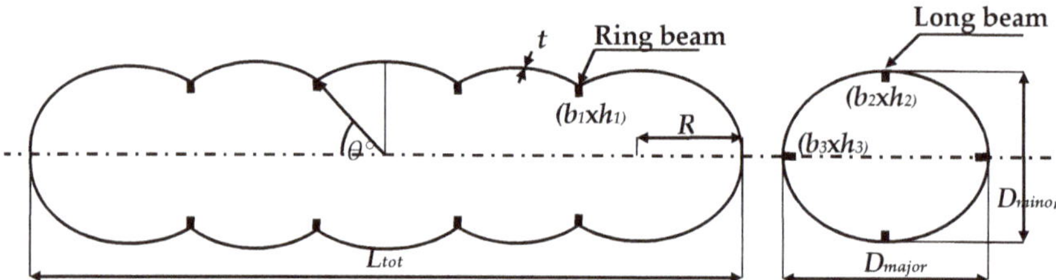

Figure 4. Miceph.

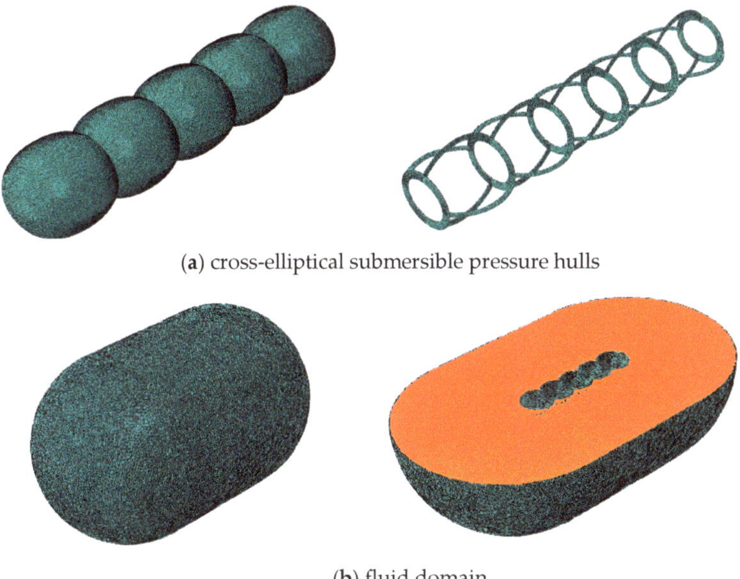

(**a**) cross-elliptical submersible pressure hulls

(**b**) fluid domain

Figure 5. Finite element modeling of: (**a**) cross-elliptical submersible pressure hulls, and (**b**) fluid domain.

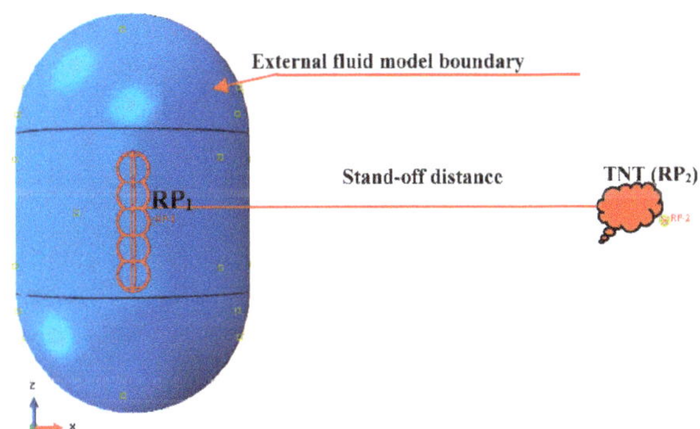

Figure 6. The location of charge (source point (RP_2)) and stand-off point (RP_1).

6. Results and Analysis

6.1. Propagation of Shock Wave

In this study, the MICEPH is subjected to shock wave and bubble pulse owing to UNDEX. Its performance depends mainly on the strength of the waves and the resilience of its structure. The stand-off point at node (A) was first struck by the shock wave. The charge is located at different stand-off distances—5.5 m, 15.5 m and 20.5 m—measured from the stand-off point. In the modelling procedures, two output variables were provided for the acoustic pressure: (i) the acoustic pressure (POR), which represents the total dynamic pressure in the wave formulation analysis including additional pressure induced by the incident and scattered waves; and (ii) the absolute acoustic pressure (PABS), which is the sum of the acoustic pressure and hydrostatic pressure. The acoustic pressure magnitudes are in Pascal [56]. Figure 7 demonstrates the POR at which the total dynamic pressure instantaneously occurs at the time of explosion (zero time instant) due to total wave formulation. Also, the figure shows the spherical shape of the shock wave and the initial instantaneous shock wave propagation at the time of the explosion due to different explosive weights and stand-off distances. It can be observed that the maximum shock wave pressure determined was 3.471×10^7 Pa due to 30 kg TNT at offset distance of 5.5 m. Figure 7b–d illustrates the computed POR field distribution due to 20 kg TNT at offset distances of 5.5 m, 15.5 m and 20.5 m. The maximum shock wave pressure measured was 2.373×10^7 Pa, 6.777×10^6 Pa and 4.853×10^6 Pa at explosive offset distance 5.5 m, 15.5 m and 20.5 m, respectively. The location of the source and standoff point greatly affect the computed POR field distribution, propagation and its magnitude.

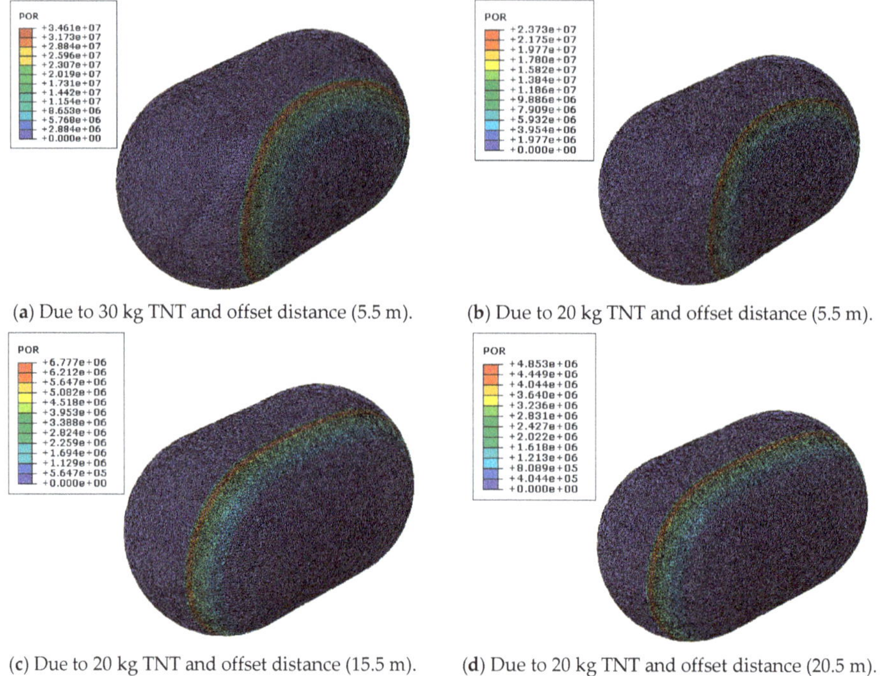

(**a**) Due to 30 kg TNT and offset distance (5.5 m).   (**b**) Due to 20 kg TNT and offset distance (5.5 m).

(**c**) Due to 20 kg TNT and offset distance (15.5 m).   (**d**) Due to 20 kg TNT and offset distance (20.5 m).

Figure 7. Computed POR field at the zero time instant due to total wave formulation.

Figures 8 and 9 present the computed POR field distribution at several time instants at the front and back sides of fluid domain. The propagation of the shockwave and the dynamic pressure in the surrounding water were clearly presented in these figures. The figures demonstrated that there are some radial POR waves that hits the structure and consequently leads to some deformations. Furthermore, it was observed that the dynamic pressure is affected by the reflections and the emissions from the pressure hull in addition to the incident field from the source point. When the shock wave hits the cross-pressure hull, it is reflected and generates negative pressure. Also, it is interaction with the incident wave decreases the dynamic pressure in the surrounding fluid. This is attributed to the fact that the water cannot withstand tension. That is why the total dynamic pressure acquired negative values as illustrated in the figures. Furthermore, cavitation occurs immediately after the incident shock wave hits the MICEPH. Additionally, local cavitation around the MICEPH was observed once the acoustic pressure declines to the steam pressure of the fluid. Subsequently, when the local cavitation disappears, the load of the fluid on the pressure hull generates vibrations. Additionally, Figure 9 illustrates that the shock wave propagates symmetrically from the stand-off point to the aft and fore, while it expands in the perpendicular direction. The aforementioned results are in accordance with the results reported in [2,43,59]. Also, Figure 9 presents the volume changes in gas bubbles. If the oscillating gas bubble is close enough to the pressure hull, the differential pressure will be created. When the bubble decreases in volume (due to resistance to water flow close to the hull), that would result in bubble collapsing onto the hull and producing high speed water jet, which in some instances is capable of destroying or holing the pressure hull.

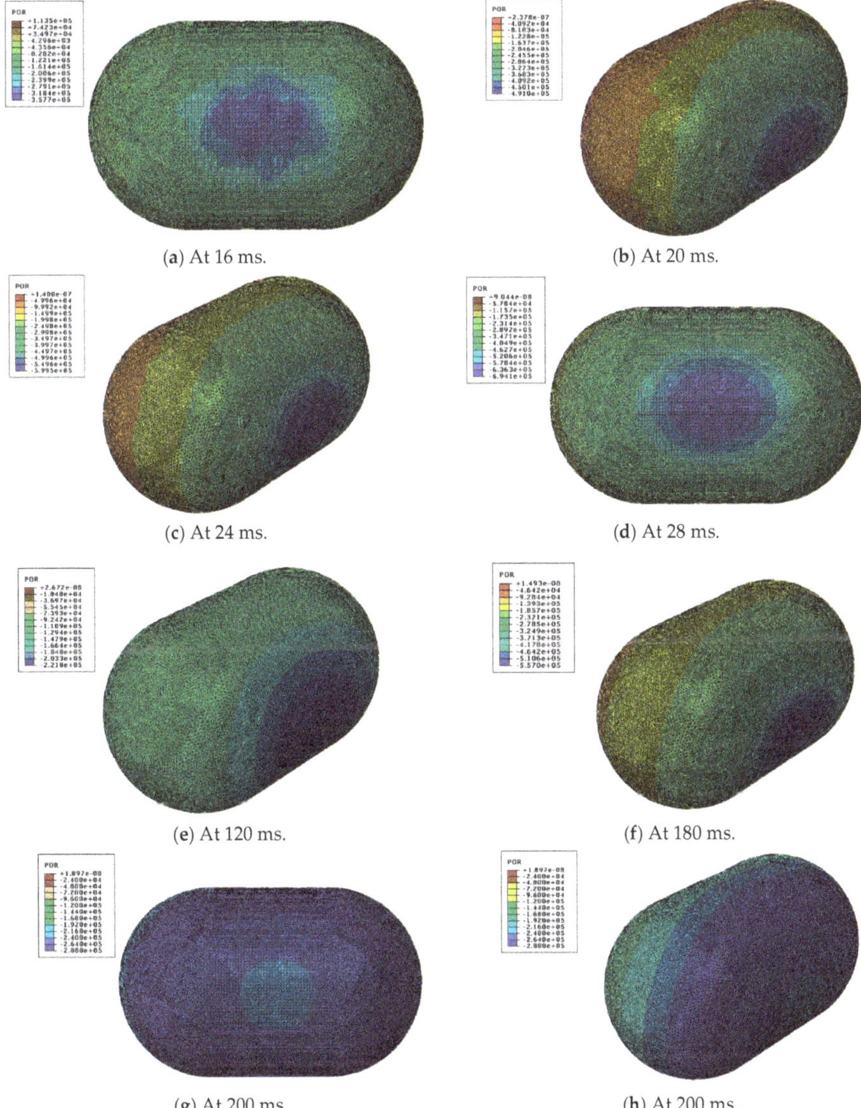

Figure 8. Computed POR field distributions at several time instants at the front side of the fluid domain (on the blast side).

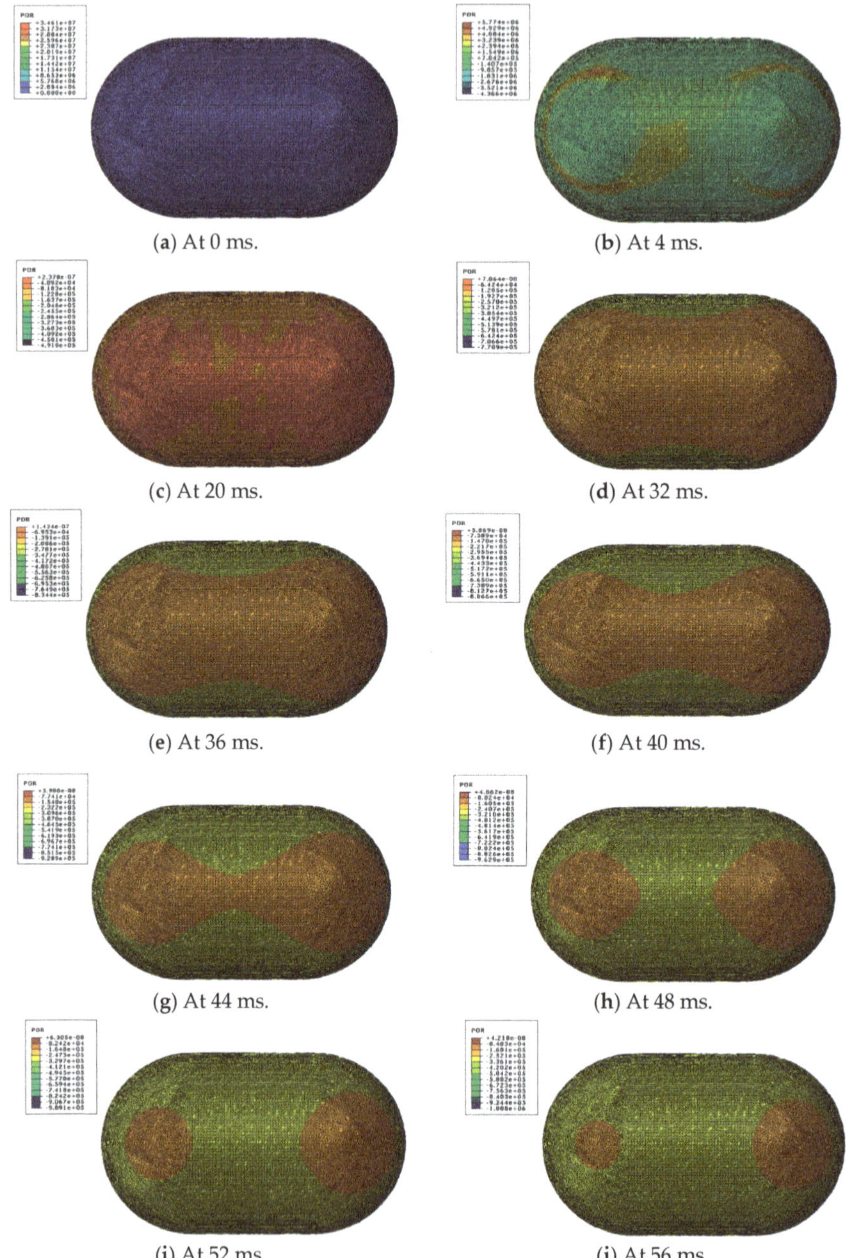

(**a**) At 0 ms.

(**b**) At 4 ms.

(**c**) At 20 ms.

(**d**) At 32 ms.

(**e**) At 36 ms.

(**f**) At 40 ms.

(**g**) At 44 ms.

(**h**) At 48 ms.

(**i**) At 52 ms.

(**j**) At 56 ms.

Figure 9. Computed POR field at several time instants at the back side of the fluid domain (the side most remote from the charge).

6.2. Responses of Submarine Pressure Hull to UNDEX

The optimum MICEPH subjected to non-contact UNDEX will show three major response modes: (i) motion in the axial direction that makes the accordion motion, (ii) motion in a direction at right angle to the fore-and-aft line of the MICEPH that makes the whipping mode parallel to the direction of the shock wave propagation, and (iii) motion in the vertical direction that makes the breathing motion perpendicular to the shock wave direction of the travel. Several locations were chosen in the model (location A, B, C, D, E and F) to demonstrate the responses in the MICEPH as illustrated in Figure 10.

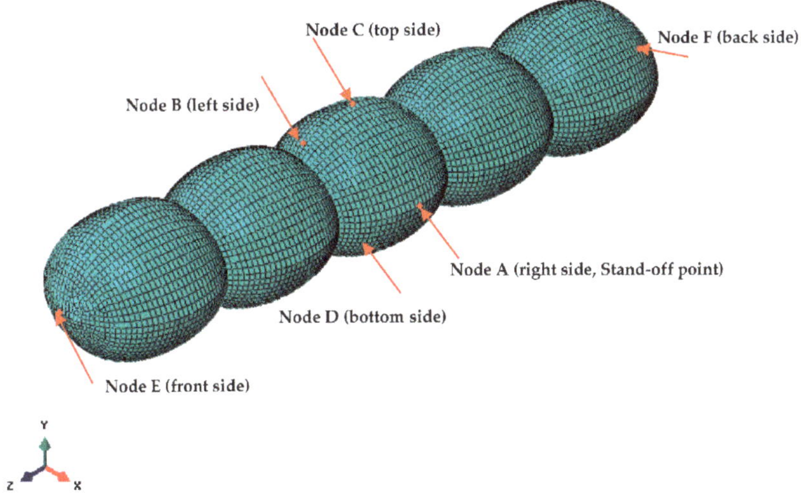

Figure 10. Finite element model of Multiple. Intersecting Cross Elliptical Pressure Hull (MICEPH) and the locations of different test points.

6.2.1. The Acceleration Response at Different Locations

First, the response of the MICEPH in the axial direction is illustrated in Figure 11. The figure presents the time history response of Z-axial acceleration (A_3). UNDEX of 20 kg TNT at different offset distances of 5.5 m, 15.5 m and 20.5 m were discussed. At points E and F, located at the center of each end of the MICEPH, it was observed that the acceleration responses (A_3) were in the opposite directions and occurs at the same instance. The peak value of the acceleration is 26×10^3 m/s^2 at an offset distance of 5.5 m, and occurs at 4 m/s. This measurement then oscillates and decays after 4 m/s. In addition, while the offset distance increases, the acceleration decreases and acquires 8.3×10^3 m/s^2 and 6.65×10^3 m/s^2 at offset distances of 15.5 m and 20.5 m, respectively. This is attributed to increasing the standoff distance of the explosive charge. Also, the figure demonstrates that the bubble pulses show a minor impact on nodes E and F.

Similarly, Figure 12 plots the time history curves of Z-axial acceleration (A_3) for nodes C and D. The curves illustrate that while the pressure hull stroked by the shock wave, node C moves to the left and node D moves to the right at the same time instance which causes the accordion motion. This is attributed to the propagated compressive pressure and its subsequent release on the MICEPH in the axial direction. The peak value at node D of the measured acceleration is 27.4×10^3 m/s^2 in the negative Z-direction, owing to a 20 kg TNT charge and an offset distance of 5.5 m. Likewise, at node C, the peak value of acceleration is 23.9×10^3 m/s^2, which occurs at 2 m/s. Also, as the offset distance increases at node C, the acceleration decreases and achieves 4.6×10^3 m/s^2 and 2.8×10^3 m/s^2 at offset distances of 15.5 m and 20.5 m, respectively. Similarly, the maximum acceleration at node D are 5.1×10^3 m/s^2 and 3.2×10^3 m/s^2 at offset distances of 15.5 m and 20.5 m, respectively. The aforementioned results demonstrate that the peak acceleration at nodes C and D were achieved before nodes E and F.

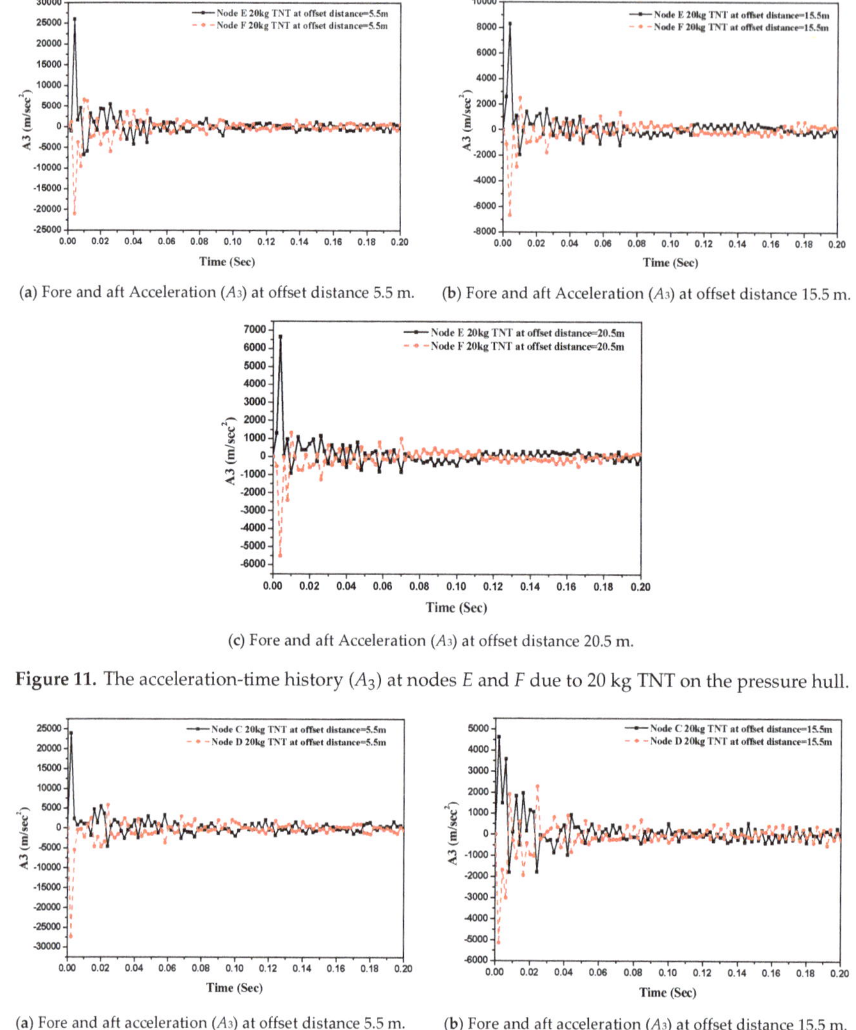

(**a**) Fore and aft Acceleration (A_3) at offset distance 5.5 m.   (**b**) Fore and aft Acceleration (A_3) at offset distance 15.5 m.

(**c**) Fore and aft Acceleration (A_3) at offset distance 20.5 m.

Figure 11. The acceleration-time history (A_3) at nodes *E* and *F* due to 20 kg TNT on the pressure hull.

(**a**) Fore and aft acceleration (A_3) at offset distance 5.5 m.   (**b**) Fore and aft acceleration (A_3) at offset distance 15.5 m.

(**c**) Fore and aft acceleration (A_3) at offset distance 20.5 m.

Figure 12. The acceleration-time history (A_3) at nodes *C* and *D* due to 20 kg TNT on the pressure hull.

On the other hand, Figure 13 illustrates the acceleration response (A_2) of the MICEPH in the vertical direction of nodes E and F under UNDEX of 20 kg TNT at offset distances of 5.5 m, 15.5 m and 20.5 m. It is observed that node E moves opposite to node F throughout the transient response of the MICEPH. Also, the figure illustrates the breathing motion caused by the expansion and subsequent contraction of the MICEPH. For instance, at node E and an offset distance of 5.5 m, the peak acceleration (A_2) is 9.45×10^3 m/s^2 in the Y-direction, while at node F, the peak acceleration is 8.96×10^3 m/s^2 in the negative Y-direction and occurs at a time instance of 8 m/s. Furthermore, at higher offset distances of 15.5 m and 20.5 m, the peak measured accelerations are 7.16×10^3 m/s^2 and 5.26×10^3 m/s^2 in negative Y-direction at node E. While at node F, the peak accelerations are 6.43×10^3 m/s^2 and 5.15×10^3 m/s^2 in the Y-direction and occur at a time instance of 2 m/s.

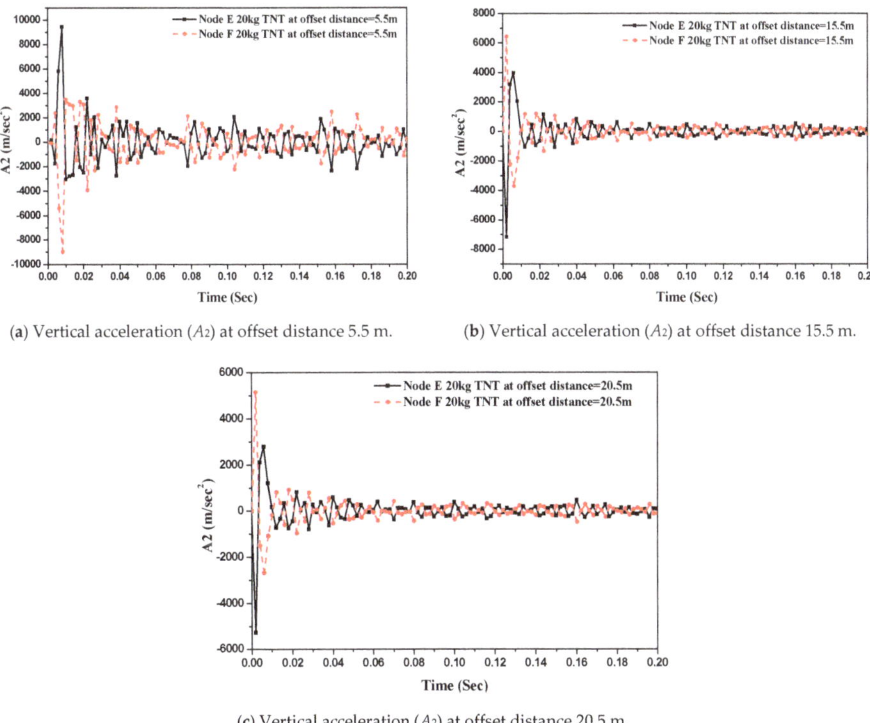

(**a**) Vertical acceleration (A_2) at offset distance 5.5 m.

(**b**) Vertical acceleration (A_2) at offset distance 15.5 m.

(**c**) Vertical acceleration (A_2) at offset distance 20.5 m.

Figure 13. The acceleration-time history (A_2) at nodes E and F due to 20 kg TNT on the pressure hull.

Furthermore, Figure 14 shows the acceleration time histories (A_2) in the vertical direction for nodes C and D which located at top and bottom of the MICEPH under the effect of 20 kg TNT charge located at offset distances of 5.5 m, 15.5 m and 20.5 m. It is revealed that the upper point (node C) moves in a direction opposite to that of the lower point (node D). The peak accelerations at node C, measured downward, are -40×10^3 m/s^2, -10.4×10^3 m/s^2, and -7.37×10^3 m/s^2 at offset distances of 5.5 m, 15.5 m, and 20.5 m, respectively. Likewise, the peak accelerations at node D, measured upward, are 49.2×10^3 m/s^2, 12.6×10^3 m/s^2, and 8.25×10^3 m/s^2 at offset distances of 5.5 m, 15.5 m and 20.5 m, respectively. The peak acceleration time histories (A_2) at node D are higher than its counterpart at node C, while the frequencies of accordion and breathing motion are nearly the same. These results ensure that the accordion and breathing motion are directly correlated as per [2].

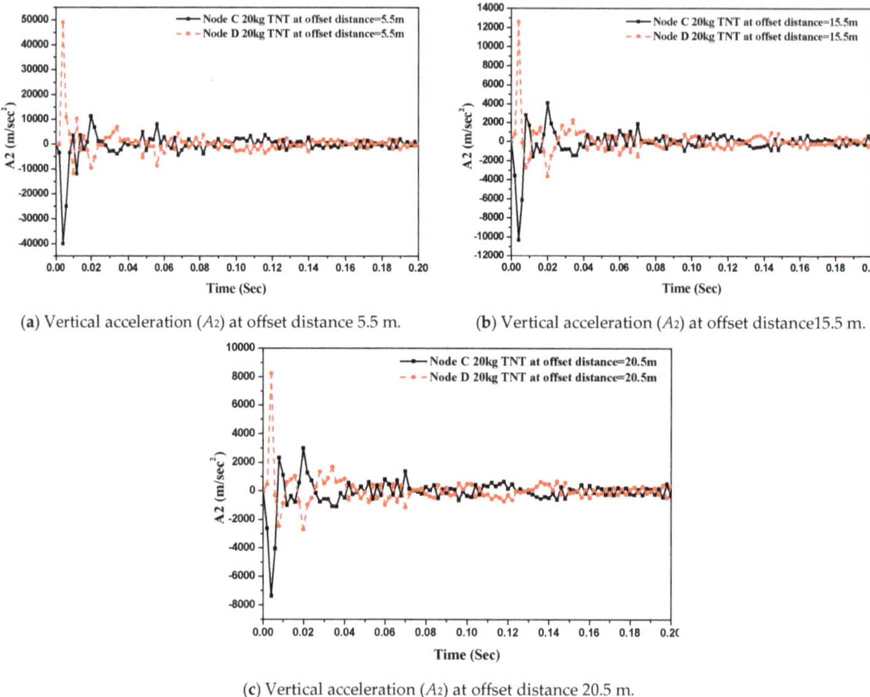

(**a**) Vertical acceleration (A_2) at offset distance 5.5 m.

(**b**) Vertical acceleration (A_2) at offset distance15.5 m.

(**c**) Vertical acceleration (A_2) at offset distance 20.5 m.

Figure 14. The acceleration-time history (A_2) at nodes C and D due to 20 kg TNT on the pressure hull.

Similarly, Figure 15 shows the acceleration time histories (A_2) at nodes A and B. It is observed that, once the shock wave struck the stand-off point at node A, the vertical movements of nodes A and B follow the same direction. Node A moves first, followed by node B. The peak acceleration at node A is -8×10^3 m/s^2, -1.8×10^3 m/s^2 and -1.3×10^3 m/s^2 at offset distance of 5.5 m, 15.5 m and 20.5 m, respectively, and occur downward. Likewise, the peak acceleration at node B is -3.5×10^3 m/s^2, -1.1×10^3 m/s^2 and -0.81×10^3 m/s^2 at offset distance of 5.5 m, 15.5 m and 20.5 m, respectively, following the same direction at point A. Figure 16 plots the athwart acceleration time histories (A_1) at nodes A and B which is the primary direction of shock wave propagation. It is revealed that the shock wave arrives at node A first then hits node B. The maximum athwart acceleration (A_1) at node A occurs in the horizontal direction and its peak is about 52.4×10^3 m/s^2 at an offset distance of 5.5 m. This is attributed to the release of the shock wave, occurrence of local cavitation and effect of bubble pulse. On the other hand, the second peak values, measured at offset distances of 5.5 m, 15.5 m and 20.5 m, occur owing to the uploading of the MICEPH. Similarly, the maximum athwart acceleration (A_1) at node B also occurs in the horizontal direction, and its peak is about 29.4×10^3 m/s^2, 10.9×10^3 m/s^2 and 7.9×10^3 m/s^2 at offset distance of 5.5 m, 15.5 m and 20.5 m, respectively. From the aforementioned analysis, it is concluded that the local cavitation has a major effect on athwart acceleration at standoff point at node A. Likewise, the uploading of the structure and the first bubble pulse also have a major effect on athwart acceleration. Additionally, for nodes A and B, the transverse responses are very severe at these locations, and the greatest acceleration occurs in the athwart direction, which is the main direction of shock wave with values of 52.3×10^3 m/s^2 and 29.4×10^3 m/s^2 at time intervals of 2 m/s and 4 m/s, respectively, followed by the vertical and longitudinal directions. While, for nodes C and D, which located at the top and bottom of MICEPH, the peak acceleration occurs in the vertical direction with values of -40×10^3 m/s^2 and 49.2×10^3 m/s^2 at time interval of 4 m/s, followed by the athwart and longitudinal directions. Furthermore, at nodes E and F, which located at the center of each end of the MICEPH (fore and aft direction), the peak acceleration occurs in the longitudinal direction followed by the vertical and athwart directions.

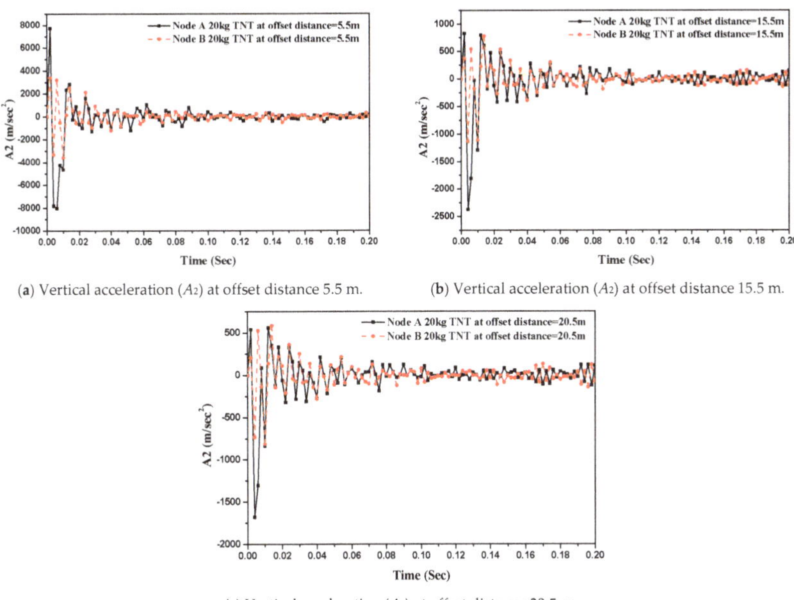

(**a**) Vertical acceleration (A_2) at offset distance 5.5 m.

(**b**) Vertical acceleration (A_2) at offset distance 15.5 m.

(**c**) Vertical acceleration (A_2) at offset distance 20.5 m.

Figure 15. The acceleration-time history (A_2) at nodes A and B due to 20 kg TNT on the pressure hull.

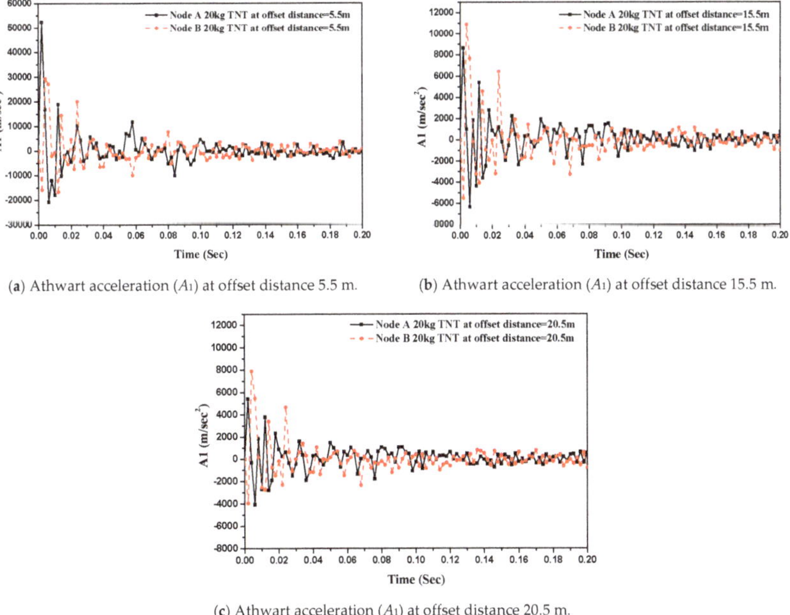

(**a**) Athwart acceleration (A_1) at offset distance 5.5 m.

(**b**) Athwart acceleration (A_1) at offset distance 15.5 m.

(**c**) Athwart acceleration (A_1) at offset distance 20.5 m.

Figure 16. The acceleration-time history (A_1) at nodes A and B due to 20 kg TNT on the pressure hull.

6.2.2. The Displacement Response at Different Locations

Figure 17 illustrates the deformation response in the translation direction (U_2) at nodes A and B owing to 20 kg TNT at various offset distances (5.5 m, 15.5 m and 20.5 m). It is observed that the vertical displacement (U_2) at nodes A and B move in opposite directions throughout the transient response and follow the same variation. In addition, the figure shows a rigid body translation of the MICEPH. Figure 17a shows the high-frequency response measured at nodes A and B, at an offset distance of 5.5 m. It is revealed that at a small offset distance, the structural wavelengths are longer than the acoustic wavelengths. Additionally, the surrounding fluid acts on the structure as a simple damping mechanism and the energy transported away from the structure through the acoustic radiation. Furthermore, Figure 17b illustrates the intermediate-frequency response at nodes A and B, at an offset distance of 15.5 m. At an intermediate offset distance, the structural wavelengths are nearly similar to the acoustic wavelengths. In addition, the surrounding fluid added mass and radiation damping on the structure of the hull. Moreover, Figure 17c demonstrates the low-frequency response at nodes A and B, at offset distance of 20.5 m. It is revealed that the structural wavelengths are nearly shorter than the acoustic wavelengths and the surrounding fluid added effective mass to the structure of the hull on the wetted interface. Figure 18 presents the translation (U_2) in the vertical direction of two points located at the top (node C) and bottom (node D) of the amid-ship section at different offset distances (5.5 m, 15.5 m and 20.5 m) due to 20 kg TNT. It is observed that the upper point (node C) is moved in a direction opposite to the lower point (node D). The figure also illustrates that the displacement at node C is higher than its counterpart at node D. In addition, the pressure hull shows an elastic deformations and rigid body motions with a significant heaving and pitching response. These results match those achieved by [60]. Similarly, Figure 19 illustrates the axial displacement response (U_2) in the vertical direction of points located at the center of each end of the MICEPH (E and F). It is observed that the two nodes move in opposite direction with almost similar frequencies forming the accordion motion. Additionally, the shape of the displacement response (U_2) is significantly affected by the offset distance. While the offset distance increases, the frequency decreases. The aforementioned results agree well with that acquired by [40,43].

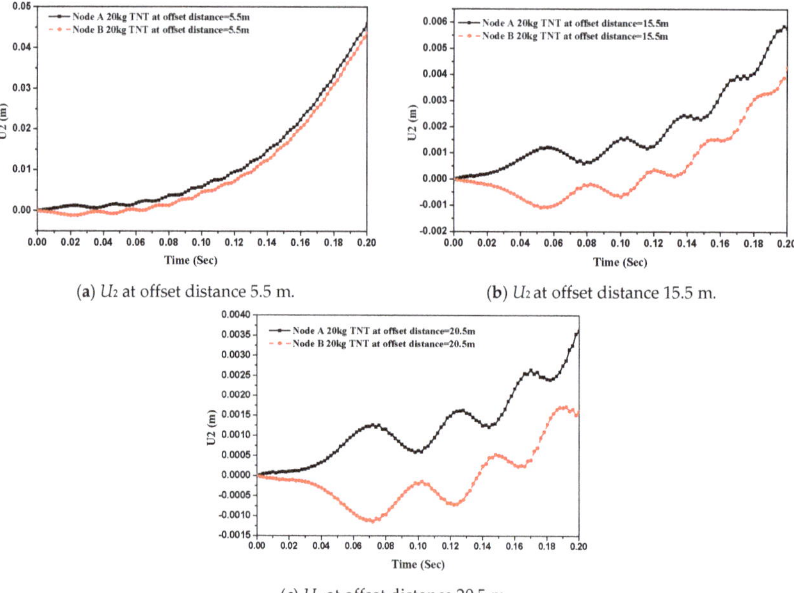

(**a**) U_2 at offset distance 5.5 m.

(**b**) U_2 at offset distance 15.5 m.

(**c**) U_2 at offset distance 20.5 m.

Figure 17. Displacement-time history responses (U_2) at nodes A and B on the MICEPH.

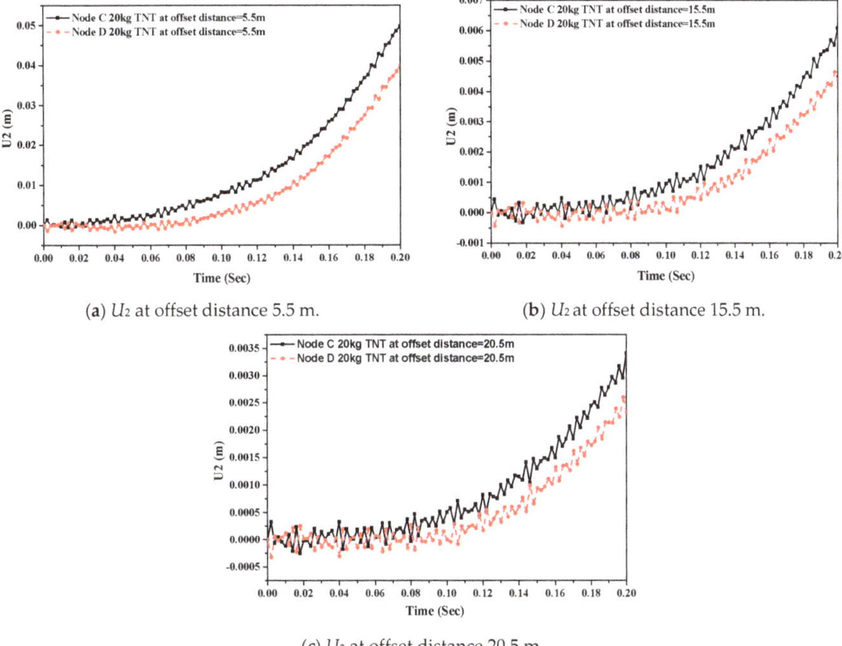

(**a**) U_2 at offset distance 5.5 m.

(**b**) U_2 at offset distance 15.5 m.

(**c**) U_2 at offset distance 20.5 m.

Figure 18. Displacement-time history responses (U_2) at nodes C and D on the MICEPH.

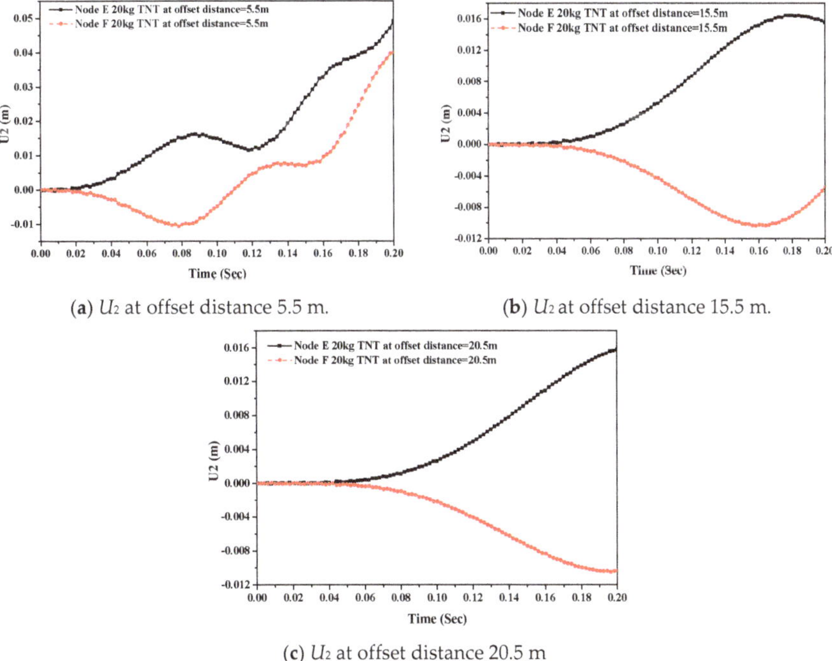

(**a**) U_2 at offset distance 5.5 m.

(**b**) U_2 at offset distance 15.5 m.

(**c**) U_2 at offset distance 20.5 m

Figure 19. Displacement-time history responses ($U2$) at nodes E, F on the MICEPH.

6.2.3. Failure Analysis of MICEPH

To evaluate the maximum allowable load that can be sustained before lamina failure, an appropriate failure criterion is needed. Therefore, the failure of the MICEPH is analyzed in this study using different failure criteria (maximum stress and Tsai-Hill failure criteria). The failure was assessed through the calculation of the failure index. If the failure index achieved unity, then the material had failed. Figures 20 and 21 demonstrate the failure index distribution on the structure of the pressure hull due to 20 kg TNT at different offset distances of 5.5 m and 15.5 m at various time instants.

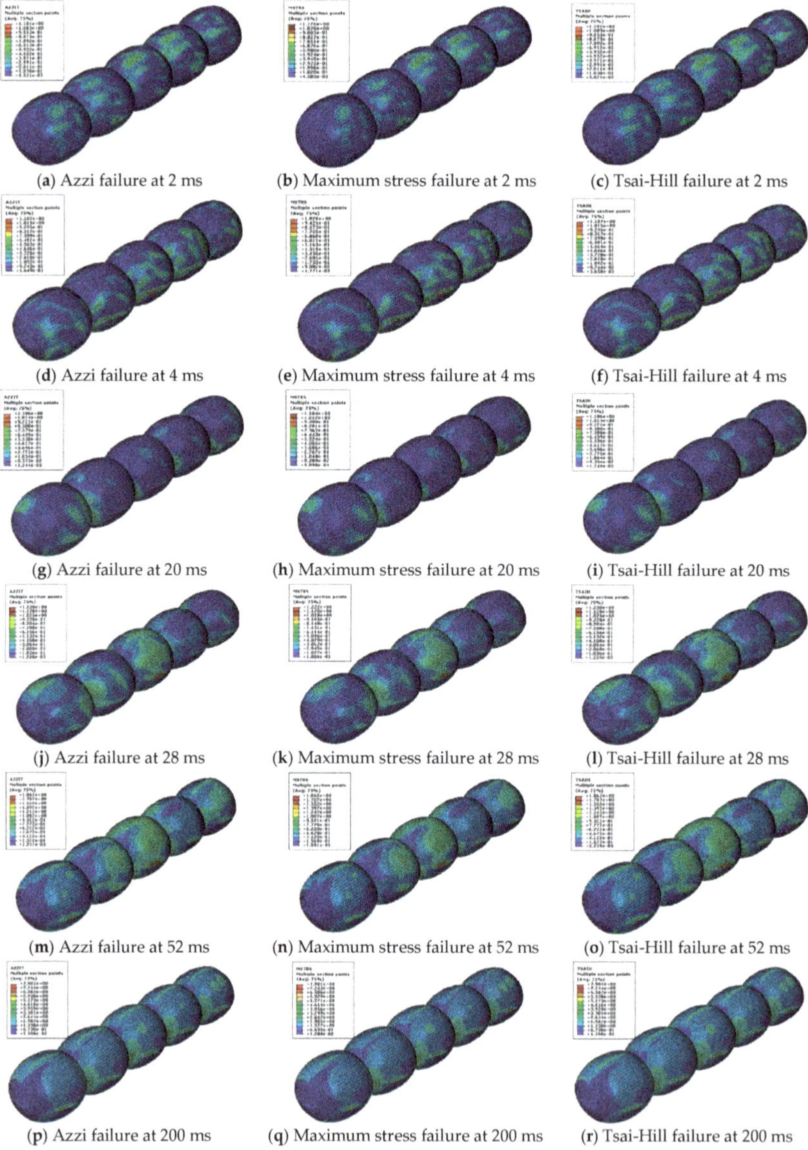

<table>
<tr><td>(a) Azzi failure at 2 ms</td><td>(b) Maximum stress failure at 2 ms</td><td>(c) Tsai-Hill failure at 2 ms</td></tr>
<tr><td>(d) Azzi failure at 4 ms</td><td>(e) Maximum stress failure at 4 ms</td><td>(f) Tsai-Hill failure at 4 ms</td></tr>
<tr><td>(g) Azzi failure at 20 ms</td><td>(h) Maximum stress failure at 20 ms</td><td>(i) Tsai-Hill failure at 20 ms</td></tr>
<tr><td>(j) Azzi failure at 28 ms</td><td>(k) Maximum stress failure at 28 ms</td><td>(l) Tsai-Hill failure at 28 ms</td></tr>
<tr><td>(m) Azzi failure at 52 ms</td><td>(n) Maximum stress failure at 52 ms</td><td>(o) Tsai-Hill failure at 52 ms</td></tr>
<tr><td>(p) Azzi failure at 200 ms</td><td>(q) Maximum stress failure at 200 ms</td><td>(r) Tsai-Hill failure at 200 ms</td></tr>
</table>

Figure 20. Failure criteria distributions on pressure hull due to 20 kg TNT and offset distance 5.5 m.

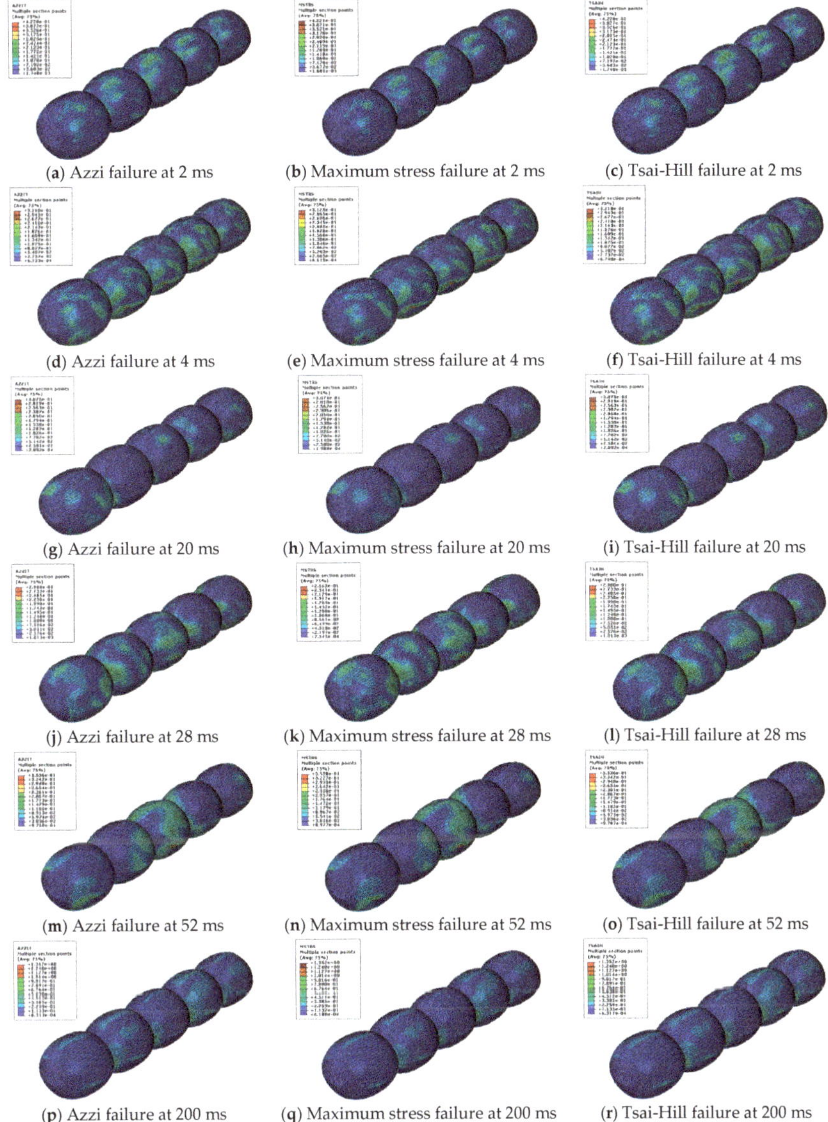

Figure 21. Failure criteria distributions on pressure hull due to 20 kg TNT and offset distance 15.5 m.

Generally, it is observed that the failure indices decrease with increasing stand-off distance. Furthermore, the maximum failure indices are measured around the ring and long beams. Moreover, it is revealed that failure is firstly initiated near the stand-off point. Figure 21 also shows the distribution of different failure criteria such as Azzi-Tsai-Hill, maximum stress, and Tsai-Hill criteria on the MICEPH. The results demonstrate that the difference among the different failure criteria is relatively too small. Figure 22 presents the maximum stress and Tsai-Hill failure-time histories for elements A, B, C, D, E and F on the structure of the pressure hull. The results illustrate that the maximum failure index occurs at offset distance of 5.5 m. The failure index owing to the maximum stress and Tsai-Hill criteria showed a similar trend with relatively small differences.

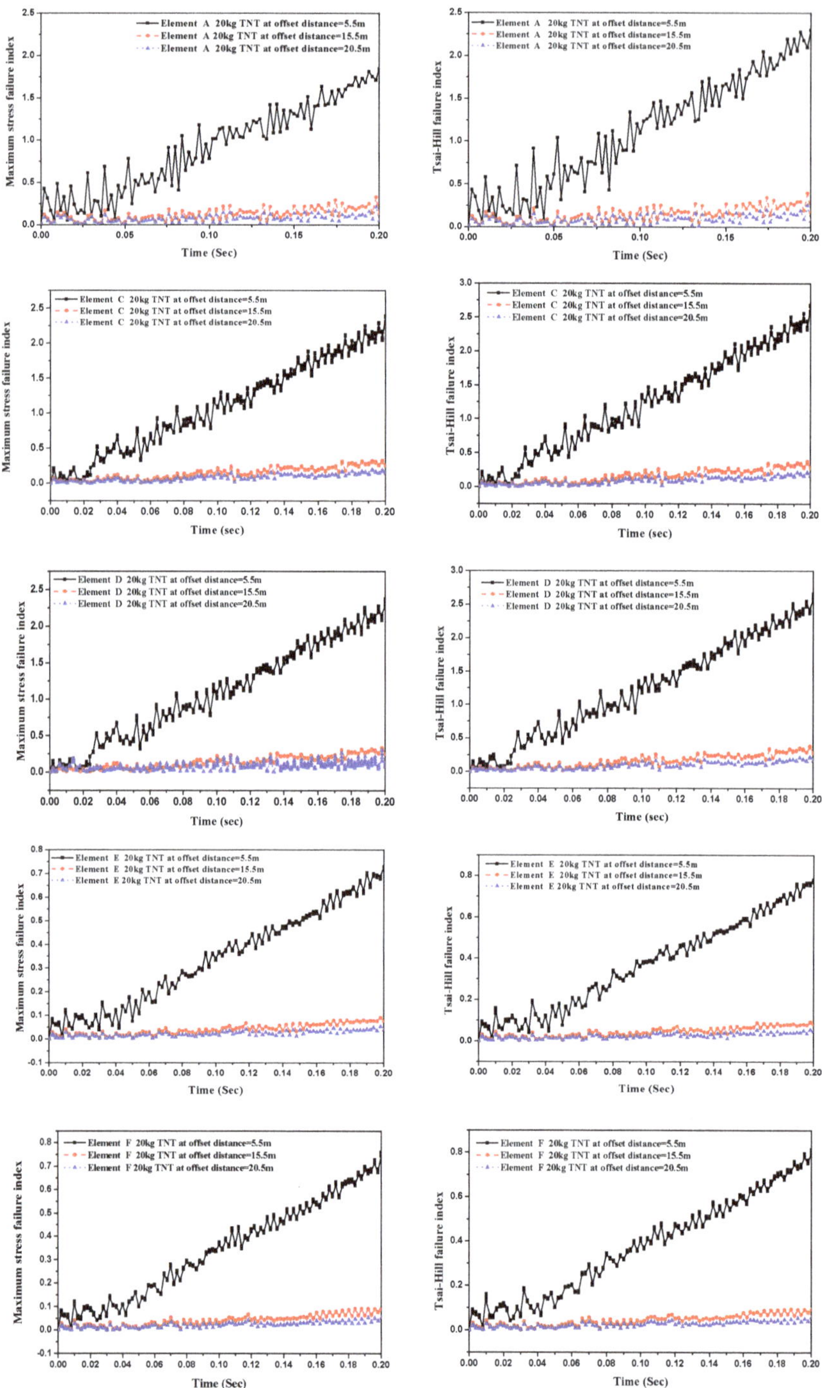

Figure 22. Maximum stress and Tsai-Hill failure-time history for elements *A*, *B*, *C*, *D*, *E* and *F* on the pressure hull.

7. Conclusions

In this study, the dynamic response of optimized multiple intersecting MICEPH under the effect of non-contact UNDEX was explored. The simulation technique was followed to avoid the expenses and complexity of physical tests. First, the multiple intersecting MICEPH subjected to hydrostatic pressure was optimized. Thereafter, using the optimum design results, the numerical simulation was carried out. Then, the response modes, breathing, accordion and whipping for the MICEPH subjected to non-contact UNDEX were discussed. Furthermore, the effects of applying various failure criterions such as maximum stress, Azzi-Tsai-Hill, and Tsai-Hill criteria for the damage initiation on failure strengths of pressure hull were studied. Based on the analysis and the simulation results, the following conclusions were drawn:

- The explosion weight and standoff distance greatly affects the computed POR field distribution, propagation and its magnitude.
- Based on the study of response modes, for nodes A and B, the greatest acceleration occurs in the athwart direction, which is the main direction of the shock wave, followed by the vertical and longitudinal directions. Meanwhile, for nodes C and D, which are located at the top and the bottom of the MICEPH, the greatest acceleration occurs in the vertical direction followed by the athwart and longitudinal directions. Likewise, for nodes E and F, which are located at the center of each end of the MICEPH (fore and aft) directions, the greatest acceleration occurs in the longitudinal direction, followed by the vertical and athwart directions.
- The cavitation occurs immediately after the incident shock wave hits the MICEPH.
- The local cavitation has a major effect on athwart acceleration. Additionally, the uploading of structure and the first bubble pulse have also, a major effect on athwart acceleration.
- This analysis can predict the failure index of the optimized pressure hull, which can be effectively used to determine the safe explosion weights and standoff distance to avoid the failure.
- The failure index of the pressure hull, caused by the shock wave due to detonation of 20 kg TNT charge at offset distance 5.5 m away from the side of the amidships of the hull is higher than one. Therefore, in this case, the pressure hull will collapse.
- On the other hand, at offset distance 15.5 m and 20.5 m, the failure index of the pressure hull is less than one, and the pressure hull will avoid the failure. Consequently, the standoff distance has a great effect on the failure index.
- The test results showed that applying different failure criterion such as maximum stress, Azzi-Tsai-Hill, and Tsai-Hill criteria for the damage initiation have a slight to no difference among them on the failure strengths of pressure hull.
- The simulation technique utilized in this study and its results can serve as a valuable reference for designers to enhance the resistance of underwater vehicles against underwater explosion.

Author Contributions: E.F. and M.H.; methodology, E.F. and M.H.; software, E.F. and M.H.; validation, E.F.; M.E.; M.A.; formal analysis, E.F. and M.H. investigation, H.H. and D.W., M.E., M.A.; resources, E.F. and M.H.; data curation, E.F.; writing—original draft preparation, all, writing—review and editing, E.F.; M.E.; M.A.; E.F.; visualization, E.F. and D.W; supervision, E.F, H.H. and D.W., project administration, D.W.; funding acquisition.

Funding: The authors are grateful for the support of this research by the 13th Five Years Key Programs for Science and Technology Development of China (Grant No. 2016YFD0701300), the Chinese Natural Science Foundation (Grant No. 51405076), and Heilongjiang Province Applied Technology Research and Development Program Major Project of China (Grant No. GA16B301). Also, the authors are grateful to Military Technical College (Cairo, Egypt), Taif University (Taif, KSA) and Mansoura University (Mansoura, Egypt) for providing all the required facilities to carry out the present research.

Acknowledgments: The authors are grateful to Military Technical College (Cairo, Egypt), Taif University (Taif, KSA) and Mansoura University (Mansoura, Egypt) for providing all the required facilities to carry out the present research.

Conflicts of Interest: The authors declare no conflict of interest

References

1. Reddy, A.C. Influence of Stiffeners on Strength of E-Glass/Epoxy Composite Submergible Hull Subjected to Shock Pressure Load using Finite Element Method. *Mater. Today Proc.* **2017**, *4*, 7507–7518. [CrossRef]
2. Jen, C.Y. Coupled Acoustic Structural Response of Optimized ring Stiffened Hull for Scaled down Submerged Vehicle Subject to Underwater Explosion. *Theor. Appl. Fract. Mech.* **2009**, *52*, 96–110. [CrossRef]
3. Cho, Y.S.; Oh, D.H.; Paik, J.K. An empirical formula for predicting the collapse strength of composite cylindrical-shell structures under external pressure loads. *Ocean Eng.* **2019**, *172*, 191–198. [CrossRef]
4. Chen, Y.; Tong, Z.P.; Hua, H.X.; Wang, Y.; Gou, H.Y. Experimental Investigation on the Dynamic Response of Scaled Ship Model with Rubber Sandwich Coatings Subjected to Underwater Explosion. *Int. J. Impact Eng.* **2009**, *36*, 318–328. [CrossRef]
5. Ramajeyathilagam, K.; Vendhan, C.P.; Rao, V.B. Non-linear Transient Dynamic Response of Rectangular Plates Under Shock Loading. *Int. J. Impact Eng.* **2000**, *24*, 999–1015. [CrossRef]
6. Liu, Y.L.; Zhang, A.M.; Tian, Z.L.; Wang, S.P. Numerical investigation on global responses of surface ship subjected to underwater explosion in waves. *Ocean Eng.* **2018**, *161*, 277–290. [CrossRef]
7. Liang, C.C.; Lai, W.H.; Hsu, C.Y. Study of the Nonlinear Responses of a Submersible Pressure Hull. *Int. J. Press. Vessel. Pip.* **1998**, *75*, 131–149. [CrossRef]
8. Kwon, Y.W.; Fox, P.K. Underwater Shock Response of a Cylinder Subjected to a Side-on Explosion. *Comput. Struct.* **1993**, *48*, 637–646. [CrossRef]
9. McCoy, R.W.; Sun, C.T. Fluid-Structure Interaction Analysis of a Thick Section Composite Cylinder Subjected to Underwater Blast Loading. *Compos. Struct.* **1997**, *37*, 45–55. [CrossRef]
10. Shin, Y.S.; Hooker, D.T. Damage Response of Submerged Imperfect Cylindrical Structures to Underwater Explosion. *Comput. Struct.* **1996**, *60*, 683–693. [CrossRef]
11. Qiankun, J.; Gangyi, D. A Finite Element Analysis of Ship Sections Subjected to Underwater Explosion. *Int. J. Impact Eng.* **2011**, *38*, 558–566. [CrossRef]
12. Adamczyk, R.; Cichocki, K. Analysis of the Shock Response of an Underwater Structure Subjected to a Far-Field Explosion. In Proceedings of the Abaqus Users, Milan, Italy, 4–6 June 1997; pp. 73–87.
13. Cichocki, K. Computer Analysis of Dynamic Response due to Underwater Explosion on Hybrid Structure. In Proceedings of the Abaqus Users, Newport, RI, USA, 1–3 June 1994; pp. 207–220.
14. Zhao, X.; Wang, G.; Lu, W.; Yan, P.; Chen, M.; Zhou, C. Damage features of RC slabs subjected to air and underwater contact explosions. *Ocean Eng. Eng.* **2018**, *147*, 531–545. [CrossRef]
15. Rajendran, R.; Narasimhan, K. Damage Prediction of Clamped Circular Plates Subjected to Contact Underwater Explosion. *Int. J. Impact Eng.* **2001**, *25*, 373–386. [CrossRef]
16. Jacinto, A.C.; Ambrosini, R.D.; Danesi, R.F. Experimental and Computational Analysis of Plates Under Air Blast. *Int. J. Impact Eng.* **2001**, *25*, 927–947. [CrossRef]
17. Kumar, P.; Stargel, D.S.; Shukla, A. Effect of Plate Curvature on Blast Response of Carbon Composite Panels. *Compos. Struct.* **2013**, *99*, 19–30. [CrossRef]
18. Guo, J.; Ji, X.-B.; Wen, Y.-Y.; Cui, X.-W. A new shock factor of SWATH catamaran subjected to underwater explosion. *Ocean Eng.* **2017**, *130*, 620–628. [CrossRef]
19. Wang, Z.; Liang, X.; Fallah, A.S.; Liu, G.; Louca, L.A.; Wang, L. A novel efficient method to evaluate the dynamic response of laminated plates subjected to underwater shock. *J. Sound Vib.* **2013**, *332*, 5618–5634. [CrossRef]
20. Zhang, A.M.; Zeng, L.Y.; Cheng, X.D.; Wang, S.P.; Chen, Y. The Evaluation Method of Total Damage to Ship in Underwater Explosion. *Appl. Ocean Res.* **2011**, *33*, 240–251. [CrossRef]
21. Gong, S.W.; Khoo, B.C. Transient response of stiffened composite submersible hull to underwater explosion bubble. *Compos. Struct.* **2015**, *122*, 229–238. [CrossRef]
22. Wang, H.; Zhu, X.; Cheng, Y.S.; Liu, J. Experimental and Numerical Investigation of Ship Structure Subjected to Close-in Underwater Shock Wave and Following Gas Bubble Pulse. *Mar. Struct.* **2014**, *39*, 90–117. [CrossRef]
23. Jun, G.; Yang, Y.; Zhang, Y.; Feng, L.-H. The Spherical Shock Factor Theory of a FSP with an Underwater Added Structure. *Shock Vib.* **2019**, *2019*, 7080941. [CrossRef]
24. Shin, Y.S.; Santiago, L.D. Surface Ship Shock Modeling and Simulation: Two-Dimensional Analysis. *Shock Vib.* **1998**, *5*, 129–137. [CrossRef]

25. Gannon, L. Submerged aluminum cylinder response to close-proximity underwater explosions—A comparison of experiment and simulation. *Int. J. Impact Eng.* **2019**, *133*, 103339. [CrossRef]
26. Ross, C.T.F. *Pressure Vessels External Pressure Technology*, 2nd ed.; British Library: London, UK, 2011.
27. Liang, C.C.; Chen, H.W.; Jen, C.Y. Optimum Design of Filament-wound Multilayer Sandwich Submersible Pressure Hulls. *Ocean Eng.* **2003**, *30*, 1941–1967. [CrossRef]
28. Craven, R.; Graham, D.; Dalzel-Job, J. Conceptual design of a composite pressure hull. *Ocean Eng.* **2016**, *128*, 153–162. [CrossRef]
29. Keil, A.H. The Response of Ships to Underwater Explosions. *Trans. Soc. Nav. Arch. Mar. Eng.* **1961**, *69*, 366–410.
30. Zhange, A.; Yao, X.; Li, J. The Interaction of an Underwatet Explosion Bubble and an Elastic-plastic Structure. *Appl. Ocean Res.* **2008**, *30*, 159–171. [CrossRef]
31. Shin, Y.S. Ship Shock Modeling and Simulation for Far-Field Underwater Explosion. *Comput. Struct.* **2004**, *82*, 2211–2219. [CrossRef]
32. Forghani, A.; McGregor, C.; McClennan, S.; Vaziri, R.; Ellyin, F.; Poursartip, A.; Bouamoul, A. Modelling of Damage Development in Blast Loaded Composite Panels. In Proceedings of the 16th International Conference on Composite Materials, Kyoto, Japan, 8–13 July 2007; pp. 1–8.
33. Kalavalapally, R.; Penmetsa, R.; Grandhi, R. Multidisciplinary Optimization of a Lightweight Torpedo Structure Subjected to an Underwater Explosion. *Finite Elem. Anal. Des.* **2006**, *43*, 103–111. [CrossRef]
34. Cole, R.H. *Underwater Explosions*; Princeton University Press: Princeton, NJ, USA, 1948.
35. Bjorno, L.; Levin, P. Underwater Explosion Research Using Small Amount of Chemical Explosives. *Ultrasonics* **1976**, *14*, 263–267. [CrossRef]
36. Hung, C.F.; Lin, B.J.; Hwang-Fuu, J.J.; Hsu, P.Y. Dynamic Response of Cylindrical Shell Structures Subjected to Underwater Explosion. *Ocean Eng.* **2009**, *36*, 564–577. [CrossRef]
37. Gupta, N.K.; Kumar, P.; Hegde, S. On Deformation and Tearing of Stiffened and Un-stiffened Square Plates Subjected to Underwater Explosion-a Numerical Study. *Int. J. Mech. Sci.* **2010**, *52*, 733–744. [CrossRef]
38. Helal, M.M.K.; Elsayed, F. Dynamic behavior of stiffened plates under underwater shock loading. *Mater. Test.* **2015**, *57*, 506–517. [CrossRef]
39. Reid, W.D. *The Response of Surface Ships to Underwater Explosions (No. DSTO-GD-0109). Defence Science and Technology Organization Canberra (Australia)*; DSTO Aeronautical and Maritime Research Laboratory: Melbourne, VIC, Australia, 1996.
40. Fathallah, E.; Qi, H.; Tong, L.; Helal, M. Numerical investigation of the dynamic response of optimized composite elliptical submersible pressure hull subjected to non-contact underwater explosion. *Compos. Struct.* **2015**, *121*, 121–133. [CrossRef]
41. Fathallah, E.; Qi, H.; Tong, L.; Helal, M. Numerical Simulation and Response of Stiffened Plates Subjected to Non-Contact Underwater Explosion. *Adv. Mater. Sci. Eng.* **2014**, *2014*, 1–18. [CrossRef]
42. Liang, C.C.; Tai, Y.S. Shock Responses of a Surface Ship Subjected to Noncontact Underwater Explosions. *Ocean Eng.* **2006**, *33*, 748–772. [CrossRef]
43. Yin, C.; Jin, Z.; Chen, Y.; Hua, H. Shock mitigation effects of cellular cladding on submersible hull subjected to deep underwater explosion. *Ocean Eng.* **2016**, *117*, 221–237. [CrossRef]
44. Soden, P.D.; Hinton, M.J.; Kaddour, A.S. A comparison of the predictive capabilities of current failure theories for composite laminates. *Compos. Sci. Technol.* **1998**, *58*, 1225–1254. [CrossRef]
45. Fathallah, E.; Qi, H.; Tong, L.; Helal, M. Design Optimization of Composite Elliptical Deep-Submersible Pressure Hull for Minimizing the Buoyancy Factor. *Adv. Mech. Eng.* **2014**, *6*, 987903. [CrossRef]
46. Fathallah, E.; Qi, H.; Tong, L.; Helal, M. Optimal Design Analysis of Composite Submersible Pressure Hull. *Appl. Mech. Mater.* **2014**, *578*, 89–96. [CrossRef]
47. Barbero, E.J. *Finite Element Analysis of Composite Materials Using Abaqus*; CRC Press: Boca Raton, FL, USA, 2013.
48. Fathallah, E.; Qi, H.; Tong, L.; Helal, M. Design optimization of lay-up and composite material system to achieve minimum buoyancy factor for composite elliptical submersible pressure hull. *Compos. Struct.* **2015**, *121*, 16–26. [CrossRef]
49. Jingxuan, H.; Mingfa, R.; Shiyong, S.; Qizhong, H.; Xiannian, S. Failure Prediction on Advanced Grid Stiffened Composite Cylinder Under Axial Compression. *Compos. Struct.* **2011**, *93*, 1939–1946. [CrossRef]

50. Lopez, R.H.; Miguel, L.F.F.; Belo, I.M.; Cursi, J.E.S. Advantages of Employing a Full Characterization Method over Form in the Reliability Analysis of Laminated Composite Plates. *Compos. Struct.* **2014**, *107*, 635–642. [CrossRef]

51. Shrivastava, S.; Mohite, P.M.; Yadav, T.; Malagaudanavar, A. Multi-objective multi-laminate design and optimization of a Carbon Fibre Composite wing torsion box using evolutionary algorithm. *Compos. Struct.* **2018**, *185*, 132–147. [CrossRef]

52. Tsai, S.W.; Wu, E.M. A General Theory of Strength for Anisotropic Materials. *J. Compos. Mater.* **1971**, *5*, 58–80. [CrossRef]

53. ANSYS Inc. *Ansys Theory Reference Release 14.5*; October ed.; ANSYS Inc.: Canonsburg, PA, USA, 2012.

54. Madenci, E.; Guven, I. *The Finite Element Method and Applications in Engineering Using ANSYS*; Springer: Berlin/Heidelberg, Germany, 2006.

55. Fathallah, E. Finite Element Modelling and Multi-Objective Optimization of Composite Submarine Pressure Hull Subjected to Hydrostatic Pressure. *Mater. Sci. Forum* **2019**, *953*, 53–58. [CrossRef]

56. Hibbitt, K.; Sorensen, I. *Abaqus Version 6.13 Abaqus Analysis User's Manual*; SIMULIA: Johnston, RI, USA, 2013.

57. Jen, C.Y.; Tai, Y.S. Deformation Behavior of a Stiffened Panel Subjected to Underwater Shock Loading Using the Non-linear Finite Element Method. *Mater. Des.* **2010**, *31*, 325–335. [CrossRef]

58. Hibbitt, K.; Sorensen, I. *AbaqusTutorial Version 6.13 Example Problems Manual*; SIMULIA: Johnston, RI, USA, 2013.

59. Panahi, B.; Ghavanloo, E.; Daneshmand, F. Transient Response of a Submerged Cylindrical Foam Core Sandwich Panel Subjected to Shock Loading. *Mater. Des.* **2011**, *32*, 2611–2620. [CrossRef]

60. Zhang, N.; Zong, Z.; Zhang, W. Dynamic Response of a Surface Ship Structure Subjected to an Underwater Explosion Bubble. *Mar. Struct.* **2014**, *35*, 26–44. [CrossRef]

Article

Issues on the Vibration Analysis of In-Service Laminated Glass Structures: Analytical, Experimental and Numerical Investigations on Delaminated Beams

Chiara Bedon

Department of Engineering and Architecture, University of Trieste, Piazzale Europa 1, 34127 Trieste, Italy; chiara.bedon@dia.units.it; Tel.: +39-040-558-3837

Received: 5 August 2019; Accepted: 7 September 2019; Published: 19 September 2019

Abstract: Load-bearing laminated glass (LG) elements take the form of simple members in buildings (i.e., columns, beams, and plates) or realize stand-alone assemblies, where glass and other traditional constructional materials can interact. Among several relevant aspects, the dynamic response of LG structures requires dedicated methods of analysis, towards the fulfilment of safe design purposes. A combination of multiple aspects must be taken into account for dynamic calculations of even simple LG elements when compared to static conditions, first of all the sensitivity of common interlayers to the imposed vibration frequency. The challenge is even more complex for the vibration serviceability assessment of in-service LG structures, where the degradation of materials and possible delamination effects could manifest, hence resulting in structural performances that can markedly differ from early-design conditions. Major uncertainties can be associated to the actual mechanical characterization of materials in use (especially the viscoelastic interlayers), as well as the contribution of restraints (as compared to ideal boundaries) and the possible degradation of the bonding layers (i.e., delaminations). All of these aspects are examined in the paper, with the support of extended analytical calculations, on-site experimental measurements, and parametric Finite Element (FE) numerical analyses. When compared to literature efforts accounting for ideal boundaries only, an analytical formulation is proposed to include the effects of flexible restraints in the dynamic performance of general (double) LG beams. Special care is also spent for the presence of possible delaminations, including size and position effects. In the latter case, existing formulations for composite laminates are preliminarily adapted to LG beams. Their reliability and accuracy is assessed with the support of test predictions and parametric FE simulations.

Keywords: laminated glass (LG); free vibrations; fundamental frequency; mechanical restraints; field experiments; analytical modelling; Finite Element (FE) numerical modelling

---

1. Introduction

Laminated glass (LG) is largely used in several ways, and mostly increasing is its application for load-bearing structural members for buildings and constructions. There, special care is required for the optimal and safe design of structural elements that are able to ensure resistance, robustness, stiffness, redundancy, etc., even under extreme loads [1,2].

From a structural point of view, critical design conditions can include explosive events [3–5] and impacts [6–8], natural hazards and earthquakes [9,10], and dynamic loads in general (including moving loads, in the case of pedestrian systems [11–16]), whose effects need dedicated calculation methods.

While the research community is spending efforts for the refinement or development of analytical/graphical design procedures that are able to capture the limit states of interest for design, the actual dynamic behaviour of even simple laminated glass elements is often described by means of approximate methods/working assumptions [17]. Such a strategy can be beneficial at the preliminary design stage. However, the same approaches are often not able to properly capture the real dynamic performance of in-service LG structures. This is especially the case of LG systems that are subjected to severe operational conditions, where mechanical properties and boundary conditions might be strongly affected by installation methods, ambient, etc. (i.e., Figure 1).

(a)   (b)

Figure 1. Example of glass structures under severe operational conditions, due to (**a**) temperature variations or (**b**) overcrowding.

In this context, the paper aims at investigating the dynamic behaviour of the in-service LG members in free vibrations. Following earlier studies of literature, special care is focused on the sensitivity analysis of their fundamental frequency to a multitude of aspects of technical interest, including the type of interlayer (and its frequency-dependent shear stiffness), the aspect ratio of LG beams, the presence of flexible restraints that can provide only partial translational/rotational constraints (with respect to ideal boundaries), or the effects of possible delaminations in the bonding interlayers.

To this aim, classical analytical methods are first recalled in Section 2, so as to introduce the literature concept of "adjusted dynamic" effective thickness for LG beams [17]. Hence, preliminary analytical calculations are carried out on a wide set of LG beams, giving evidence of the expected sensitivity of frequency estimates to some variations in the LG beam composition and/or geometrical configuration.

As shown, major uncertainties in frequency estimates can derive from the actual stiffness of restraints, as well as from the dynamic response of the interlayers in use, to the imposed vibration frequency. Even further sensitivity is related—for in-service structures—to the possible presence of damage, like delaminations or material degradations, which should be properly taken into account for safe vibration assessments, with respect to early-stage design calculations. Such an aspect is further explored with the support of on-site vibration experiments (Section 3) that were carried out on selected LG beams belonging to an existing structure. The sensitivity of test predictions to major influencing parameters is discussed in Section 4, with the support of refined Finite Element (FE) numerical models (ABAQUS [18]) and parametric analytical calculations. Based on experimental observations and refined FE numerical models (ABAQUS [18]), practical analytical expressions are proposed in the paper, so that the expected vibration frequencies of a general (double) LG beam with flexible restraints could be rationally calculated. Dedicated analytical methods are also adapted to LG members and proposed to include possible delaminations in the bonding interlayers, thus resulting in more accurate dynamic estimations for in-service glass structures, based on past literature studies on composite laminates.

2. Classical Analytical Formulation for Frequency Calculations

2.1. Reference System

The attention of the current study is focused on the vibration performance of double LG beams agreeing with Figure 2. For simplicity, the reference cross-section is symmetrical, inclusive of two glass layers (with thickness $h_1 = h_3$) and a middle viscoelastic foil (h_2), providing a certain shear coupling.

The resisting LG member has total width b, with L the span, and $d \geq 0$ representing the distance (if any) between each restraints (from the middle axis) and the beam end section. Accordingly, $L_0 >> 2d$ is the actual bending span. The thickness h_i of each layer is relatively small when compared to the global dimensions $b \times L$.

In addition, $E_1 = E_3 = 70$ GPa is the nominal modulus of elasticity (MoE) of glass, with $\rho_1 = \rho_3 = 2500$ kg/m^3 the density and $v_1 = v_3 = 0.23$ the Poisson' ratio [19]. Disregarding the interlayer type and composition, the bonding foil has generally a relatively low density ρ_2 as compared to glass ($\rho_2 \approx 1000$ kg/m^3), while its stiffness can strongly modify with operational conditions, see Section 2.2.

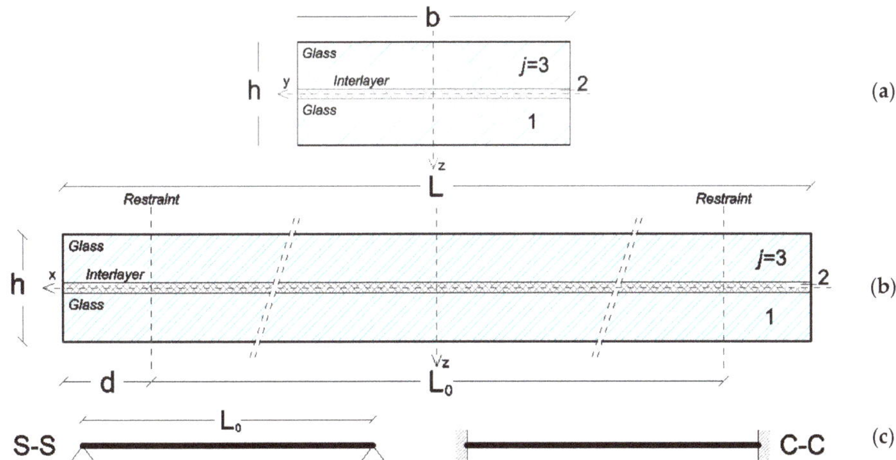

Figure 2. Double LG beam in free vibrations: (**a**) transversal and (**b**) longitudinal cross-sections, with (**c**) selected ideal restraints.

2.2. Existing Closed-Form Solutions

Assuming the ideal restraint configurations that were schematized in Figure 2c for simply supported (S-S) or clamped (C-C) members, classical theories for slender beams can be taken into account for frequency analyses of LG elements, as far as the dynamic mechanical properties of the interlayer in use are properly described.

According to Figure 2, a given LG beam in free vibrations must, in fact, satisfy the well-known Euler–Bernoulli differential equation of motion, that for a given a monolithic ($A = b \times h) \times L_0$ member is given by [20]:

$$\frac{\partial^2}{\partial x^2} EI(x)\left(\frac{\partial^2 v(x,t)}{\partial x^2}\right) + \rho A \frac{\partial^2 v(x,t)}{\partial t^2} = 0 \tag{1}$$

Moreover, in Equation (1), $v(x,t)$ is the vertical displacement at the abscissa $0 \leq x \leq L_0$ and time instant t; E and ρ the MoE and density of the material in use; and, I the second moment of area. Thus, the fundamental frequency is conventionally given by the compact expression [20]:

$$f_n = \frac{\omega_n}{2\pi} = \frac{1}{2\pi}\sqrt{\frac{\beta_n^4 E}{12\overline{m}}h^3} \tag{2}$$

with n the mode order and β the wavenumber (Table 1).

Table 1. Reference wavenumbers β_n and shape coefficients Ψ for beams with simple restraints (Figure 2c) and bending span L_0.

Beam Restraints	β_n			Ψ		
	Mode Order n			Mode Order n		
	1	2	3	1	2	3
Simply supports (S-S)	π/L_0	$2\pi/L_0$	$3\pi/L_0$	π/L_0^2	$(2\pi)^2/L_0$	$(3\pi)^2/L_0$
Clamps (C-C)	$4.73/L_0$	$7.8532/L_0$	$10.996/L_0$	$40.7/L_0^2$	$82.6/L_0^2$	$148/L_0^2$

According to several studies of literature, the challenge for a given LG member lies in the estimation of the actual composite stiffness, being strongly related to the shear coupling effect of the bonding interlayer. Besides the availability of simplified analytical approaches that are based on the use of an equivalent, monolithic glass thickness $h_{ef} = h$ for sandwich sections, according to Figure 2, it was shown in [17] that the "adjusted dynamic" effective thickness (adapted from [21] for modal analysis purposes) is able to offer reliable frequency estimates for double LG beams with ideal boundaries. In particular, such an "adjusted" thickness is given by [17]:

$$h_{ef} = \sqrt[3]{\cfrac{1}{\cfrac{\eta}{h_1^3+h_3^3+12I_s} + \cfrac{1-\eta}{h_1^3+h_3^3}}} \tag{3}$$

where η represents the shear coupling of the composite section:

$$0 \leq \eta = \cfrac{1}{1 + \frac{E_1 h_2}{G_2 b}\cdot\frac{I_1+I_3}{I_{tot}}\cdot\frac{A_1 A_3}{A_1+A_3\cdot\Psi}} \leq 1 \tag{4}$$

and the other relevant terms are given by:

$$I_i = \frac{bh_i^3}{12} \quad A_i = bh_i \tag{5}$$

$$I_s = \frac{h_1 h_3}{h_1 + h_3}\cdot[h_2 + 0.5(h_1 + h_3)]^2 \tag{6}$$

$$I_{tot} = I_1 + I_3 + \frac{A_1 A_3}{A_1 + A_3}\cdot[h_2 + 0.5(h_1 + h_3)]^2 \tag{7}$$

The coefficient Ψ in Equation (4) depends on the normalized shape of deflections, for a given homogeneous beam. For basic boundary conditions and several mode orders n, Ψ can be calculated from Table 1.

Finally, η in Equation (3) is strictly affected by the shear modulus $G_2 = G_2(\omega)$ of the bonding layer. Given that the common materials for LG applications have a viscoelastic behaviour that depends on the material composition and its vibration frequency and/or ambient conditions, this turns out in an effective thickness $h_{ef} = h_{ef}(\omega)$, which explicitly reflects the dynamic response of the interlayer itself, as a part of a composite system it belongs. However, according to Equation (2), it is also $\omega = \omega(h_{ef})$, and hence an iterative calculation approach is required for accurate thickness/frequency estimates.

Both the real, frequency-independent term (storage modulus $G_{2,0}$) and the imaginary part (loss modulus, $G_{2,\omega}$), are in fact involved in the frequency domain, where:

$$G_2(\omega) = G_{2,0} + G_{2,\omega}(\omega) \tag{8}$$

and their typically high sensitivity to frequency is shown in Figure 3 (selected examples reproduced from [7,17,22]).

In this regard, literature projects have been dedicated to the mechanical characterization of interlayers in use for LG systems, under the assumption of various severe conditions of temperature or time loading (i.e., [23–30]). As far as the interlayer composition and the test method both modify; however, different mechanical properties can be derived for a given interlayer material [31]. The study reported in [32] also emphasized that the mechanical properties of the interlayer samples (i.e., material test), or interlayer foils belonging to small portions of LG sections (i.e., section test), can result in markedly different stiffness estimates, due to variation of the actual boundary conditions. Finally, for in-service glass structures, it is generally recognized that the degradation of interlayers can affect several material features, including the shear stiffness, but also the adhesion properties, and other thermo-mechanical parameters that could indirectly affect the overall structural performance of a given LG section (see [33–35]).

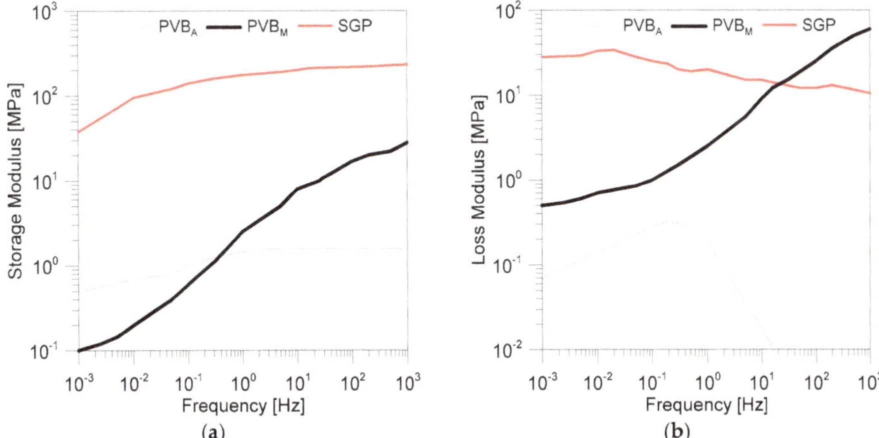

Figure 3. Examples of (**a**) storage and (**b**) loss moduli variation with frequency, for Polyvinyl Butyral (PVB) or SentryGlas Plus (SGP) interlayers at 25 °C, according to [7,17,22].

In other words, Equation (3) represents a practical tool for design, being able to simplify the original dynamic problem of composite beams with flexible, viscoelastic connection and ideal boundaries. On the other side, the reference $G_2 = G_2(\omega)$ value for dynamic estimates should be properly assessed, including possible delaminations (Section 4).

In Figure 4, frequency calculations are proposed for selected LG beams in the S-S or C-C conditions, as a function of G_2. The collected frequency values are derived from Equation (2), while assuming that h_{ef} (Equation (3)) modifies with G_2, and 10^{-4} MPa $< G_2 < 10^5$ MPa. In the figures, both the limit *"layered"* and *"monolithic"* conditions can be easily detected. In addition, the grey regions represent all the possible frequency values that could characterize the dynamic performance of a given LG beam geometry, as far as its end restraints are characterized by a certain translational/rotational stiffness that can be comprised within the limit conditions of ideal simple supports (S-S) or clamps (C-C).

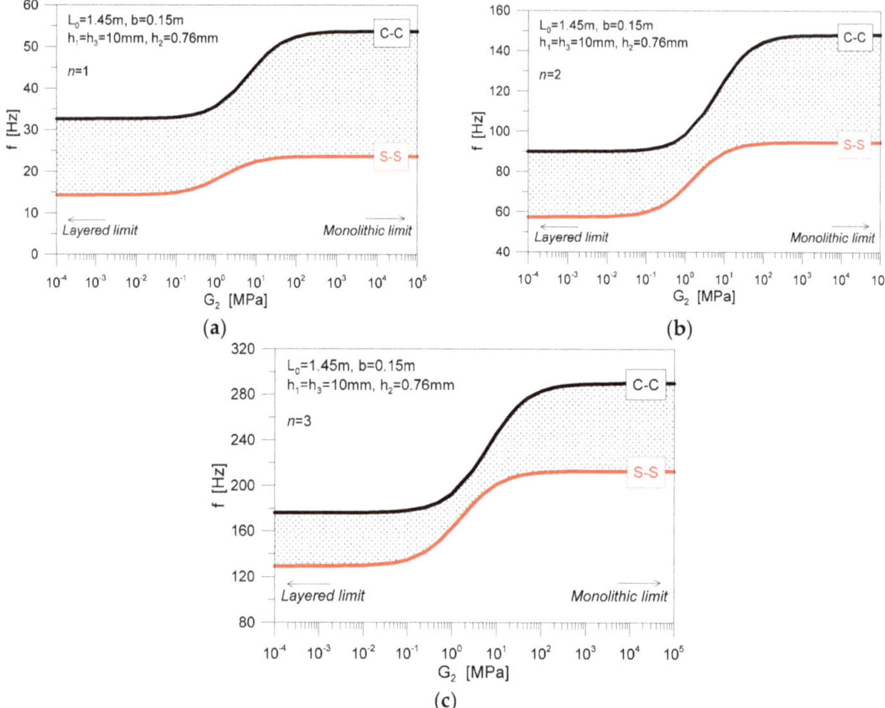

Figure 4. Example of analytical frequency estimations for double laminated glass (LG) beams, as a function of the restraint type (S-S or C-C) and shear stiffness of the interlayer. The grey region denotes the sensitivity of analytical frequency estimates to the restraints and interlayer stiffnesses.

As far as higher vibration modes are taken into account in Figure 4, it is possible to notice that the grey region progressively minimizes for $n = 2$ and $n = 3$. As such, major uncertainties for simplified analytical calculations can be expected, especially for $n = 1$, that in most of the cases is a key parameter for design purposes.

2.3. Restraints and Delaminations for In-Service LG Systems

A relevant influencing parameter for the vibrational analysis of LG beams is certainly represented by the effect of real restraints, with respect to the ideal supports (Figure 2c).

It was shown in [12], for example, that the restraints characterized by a certain flexibility (i.e., axial (K_s) and rotational (K_r) stiffness due to the presence of soft layers, gaskets, etc., see Figure 5) should be properly taken into account for the dynamic analysis of even monolithic glass members with cantilever or beam behaviour, affecting both frequency and damping calculations.

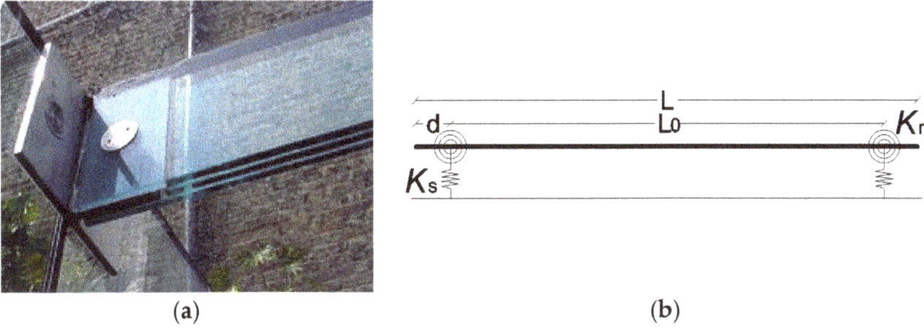

(a)   (b)

Figure 5. Example of LG beams with end mechanical restraints: (**a**) real system and (**b**) corresponding mathematical model.

The presence of delaminations and their effects on the flexural stiffness of the resisting sandwich section can represent another crucial aspect to account for dynamic analysis of LG members. Critical regions for delaminations are commonly represented by restraints and edges (see, for example, Figure 6 and [36,37]). However, research studies on the structural performance of delaminated LG sections are still limited (i.e., [38–41]). In addition, literature investigations are focused on the stress response of simple LG members under laboratory conditions, which is with artificially imposed delaminations.

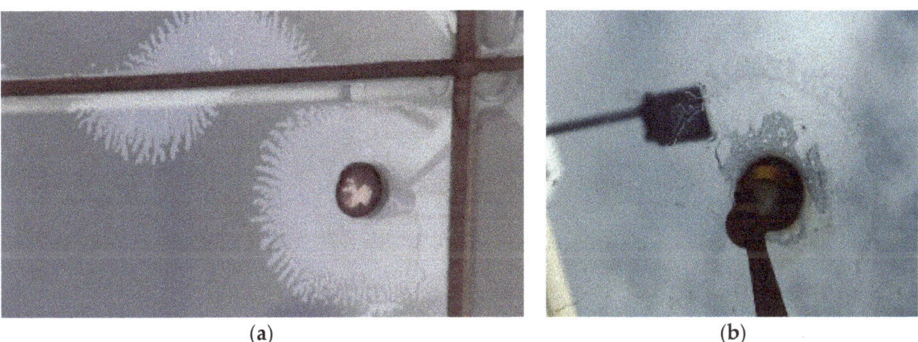

(a)   (b)

Figure 6. Examples of severe delamination in LG member, in the region of point mechanical restraints. Reproduced with permission from (**a**) [36] and (**b**) [37].

In this regard, Equation (2) is expected to provide only approximate estimates of the actual dynamic performance of in-service LG members, thus suggesting the need of more refined methods of analysis.

3. Experimental Study on In-Service LG Beams

A series of field experiments was carried out on existing LG member in order to further explore the actual dynamic behaviour of glass structures in operational conditions.

3.1. Specimens and Test Methods

The experimental study was carried out in May 2019, on a selection of 32 LG beams composed of two, $h_1 = h_3 = 10$ mm thick fully tempered glass layers and a middle PVB foil ($h_2 = 0.76$ mm its thickness). The width of LG beams was fixed ($b = 0.14$ m), while the variations were represented by the total span L. For the majority of them, L was in the range of 2.4 m and 2.7 m. The minimum span—even with identical nominal section properties and restraints—was in the order of 1.45 m. Such a marked variability in the span was required—at the design stage—to accommodate some geometrical irregularities of the primary structure hosting the specimens. The examined LG beams are, in fact, currently part of an in-service glass walkway (in the form of handrails, see Figure 7), being constructed in the early 2000 in the context of a Roman age Basilica monument in Aquileia, Italy (see also [15,16] for further details on the pedestrian system).

Figure 7. Example of service loads for the tested LG beams (photos by C. Bedon, courtesy of So.Co.Ba.).

For all of the LG members, the end restraints were realized in the form of stainless steel point-fixings according to Figure 8a, with holes having 42 mm nominal diameter and positioned at a distance $d = b/2 = 70$ mm from the edges (see Figure 8b).

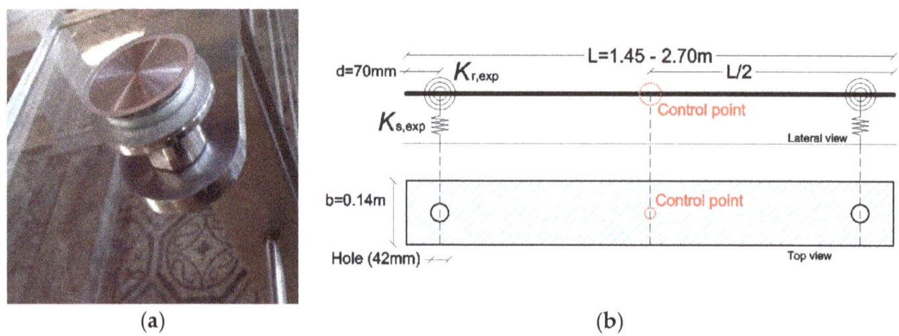

Figure 8. Experimental LG specimens: (**a**) detail of the typical restraint (photo by C. Bedon, courtesy of So.Co.Ba.) and (**b**) schematic representation of the test setup test setup.

At the time of the on-site experiments, the selected LG beams were subjected to induced vibrations and their acceleration in time was monitored via a single tri-axial sensor [42], glued on the top surface of the mid-span section (Figure 8b). All of the tests were carried out with a mean temperature of 23 °C and a 62% relative humidity.

For each specimen, multiple measurements were collected (minimum three test repetitions), and then post-processed to predict the corresponding (mean) frequency.

In this regard, the advantage and potential of Operational Modal Analysis techniques is represented by the possibility to derive even relevant mechanical parameters for in-service structures that cannot be subjected to destructive (or laboratory) experiments. Otherwise, for the LG structure object of analysis, the serviceability of the religious monument did not allow for performing massive experimental measurements, and required the use of a minimum number of instruments, thus involving some further uncertainties for the interpretation of test results.

A key influencing parameter, for example, was represented by a certain degradation of the PVB interlayers (due to repeated non-controlled ambient conditions and time), with visible delaminated regions, close to the restraints and along the edges of each beam (detail views are proposed in Figure 9 for some of the tested specimens).

Another major issue in the experimental study and vibration serviceability assessment consisted in the actual life-time of the selected LG specimens, thus in additional difficulties for the reliable estimate of PVB mechanical properties. Most of them were characterized by a mean service life of ≈15 years at the time of the research study. However, some of them have been replaced during the years, without any track of maintenance/replacement interventions.

Finally, a further uncertainty was represented by the actual stiffness contribution of the steel point-fixings in use (Figure 8), thus its effects on the overall dynamic response of the LG members.

Figure 9. Example of delaminations (i.e., bubbles and shadows) for a selection of tested LG beams (photos by C. Bedon, courtesy of So.Co.Ba.).

3.2. Derivation of Experimental Fundamental Frequencies

The analysis of the experimental results was based on post-processing of the collected acceleration-time data, see Figure 10. Given the availability of a single control point only for each test specimen, special care was spent for the fundamental frequency of LG beams, disregarding higher experimental modes, or vibration shapes and damping related issues.

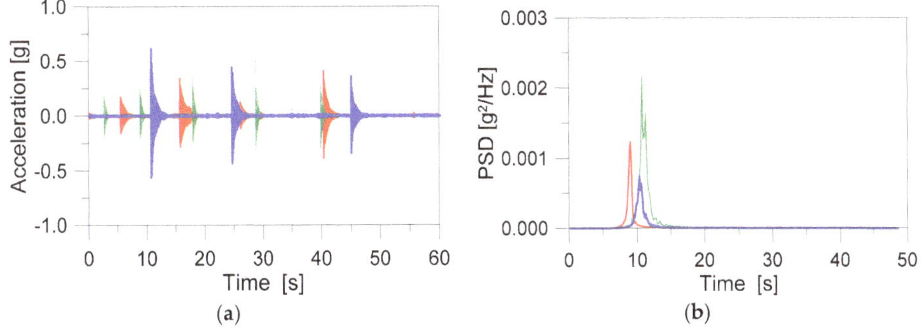

Figure 10. Examples of test records: (**a**) time-acceleration data and (**b**) Power Spectral Density (PSD) function for selected specimens.

All of the LG specimens proved to offer a beam-like behaviour, but strictly related to the effect of mechanical supports and PVB layers. In Figure 11, for example, the experimental frequencies are proposed for the tested beams, as a function of L_0. Analytical estimates from Equation (2) are also collected, as obtained for LG members having nominal experimental dimensions, but classical boundaries (S-S or C-C). Disregarding the uncertain PVB shear stiffness of the test specimens, moreover, three configurations are analytically taken into account in Equation (2) and Figure 11, that is the limit *"layered"* and *"monolithic"* theoretical configurations, and the PVB_A properties of Figure 3.

It is worth of interest that most of the test predictions are comprised within the lower limit of S-S beams with "rigid" shear connection (*"monolithic"* curve of Figure 11a) and the upper limit of C-C beams with "weak" mechanical bonding between the glass layers (*"layered"* curve of Figure 11b).

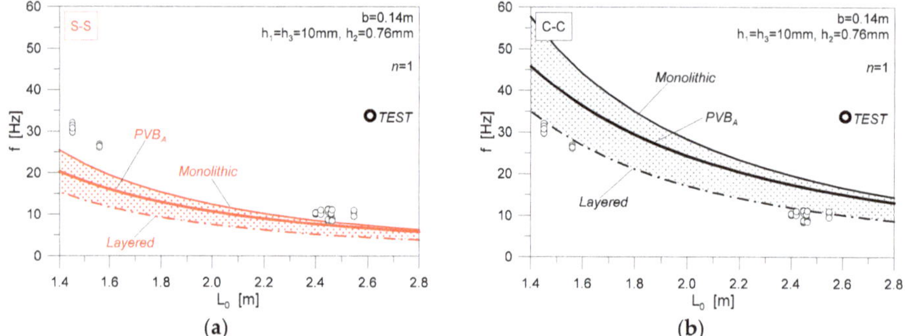

Figure 11. Experimental and analytical frequency estimates for the examined LG specimens, grouped by L_0, with (S-S) or (**b**) clamped (C-C) configurations.

4. Analysis of Relevant Influencing Parameters

4.1. Stiffness Contribution of Point-Fixings

The actual stiffness contribution of the joints in use was first assessed, with the support of FE numerical models (ABAQUS). In Figure 12, the reference numerical model is shown, being representative of the nominal geometry for a connection detail. A small portion of glass (0.14×0.14 m) was also taken into account to reproduce the actual joint region and interactions. In doing so, a set of three-dimensional (3D) solid brick elements was used to describe the steel restraint (C3D8R type from ABAQUS element library). The mesh size and pattern was chosen to ensure a refined description of the examined system, with 3300 solid elements for the steel joint and the portion of LG plate.

The mechanical interaction between the steel connector and the LG portion was accounted in the form of a *penalty* & *normal behaviour* surface-to-surface contact algorithm (ABAQUS library), so as to allow for possible relative sliding between the steel and glass components in the region of glass hole, but also the possible separation under tensile loads. Rigid nodal restraints were applied at the bottom face of the steel connector. At the same time, the LG plate was restrained in its thickness and width (end section), so as to avoid possible rotations, being a part of a full beam in bending about the minor axis.

A preliminary static nonlinear analysis was carried out to estimate the expected stiffness contributions under the assumption that the so assembled FE model can be representative of the actual end region for one of the tested specimens. Based on Figure 12, the bending performance of the small scale FE model was, in fact, explored by imposing a linearly increasing bending moment M_z at the top face of the steel point-fixing (via a reference RP node, and hence distributed on the full surface of steel with a *"coupling"* constraint).

Hence, the longitudinal (i.e., x direction) and vertical (i.e., y direction) displacements of four selected control points were monitored, so that the corresponding elastic stiffnesses (both rotational and translational/axial) could be properly calculated. Two of these control points, see Figure 12, were set at the edges of the hole, to monitor the bending rotation of the portion of LG plate, with respect to the steel connector.

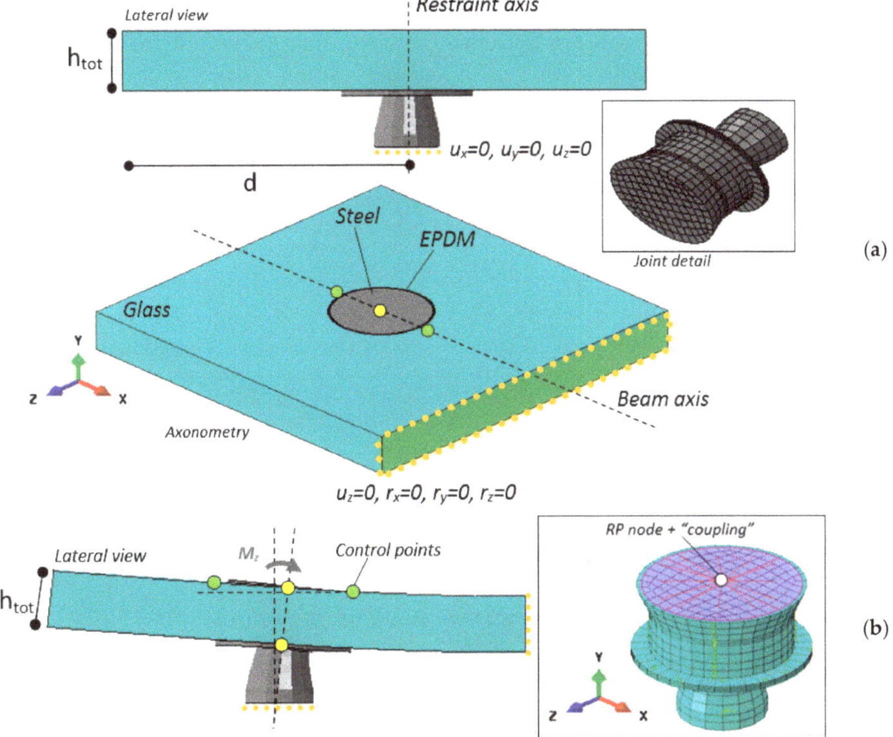

Figure 12. Local numerical model of the reference joint (ABAQUS). (**a**) Assembly details (with hidden mesh pattern) and (**b**) bending deformation.

Globally, the translational stiffness K_s was found to be relatively high, hence justifying the assumption of an ideal rigid joint for comparative purposes (i.e., $K_s = \infty$). Given the reference system of Figure 12, otherwise the rotational stiffness was estimated in $K_r \approx 150$ kN m/rad.

The so-calculated stiffness was first used form some preliminary FE calculations, carried out in ABAQUS with the support of Equation (3), to explore the sensitivity to K_r of the experimentally predicted frequencies. In Figure 13a, comparative estimates are shown for selected LG specimens. Frequency estimates are also proposed for the limit configurations of C-C and S-S beams. In general, it is possible to notice that "ideal" analytical calculations tend to result in marked percentage scatter, with respect to on-site experiments (Figure 13b).

For the S-S calculations, the FE predictions resulted in a mean $\Delta = -29\%$, as compared to the test data (with $\Delta_{min} = -14\%$, $\Delta_{max} = -40\%$, Dev.St = ±7). Given that the S-S boundary fully disregards the rotational role of the point-fixings in use, such an assumption is generally expected to severely underestimate the actual bending stiffness of a given composite system, thus the corresponding frequency. On the other side, see Figure 13b, the C-C assumption gave evidence of a marked overestimation of test results, with a mean $\Delta = +59\%$ ($\Delta_{min} = +35\%$, $\Delta_{max} = +92\%$, Dev.St = ±11).

For the calculations that were carried out with partially rigid restraints, finally the mean percentage scatter was found in the order of $\Delta = +50\%$ (with $\Delta_{min} = +27\%$, $\Delta_{max} = +82\%$, Dev.St = ±10), thus suggesting an improved agreement with, but still recommending more detailed analyses to assess, the actual dynamic performance of the tested LG specimens. The poor correlation of Figure 13 could be, in fact, justified by the actual shear contribution of the PVB layers, which, besides the availability of shear/loss moduli of literature (i.e., Figure 3), can be strongly affected by severe operational conditions, including high temperature and humidity variations, and medium/long-term degradation phenomena. The presence of visible delaminations that were visually detected for most of the tested LG beams (i.e., Figure 9) is another key parameter that affects the composite stiffness of the specimens, but it can be hardly quantified with detail.

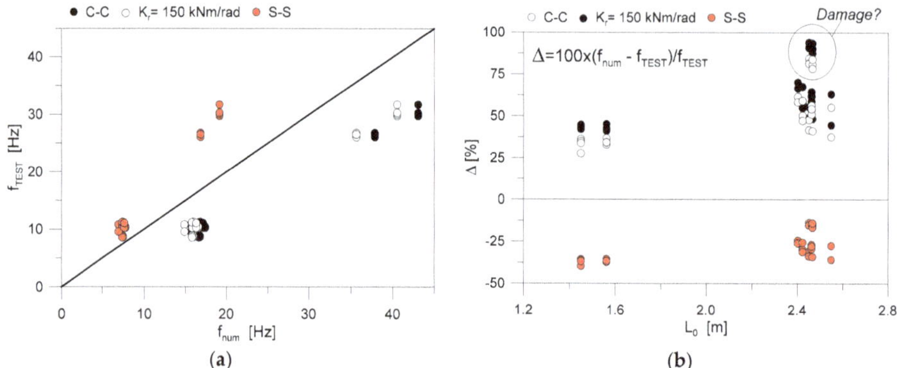

Figure 13. Numerical (with PVB$_A$ interlayer properties) and experimental frequencies for the tested LG beams: (**a**) comparative frequency values and (**b**) corresponding percentage scatter Δ, as a function of L_0.

4.2. Derivation of Practical Fitting Curves for LG Members with Flexible Restraints

According to literature, the definition of closed-form formulations accounting for the actual mechanical restraint in viscoelastic LG beams in free vibrations can involve complex mathematical problems (i.e., [43]). Otherwise, simple analytical fitting curves can offer robust support for reliable frequency estimations.

Assuming that a given glass member is restrained via mechanical supports having partial rotational stiffness $K_r \neq 0$ (in kNm/rad) and $K_s = \infty$, the restraints act as flexible clamps that must be properly taken into account for design. Moreover, following Section 4.1, it is convenient to express these rotational and axial stiffness contributions as:

$$R_r = \frac{K_r L_0}{EI} \tag{9}$$

and

$$R_s = \frac{K_s L_0^3}{EI} \tag{10}$$

where $R_r = 0$ corresponds to the limit condition of a S-S beam and $R_r = \infty$ denotes the C-C configuration. As far as a monolithic or LG beam is taken into account in Equations (9) and (10), its bending stiffness EI can be expressed as a function of the total $b \times h$ or equivalent $b \times h_{ef}$ section (Equation (3)).

Thus, from a practical point of view, the vibration frequency of a given glass beam with non-ideal restraints can be conveniently calculated as (with $R_r \geq 0$ and $R_s = \infty$):

$$\bar{f}_n = k_f f_n \tag{11}$$

where f_n is given by Equation (2) for a S-S beam and k_f represents a magnification factor, depending on R_r.

In this paper, a series of parametric FE numerical simulations was carried out in ABAQUS for glass beams with different geometrical features (b = 0.1–0.5 m, h = 0.005–0.04 m, $L = L_0$ = 1–4 m) and variable rotational stiffness K_r (with $K_s = \infty$). In doing so, beam (B31) type elements were used, with equivalent springs being able to reproduce the desired rotational stiffness K_r (Figure 14).

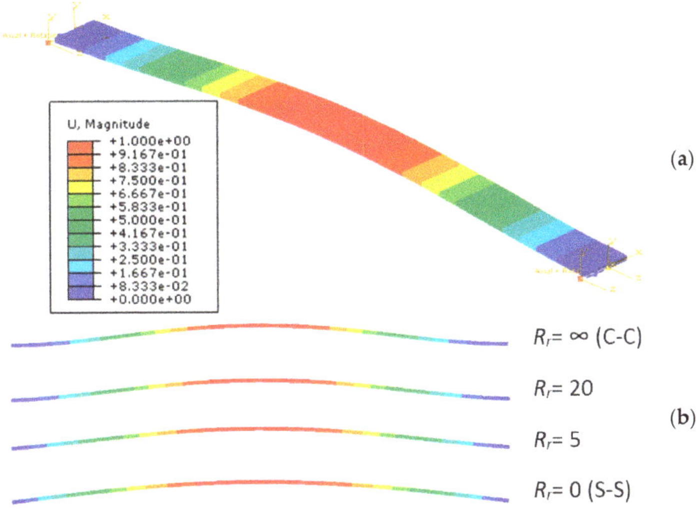

Figure 14. Parametric numerical analysis of monolithic glass beams (ABAQUS): (**a**) axonometric view of the expected vibration shape ($n = 1$) and (**b**) qualitative shape variation, as a function of R_r.

Analytical fitting curves are thus proposed in this paper, in support of practical calculations (i.e., based on Equation (11)) on general LG beams with flexible restraints, see Figure 15 and Table 2.

As shown in Figure 15, in particular, it was observed that the use of flexible restraints and their combination with different geometrical/mechanical LG properties can result in even marked variations of the expected vibration frequencies. This is especially the case of the fundamental mode ($n = 1$), where relatively stiff restraints can amplify up to ≈ 2.25 times the S-S estimations. Certainly, the vibration shape is also expected to modify with the stiffness variation of restraints (i.e., Figure 14b). However, the current investigation was specifically focused on frequency estimates, and the accurate analysis of shape sensitivity would require more detailed experimental methods, as compared to the available test predictions.

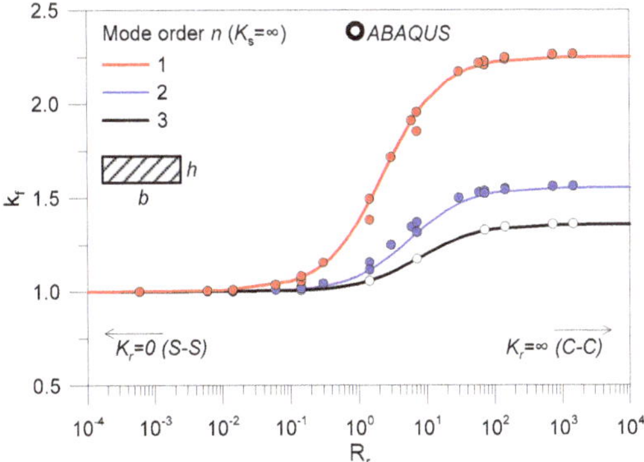

Figure 15. Analytical fitting curves (see also Table 2) for the frequency estimation of glass beams with flexible restraints ($K_r > 0$ and $K_s = \infty$).

Table 2. Key input parameters for the magnification factor k_f (Equation (11)), given that $K_r > 0$, $K_s = \infty$, and R_r in Equation (9)).

	Mode Order n		
	1	2	3
k_f		$\frac{R_r}{A+BR_r} + C$	
A	5.4	9.9	21.5
B	0.8	1.8	2.8
C	1.0	1.0	1.0

Given that h_{ef} in Equation (3) modifies with $G_2(\omega)$, another relevant outcome of Figure 15 is that the role of real restraints can be properly taken into account in the iterative analytical procedure recalled in Section 2. Marked variations of h_{ef} (and thus frequency estimates), as in Figure 16, can in fact be obtained within the "*layered*" and "*monolithic*" limit conditions, as far as R_t modifies.

Figure 16a, in particular, shows the evolution of h_{ef} (Equation (3) for selected LG members bonded with PVB_M foils. The fundamental mode ($n = 1$) is the most sensitive to restraint parameters. As far as stiff interlayers are used for a given LG beam, a mostly "*monolithic*" equivalent section h_{ef} can be also expected, with minimum sensitivity to restraints, as in Figure 16b. On the other side, the thickness and frequency sensitivity to the restraints in use progressively increases in presence of weak interlayers (as it is in presence of severe operational conditions). Moreover, from Figure 16, it is also possible to perceive that frequency estimates based on the simplified assumption of "*layered*" or "*monolithic*" thicknesses with ideal supports ($R_R = 0$ or $R_R = \infty$) can result in mostly rough (and even unconservative) calculations, with respect to the real performance of a given LG system.

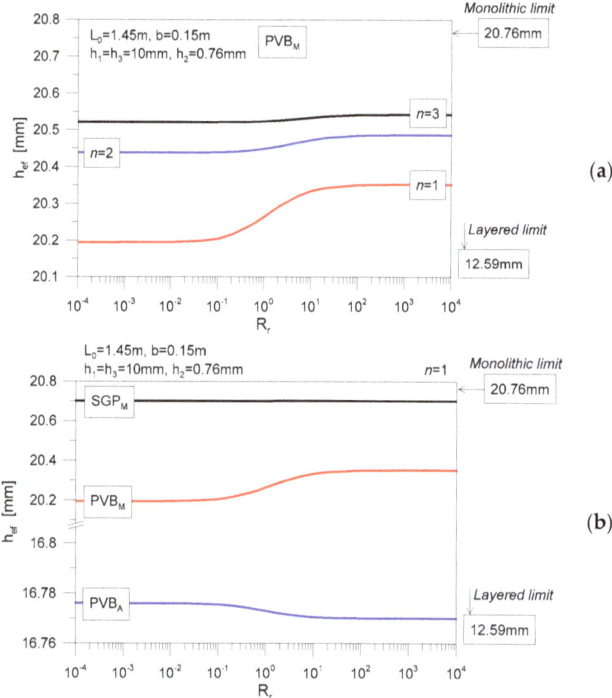

Figure 16. Variation of the dynamic thickness h_{ef} (Equation (3)) of LG beams, as a function of (**a**) the order of the vibration mode n or (**b**) by changing the interlayer properties.

4.3. Effect of Delaminations

The vibration analysis of composite beams with delaminations is another complex issue, which can be generally solved with the use of advanced FE models (i.e., [44–47]), or with the support of coupled numerical-experimental studies on artificially damaged specimens (see, for example [48–50]). In both the cases, besides different composite laminates are taken into account, a common outcome of existing literature projects is represented by the high sensitivity of vibration frequencies to delaminations. Otherwise, dedicated studies for delaminated LG structures are still missing.

4.3.1. Analytical Description of the Problem

Given a composite beam with delaminations, the mathematical problem can be solved in accordance with [51], i.e., in the form of an effective longitudinal modulus (E_{ef}) representative of the actual mechanical interaction between the constituent layers. For a laminated section. according to Figure 17a, such a modulus is given by:

$$E_{ef} = \frac{8}{h^3} \sum_{j=1}^{m/2} (E_x)_j \left(z_j^3 - z_{j-1}^3 \right) \tag{12}$$

with E_x the MoE of the j-th layer, m is the total number of layers composing the beam section (with $b \times h$ total dimensions); z_j is the distance between the outer face of the j-th layer and the neutral plane of the section. When m is an odd number—as in the case of the examined double LG sections, see Figures 2 and 17b—Equation (12) is still valid, given that the external j-th layer is subdivided into two symmetric parts.

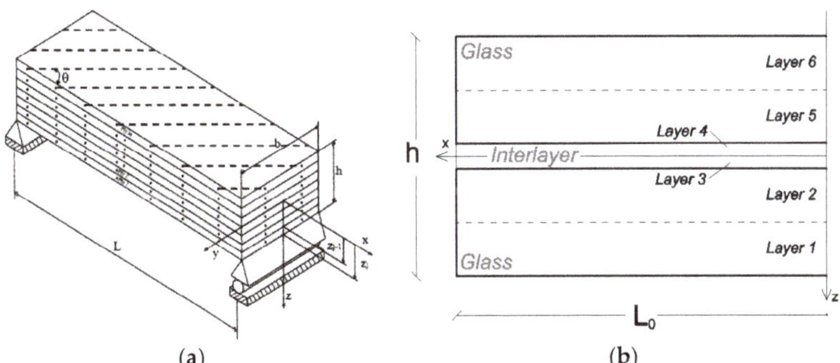

Figure 17. LG beams with delaminations: (**a**) reference analytical model for simply supported, delaminated composite beams (reproduced from [51]) and (**b**) detail example for a double LG section.

The effect of delaminations is, in fact, accounted for by reducing E_{ef}, as a function of the number and size of delaminated regions. Based on [51], the resulting longitudinal MoE of a partially delaminated composite section is, in fact, given by:

$$E_d = \left(E_{zd} - E_{ef} \right) \frac{A_d}{A_t} + E_{ef} \tag{13}$$

with:

$$E_{zd} = \frac{\sum_{j=1}^{s} E_{ef} z_j}{z} \tag{14}$$

Moreover:

E_{zd}—is the longitudinal MoE of a totally delaminated section (along one or more interfaces), representative of the so-called imperfect effective MoE;

S—is the number of sub-layers detected by the delamination;

z_j—represents the thickness of the j-th sub-layers; and,

A_d, A_t—are respectively the delaminated and total interfacial area between the bonded layers.

Key assumptions of Equations (12)–(14) are that:

(1)   Plane sections are initially normal to the longitudinal axis of the glass beam, and remain plane and normal also during flexure;

(2)   The beam has symmetrical properties about the neutral axis (both geometrical and mechanical);

(3)   The sandwich beam section is composed of layers with a linear elastic behaviour; and,

(4)   Shear coupling between each ply can be disregarded.

Based on point (d), it is expected that the analytical method herein recalled could overestimate the bending stiffness decrease of the delaminated LG beams, thus resulting in conservative frequency predictions. Another issue can be related to the location of delaminated regions (disregarded by the surface parameter A_d).

For general LG members, which were included the tested specimens, delamination phenomena typically occur along the edges and close to restraints (Figure 9), where the LG layers are not protected and/or properly sealed. In some other cases, the spotted delaminated regions can be recognized over the surface of a given LG member. However, in both the cases, natural delaminated regions (and their quantitative effects for structural calculations) can be hard to visually detect and properly quantify.

4.3.2. Reliability of Frequency Calculations for Delaminated LG Specimens

The application of Equations (12)–(14) to the examined LG beams, where even marked frequency reductions can be expected, as compared to the theoretical values of LG beams with uniform PVB bonding.

Generally, the MoE of partially delaminated beams (E_d) is linearly dependent on the delaminated area, when compared to the total bonding surface (see Equation (13)). As far as the effective MoE is taken into account for the examined LG beams (i.e., Figure 17b), it is thus possible to expect a MoE decrease in the order of 30%, for even a limited portion of delaminated interlayer. Such a MoE variation turns out in a marked variation of the bending stiffness for the composite LG beam, and at the same time reflects on a different shear stiffness from the interlayer in use (i.e., Figure 3), thus representing a relevant influencing parameter for calculations.

A parametric study was hence carried out on selected configurations in order to assess the reliability of Equations (12)–(14) for simple analytical calculations on delaminated LG beams. In accordance with Figure 18, more in detail, schematic delaminations were defined, to reproduce—even in a simplified way—some of the on-site qualitative observations from the experimental tests.

In doing so, symmetry was taken into account for all of the possible configurations, thus resulting in:

-    Scheme D1: $A_{d,tot} = 2A_{d,1}$ delaminated surface close to each restraint, where $A_{d,1} = b \times d$;
-    Scheme D2: similar to D1, but with $A_{d,tot} = 2A_{d,2}$ and $A_{d,2} = b \times 2d$ close to each restraint;
-    Scheme D3: like D2, with $A_{d,tot} = 2A_{d,2} + 2A_{d,3}$ and $A_{d,3} = s \times b$ ($s = 30$, 60 and 90 mm); and,
-    Scheme D4: inclusive of delamination along the longitudinal edges, thus $A_{d,tot} = 2A_{d,2} + 2A_{d,4}$, with $A_{d,4} = t \times L_0$ ($t = 15$, 30, 45 mm, that is $\approx b/10$, $\approx b/5$ and $\approx b/3$ for the selected specimens).

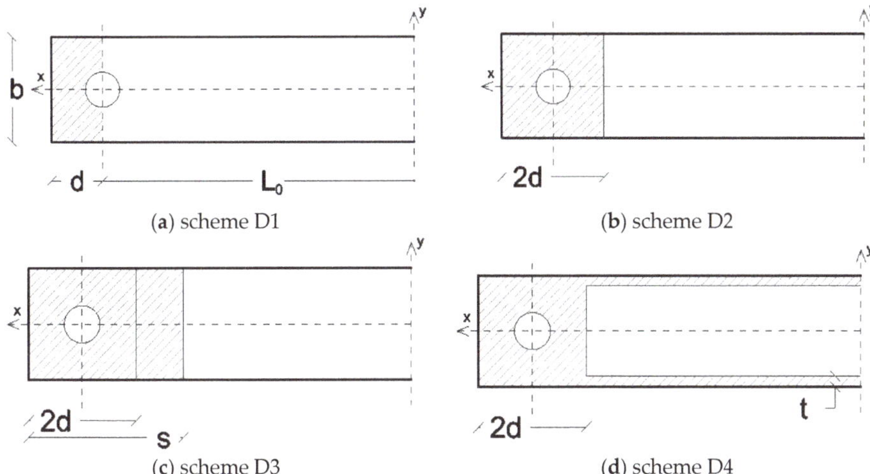

Figure 18. Selected delamination schemes for the parametric study on LG beams. Dashed area are representative of delaminations.

The analytical estimates of fundamental frequencies were carried out with the dynamic thickness h_{ef} and iterative approach of Section 2, with the additional set of iterations due to a decreasing MoE (with increasing the delamination surface A_d). The same iterative approach includes the effect of mechanical restraints, in accordance with Equation (11).

The support of full 3D solid models from ABAQUS was also taken into account. In the latter case, the choice of full 3D models was suggested by the need of including possible delaminations in several regions of interest (see Figure 19). The glass and interlayer foils were rigidly connected via *"tie"* constraints, in order to ensure a rigid mechanical connection for the involved surfaces. In the presence of delaminated regions, moreover, additional contact interactions were used between the involved layers. In Figure 19a, an example is proposed for the D4 scheme. The mechanical restraints were then reproduced in the form of equivalent axial and rotational springs (with $K_r > 0$ and $K_s = \infty$). These springs were connected to reference RP nodes, and then restrained to each LG beam with the use of additional *"coupling"* constraints, thus reproducing the actual effect of the steel point-fixings in use (see Figure 19b)). Frequency analyses were thus carried out on a wide set of configurations of LG beams. In Figure 19c, an example of typical deformed shape is shown ($n = 1$).

Through the FE parametric calculations, variations were made in terms of span (L) and delaminated region (A_d).

Figure 20 reports some of the collected results, for a LG beam with $L = 2.65$ m and $d = 0.07$ m, 10 mm and 0.76 mm the thicknesses of glass and PVB layers, respectively. The interlayer properties were defined as for the PVB_A material law reported in Figure 3.

From Figure 20a, in particular, it is possible to notice that delaminations close to the restraints (i.e., schemes D1 to D3 of Figure 18) can be mostly disregarded. Frequency variations with respect to the vibration frequency of the undamaged (fully bonded) LG members are, in fact, proposed as a function of A_d/A_t, as obtained from Equations (12)–(14), giving evidence of an expected frequency decrease in the order of -0.2%. This finding is also in line with literature efforts (for different typologies of composite laminates, see, for example [48–50]), where it was proven that (artificially imposed) delaminated regions close to restraints have minimum effects.

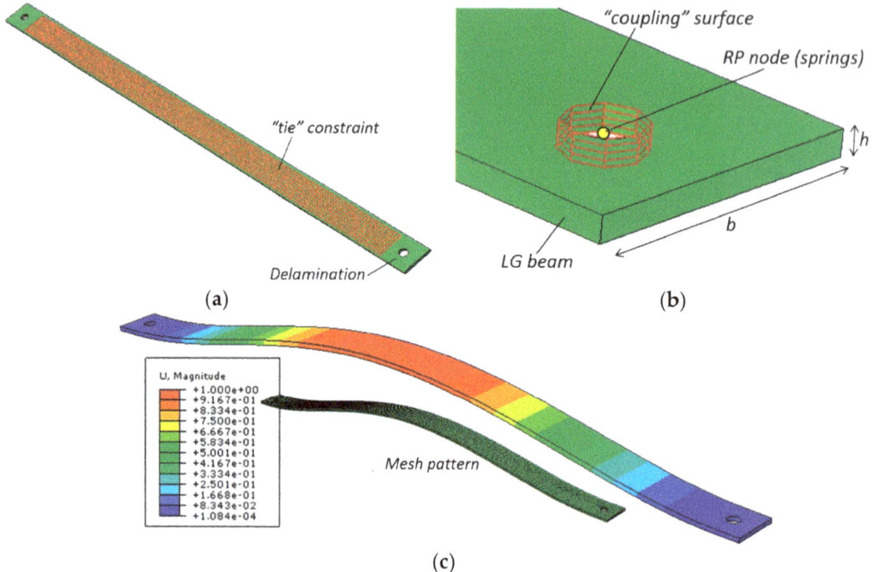

Figure 19. Numerical modelling of LG beams with delaminations: (**a**) delaminated regions, (**b**) end detail, and (**c**) typical deformed shape (ABAQUS).

Otherwise, as far as the damage location moves along the LG beam span (i.e., scheme D4), a mostly linear frequency decrease was observed from the collected parametric results, see Figure 20b. In this sense, the parametric investigation summarized herein proved that:

- The analytical method recalled from [51] and extended to the adjusted dynamic thickness for viscoelastic LG beams could be rationally used for preliminary frequency estimates, especially when refined methods of analysis or dedicated experimental investigations are not available;
- The presence of even slight delaminations along the edges of LG beams (i.e., with limited thickness, with respect to the beam width b) can have marked effects on the bending stiffness of the composite LG sections, thus on the corresponding frequency calculations; and,

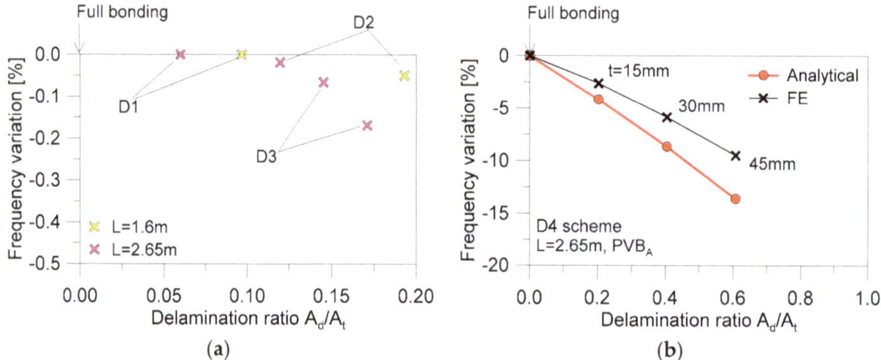

Figure 20. Frequency variation of delaminated LG beams with flexible mechanical restraints (with PVB$_A$ interlayer, $K_r = 150$ kNm/rad and $K_s = \infty$). (**a**) D1-to-D3 or (**b**) D4 scheme results, for selected LG beams.

- On the other side, the simplified assumptions of Equations (12)–(14) gave evidence of a certain scatter from the corresponding FE calculations, with respect to a given A_d/A_t ratio. Such an effect can be also observed in Figure 20b, for selected LG configurations. As a general trend, the analytical formulation for delaminated LG beams was found to clearly overestimate the FE frequency variations, thus providing even more conservative predictions.

4.3.3. Final Remarks on Practical Analytical Calculations for Design

In conclusion, some final analytical calculations were carried out, in order to assess—even in the lack of more detailed experimental methods—the actual bonding effect of the PVB foils in use for the tested LG specimens, with respect to practical literature recommendations. Given that the fundamental frequency of the selected LG beams is strictly related to several parameters, the degradation of shear stiffness for the PVB foils in use—as a direct effect of long-term performances—represents another relevant parameter, as also discussed in the previous sections.

Literature contributions addressing the actual dynamic shear modulus of PVB foils (including both the storage and loss moduli), however, are available for dynamic estimates at different vibration frequencies (i.e., Figure 3), or different temperature scenarios, which can be hardly adapted to specific case-studies. Long-term effects of design loads (up to 50 years) are mostly referred to the relaxation of the storage modulus of PVB (and other common interlayers) under permanent loads only (see for example [23] and others).

In this regard, Figure 21 collects some analytical estimates for the tested LG beams. Following the research outcomes that are partly summarized in Figure 20, and the real experimental scenarios of Figure 9, three levels of delamination were taken into account, i.e., corresponding to (i) un-delaminated LG beams, or to a MoE degradation up to (ii) 5% and (iii) 15% the nominal value (based also on Equation (13)). In addition, two tentative storage moduli were considered for the PVB foils in use, namely represented by:

- $G_2 = 0.50$ MPa, as derived from technical data sheets, for PVB foils at 25 °C and 15 years of time loading (see [52]), and
- $G_2 = 0.07$ MPa, corresponding to the recommended value for the design of LG elements under permanent, dead loads [23].

As shown in Figure 21 and Table 3, the best correlation between analytical frequencies and most of the experimental results was obtained in the presence of a relatively weak PVB layer ($G_2 = 0.07$ MPa).

Table 3. Percentage scatter Δ (mean value) of analytical frequency estimates, compared to experimental data.

| | Delamination Severity (Equation (13))—E_d/E | | | | | | | | |
| | 1 | | | 0.95 | | | 0.85 | | |
Restraint	C-C	K_r	S-S	C-C	K_r	S-S	C-C	K_r	S-S
$G_2 = 0.5$ MPa	+40.8	+35.6	−37.4	+32.2	+30.1	−38.7	+38.1	+27.8	−41.2
$G_2 = 0.07$ MPa	+20.6	+17.7	−46.2	+17.9	+15.3	−47.5	+15.5	+10.1	−49.9

Besides that, the collected comparisons allow for further emphasizing the key role of multiple input parameters on the reliable assessment of in-service glass structures, and, in particular, the role of restraints (compared to C-C or S-S conditions) and the presence of possible delaminations, which should be properly taken into account.

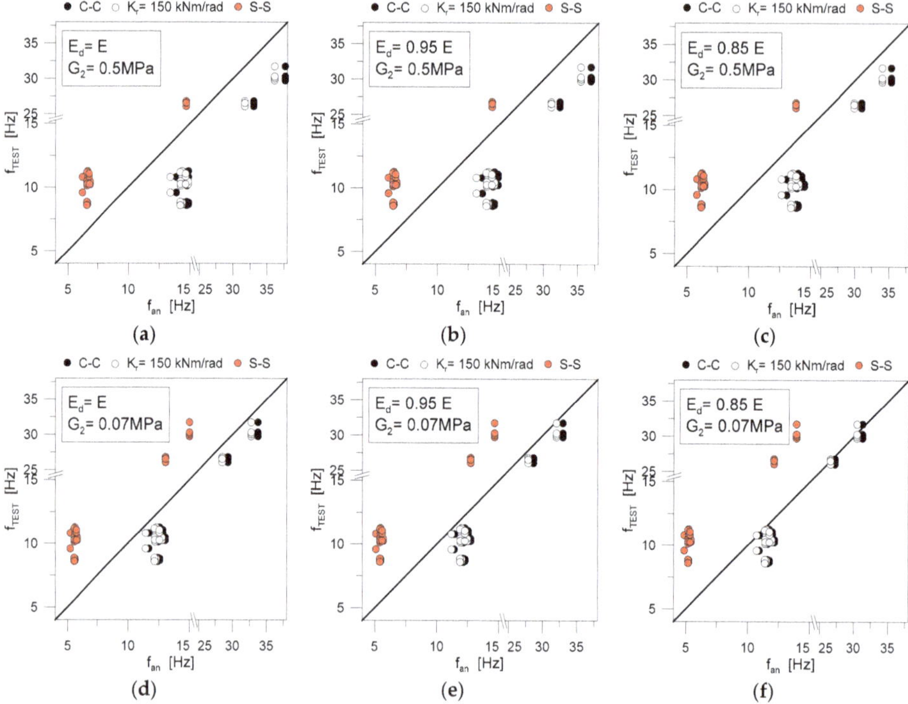

Figure 21. Comparison of experimental and analytical frequencies for the tested LG beams, as a function of deamination severity and interlayer stiffness (selection).

5. Conclusions

Laminated glass (LG) elements are largely used in buildings and civil engineering infrastructures, in the form of simple members (i.e., columns, beams and plates), but also combined together to realize stand-alone assemblies, where glass can interact with other traditional constructional materials. Besides the need of safe design methods for glass structures under ordinary loads, special care is increasingly spent by researchers in the dynamic response of LG systems under impact, moving loads due to pedestrians (in the case of walkways and roofs) or seismic events. As a part of buildings or complex systems, the dynamic parameters of LG elements should be properly taken into account, for the design of independent members or complex systems.

In this regard, several analytical methods available in the literature, for practical estimates of the fundamental vibration frequencies of simple LG members with viscoelastic interlayers. Otherwise, in most of the cases, reliable dynamic predictions can be obtained for the early-design stage. The intrinsic limit of literature methods is, in fact, represented by the assumption of ideal theoretical boundaries that often do not capture the mechanical effect of typical restraints in in use for LG systems. At the same time, they do not include the potential effects due to material degradations that can often manifest in in-service LG structures, like delaminations of the bonding interlayers.

In this paper, major issues for the vibration analysis of in-service (double) LG members were discussed and explored, with the support of analytical calculations, on-site experimental tests, and parametric Finite Element (FE) numerical analyses. As shown—as compared to laboratory research studies or early-stage design calculations—major uncertainties can be represented by the lack of accurate input material properties, especially for characterizing the shear stiffness of the bonding interlayers. The latter is strictly affected by the vibration frequency, but also by operational and ambient conditions to properly assess. Such an issue can in fact reflect on unsafe equivalent thickness assumptions for analytical and numerical calculations, thus dynamic estimates. Further relevant influencing parameters that can magnify the material uncertainties and approximate assumptions are then represented by the actual boundary condition of real LG structures, which can be markedly different with respect to the ideal restraints of practical use for design. Accordingly, it was shown that even minor flexibility contributions of joints can turn out in marked stiffness for the examined LG members, thus requiring separate calculations. The restraints themselves can, in fact, modify the vibration frequency of a given LG members, thus further affecting the PVB stiffness and require iterative calculations. Finally, another relevant issue for in-service LG members derives from the degradation of common interlayers in use, especially delaminations that can be hardly quantified, but having marked effects on dynamic estimates.

Based on on-site experimental tests on in-service LG beams, these influencing parameters were separately explored in the paper. Analytical methods were then proposed for reliable estimates on general LG beams. Their advantage is that both restraint features and delaminations can be taken into account when compared to classical theories for slender composite beams of literature proposals for LG members in ideal conditions.

Author Contributions: The paper results from original research investigations carried out by the author.

Funding: No external financial funding was received for the study.

Acknowledgments: The So.Co.Ba Foundation "Società per la conservazione della Basilica" (Bergamin) is gratefully acknowledged for facilitating the on-site experimental measurements.

Conflicts of Interest: The author declares no conflict of interest.

References

1. Haldimann, M.; Luible, A.; Overend, M. *Structural Use of Glass*; IABSE: Zurich, Switzerland, 2008; ISBN 978-3-85748-119-2.
2. Feldmann, M.; Kasper, R.; Abeln, B.; Cruz, P.; Belis, J.; Beyer, J.; Geßler, A.; Colvin, J.; Ensslen, F.; Eliasova, M.; et al. *Guidance for European Structural Design of Glass Components—Support to the Implementation, Harmonization and Further Development of the Eurocodes*; Dimova, P., Feldmann, D., Eds.; Report EUR 26439; European Union: Luxembourg, 2014.
3. Peroni, M.; Solomos, G.; Pizzinato, V.; Larcher, M. Experimental investigation of high strain-rate behaviour of glass. *Appl. Mech. Mater.* **2011**, *82*, 63–68. [CrossRef]
4. Del Linz, P.; Wang, Y.; Hooper, P.A.; Arora, H.; Smith, D.; Pascoe, L.; Cormie, D.; Blackman, B.R.K.; Dear, J.P. Determining material response for Polyvinyl Butyral (PVB) in blast loading situations. *Exp. Mech.* **2016**, *56*, 1501–1517. [CrossRef]
5. Zhang, X.; Hao, H.; Wang, Z. Experimental study of laminated glass window responses under impulsive and blast loading. *Int. J. Impact Eng.* **2015**, *78*, 1–19. [CrossRef]
6. Bedon, C.; Kalamar, R.; Eliasova, M. Low velocity impact performance investigation on square hollow glass columns via full-scale experiments and Finite Element analyses. *Compos. Struct.* **2017**, *182*, 311–325. [CrossRef]
7. Mohagheghian, I.; Wang, Y.; Jiang, L.; Zhang, X.; Guo, X.; Yan, Y.; Kinloch, A.; Dear, J. Quasi-static bending and low velocity impact performance of monolithic and laminated glass windows employing chemically strengthened glass. *Eur. J. Mech.* **2017**, *63*, 165–186. [CrossRef]
8. Biolzi, L.; Bonati, A.; Cattaneo, S. Laminated glass cantilevered plates under static and impact loading. *Adv. Civ. Eng.* **2018**, *2018*, 7874618. [CrossRef]
9. Xue, L.; Coble, C.R.; Lee, H.; Yu, D.; Chaparala, S.; Park, S. Dynamic analysis of thin glass under ball drop impact with new metrics. In Proceedings of the ASME 2013—International Technical Conference and Exhibition on Packaging and Integration of Electronic and Photonic Microsystems, Burlingame, CA, USA, 16–18 July 2013. [CrossRef]

10. Behr, R.; Minor, J.; Kremer, P. Effects of Accelerated Weathering on Architectural Laminated Glass in a Windstorm Environment. In *Science and Technology of Building Seals, Sealants, Glazing, and Waterproofing: Sixth Volume*; Myers, J., Ed.; ASTM International: West Conshohocken, PA, USA, 1996; pp. 27–45.

11. Santarsiero, M.; Bedon, C.; Moupagitsoglou, K. Energy-based considerations for the seismic design of ductile and dissipative glass frames. *Soil Dyn. Earthq. Eng.* **2019**, *125*, 105710. [CrossRef]

12. Lenci, S.; Consolini, L.; Clementi, F. On the experimental determination of dynamic properties of laminated glass. *Ann. Solid Struct. Mech.* **2015**, *7*, 27–43. [CrossRef]

13. Bedon, C.; Fasan, M.; Amadio, C. Vibration analysis and dynamic characterization of structural glass elements with different restraints based on Operational Modal Analysis. *Buildings* **2019**, *9*, 13. [CrossRef]

14. Aenlle, M.L.; Pelayo, F.; Ismael, G. An effective thickness to estimate stresses in laminated glass beams under dynamic loadings. *Compos. Part B Eng.* **2015**, *82*, 1–12. [CrossRef]

15. Bedon, C. Diagnostic analysis and dynamic identification of a glass suspension footbridge via on-site vibration experiments and FE numerical modelling. *Compos. Struct.* **2019**, *216*, 366–378. [CrossRef]

16. Bedon, C.; Fasan, M. Reliability of field experiments, analytical methods and pedestrian's perception scales for the vibration serviceability assessment of an in-service glass walkway. *Appl. Sci.* **2019**, *9*, 1936. [CrossRef]

17. Zemanova, A.; Zeman, J.; Janda, T.; Schmidt, J.; Sejnoha, M. On modal analysis of laminated glass: Usability of simplified methods and Enhanced Effective Thickness. *Compos. Part B* **2018**, *151*, 92–105. [CrossRef]

18. *ABAQUS Computer Software*; Simulia: Dassault, RI, USA, 2019.

19. EN 572–2:2004. *Glass in Buildings-Basic Soda Lime Silicate Glass Products*; CEN: Brussels, Belgium, 2004.

20. Clough, R.W.; Penzien, J. *Dynamics of Structures*; McGraw-Hill: New York, NY, USA, 1993; ISBN 0-07-011394-7.

21. Galuppi, L.; Royer-Carfagni, G. Effective thickness of laminated glass beams: New expression via a variational approach. *Eng. Struct.* **2012**, *38*, 53–67. [CrossRef]

22. Andreozzi, L.; Bati, S.B.; Fagone, M.; Ranocchiai, G.; Zulli, F. Dynamic torsion tests to characterize the thermo-viscoelastic properties of polymeric interlayers for laminated glass. *Constr. Build. Mater.* **2014**, *65*, 1–13. [CrossRef]

23. Bennison, S.J.; Jagota, A.; Smith, C. Fracture of glass/poly(vinyl butyral) (Butacite®) laminates in biaxial flexure. *J. Am. Ceram. Soc.* **1999**, *82*, 1761–1770. [CrossRef]

24. Hooper, P.A.; Blackman, B.R.K.; Dear, J.P. The mechanical behaviour of poly(vinyl butyral) at different strain magnitudes and strain rates. *J. Mater. Sci.* **2012**, *47*, 3564–3576. [CrossRef]

25. Ensslen, F. Influences of laboratory and natural weathering on the durability of laminated safety glass. In Proceedings of the Glass Performance Days, GDP 2007, Tampere, Finland, 15–18 June 2007.

26. Hana, T.; Janda, T.; Schmidt, J.; Zemanova, A.; Sejnoha, M.; Eliasova, M.; Vokac, M. Experimental and numerical study of viscoelastic properties of polymeric interlayers used for laminated glass: Determination of material parameters. *Materials* **2019**, *12*, 2214. [CrossRef]

27. Froli, M.; Lani, L. Adhesion, creep and relaxation properties of PVB in laminated safety glass. In Proceedings of the Glass Performance Days, GPD 2011, Tampere, Finland, 17–20 June 2011.

28. Serafinavicius, T.; Lebet, J.P.; Louter, C.; Lenkimas, T.; Kuranovas, A. Long-term laminates glass four point bending test with PVB, EVA and SG interlayers at different temperatures. *Procedia Eng.* **2013**, *57*, 996–1004. [CrossRef]

29. Xu, X.; Liu, B.; Li, Y. Experimental studies on viscoelasticity of film materials in laminated glass sheets. *Sae Int. J. Mater. Manuf.* **2015**, *8*, 922–931. [CrossRef]

30. Rodrigues, T.; Jordão, S.; Bedon, C. Long-term effects on structural glass beams. In Proceedings of the XI CMM Conference—Congresso de Construção Metálica e Mista, Coimbra, Portugal, 23 November 2017; pp. 933–942.

31. Stevels, W.; D'Haene, P.; Zhang, P.; Haldeman, S. A comparison of different methodologies for PVB interlayer modulus characterization. In Proceedings of the Challenging Glass 5—Conference on Architectural and Structural Applications of Glass, Ghent, Belgium, 16–17 June 2016; ISBN 978-909-825-2680-6.

32. Lopez-Aenlle, M.; Noriega, A.; Pelayo, F. Mechanical characterization of polyvinyl butyral from static and modal tests on laminated glass beams. *Compos. Part B Eng.* **2019**, *169*, 9–18. [CrossRef]

33. Carrot, C.; Bendaoud, A.; Pillon, C. Polyvinyl Butyral. In *Handbook of thermoplastics*; Olabisi, O., Adewale, K., Eds.; CRC Press: Boca Raton, FL, USA, 2016.

34. Chapuis, V.; Pelisset, S.; Raeis-Barneoud, M.; Li, H.Y.; Ballif, C.; Perret-Aebi, L.E. Compressive-shear adhesion characterization of polyvinyl-butyral and ethylene-vinyl acetate at different curing times before and after exposure to damp-heat conditions. *Prog. Photovolt. Res. Appl.* **2014**, *22*, 405. [CrossRef]
35. Liang, R.H.; Gupta, A.; di Stefano, S. Photothermal Characterization of Encapsulated Materials for Photovoltaic Modules. 1982. Available online: https://www2.jpl.nasa.gov/adv_tech/photovol/2016ENG/Phototherm%20Char%20of%20Encap%20Mtls_5101-210_JPL1982.pdf (accessed on 13 September 2019).
36. Kuraray. Edge Stability, Durability and Weathering. Document Ref. GLS-TECBU-2014-04. 2014. Available online: https://www.trosifol.com/fileadmin/user_upload/Kuraray_4_4_Edge_Stability.pdf (accessed on 13 September 2019).
37. CPNI EPB 04/13. Guidance note—Influence of Delamination of Laminated Glass on Its Blast Performance. Centre for the Protection of National Infrastructure. 2013. Available online: https://www.cpni.gov.uk/system/files/documents/4d/8f/Delamination-of-laminated-glass.pdf (accessed on 13 September 2019).
38. Dural, E. Experimental and numerical treatment of delamination in laminated glass plate structures. *J. Reinf. Plast. Compos.* **2016**, *35*, 56–70. [CrossRef]
39. Dural, E. Analysis of delaminated glass beams subjected to different boundary conditions. *Compos. Part B* **2016**, *101*, 132–146. [CrossRef]
40. Jaskowiec, J. Numerical modelling mechanical delamination in laminated glass by XFEM. *Procedia Eng.* **2015**, *487*, 181–184.
41. Vedrtnam, A. Experimental and simulation studies on delamination strength of laminated glass composites having polyvinyl butyral and ethyl vinyl acetate inter-layers of different critical thicknesses. *Def. Technol.* **2018**, *14*, 313–317. [CrossRef]
42. Bedon, C.; Bergamo, E.; Izzi, M.; Noè, S. Prototyping and validation of MEMS accelerometers for structural health monitoring—The case study of the Pietratagliata cable-stayed bridge. *J. Sens. Actuator Netw.* **2018**, *7*, 18. [CrossRef]
43. De Rosa, M.A.; Auciello, N.M. Free vibrations of tapered beams with flexible ends. *Comput. Struct.* **1996**, *60*, 197–202. [CrossRef]
44. Lee, J. Free vibration analysis of delaminated composite beams. *Comput. Struct.* **2000**, *74*, 121–129. [CrossRef]
45. Della, C.N.; Shu, D. Free vibration analysis of composite beams with overlapping delaminations. *Eur. J. Mech. A/Solids* **2005**, *24*, 491–503. [CrossRef]
46. Ramtekkar, G.S. Free vibration analysis of delaminated beams using mixed finite element model. *J. Sound Vib.* **2009**, *328*, 428–440. [CrossRef]
47. Callioglu, H.; Atlihan, G. Vibration analysis of delaminated composite beams using analytical and FEM models. *Indian J. Eng. Mater. Sci.* **2011**, *18*, 7–14.
48. Zhang, Z.; Shankar, K.; Morozov, E.V.; Tahtali, M. Vibration-based delamination detection in composite beams through frequency changes. *J. Vib. Control* **2016**, *22*, 496–512. [CrossRef]
49. Krawczuk, M.; Ostachowicz, W.; Zuk, A. Analysis of natural frequencies of delaminated composite beams based on finite element method. *Struct. Eng. Mech.* **1996**, *4*, 243–255. [CrossRef]
50. Tate, I.V.; Roy, S.; Jagtap, K.R. Delamination detection of composite cantilever beam coupled with piezoelectric transducer using natural frequency deviation. *Procedia Eng.* **2014**, *97*, 1293–1304. [CrossRef]
51. Gibson, R.F. *Principles of Composite Material Mechanics*; CRC Press: Philadelphia, PA, USA, 1994.
52. AGC Glass Europe. Material Properties of PVB Interlayers Used in Stratobel and Stratobel Strong Laminated Glass. 2018. Available online: https://www.agc-yourglass.com/sites/default/files/technical_documents/original/STRATOBEL_PROPERTIES44907Material%20Properties%20of%20PVB_Stratobel_June2018_ENG.pdf (accessed on 13 September 2019).

Article

Evolution Mechanism of Wind Vibration Coefficient and Stability Performance during the Whole Construction Process for Super Large Cooling Towers

Shitang Ke [1,2,*], Peng Zhu [3], Lu Xu [4] and Yaojun Ge [2]

[1] Department of Civil Engineering, Nanjing University of Aeronautics and Astronautics, Nanjing 210016, China

[2] State Key Laboratory for Disaster Reduction in Civil Engineering, Tongji University, Shanghai 200092, China; yaojunge@tongji.edu.cn

[3] China Information Consulting & Designing Institute Co., LTD, Nanjing 210000, China; zhupeng.cicdi@chinaccs.cn

[4] China Energy Engineering Group Guangdong Electric Power Design Institute Co., LTD, Guangzhou 510663, China; luxunuaa@163.com

* Correspondence: keshitang@163.com

Received: 28 August 2019; Accepted: 29 September 2019; Published: 9 October 2019

Abstract: Wind-induced damage during the construction process and the evolution of damage over time are important reasons for the wind-induced destruction of large cooling towers. In fact, wind vibration coefficient and stability performance will evolve with the construction height and material properties over time. However, the existing studies generally ignore the impact of wind load and structural performance during the construction period. In this study, we built the 3D physical model separately for all eight construction stages a super large cooling tower which is being currently constructed and stands 210 m. The dynamic characteristics of the cooling tower were analyzed in each stage. First, the flow field information and 3D time history of aerodynamic forces were obtained for the whole construction process using large eddy simulation (LES). Full transient dynamic finite element analysis was used to calculate the dynamic responses of the tower under the real-time changes of wind loads during the whole construction process. Five calculation methods were used to trace the evolution of wind vibration coefficient during the whole construction process of the super large cooling tower. Then the formula for wind vibration coefficient changing with the construction height was fitted. The differential values of wind vibration coefficient during the whole construction process of the cooling tower were discussed by taking the meridional axial force as the objective function. On this basis, the influence and working mechanism of wind vibration coefficient, concrete age, construction load, geometric nonlinearity, internal suction force on buckling stability, and ultimate bearing capacity of the cooling towers were investigated. This research provides an enhanced understanding on the evolution of wind-induced stability performance in super large cooling towers and a methodology to prevent wind-induced damage during the construction process.

Keywords: super large cooling tower; whole construction process; wind vibration coefficient; buckling stability; ultimate bearing capacity

---

1. Introduction

After Ferrybridge Cooling Tower failures in the UK in 1965 [1], the international wind engineering circle began to conduct studies in the following topics: influence of tower group and surrounding structures on wind pressure distribution on the surface of the tower body [2,3], buckling stability, and ultimate bearing capacity of the tower body under wind load [4,5], finite element analysis of responses of large cooling towers considering the tower defect and soil–structure interaction [6,7], and random dynamic responses of tower body induced by pulsating wind pressure [8]. Wind-induced damage and subsequent evolution of the damage during the construction process are considered [9] responsible for the collapse of three cooling towers (at Ardeer Power Station in Scotland in 1973, power plant in Bouchain, Franch in 1979, and Fiddlers Ferry Power Station in 1984). This is closely related to the wind loads, concrete performance, and crack evolution during the construction process of the cooling tower. We have also found through the overall and local stability performance of China's tallest exhaust cooling tower during the construction process that the wind vibration coefficient changes with the construction height and evolution of material properties. Moreover, dynamic wind pressure inside the cooling tower also has a non-negligible impact on the wind-induced stability performance during the construction process.

In China, the height of newly built thermal and nuclear power plants has far exceeded the upper limit of standard or broken the world's record. This directly leads to substantial 3D dynamic wind load effect [10,11]. The construction period of the main structure and the construction difficulty also increase [12]. For the template of the cooling tower, the concrete strength may be insufficient before the concrete pouring of the cooling tower is complete. The concrete, though having a relatively low strength during the construction process, is subjected to dead load, wind load, and construction load. The strength and modulus of elasticity of concrete will increase with the construction height, which results in the constant evolution of stiffness and stress performance of the overall tower. Changes in the morphology and mechanical performance of the cooling tower during this process will further lead to alterations of static and dynamic wind pressure distribution on tower surface, wind-induced response, and wind vibration coefficient. As a result, the calculation of internal force of the structure and analysis of stability performance and ultimate bearing capacity will be also affected. In light of this, it is of high importance to discuss the evolution and non-linear influence of wind-induced stability of super large cooling towers during the construction process.

Few studies have been devoted to the wind-induced stability performance of large cooling towers during the construction process so far. In one literature report [13], buckling failure and ultimate bearing capacity of cooling tower during the whole construction process were analyzed based on secondary development of the ANSYS and wind tunnel test. Ke employed self-written preprocessing and post-processing programs for checking computation of the local and overall ultimate bearing capacity of the exhaust cooling tower. The variation of critical wind speed with construction height was also discussed [14].

In this study, we focused on a 210 m super large cooling tower under construction, the tallest tower ever built in the world. We built the 3D physical model separately for all eight construction stages. The dynamic characteristics of the cooling tower were analyzed in each stage. First, the flow field information and 3D time history of aerodynamic force were obtained for the whole construction process using large eddy simulation. The wind pressure distributions of the constructed tower were compared against the standard and measured curves to validate the numerical simulation. Full transient dynamic finite element analysis was used to calculate the dynamic responses of the tower under the real-time changes of wind load during the whole construction process. Five calculation methods were used to trace the evolution of wind vibration coefficient during the whole construction process of the super large cooling tower. Then the formula for wind vibration coefficient changing with the construction height was fitted. The differential values of wind vibration coefficient during the whole construction process of the cooling tower were discussed by taking the meridional axial force as the objective function. On this basis, the influence and working mechanism of wind vibration coefficient, concrete age, construction load, geometric nonlinearity, and internal suction force on buckling stability and ultimate bearing capacity of the cooling tower were investigated.

2. An Illustrative Example

2.1. An Overview of the Project

This super large cooling tower stood 210.0 m, with a throat height of 157.5 m and air inlet height of 32.5 m. The top section had a diameter of 115.8 m, and the throat section in the middle portion had a diameter of 110 m. The zero-meter diameter was 180. The tower body was connected to the annular plate resting on the foundation with 52 pairs of X-shaped pillars. The X-shaped pillars had a rectangular cross section, which measured 1.2 m × 1.8 m. The foundation was a cast-in-situ reinforced concrete structure with a width of 12.0 m and a height of 2.5 m. The terrain category was B, with a basic wind velocity of 23.7 m/s. Table 1 shows the main parameters of the super large cooling tower.

Table 1. Main structural parameters of super large cooling tower.

Component	Height/m	Wall Thickness/m	Radius/m	Concrete Grade	Structural Schematic
Tower body	32.5	2.00	80.5	C40	
	44.7	0.58	76.8		
	69.8	0.46	69.8		
	82.3	0.43	66.5		
	107.7	0.39	60.7		
	120.4	0.39	58.4		
	133.3	0.38	56.7		
	159.2	0.37	55.2		
	172.2	0.37	55.6		
	185.2	0.37	56.4		
	210.0	0.50	58.5		
Pillar	52 pairs of X-shaped pillars with a cross-sectional area of 1.8 m × 1.2 m			C45	
Foundation	Circular plate foundation with a cross-sectional area of 12 m × 2.5 m			C35	

2.2. Modeling of the Whole Construction Process

The construction process was divided into eight stages based on the progress of construction and calculation precision. Evolution of wind-induced stability performance was analyzed for each stage. Table 2 shows the parameters of each working condition.

Table 2. Parameters of super large cooling tower under typical working conditions.

Schematic of Working Conditions				
No. of working condition	Working condition 1	Working condition 2	Working condition 3	Working condition 4
Template number	10	30	50	70
Height/m	44.1	69.8	94.9	120.4
Air inlet diameter/m	154.41	140.31	127.61	117.21
Minimum shell thickness/m	0.590	0.460	0.405	0.385
Schematic of Working Conditions				

Table 2. *Cont.*

No. of working condition	Working condition 5	Working condition 6	Working condition 7	Working condition 8
Template number	90	105	120	139
Height/m	146.2	165.7	185.2	210.0
Air inlet diameter/m	111.21	110.61	112.61	115.81
Minimum shell thickness/m	0.375	0.370	0.370	0.370

3. Numerical Simulation

3.1. Methodology

Fluid is considered to be incompressible viscous flow in anti-wind design of structures. Spatial averaging of transient N–S equation can yield the governing equation in large eddy simulation (LES):

$$\frac{\partial \overline{\mu_i}}{\partial x_i} = 0 \tag{1}$$

$$\frac{\partial \overline{\mu_i}}{\partial t} + \frac{\partial (\overline{\mu_i \mu_j})}{\partial x_j} = -\frac{1}{\rho}\frac{\partial \overline{p}}{\partial x_i} + v\frac{\partial^2 \overline{\mu_i}}{\partial x_j \partial x_j} - \frac{\partial \tau_{ij}}{\partial x_j} \tag{2}$$

where ρ is air density, t is time, v is kinematic viscosity coefficient of air, $\overline{\mu_i}$ and $\overline{\mu_j}$ are velocities in three directions after filtering, and τ_{ij} is non-closed term in the N–S equation after spatial averaging, i.e., subgrid scale stress.

$$\tau_{ij} = \overline{\mu_i \mu_i} - \overline{\mu_i}\,\overline{\mu_j} \tag{3}$$

Boussinesq approximation was introduced according to Smagorinsky subgrid–scale model based on eddy viscosity assumption. Thus, the subgrid–scale stress is written as

$$\tau_{ij} - \frac{1}{3}\tau_{ij}\delta_{ij} = -2\mu_t \overline{S_{ij}} = -\mu_t\left(\frac{\partial \overline{\mu_i}}{\partial x_j} + \frac{\partial \overline{\mu_j}}{\partial x_i}\right) \tag{4}$$

e tensor of solvable scale, τ_{kk} is the isotropic component of subgrid–scale stress, which is contained in the pressure item after filtering, δ_{ij} is Kronecker delta function, μ_t is subgrid–scale turbulence eddy viscosity coefficient, generally using the Smagorinsky assumption:

$$\mu_t = (C_s\Delta)^2|\overline{S}| \tag{5}$$

where C_s is Smagorinsky constant, generally taken as 0.1–0.23 and being 0.1 in this study. Strain rate tensor $|\overline{S}| = \sqrt{2S_{ij}S_{ij}}$. Δ is grid scale, $\Delta = (\Delta_x\Delta_y\Delta_z)^{1/3}$, where Δ_x, Δ_y, and Δ_z are the grid size in x, y, and z directions, respectively. This is the standard Smagorinsky subgrid–scale model. Some researchers propose dynamic determination of C_s value to better characterize collision, separation, free shear layer, and vortex shedding of the flow field around the blunt body. It is known as the dynamic Smagorinsky model.

However, both the Smagorinsky model and dynamic Smagorinsky model are algebraic models, which assume a local equilibrium between the generation and dissipation of subgrid–scale turbulent kinetic energy. These models are not fit for simulating structures with a high Reynolds number, such as the cooling towers. To solve this problem, we proposed a new form of subgrid–scale model based on literature [15]:

$$\frac{\partial k_{sgs}}{\partial t} + \frac{\partial \mu_j k_{sgs}}{\partial x_j} = -\tau_{ij}S_{ij} - C_\varepsilon\frac{k_{sgs}^{3/2}}{\Delta} + \frac{\partial}{\partial x_j}\left[(C_d\Delta v\sqrt{k_{sgs}} + v)\frac{\partial k_{sgs}}{\partial x_j}\right] - \varepsilon_w \tag{6}$$

where, K_{sgs} is the kinetic energy of transportation equation SGS, C_s is Smagorinsky constant, and v is positive. Here, C_s will reduce the amplitude of K_{sgs}, and a Gaussian filter is needed for finite difference.

This model is better applied to engineering applications. There is no need for experimental filtering, and the computational load is small. So, it can be used to simulate structures with a high Reynolds number, such as cooling towers [10].

3.2. Parameter Configuration and Grid Generation

A physical model of the super large cooling tower was built according to original size, so that the Reynolds number used in numerical simulation would be comparable to that in the actual project. The size of the computational domain was X × Y × Z = 6000 m × 4000 m × 1000 m (X is across-wind direction, Y is along-wind direction, and Z is height direction). The blocking rate of the model was below 1%. The computational domain was divided into dense region and peripheral region so as to ensure both computational efficiency and precision. Non-structured grids with a high adaptability were used for the region near the cooling tower; for the peripheral space further away from the cooling tower, structured grids having a regular topology were used for the discretization. Therefore, the total number of grids was reduced, which improved the computational efficiency. The minimum grid size was 0.2 m in the core region. The model of the built-up tower had approximately 12.8 million grids.

Boundary conditions were defined using UDF (User define function) file. Inlet boundary condition was velocity inlet, and the outlet boundary condition was pressure outlet. The top and sides of the computational domain were equivalent to free slip walls, the symmetry boundary conditions (Symmetry). The floor and structure surface were equivalent to no-slip wall boundary condition. The wind field was considered as incompressible flow field. Discrete equations were solved using SIMPLEC. This calculation method has good convergence performance and applies to LES (Large eddy simulation) with small time step. The time step of LES was set to 0.05 s. Here, only the schematic for the computational domain and grid generation of the model of the built-up tower is provided, as shown in Figure 1.

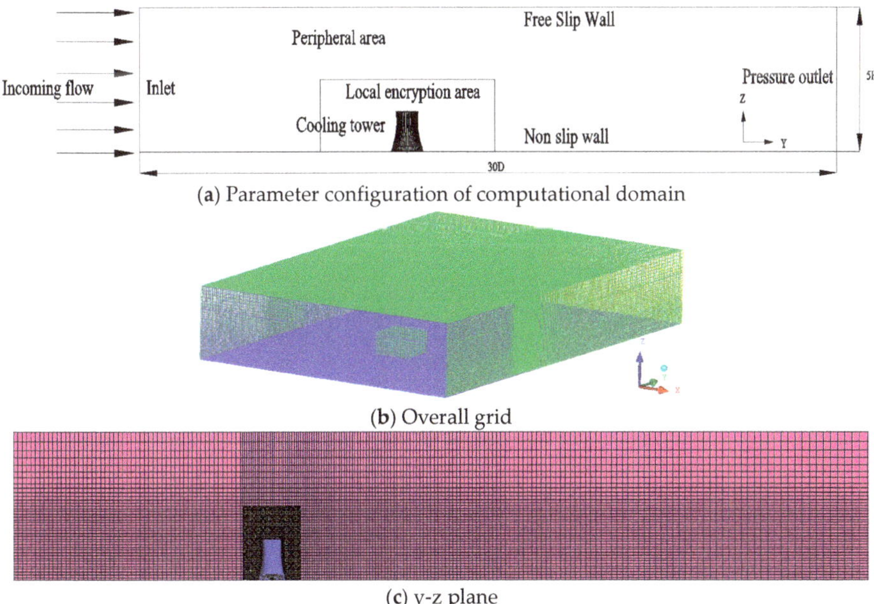

(**a**) Parameter configuration of computational domain

(**b**) Overall grid

(**c**) y-z plane

Figure 1. *Cont.*

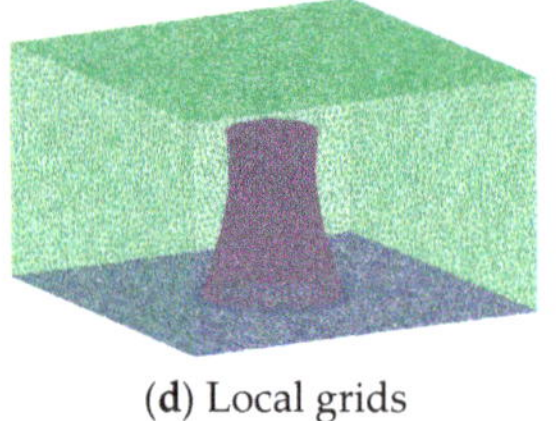

(**d**) Local grids

(**e**) Dense region

Figure 1. Computational domain and grid generation for the super large cooling tower.

3.3. Validation

The design code for cooling towers [16–19] only provides the average and pulsating wind pressure distribution curves of the built-up towers. Therefore, we only validated the numerical simulation for the built-up tower.

Figure 2 shows the comparison of simulated average and pulsating wind pressure distributions of a cross section of the built-up tower against the measured and standard curves. It can be seen that the simulated average wind pressure distribution curve agreed well with the standard curve. The pressure coefficients in the upwind side, region of extreme negative pressure, and at the separation point on the leeward side were consistent with the standard curve. This validated the average wind pressures obtained by LES.

Moreover, the simulated pulsating wind pressure distribution curve basically coincided with the measured curve and the curve of wind tunnel test. The values lay between the results measured at home and abroad. Five curves are presented respectively in Figure 2b, among which three are the measured pulsating wind pressure curves in domestic and foreign regulations (VGB–R 610Ue 2005, Blanchette et al. 2013, DL/T 5339–2006, GB/T 50102–2014), and two are the measuring points of the third and ninth floors of the full-size cooling tower in this paper. Pulsating wind pressure distribution is closely associated with the terrain, incoming turbulent flow, and surrounding interference. The trend and values of pulsating wind pressure distribution estimated by LES were close or fell into the range of the measured values. Therefore, the pulsating wind pressures simulated by LES were reliable and suitable for subsequent analysis of wind-induced vibration and stability performance.

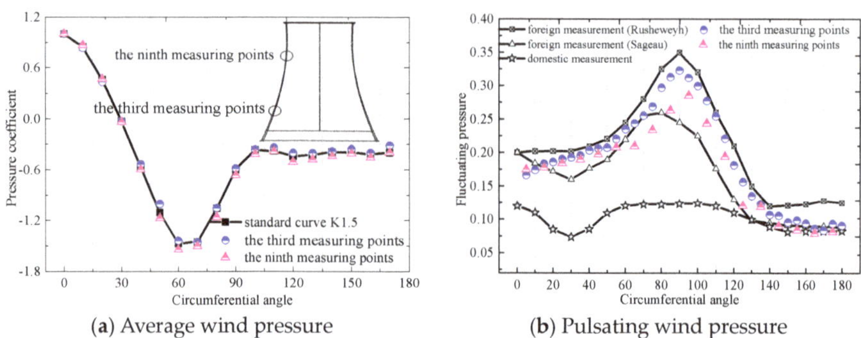

(**a**) Average wind pressure

(**b**) Pulsating wind pressure

Figure 2. Comparison of numerical calculation, field measurement, and wind tunnel test.

3.4. Simulation Results

3.4.1. Pulsating Wind Pressure

The same parameter configuration as the numerical simulation for the built-up tower was used for LES under the other seven working conditions. Thus, the flow field on the surface of the tower and the time history of pulsating wind pressure were obtained for the whole construction process. Due to limited space, Figure 3 only provides the time history of pressure coefficient at the upwind side, crosswind side, separation point and leeward side for the built-up model.

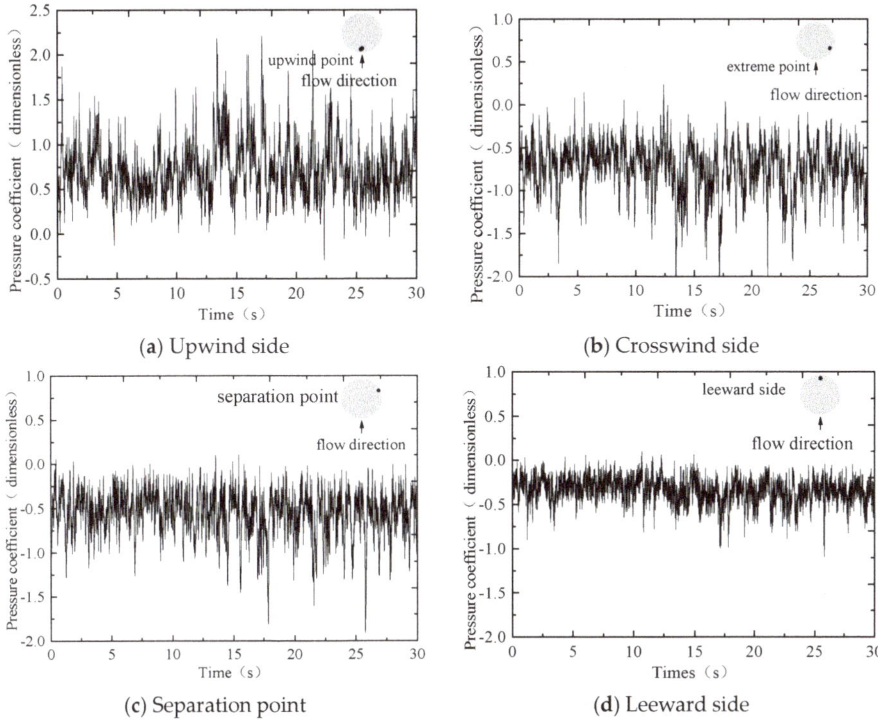

Figure 3. Time history of pressure coefficient at representative measuring points of the built-up tower.

Figure 4 provides the power spectral density curves of pulsating wind pressure at the representative measuring points. Comparison showed that for different positions, the peaks of the power spectral density function all occurred in the low frequency range. The energy of the pulsating wind pressure was mainly concentrated in low frequencies. The values of the power spectral density function were slightly higher on the upward side than at the separation point and leeward side.

3.4.2. Pressure Coefficients on the Tower Surface

Figure 5 shows the nephograms of pressure coefficients on tower surface under eight typical working conditions. The distributions of pressure coefficient were basically consistent under different working conditions. The increase of construction height did not change the wind field characteristics of cylindrical structure. The flow separated in front of the tower body due to collision, which resulted in the separation bubble, and the bubble shedded off from the crosswind side. Consequently, there were positive and negative pressure distributions on the upward and crosswind sides of the tower body. The increase in the construction height greatly decreased the negative pressure on the crosswind and leeward sides, especially for the built-up tower.

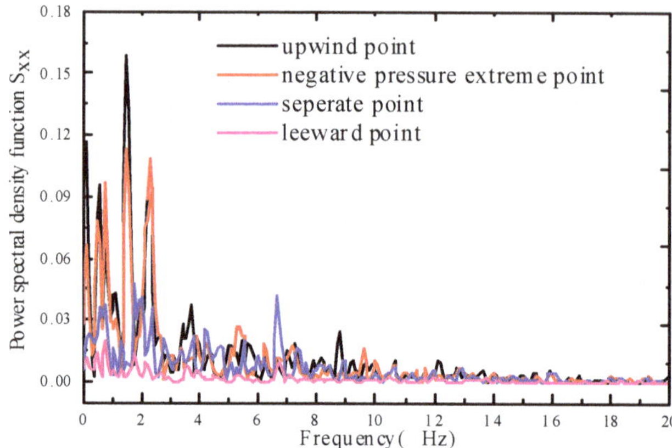

Figure 4. Power spectral density curves of pulsating wind pressure at the representative measuring points.

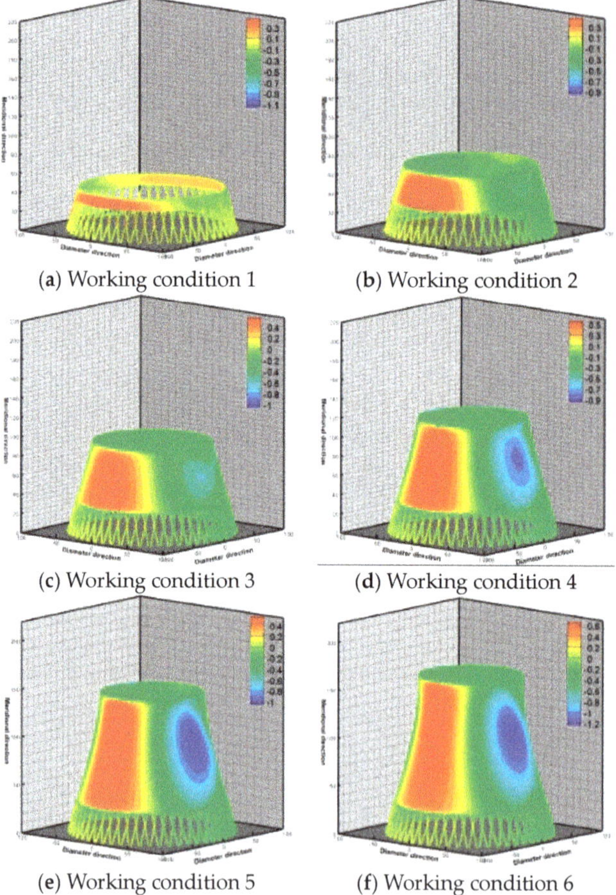

(**a**) Working condition 1

(**b**) Working condition 2

(**c**) Working condition 3

(**d**) Working condition 4

(**e**) Working condition 5

(**f**) Working condition 6

Figure 5. *Cont.*

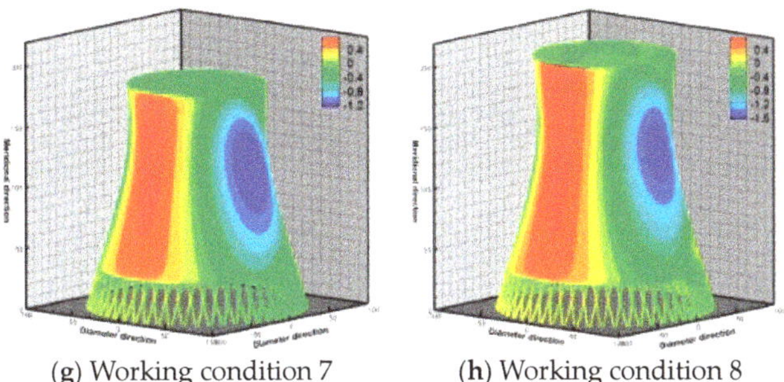

(**g**) Working condition 7　　　　(**h**) Working condition 8

Figure 5. Nephograms of pressure coefficients on tower surface under eight typical working conditions.

3.4.3. Turbulent Kinetic Energy

Figure 6 is the schematic for the turbulent kinetic energy distribution under the eight working conditions. There were significant differences between the eight working conditions. As the construction height increased, the scale component of maximum turbulent kinetic energy deviated from the upper end of the air inlet to the wake stream. In addition, as the construction height increased, the turbulent kinetic energy of fluid inside the tower decreased. This resulted in uniform distribution of internal pressure of the built-up tower along the circumferential and meridional directions.

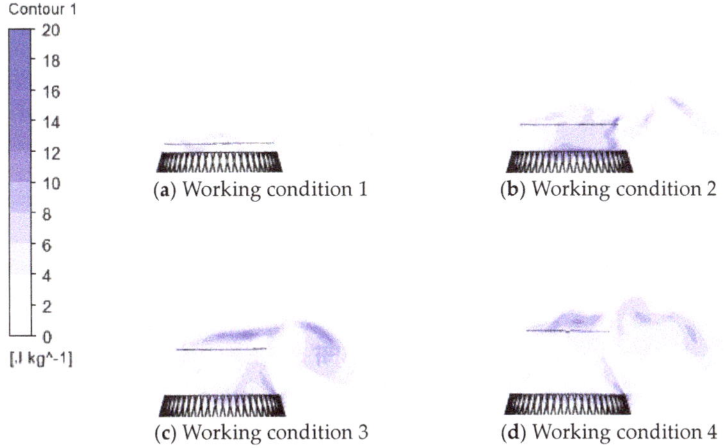

(**a**) Working condition 1　　　　(**b**) Working condition 2

(**c**) Working condition 3　　　　(**d**) Working condition 4

Figure 6. *Cont.*

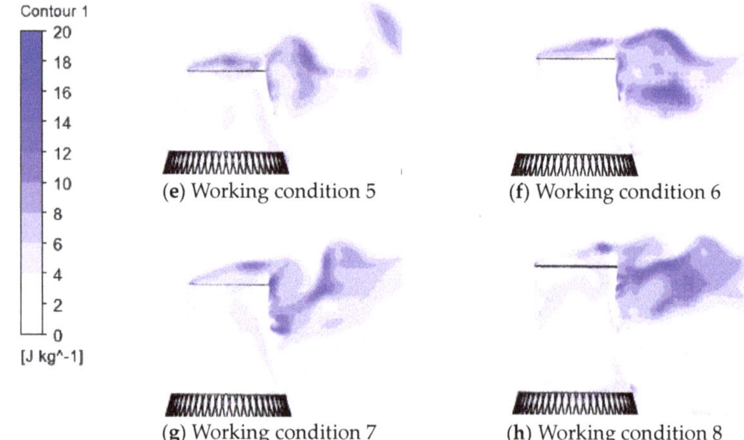

Figure 6. Schematic for turbulent kinetic energy distribution under working condition 8.

4. Analysis of Dynamic Characteristics

The integrated model of the tower body–pillar–circular foundation was built using ANSYS software. The tower body, pillar, circular foundation, and elastic foundation were simulated with shell element, beam element, and spring element, respectively. The connections of circular foundation to the tower body and pillars were simulated using multi-point constraint and rigid domain, respectively. Block Lanczos method was employed to analyze the dynamic characteristics of finite element (FE) model of the cooling tower through the whole construction process. Table 3 shows the fundamental frequencies and vibration mode distribution under each working condition. Figure 7 is the distribution curve of natural frequencies of the first fifty modes under different working conditions. It can be seen that construction height had a significant impact on lower-order frequencies, but a lesser impact on higher-order frequencies. As the construction height increased, the fundamental frequencies of the cooling tower decreased. The fundamental frequency was the largest under working condition 1, which was 0.963 Hz.

Table 3. Vibration modes at fundamental frequency under each working condition.

Working Condition	Working Condition 1	Working Condition 2	Working Condition 3	Working Condition 4
Vibration mode				
Frequency/Hz	0.963	0.739	0.649	0.608
Working Condition	**Working Condition 5**	**Working Condition 6**	**Working Condition 7**	**Working Condition 8**
Vibration mode				
Frequency/Hz	0.598	0.596	0.580	0.572

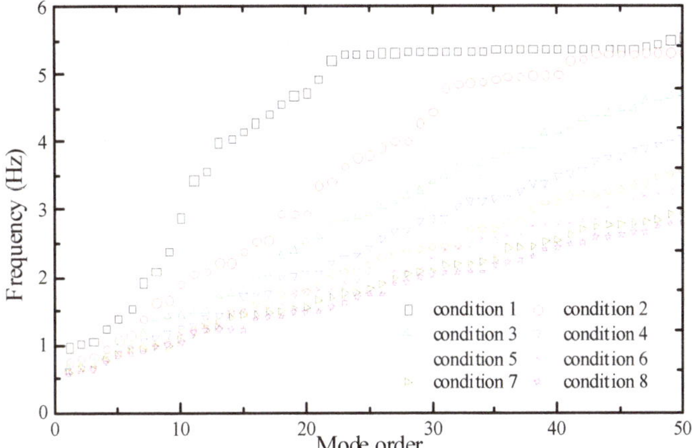

Figure 7. Frequency distribution of the first fifty modes under each working condition.

Figure 8 shows the changes of natural frequencies of the first ten modes with the number of circumferential harmonics under each working condition. Comparison shows that the minimum natural frequencies occurred when the number of circumferential harmonics was 4 under different working conditions. As the natural frequencies increased, the number of circumferential harmonics increased as well.

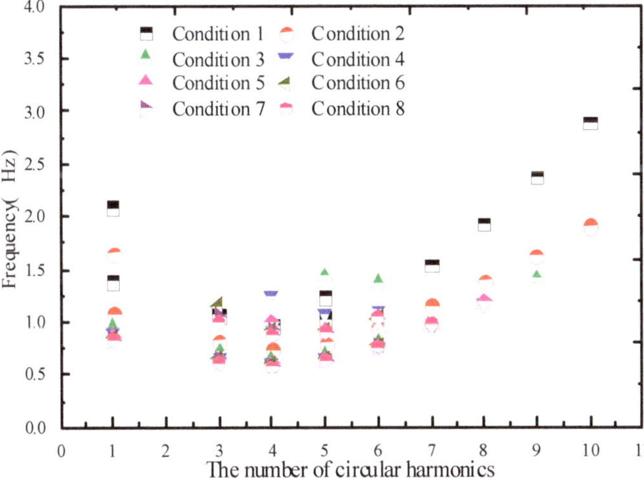

Figure 8. Changes of frequencies with the number of circumferential harmonics under each working condition.

Table 4 shows the capsize modes of the cooling tower for the whole construction process. As the construction height increased, the order of capsize mode also increased, while the frequency corresponding to the excitation mode decreased.

Table 4. Capsize modes of the cooling tower for the whole construction process.

Working Condition	Working Condition 1	Working Condition 2	Working Condition 3	Working Condition 4
Capsize mode				
Frequency/Hz	2.09	1.66	1.49	1.39
Working Condition	Working Condition 5	Working Condition 6	Working Condition 7	Working Condition 8
Capsize mode				
Frequency/Hz	1.33	1.29	1.29	1.21

5. Analysis of Differential Wind Vibration Coefficient for the Whole Construction Process

5.1. Methods and Parameter Explanations

Full transient dynamic analysis was performed to solve the dynamic equilibrium equations. The core principle was to use implicit methods, such as the Newmark method and HHT (Hilber–Hughes–Taylor) to directly solve the transient problems. The Newmark method uses the finite difference method, ad within one time interval, there are

$$[M]\{\ddot{u}\} + [C]\{\dot{u}\} + [K]\{u\} = \{F^a\} \tag{7}$$

$$\{\dot{u}_{n+1}\} = \{\dot{u}_n\} + [(1-\delta)\{\ddot{u}_n\} + \delta\{\ddot{u}_{n+1}\}]\Delta t \tag{8}$$

$$\{u_{n+1}\} = \{u_n\} + \{\dot{u}_n\}\Delta t + [(\frac{1}{2}-\alpha)\{\ddot{u}_n\} + \alpha\{\ddot{u}_{n+1}\}]\Delta t^2 \tag{9}$$

where α and δ are integral parameters. However, the use of the Newmark method for the calculation of discrete spatial domain in FE model cannot satisfy the requirement (that is, numerical damping at high frequencies should not be introduced at the expense of precision, and not too many values of numerical damping should be generated at low frequencies). This defect can be compensated by combining with HHT method.

The basic HHT has the following expression:

$$[M]\{\ddot{u}_{n+1-\alpha_m}\} + [C]\{\dot{u}_{n+1-\alpha_f}\} + [K]\{u_{n+1-\alpha_f}\} = \left\{F^a_{n+1-\alpha_f}\right\} \tag{10}$$

where $\{\ddot{u}_{n+1-\alpha_m}\} = (1-\alpha_m)\{\ddot{u}_{n+1}\} + \alpha_m\{\ddot{u}_n\}$; $\{\dot{u}_{n+1-\alpha_f}\} = (1-\alpha_f)\{\dot{u}_{n+1}\} + \alpha_f\{\dot{u}_n\}$; $\{u_{n+1-\alpha_f}\} = (1-\alpha_f)\{u_{n+1}\} + \alpha_f\{u_n\}$; $\left\{F^a_{n+1-\alpha_f}\right\} = (1-\alpha_f)\{F^a_{n+1}\} + \alpha_f\{F^a_n\}$.

In order to ensure unconditional stability of the second-order system without reducing the accuracy of time integral, four parameters α, δ, α_f, and α_m should satisfy the following relationships:

$$\delta \geq \frac{1}{2}; \ \alpha = \frac{1}{2}\delta; \ \delta = \frac{1}{2} - \alpha_m - \alpha_f; \ \alpha_m \leq \alpha_f \leq \frac{1}{2} \tag{11}$$

Combining Formulae (2), (4), and (6),

$$
\begin{aligned}
(a_0[M] + a_1[C] + (1-\alpha_f)[K])\{u_{n+1}\} &= (1-\alpha_f)\{F^a_{n+1}\} + \alpha_f\{F^a_n\} - \alpha_f\{F^{int}_n\} \\
&+ [M](a_0\{u_n\} + a_2\{\dot{u}_n\} + a_3\{\ddot{u}_n\}) + [C](a_1\{u_n\} + a_4\{\dot{u}_n\} + a_5\{\ddot{u}_n\})
\end{aligned}
\tag{12}
$$

where $a_0 = \frac{1-\alpha_m}{\alpha\Delta t^2}$, $a_1 = \frac{(1-\alpha_f)\delta}{\alpha\Delta t}$, $a_2 = \frac{1-\alpha_m}{\alpha\Delta t}$, $a_3 = \frac{1-\alpha_m}{2\alpha} - 1$, $a_4 = \frac{(1-\alpha_f)\delta}{\alpha} - 1$, $a_5 = (1-\alpha_f)(\frac{\delta}{2\alpha} - 1)\Delta t$.

For terrain category B, the basic wind speed was 23.7 m/s and the damping ratio of the structure was 5%. Then the wind vibration coefficient was calculated as follows:

$$
\beta_{Ri} = \frac{R_i}{\overline{R}_i} = 1 + \frac{g\sigma_t}{\overline{R}_i}
\tag{13}
$$

where β_{Ri} is the wind vibration coefficient of node i; R_i, $\overline{R}_I$, and σ_t are the overall response, average response, and pulsating response of node I, respectively; g is peak factor of node I, taken as 3.0 (Ke et al. 2012).

5.2. Distribution of Wind Vibration Coefficient

Based on the time history under eight working conditions, the wind vibration coefficient was calculated dynamically. The distributions of wind vibration coefficient for the whole construction process were discussed under five equivalent targets, as shown in Table 5.

Table 5. Five equivalent targets for value determination of wind vibration coefficient.

Equivalent Target	No.
Meridional axial force on the upwind side	Equivalent target 1
Von Mises stress on the upwind side	Equivalent target 2
Average absolute value of the mean response	Equivalent target 3
Maximum absolute value of the mean response	Equivalent target 4
Maximum pressure coefficient *	Equivalent target 5

Note: Maximum pressure coefficient * refers to wind pressure at the measuring point multiplied by the corresponding wind vibration coefficient.

The equivalent target 3–1 is the wind vibration coefficient and its mean value of the meridional axial force, the equivalent target 3–2 is the wind vibration coefficient and its mean value of the toroidal bending moment, and the equivalent target 3–3 is the wind vibration coefficient and its mean value of the radial displacement.

Equivalent goals 4–1, 4–2, 4–3 and equivalent goals 5–1, 5–2, 5–3 have the same representative meaning as above, which is not repeated here.

Figure 9 shows the distributions of wind vibration coefficient with construction height under five equivalent targets. It can be found that the wind vibration coefficient decreased with height under the eight working conditions. For the same construction model, the wind vibration coefficient was the largest under equivalent target 5, and that under equivalent target 1 was the smallest. Figure 10 shows the recommended values of wind fluttering coefficient under the eight working conditions for the five equivalent targets. Equivalent target 1 was the value of wind vibration coefficient, while all other equivalent targets were increments relative to target 1.

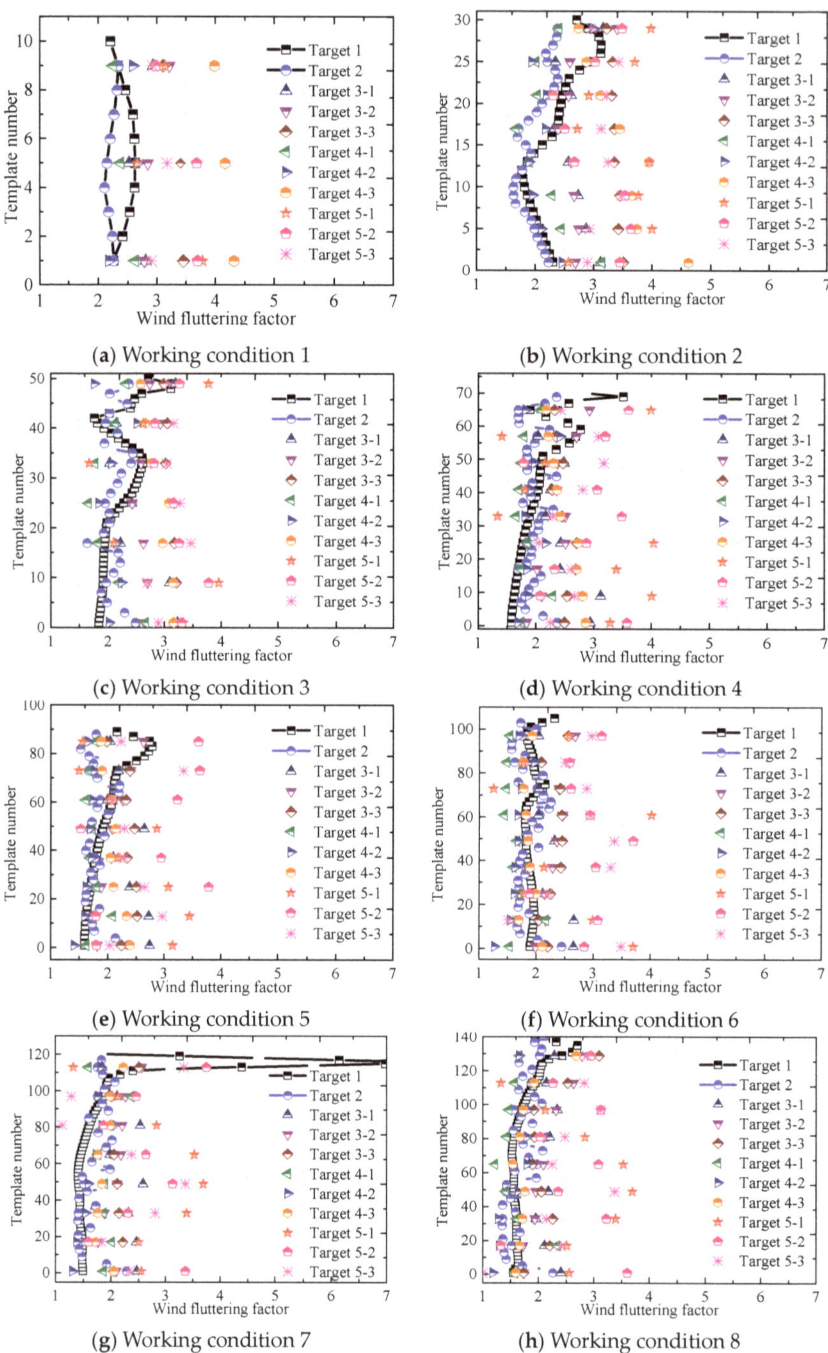

(**a**) Working condition 1

(**b**) Working condition 2

(**c**) Working condition 3

(**d**) Working condition 4

(**e**) Working condition 5

(**f**) Working condition 6

(**g**) Working condition 7

(**h**) Working condition 8

Figure 9. Wind vibration coefficient with meridional heights under five equivalent targets for each working condition.

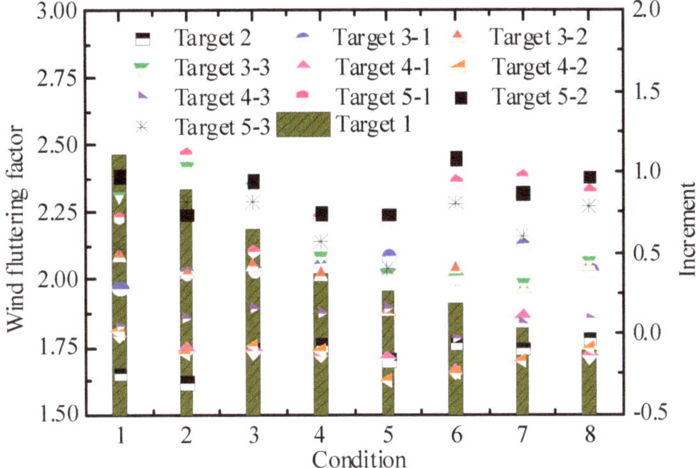

Figure 10. Comparison of wind vibration coefficient under the five equivalent targets between eight working conditions.

5.3. Fitting Formula of Wind Vibration Coefficient

Wind vibration coefficient of super large cooling towers is greatly affected by structural performance and wind pressure distribution during the whole construction process. However, the wind vibration coefficient does not linearly increase with the height significantly, and there is considerable discreteness of wind vibration coefficient for different equivalent targets. Therefore, the control of internal force, safety, and economic performance of the cooling towers were the major considerations for the determination of wind vibration coefficient in this study. We proposed the following formula of wind vibration coefficient by taking the meridional axial force as the target (equivalent target 1, GB/T 50102–2014, 2014):

$$y = \frac{m - \beta_0}{1 + \left(\frac{x}{n}\right)^k} + \beta_0 \tag{14}$$

where β_0 is the wind vibration coefficient of the built-up tower, $\beta_0 = 1.74$, m, n, and k are calculation parameters, x is the template number, and y is the wind vibration coefficient for the corresponding template number. After several iterations, the calculation parameters in the fitting formula were as follows: $m = 2.526$, $n = 116.511$, $k = 1.320$.

Figure 11 shows the fitted curve and the comparison of wind vibration coefficients under the five equivalent targets. The fitted curve could well reflect the differential values of wind vibration coefficient during the whole construction process when the meridional axial force was taken as the target. Table 6 shows the recommended values of wind vibration coefficient under each working condition.

Table 6. Recommended values of wind vibration coefficient under each working condition.

Working Condition	Working Condition 1	Working Condition 2	Working Condition 3	Working Condition 4	Working Condition 5	Working Condition 6	Working Condition 7	Working Condition 8
Wind vibration coefficient	2.47	2.34	2.19	2.02	1.96	1.91	1.82	1.74

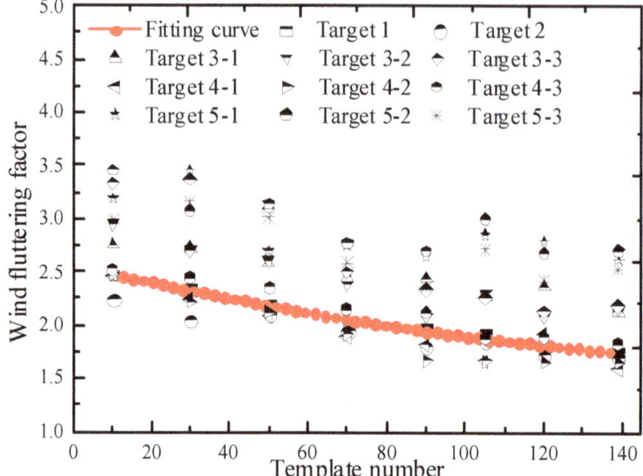

Figure 11. Fitted curve of wind vibration coefficients under the five equivalent targets for eight working conditions.

6. Stability Performance of the Cooling Tower during the Whole Construction Process

We then analyzed the evolution of wind-induced stability performance during the whole construction process of the super large cooling tower. The effects of wind vibration coefficient, concrete age, construction load, geometric non-linearity, and internal suction force on the buckling stability and ultimate bearing capacity of the cooling tower were discussed.

6.1. Influence of Concrete Age and Construction Load

The elastic modulus of concrete with different age was calculated under each working condition:

$$E_c(t) = E_c \sqrt{\beta_t} \tag{15}$$

where $E_c(t)$ is the elastic modulus (kPa) of C40 concrete with an age of t days, E_c is the elastic modulus of concrete shell with an age of 28 days, β_t is a coefficient, $\beta_t = e^{s(1-\sqrt{28/t})}$, S depends on the type of concrete, the value is 0.25 for ordinary cement and rapid-hardening cement, and t is the age of concrete (day). The Poisson's ratio and linear expansion coefficient of concrete with an age of t days were the same as those of concrete with an age of 28 days. Shear modulus was 0.4 times that of the elastic modulus.

Figure 12 shows the distribution of elastic modulus of concrete for different template number under working condition 2 (30 templates).

The construction loads were determined based on the following criteria: (1) the uniformly distributed load imposed by templates, slidewalk, scaffold, hanging basket, railings, a-frame, and supporting system to the shell below along the circumferential direction was about 3.6 kN/m, (2) the newly cast concrete exerted a uniformly distributed load along the circumferential was calculated as 25 × template height (1.277 m) × average thickness of the plate (m) kN/m, (3) the uniformly distributed load exerted by the construction workers turning over the template to the shell below along the circumferential direction was about 0.75 kN/m, and (4) concentrated load would be generated by the reinforcing bars stacked on the slidewalk. The maximum concentrated load produced this way was 18 kN; (5) Concentrated load would be generated by the weight of electric welder and switchboard and acted on the slidewalk, reaching a level of about 3.6 kN.

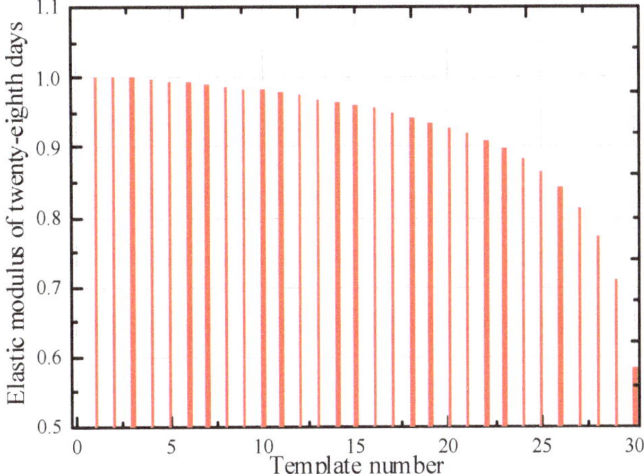

Figure 12. Distribution of elastic modulus of concrete under working condition 2.

Figure 13 shows the typical construction conditions of cooling tower under two kinds of wind loads during the whole construction process (wind pressure of standard wind vibration coefficient and wind pressure of actual wind vibration coefficient).The buckling coefficients and displacements of cooling towers under typical construction conditions during the whole construction process are analyzed under the two conditions of concrete age change and concrete age change. Comparison reveals that:

(1) As the construction height increased, the buckling coefficient decreased, and the rate of decrease became smaller over time. Buckling displacement showed a discrete distribution, and no consistent variation trend was observed. Thus, wind vibration coefficient and whether the concrete age and construction load were considered had little impact on buckling mode and buckling displacement under the eight working conditions.

(2) Buckling coefficient decreased if the concrete age and construction load were considered. The influence of wind vibration coefficient referred from the specification and the actual wind vibration coefficient on the buckling stability was much smaller than the influence of whether the concrete age and construction load were considered or not.

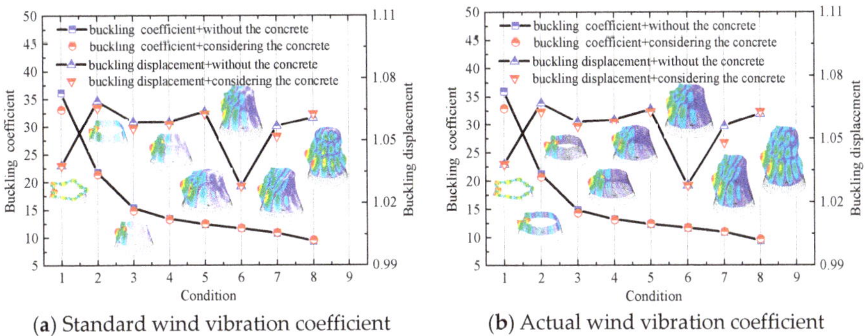

(a) Standard wind vibration coefficient   **(b)** Actual wind vibration coefficient

Figure 13. Changes of buckling coefficient and buckling displacement under eight working conditions.

6.2. Analysis of Geometric Non-Linearity

Figures 14 and 15 respectively show the typical construction conditions of cooling tower under two kinds of wind loads (actual wind-induced coefficient wind load, buckling wind speed wind load).Considering the concrete age change and not considering the concrete age change, the linear and non-linear calculation of the typical construction conditions during the whole construction process of the cooling tower is carried out, and the variation law of the maximum displacement of the structure is analyzed.

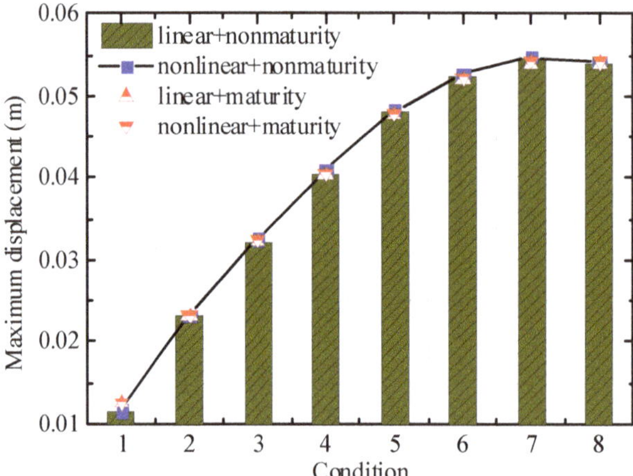

Figure 14. Comparison of maximum displacement under linearity and non-linearity of the structure for basic wind speed during the whole construction process.

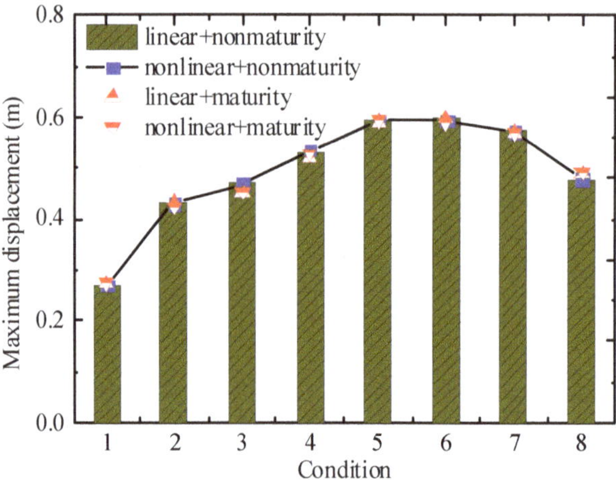

Figure 15. Comparison of maximum displacement under linearity and non-linearity of the structure for critical wind speed of buckling during the whole construction process.

It can be seen from the Figure 14, as the construction height increased, the maximum displacement of the tower increased constantly, but the amplitude decreased. The maximum displacement under working condition was smaller than that under working condition 7 due to the rigid ring constraint. The maximum displacement response was consistent under linearity and non-linearity of the cooling tower under different working conditions; the values differed very slightly.

It can be seen from the Figure 15 that the distribution pattern of maximum displacement changed under the critical wind speed of buckling when considering geometric non-linearity. Below is the linear analysis of wind-induced maximum displacement under some working conditions.

6.3. Influence of Internal Suction Force

The influence of internal suction force on stability performance of the cooling tower during the whole construction process was further analyzed. Figure 16 is the comparison of buckling coefficient and buckling displacement with or without internal suction force under different working conditions. There was an increment without internal suction force as compared with the condition with internal suction force. An absence of internal suction force caused a significant increment in the buckling coefficient during the whole construction process, but it had a lesser impact on buckling displacement. The maximum increment of buckling coefficient was 16.2%, and the maximum difference in buckling displacement was 1.2% in an absence of internal suction force without considering concrete age and construction load. The maximum increment of buckling coefficient was 16.6%, and the maximum difference in buckling displacement was −0.8% in an absence of internal suction force considering concrete age and construction load.

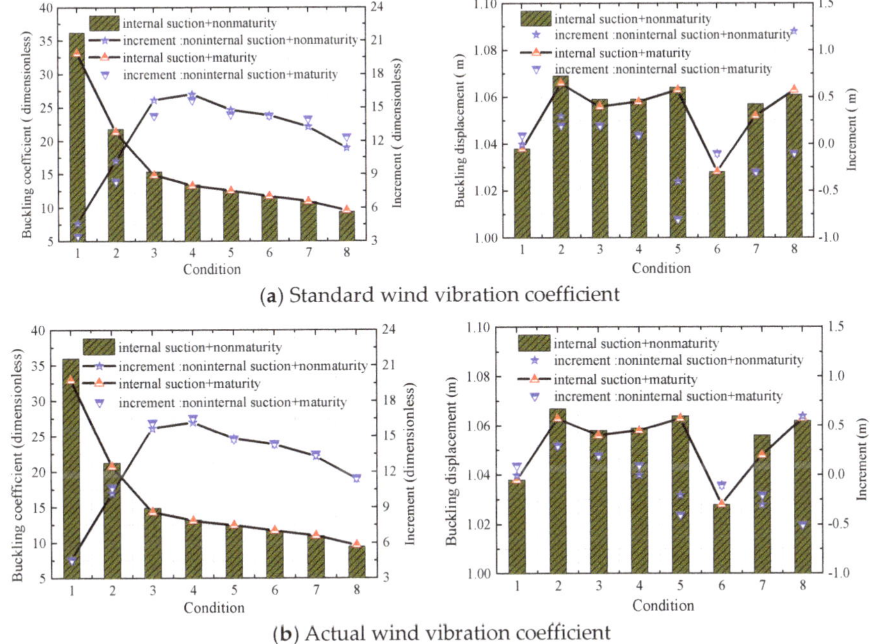

(**a**) Standard wind vibration coefficient

(**b**) Actual wind vibration coefficient

Figure 16. Comparison of buckling coefficient and displacement under standard and actual wind vibration coefficients.

Figures 17 and 18 are the comparisons of maximum displacement in linear and non-linear analyses under basic wind speed and critical wind speed of buckling for each working condition. It is easy to see that the internal suction force under the basic wind speed had less impact on the maximum displacement increment in the presence of internal suction force. In contrast, under the critical wind speed of buckling, the internal suction force caused a significant increment in maximum displacement under each working condition.

The comparison shows that when the wind load is the basic design wind speed, considering the influence of internal suction, the maximum displacement increment caused by wind has little influence on each working condition, and the influence is positive or negative.

However, when the wind load is buckling wind speed, considering the influence of internal suction, the maximum displacement of wind-induced displacement increases significantly under different working conditions.

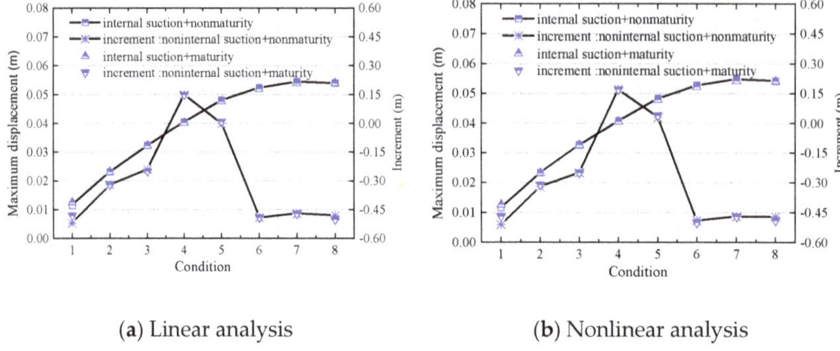

(**a**) Linear analysis   (**b**) Nonlinear analysis

Figure 17. Comparison of maximum displacement in linear and non-linear analyses under basic wind speed.

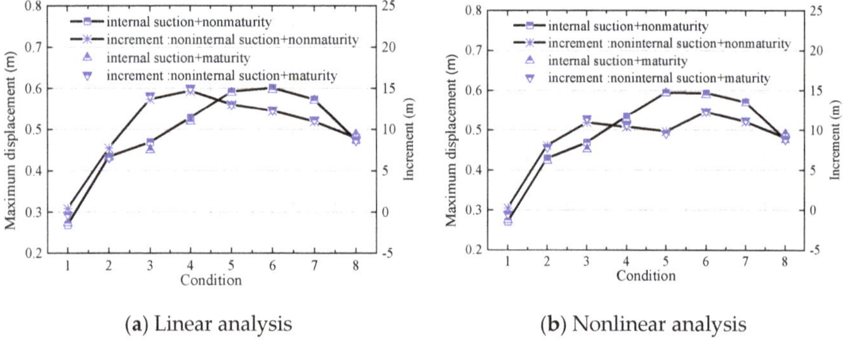

(**a**) Linear analysis   (**b**) Nonlinear analysis

Figure 18. Comparison of maximum displacement in linear and non-linear analyses under critical buckling wind speed.

6.4. Ultimate Bearing Capacity

Figure 19 provides the curve of maximum displacement with wind speed during the whole construction period for each working condition. The histogram indicates the changes of maximum displacement with wind speed under the standard wind vibration coefficient. The displacement under actual wind vibration coefficient considering concrete age, construction load, geometric non-linearity, and internal suction force was the increment relative to the condition of standard wind vibration coefficient. Stepwise loading was performed using the initial wind speed of 23.7 m/s at the height of 10 m as the baseline. The step length was 1–20 m/s. Comparison indicated that the increase of construction height greatly reduced the ultimate bearing capacity of the cooling tower. The critical wind speed of buckling decreased from 350 (±20) m/s to 100 (±20) m/s, and the decrease rate became smaller over time. No consistent variation trend was observed for the maximum displacement upon buckling under each working condition.

Ultimate bearing capacity of the cooling tower increased when considering the geometric non-linearity, and decreased when considering the concrete age and construction load. The ultimate bearing capacity during the whole construction process was sensitive to wind vibration coefficient. The ultimate bearing capacity was much lower under actual wind vibration coefficient at a low construction height. As the construction height increased, the ultimate bearing capacity of the structure calculated with the actual wind vibration coefficient increased gradually.

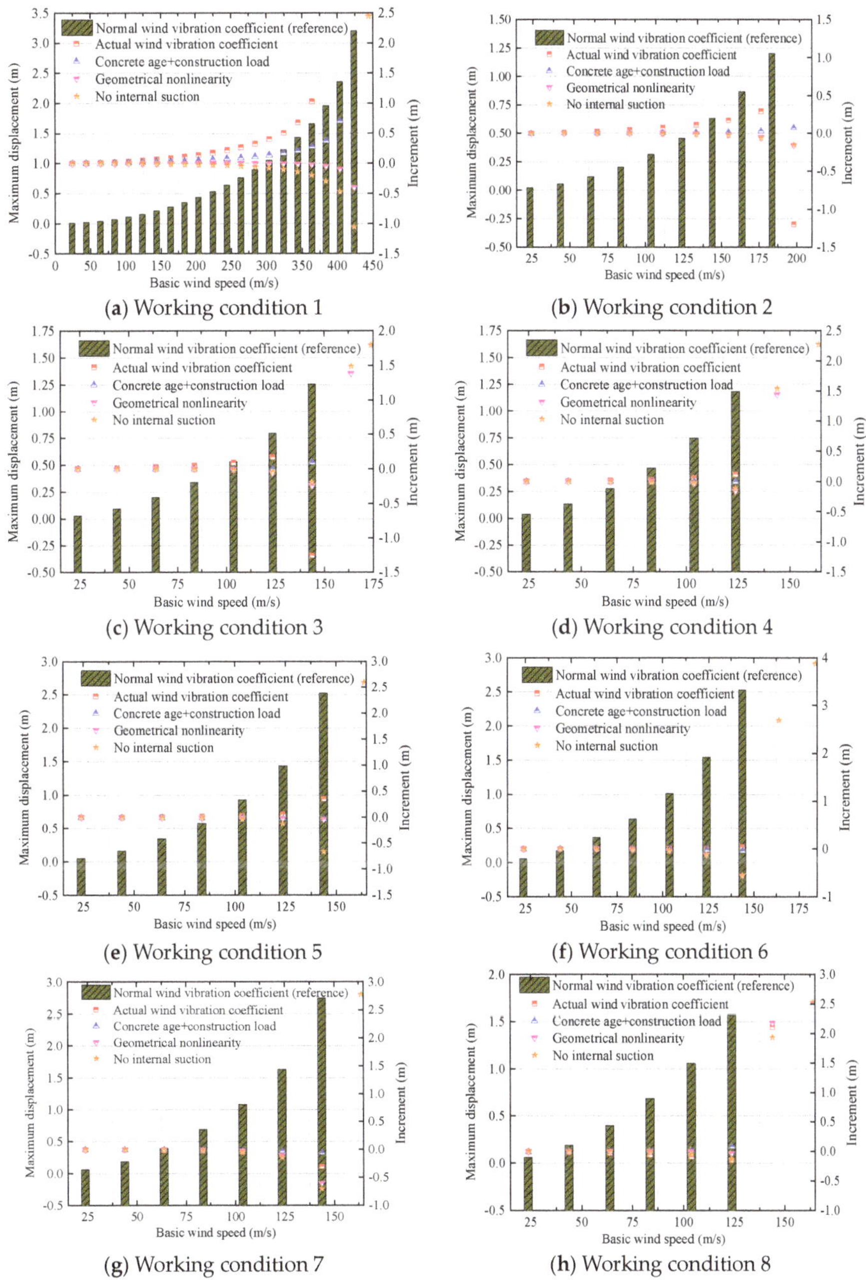

Figure 19. Changes of maximum displacement and increment with wind speed under eight working conditions.

7. Conclusions

We discussed the evolution of wind-induced stability performance and performed parameter analysis for the whole construction process of super large cooling towers. The contents of research included dynamic characteristics, wind vibration coefficient, wind-induced response, buckling instability, ultimate bearing capacity, and geometric non-linearity of the tower. The following conclusions were reached:

(1) The fundamental frequency of the built-up tower was 0.57 Hz. As the construction height increased, the fundamental frequency decreased. Construction height had a significant impact on the lower-order frequencies, but a lesser impact on the higher-order frequencies. The order of capsize mode increased with the increase of construction height, while the frequency of the excitation mode decreased gradually.

(2) The wind vibration coefficient decreased with construction height during the whole construction process. For the same construction model, the wind vibration coefficient was the maximum under equivalent target 5, and it was the smallest under equivalent target 1. Based on the calculation results, we proposed the formula of wind vibration coefficient by taking the meridional axial force as the target for the tower, as shown below. In the formula, x is the template number, and y is the wind vibration coefficient for the corresponding template number.

$$y = \frac{0.786}{1 + (\frac{x}{116.511})^{1.32}} + 1.74 \tag{16}$$

(3) The buckling coefficient of the cooling tower decreased, and the maximum displacement increased gradually as the construction height increased. Buckling displacement showed a discrete distribution, without a consistent variation trend. In addition, the ultimate bearing capacity of the tower decreased with the construction height. The critical wind speed of buckling decreased from 350 ($\pm$20) m/s to 100 ($\pm$20) m/s, and the decrease trend slowed down over time.

(4) The buckling coefficient of the tower during the whole construction period decreased when considering the concrete age and construction load. Geometric non-linearity had mild impact on the maximum displacement under the basic wind speed, but the impact was higher under the critical wind speed of buckling. The presence of internal suction force caused a reduction in the buckling coefficient of the cooling tower. The buckling coefficients calculated from standard or actual wind vibration coefficient differed little. The influence factors of bucking stability of the cooling tower can be ranked in a decreasing order as follows: internal suction force > geometric non-linearity > concrete age and construction load > wind vibration coefficient. The degree of influence of these factors did not show a consistent variation trend over the construction height.

(5) The ultimate bearing capacity of the cooling tower during the construction period increased when considering geometric non-linearity, and decreased when considering concrete age and construction load. The ultimate bearing capacity during the whole construction process was sensitive to the wind vibration coefficient. As the construction height increased, the ultimate bearing capacity calculated with the actual wind vibration coefficient increased gradually.

To conclude, the checking computation of stability performance of the super large cooling tower during the whole construction period should consider the effect of differential values of wind fluttering coefficient, as well as the influence of concrete age, construction load, and internal suction force. The effect of geometric non-linearity is negligible.

Author Contributions: Conceptualization, S.K.; Data curation, S.K.; Formal analysis, P.Z.; Funding acquisition, S.K.; Investigation, P.Z.; Methodology, Y.G.; Software, L.X.; Writing–original draft, S.K.; Writing-Review & Editing, S.K.

Funding: This project is jointly supported by National Natural Science Foundation (51878351; 51208254; U1733129, 51878351, 51761165022), Jiangsu Province Natural Science Foundation (BK2012390), and Postdoctoral Science Foundation (2013M530255; 1202006B), which are gratefully acknowledged.

Conflicts of Interest: The authors declare no conflict of interest.

References

1. Niemann, H.J.; Kopper, H.D. Influence of adjacent buildings on wind effects on cooling towers. *Eng. Struct.* **1998**, *20*, 874–880.
2. Orlando, M. Wind-induced interference effects on two adjacent cooling towers. *Eng. Struct.* **2001**, *23*, 979–992. [CrossRef]
3. Ke, S.; Liang, J.; Zhao, L. Influence of ventilation rate on the aerodynamic interference for two IDCTs by CFD. *Wind Struct. Int. J.* **2015**, *20*, 449–468. [CrossRef]
4. Radwańska, M.; Waszczyszyn, Z.; Radwańska, M. Buckling analysis of a cooling tower shell with measured and theoretically-modelled imperfections. *Thin-Walled Struct.* **1995**, *23*, 107–121. [CrossRef]
5. Noh, H.C. Nonlinear behavior and ultimate load bearing capacity of reinforced concrete natural draught cooling tower shell. *Eng. Struct.* **2006**, *28*, 399–410. [CrossRef]
6. Ke, S.; Ge, Y.; Zhao, L. Evaluation of Strength and Local Buckling for Cooling Tower with Gas Flue. *Comput. Struct. Eng.* **2009**, 545–551. [CrossRef]
7. Karisiddappa; Viladkar, M.N.; Godbole, P.N.; Krishna, P. Finite element analysis of column supported hyperbolic cooling towers using semi-loof shell and beam elements. *Eng. Struct.* **1998**, *20*, 75–85. [CrossRef]
8. Waszczyszyn, Z.; Pabisek, E.; Pamin, J.; Radwańska, M. Nonlinear analysis of a RC cooling tower with geometrical imperfections and a technological cut-out. *Eng. Struct.* **2000**, *22*, 480–489. [CrossRef]
9. Witasse, R.; Georgin, J.F.; Reynouard, J.M. Nuclear cooling tower submitted to shrinkage: behavior underweight and wind. *Nucl. Eng. Des.* **2002**, *21*, 247–257. [CrossRef]
10. Ke, S.; Ge, Y. Extreme wind pressures and non-Gaussian characteristics for super-large hyperbolic cooling towers considering aero-elastic effect. *J. Eng. Mech. ASCE* **2015**, *141*, 04015010. [CrossRef]
11. Ke, S.; Zhu, P. Impact study of the different air-deflectors on the wind pressure for super-large cooling towers. *J. Vib. Shock* **2016**, *35*, 136–141. (In Chinese)
12. Zhang, M.; Wang, F.; Li, Q.; Tang, D. Design wind loads on hyperbolic cooling towers during construction. *J. Tsinghua Univ. (Sci. Technol.)* **2015**, *12*, 1281–1288. (In Chinese)
13. Du, L.; Ke, S. Wind-induced Limit Bearing Capacity in Whole Construction Process for Cooling Tower based on ANSYS Secondary Development. *J. Vib. Shaock* **2016**, *35*, 170–175. (In Chinese)
14. Ke, S.; Zhao, L.; Zhang, J. Wind tunnel test and stability performance analysis of super large cooling tower with gas flue in power plant. *J. Harbin Inst. Technol.* **2011**, *43*, 114–118. (In Chinese)
15. Lu, C.; Li, Q.; Huang, S.; Chen, F.; Fu, X.; Guo, M. Simulation of large eddy of wind load on a long-span complex roof structure. *China Civ. Eng. J.* **2011**, *44*, 1–9.
16. VGB-R610Ue. *VGB-Guideline: Structural Design of Cooling Tower-Technical Guideline for the Structural Design, Computation and Execution of Cooling Tower*; BTR Bautechnik Bei Kühltürmen: Essen, Germany, 2005.
17. Blanchette, J.; Nelson, T.F.; Xuan, Z. The British Standard Institution. In Proceedings of the Apha Meeting and Exposition, New Orleans, LA, USA, 2013; pp. 633–634.
18. DL/T 5339-2006. *Code for Hydraulic Design of Fossil Fuel Power Plants*; The Ministry of Construction of China: Beijing, China, 2006. (In Chinese)
19. GB/T 50102-2014. *Code for Design of Cooling for Industrial Recirculating Water*; The Ministry of Construction of China: Beijing, China, 2014. (In Chinese)

Article

Research on Snow Load Characteristics on a Complex Long-Span Roof Based on Snow–Wind Tunnel Tests

Guolong Zhang [1,2], Qingwen Zhang [1,2,*], Feng Fan [1,2] and Shizhao Shen [1,2]

[1] Key Laboratory of Structures Dynamic Behavior and Control of China Ministry of Education, Harbin Institute of Technology, Harbin 150090, China; zhanggl_hit@126.com (G.Z.); fanf@hit.edu.cn (F.F.); szshen@hit.edu.cn (S.S.)

[2] Key Laboratory of Smart Prevention and Mitigation of Civil Engineering Disasters of the Ministry of Industry and Information Technology, Harbin Institute of Technology, Harbin 150090, China

* Correspondence: zhangqw@hit.edu.cn; Tel.: +86-1504-581-1536

Received: 2 September 2019; Accepted: 14 October 2019; Published: 16 October 2019

Featured Application: **This study provides a simplified method that can be applied to help preliminarily estimate the snow load on a complex roof by combining the snow loads on several simple roofs, which have been provided in the load code or obtained by numerical studies.**

Abstract: A considerable number of studies have been carried out for predicting snowdrifts on roofs over the years. However, few studies have focused on snowdrifts on complex long-span roofs, as the complex shape and fine structure pose significant challenges. In this study, to simplify the calculation requirements of snow load on such roofs, work was conducted to decompose the snowdrift on a complex roof into snowdrifts on several simple roofs. First, the snow–wind tunnel test similarity criteria were investigated based on a combined air–snow–wind experimental system. Thereafter, with reference to the validated experimental similarity criteria, a series of snow–wind tunnel tests were performed for snowdrifts on a complex long-span structure under the conditions of different inflow directions. Finally, based on empirical orthogonal function (EOF) analysis, the snowdrifts on the complex roof were decomposed into basic characteristic distribution modes, including snowdrifts caused by the local and overall roof forms. The snow distribution under a specific inflow direction could be derived from the weighted combination of the basic characteristic modes, based on the wind direction coefficients. Therefore, it is possible for the snow load on a complex roof to be estimated preliminarily based on the snow distributions on several simple roofs.

Keywords: snow load; complex roof; snowdrift; EOF analysis; characteristics decomposition

---

1. Introduction

Snow-induced building collapse occurs frequently in long-span structures; for example, the collapse of the Katowice Trade Hall (Poland, 2006), the collapse of an ice rink in Bad Reichenhall (Germany, 2006), and the collapse of the Minnesota Vikings membrane stadium (USA, 2010). The sudden increase in the snowfall and unbalanced snow distribution, which leads to an excessive snow load in a local area, are the main reasons for such collapses [1]. Unfortunately, the snow load requirements are only provided for the design of buildings with certain simple roofs (e.g., flat roof, pitch roof, and gable roof, among others) in different national load codes [2,3], and few requirements are available for long-span structures. It is usually suggested that the snow load distribution coefficients for large or special-shaped buildings be defined following specific research or experiments in certain national codes [4,5].

In terms of the related special research methods, the computational fluid dynamics (CFD) method and wind tunnel experiment are usually adopted for the practical design of snowdrifts on or around buildings [6,7]. In the field of CFD research, Tominaga et al. [6] analyzed the snowdrift around an actual apartment building based on the revised k-ε model. In this study, the transport equation of snow concentration was solved only for snow particles suspended in the air. Thiis et al. [8,9] predicted the snow distribution on the curved roof of a sports hall under specific weather conditions. Generally, the overall snowdrift pattern fitted the measured result well and the location of end effects was reproduced close to the side edge of the roof. Beyers et al. [10] simulated the snowdrift development around a group of surface-mounted buildings and elevated structures, respectively. The analysis assisted the conceptual building design process to manage potential snowdrifts on and around these structures. The application of the CFD technique to snowdrift problems can provide detailed information on the relevant flow and phase variables in the entire calculation domain under well-controlled conditions and at a low cost. Unfortunately, the numerical technique is mainly applied to the snowdrift on small-scale roofs with usual shapes or simplified long-span roofs, as the large scale and fine structural details of complex roofs require higher resolution and higher quality grids. Cases with finer grids would be more time consuming, and the accuracy will be reduced if the grid quality is poor. Therefore, rare work was carried out on the snowdrift on complex long-span roofs by using the CFD method.

In the field of experimental research, Isyumov et al. [11] examined snowdrift formation on the lower level of a large-area two-level roof in different-surface shear stress and terrain roughness conditions. Delpech et al. [12] explored the hazard and alleviating measures of the snowdrift around the Concordia Antarctic research station by using real snow particles in the Jules Verne climatic wind tunnel. Flaga et al. [13] performed a series of snow load tests for three different complex roofs of sports facilities. Snow precipitation and wind-induced redistribution were simulated by using powdered polystyrene foam as artificial snow. The results of snow load distributions were presented for the practical structural design. Compared with the CFD method, this experiment can reproduce the snowdrift mechanism to the greatest extent and restore every detail of the building. Unfortunately, the required equipment is not always available and is usually expensive for experimental preparation and model creation. Therefore, it greatly limits the large-scale application of the experimental method in the study of snowdrift problems.

In order to analyze and summarize the snow load information on complex long-span roofs more efficiently and systematically, assisting the design process of the structure of a building, it is necessary to combine the advantages of CFD and experiment methods and avoid their limitations. Specifically, the distribution of snow load on a complex long-span roof should be analyzed in detail by using experiments. Based on the results, the distribution characteristics could be summarized systematically, before helping reduce the computing requirements for the snowdrifts on complex roofs and the large-scale analytical studies by using the CFD method. As a preliminary step, it is important to make clear the snow distribution characteristics on a complex long-span roof.

In order to express the distribution characteristics in detail, this study firstly presents the validation of the similarity criterion to figure out the experimental theory, based on an air–snow–wind combined experimental system. Thereafter, based on the similarity criterion, a description of the wind-induced snow drifting on a long-span structure with a membrane roof under different wind direction conditions is provided. Finally, according to the experimental results, the snowdrifts' characteristics on the complex membrane roof are analyzed. Three basic characteristic distribution modes and wind direction series are decomposed and derived from the results. The distribution pattern and wind direction coefficients of each mode are investigated with reference to the roof form.

2. Validation of Experimental Approach

2.1. Experimental Facility

Experimental works were carried out based on an air–snow–wind combined experimental system, which allows for the appropriate creation of natural wind velocity and turbulence profiles, as well as the precipitation environment. The snow–wind tunnel facility used belongs to the Key Laboratory of Structures Dynamic Behavior and Control of China Ministry of Education, Harbin Institute of Technology, Harbin, China, as shown in Figure 1. The facility was improved on the basis of previous research [14]. The snow–wind tunnel test chamber was closed, and the dimensions were 10 × 4.5 × 3 m. The temperature inside the chamber was the same as the outdoor air temperature, which ranged from −30 °C to −10 °C during winter. The wind velocity was measured with a hot-wire anemometer.

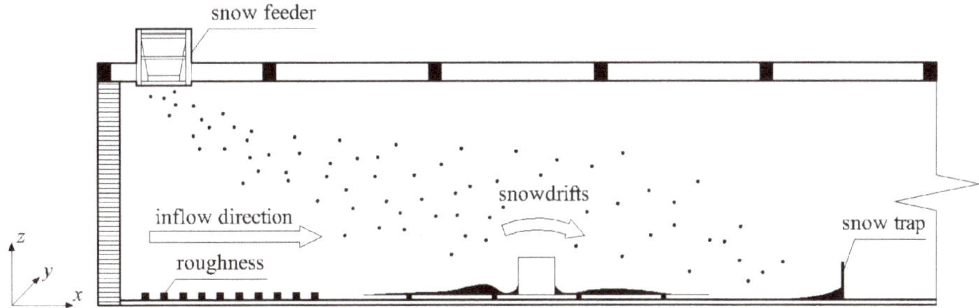

Figure 1. Schematic view of the experimental facility.

Artificial snow particles were used to simulate the snowdrift environment to deal with the combined wind–snow engineering problems (Figure 2a). The artificial snow particles were generated by spraying water droplets into freezing air with a snowmaker. A particle feeder was fixed at the top of the chamber to simulate precipitation. The snow particle feeder (Figure 2b) consisted of a stable steel frame and vibrating sieve with a perforated bottom (bottom size: 4.5 m long, 0.4 m wide; hole diameter: 5 mm). The sieve was moved by a motor with continuous speed regulation. The snowdrift flux profile was measured with a snow particle counter [15].

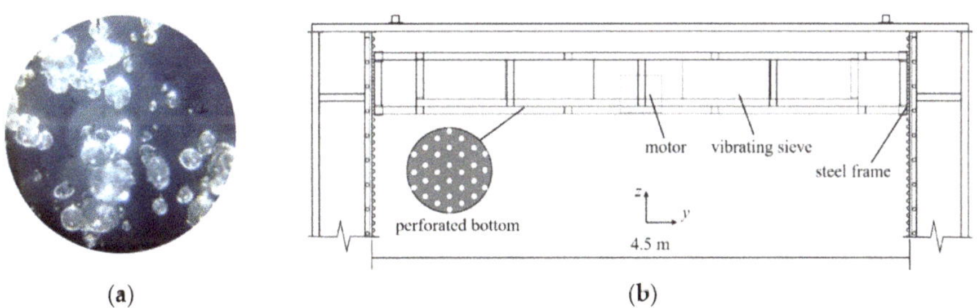

Figure 2. Snowfall simulator: (**a**) artificial snow particles; (**b**) snow particle feeder.

2.2. Similarity Criterion

The snowdrift phenomenon can be explained as solid particle transport. Depending on whether the friction velocity reaches a threshold, three particle transport mechanisms can be observed: creep, saltation, and suspension. Creep is a phenomenon in which snow particles move by rolling, sliding, or creeping at the surface; saltation is a process in which snow particles move with repeated leaping up or jumping and colliding with a snow surface; and at a higher wind velocity, particles are transported upwards by turbulent eddies and transported far downwind. Among the mechanisms, saltation has been identified as the major particle transport process, which causes approximately 67% of the total drifting mass [12]. For reliable modeling of these transport mechanisms, a reasonable similarity criterion is necessary for reduced-scale experiments. As the similarity criterions are incompatible with one another, compromises have to be made according to their relevance [12,16]. Even so, none of the models cited in the literature are able to provide a satisfying interpretation of experimental results.

In this study, considering that the experimental modeling was focused on the reliable preproduction of snowdrift on a long-span roof, the similarity number based on the similarity of the drifting snow flux in the saltation layer was selected. These similarity criteria were first introduced by Iversen [17] and also adopted by Delpech [12]. The similitude criterion based on the drifting flux is shown in Equation (1).

$$(\rho/\rho_p)(U^2/2g)(1 - U_0/U) \tag{1}$$

where ρ and ρ_p are the air and particle densities, respectively; U and U_0 are the reference wind velocity and threshold reference velocity, respectively; and g is the gravitational acceleration. Furthermore, similarity criteria related to geometric, dynamic, and kinetic criteria were also considered.

To validate the similarity criterion, snowdrift around a surface-mounted cubic model, employed in detailed field measurements carried out by Oikawa et al. [18] in Sapporo, Hokkaido, Japan, was adopted as the analyzed prototype. The snowdrift was observed for only one day by cleaning up the snow around the model following each drifting snow event. The model was 1.0 m on each edge. The averaged wind velocity was approximately 1.7 m/s at a 1.0 m height, and the maximum wind velocity was close to 4.3 m/s (Figure 3). The snow depth at a reference point was 20 cm. The other measurement parameters are summarized in Table 1. The smallest undulation of snowdrift due to the weak wind velocity and the large snowfall make it the most suitable measurement data for validation of the similarity criteria.

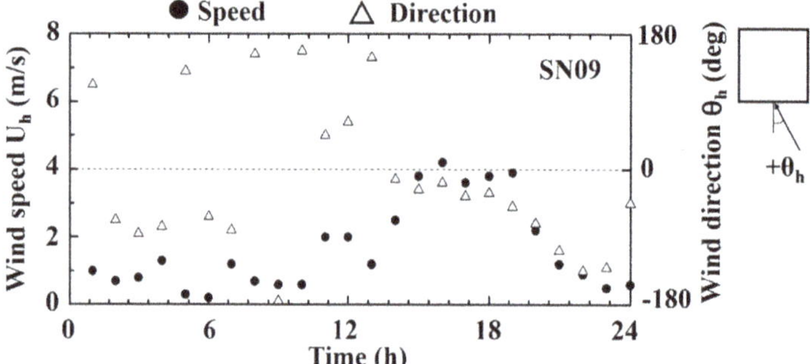

Figure 3. Wind speed U_h and direction θ_h at 1.0 m height for the entire measuring period on SN09 [18].

Table 1. Prototype drifting parameters.

Particle Parameters	Values	Particle Parameters	Values
Particle diameter D_p	150 to 200 μm	Threshold friction velocity U_t^*	0.15 to 0.36 m/s
Snowfall velocity W_f	0.2 to 0.5 m/s	Accumulated snow density ρ_p	50 to 700 kg/m^3
Air density ρ	1.22 kg/m^3	Repose angel	30° to 50°

The artificial snow particles were selected for the validation experiment. The air/water ratio in the snowmaker was set to make sure the artificial snow was dry enough to prevent the particles from sticking together. The accumulated snow density was measured with a scale and cylinder; whereas, the particle diameter was measured with a microscope. The snowfall velocity was obtained by measuring the artificial particle falling times from 1.0 m height [8]. The threshold friction velocity was estimated according to an empirical formula [1]. The parameters of the artificial snow particles are summarized in Table 2.

Table 2. Parameters relating to the physical properties of artificial snow.

Particle Parameters	Values	Particle Parameters	Values
Particle diameter D_p	100 to 150 µm	Threshold friction velocity U_t^*	0.3 to 0.6 m/s
Snowfall velocity W_f	2.0 to 2.4 m/s	Accumulated snow density ρ_p	386 to 447 kg/m^3

The selected model scale of 1/2 satisfied the blockage effects. The vertical profiles of the flow field identified over the snow mantle in the test section were similar to these over the measured field in Sapporo. In order to assess the modeling reliability, the results of the typical similarity numbers for both the prototype and model are summarized in Table 3. The experimental wind velocity (2.3 m/s at the height of the model top) used for calculating the scaled model values was determined according to Equation (1). Requirement 2 in Table 3 represents the geometric similarity of both model and prototype. The correct modeling of the ejection process of particles is realized by satisfying requirements 3, 4, and 5. Requirement 5 is a roughness–height Reynold number, where ν is the fluid's kinematic viscosity. The basic requirements of dynamic similarity could be measured with densimetrical Froude numbers 6 and 7 (ratio between the inertia force and gravity force), and similarity number 8.

Table 3. Similitude parameters.

Dimensionless Parameters	Prototype Value	Model Value	Number
D_p/L	1.5×10^{-4} to 2×10^{-4}	2×10^{-4} to 3×10^{-4}	(2)
U/U_t^*	4.7 to 11.3	3.8 to 7.7	(3)
ρ_p/ρ	41 to 574	316 to 366	(4)
$U_t^{*3}/2g\nu > 30$	11.4 to 157	91 to 730	(5)
$\dfrac{U_t^{*2}}{D_p g}\left(\dfrac{\rho}{\rho_p - \rho}\right)$	0.02 to 2.21	0.17 to 1.17	(6)
$\dfrac{U^2}{D_p g}\left(\dfrac{\rho}{\rho_p - \rho}\right)$	2.58 to 49.38	9.89 to 17.19	(7)
W_f/U	0.1 to 0.3	0.9 to 1.0	(8)

In order to model the fully rough saltation flow, it is desirable to guarantee the lower limit of the Reynolds number $U^{*3}/2g\nu > 30$. If the saltation mechanism occurs, the fully rough flows will be satisfied if $U_t^{*3}/2g\nu > 30$ [16]. This lower limit was satisfied in this experimental case as shown in Table 3. Except for this, the noticeable mismatches are found in the particle ejection process scaling and dynamic similarity. For the particle ejection scaling, the gravitational force is overestimated in requirement 4 with a greater particle density, which is also observed in requirement 8 with a higher snowfall velocity. This indicates that the trajectory of the artificial snow particle is smaller than the natural snow particle one [19]. However, a higher particle/air density ratio is usually required [19] and the saltation trajectory for the scale model is small in comparison with the actual snowdrift observed [20]. For the dynamic similarity, the particulate Froude number weighted by the density ratio (requirement 7) is the parameter that allows for assessing the similarity of the transport mechanism of suspended particles [12]. The evaluation of the particulate Froude number based on the threshold friction velocity (requirement 6) is linked to surface transport. According to comparisons of requirement 6 and requirement 7, the saltation mechanism, which has been identified as the major particle transport process, is better reproduced than other transport processes [12].

The prototype snowstorm duration was assumed to be 24 h according to the observation duration [18]; whereas, the experimental snowstorm duration was set to approximately 8 h according to the definition for the dimensionless time used by Delpech [12]. As the experiment would have been excessively lengthy, the test was divided into four stages to allow the fan to rest and refill the snow particles into the feeder.

2.3. Results and Discussion

Figure 4 compares the dimensionless snow depth distributions obtained from the field measurement and experiment. The snow depths were normalized by the reference snow depth far from the model, which was not affected by the flow around the cube. The deep-colored part in the figure indicates greater snow coverage. The deposition areas in the upwind region ahead of the cube and erosion areas near the upwind corners of the model in the experiment correspond strongly to those of the prototype. However, as the snow particle feeder was set in the upwind region, the air–snow flow originated from the upwind direction. The incoming particles bypassed the model and moved downstream following the separating airflow. Few particles could enter the wake region behind the model with the aid of the vortex. Therefore, the deposition region behind the model was not reproduced in the experiment. Therefore, the building size should be limited to ensure that the entire building was covered by the stable air–snow flow field, on the premise of satisfying the drifting similitude.

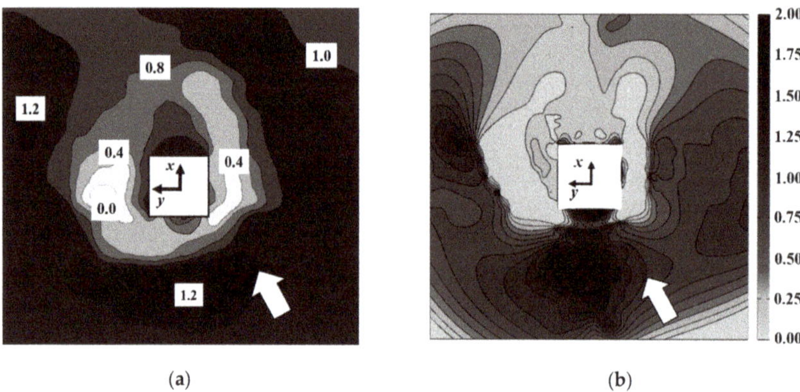

(a) (b)

Figure 4. Comparison of horizontal distributions of normalized snow depths: (**a**) prototype result [18]; (**b**) experimental result.

3. Experimental Research on Snowdrifts on Complex Long-Span Roof

3.1. Experimental Setup

3.1.1. Test Model

For buildings with a special shape, complex separation and reattachment occur when airflow passes over the roof. The snow load on the roof will be redistributed easily, which may change the stress inside the roof structures, thereby leading to building collapse, especially for long-span lightweight roofs. Therefore, significant attention should be paid to the snow distribution on complex-shaped buildings. Based on the advantages of the experimental method and the validated similarity criterion, a series of snow–wind tunnel tests were conducted to investigate the snowdrift on such complex long-span roofs. In this study, Tongren Olympic Sports Center Stadium was selected as the target building. As the stadium is located on the Yunnan-Guizhou Plateau, China, the frequent snowfall requires additional consideration. The geometric shape of the roof is illustrated in Figure 5a. The cable membrane structure was adopted for the prototype roof. In order to avoid the excessive deformation of the roof, which may affect the safety of the structure and comfortableness of the space, the roof deformation should be strictly controlled at the structural design stage. Therefore, the roof deformation under the action of snow load was assumed to be negligible. Furthermore, considering that the unique roof shape substantially affects the shape of the snow cover, acrylonitrile butadiene styrene (ABS) plastic was used for creating the model roof to reproduce the complex shape as far as possible. Similar hard material was also used by Flaga et al. [13] to make the membrane structure models. Perforated plates were adopted for the sidewall, with an air permeability of approximately 40% to realize the simulation of the wall air permeability (Figure 5b). Based on the above conclusion, the model scale of 1/150 was selected with model size limited to 1.81 (x) × 1.69 (y)0.33 × (z) m.

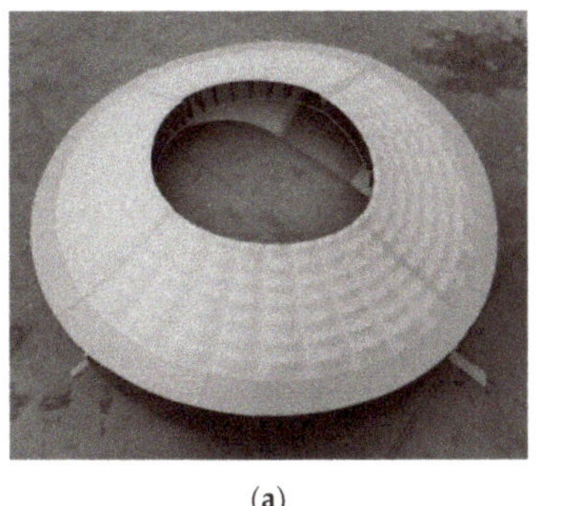

(a)

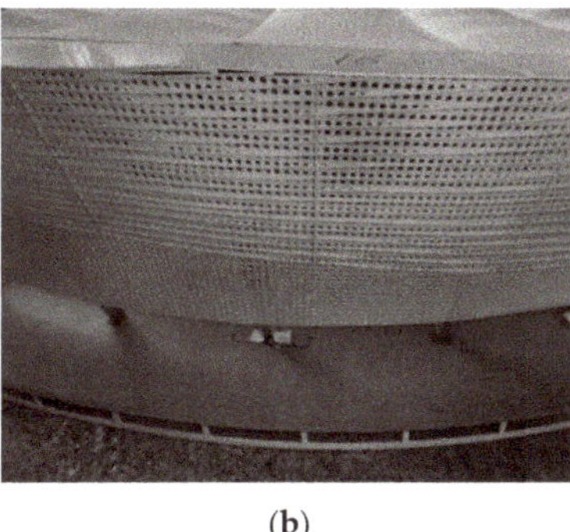

(b)

Figure 5. Model of membrane structure gymnasium: (**a**) Tongren Olympic Sports Center Stadium model (made of acrylonitrile butadiene styrene (ABS) plastic); (**b**) wall of membrane structure (40% air permeability).

The measuring points were designed according to the special shape of the roof (Figure 6a), combined with the distribution characteristics whereby snow would be deposited in the concave region more easily. Two types of concave regions were formed on the roof: along the radial cable and near the center of each membrane piece. For the concaves along the radial cables, 36 lines were set along the cables, with 16 measuring points set along each line (Figure 6b). For the concaves formed on the membrane, 36 lines were set along the circumferential direction, with an additional 9 measuring points set at the center of each membrane piece (Figure 6b). In total, 900 measuring points were set on the model roof. The snow depth at the measuring point was measured with a snow stick. Furthermore, a snowdrift experiment with an empty field was conducted to simulate the snow distribution on the ground. The shape coefficient μ of the snow distribution on the roof was calculated as the ratio of the snow depth on the model roof to the snow height on the ground.

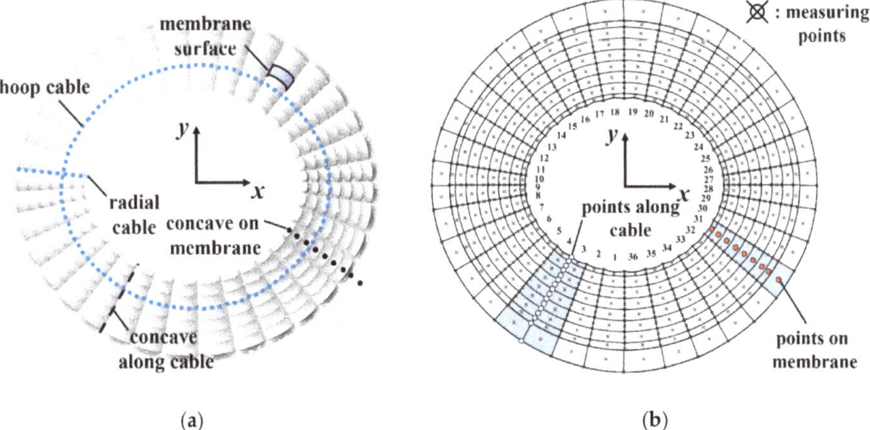

(a)

(b)

Figure 6. Arrangement of measuring points for snow depth: (**a**) roof form; (**b**) arrangement of measuring points.

3.1.2. Setup Parameters

The prototype stadium was constructed in a suburban area of Tongren City, Guizhou province, China. The reference wind pressure is 0.35 kN/m^2 (100 year return period); whereas, the reference snow pressure is 0.35 kN/m^2 (100 year return period) [2]. The prototype wind velocity at a standard height (10 m) during a snowstorm is assumed as 0.45 of the reference wind velocity, calculated based on the reference wind pressure [21]. The air density there is about 1.22 kg/m^3, and hence the reference wind velocity is close to 24 m/s. The snow density is about 150 kg/m^3, and the maximum snow depth during a snowstorm could reach 0.23 m [2]. The threshold friction velocity and snowfall velocity are set to 0.15 m/s and 0.2 m/s, respectively. Regarding the snowfall duration, as no outdoor measuring record was available; the experimental duration was provided based on an eight-year-long outdoor measurement in China by the authors. This work was carried out from 2010 to 2017 and is ongoing. The measurements included the snowfall duration, interval time between two snowfalls, accumulated snow density, and snow depth. Based on the measurement data, a single snowfall usually lasted for nearly 6 to 7 h. The probability of snowfall lasting for less than 12 h was approximately 72% (Figure 7). Therefore, the prototype duration was set as 12 h.

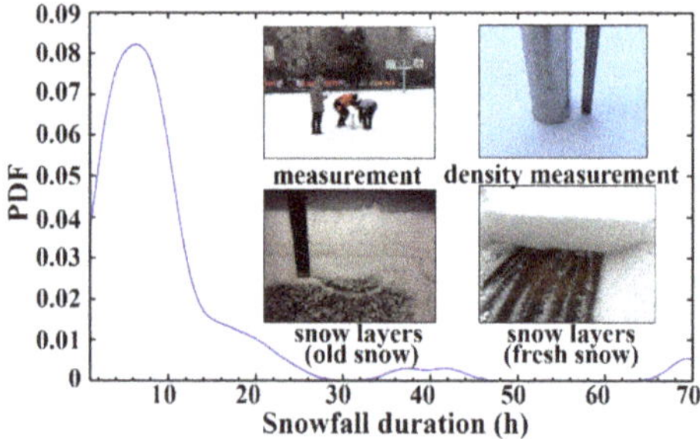

Figure 7. Probability density curve of single snowfall duration, derived from an eight-year field measurement result.

The artificial snow particles were selected as the experimental particles. Prior to the experiment, the snow was sieved into uniform sphere particles. Based on the similarity criterion introduced in Section 2.2, the experiment reference velocity was 1.2 m/s at 0.06 m, and the experimental snowstorm duration was set to 0.6 h. A total of 700 L of artificial snow was poured into the sieve-like feeder during the experimental duration, and the snowfall flux was approximately 0.007 kg/m^{2}s. As this stadium exhibited biaxial symmetry and the wind direction frequency was provided with a 22.5° interval in the annual wind rose diagram, five cases were designed and conducted for different inflow directions at 22.5° intervals, namely 0°, 22.5°, 45°, 67.5°, and 90°, from the x-axis to y-axis (Figure 8).

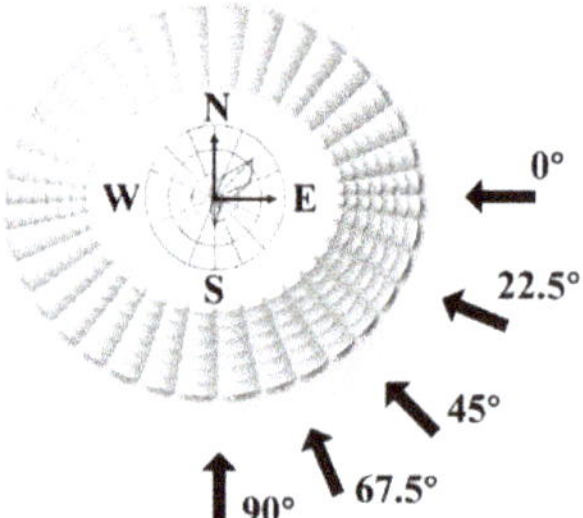

Figure 8. Study cases conducted for different inflow directions at 22.5° intervals (0°, 22.5°, 45°, 67.5°, and 90°).

3.2. Results and Discussion

The snowdrift shape coefficients μ on the membrane roof in different inflow directions are illustrated in Figure 9. The snowdrifts are extremely complex. For different inflow conditions, depositions occur on the windward and leeward surfaces along the air stream. The overall packing forms are approximately fan-shaped. When the inflow direction moved from 0° to 90°, the angle between the windward and leeward depositions varies from 90° to 140°. The results are similar to the unbalanced snow distribution on the umbrella-shaped roof, as indicated in the technical specification for cable structures (Figure 10a) [22]; whereas, in the experimental case, the angle between the depositions changed under asymmetrical conditions because of the non-central symmetrical roof form. Owing to the existence of the concave regions along the radial cables and at the center of each membrane piece, a large amount of snow is accumulated locally and a peak value of approximately 1.8 is generated. This is substantially larger than the peak value of 1.0 for the umbrella-shaped roof, but much closer to the maximum coefficient of the unbalanced distribution on the multi-span gable roof (Figure 10b) [2]. Therefore, the local accumulations that formed on the roof are likely caused by the local form of the roof itself. Evidently, the snow distribution pattern is closely related to the overall and local roof shapes.

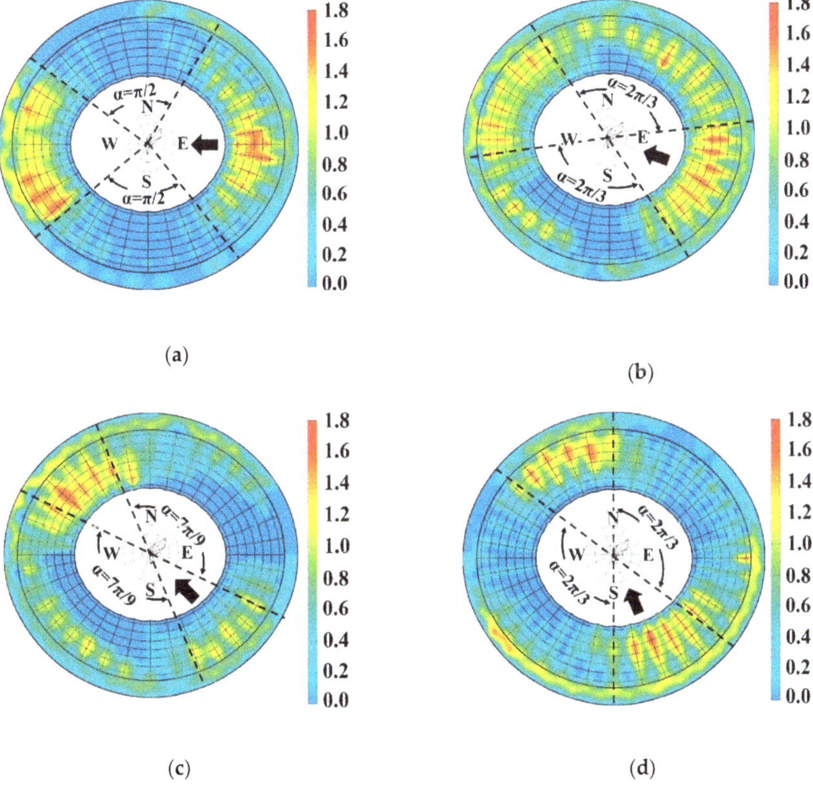

Figure 9. *Cont.*

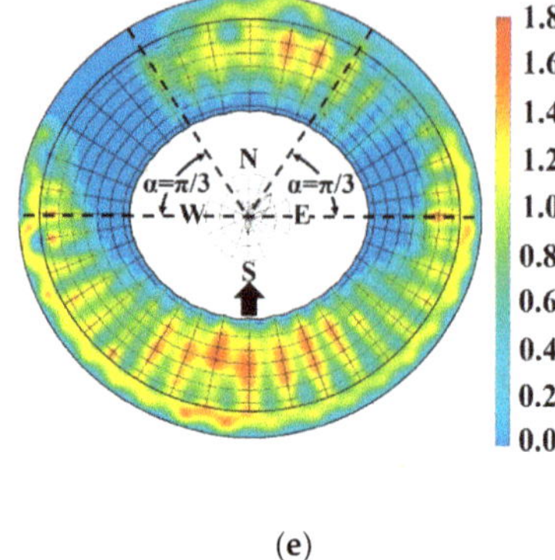

(**e**)

Figure 9. Snow distribution shape coefficient μ: (**a**) 0°; (**b**) 22.5°; (**c**) 45°; (**d**) 67.5°; and (**e**) 90°.

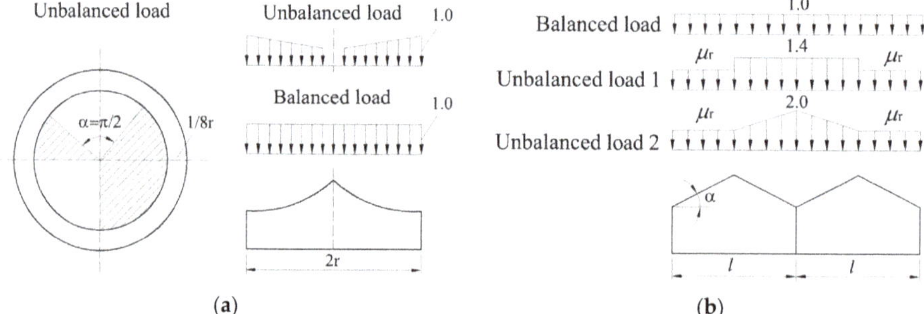

Figure 10. Unbalanced distributed shape coefficient on building roofs: (**a**) shape coefficient for umbrella-shaped roof indicated in the technical specification for cable structures; (**b**) shape coefficient for multi-span gable roof indicated in load code for design of building structures.

4. Analysis of Snow Distribution Characteristics

4.1. Analytical Method

To deeply analyze the relationship between the snow distribution pattern and the shape characteristic of the roof, the empirical orthogonal function (EOF) analysis was adopted. This approach is generally used to analyze the structural features in data and extract major data feature quantities. Specifically, EOF analysis can decompose a variable field matrix **X** that changes with time into a time-independent spatial matrix **EOF** and a time matrix **PC**, as shown in Equation (9). In this study, the time-dependent field matrix was replaced by a field matrix that changed with the wind direction, and a wind direction related matrix could be obtained instead of the time matrix. The spatial matrix **EOF** generalizes the geographical distribution characteristics of the field; whereas, the wind direction matrix **PC** is composed of the linear combination of coefficients of the spatial points of the field. As the main information of the original field **X** is concentrated in the first several components of the spatial matrix **EOF**, the study of the original field **X** changing with wind direction can be converted into the investigation of the first components of the spatial matrix **EOF** and their wind direction matrix **PC**, respectively.

$$\mathbf{X} = \mathbf{EOF} \times \mathbf{PC} \tag{9}$$

The specific method is illustrated in Equation (10)–(14). Firstly, a field matrix $\mathbf{X}_{m \times n}$ is obtained by collating the measured data, where the subscript "m" represents the number of spatial points under a certain inflow direction condition, namely 900 and the subscript "n" represents the number of inflow cases, namely five. Based on the orthogonal decomposition theory, the spatial matrix **EOF** and its eigenvalue matrix Λ can be derived, as shown in Equation (11)–(13), where **C** is the cross-product of the field matrix **X** and its transpose matrix $\mathbf{X}^T$. The kth column of the spatial matrix **EOF** represents the kth spatial mode $\mathbf{EOF}_k$. Finally, the wind direction matrix **PC** can be derived from the spatial matrix **EOF** and original field matrix **X**, as indicated in Equation (14).

$$\mathbf{X}_{m \times n} = \begin{bmatrix} x_{11} & x_{12} & \cdots & x_{13} \\ x_{21} & x_{22} & \cdots & x_{23} \\ \cdots & \cdots & \cdots & \cdots \\ x_{m1} & x_{m2} & \cdots & x_{mn} \end{bmatrix} \tag{10}$$

$$\mathbf{C}_{m \times n} = \frac{1}{n} \mathbf{X} \times \mathbf{X}^T \tag{11}$$

$$\mathbf{C}_{m \times n} \times \mathbf{EOF}_{m \times m} = \mathbf{EOF}_{m \times m} \times \Lambda_{m \times m} \tag{12}$$

$$\Lambda_{m \times m} = \begin{bmatrix} \lambda_1 & 0 & \cdots & 0 \\ 0 & \lambda_2 & \cdots & 0 \\ \cdots & \cdots & \cdots & \cdots \\ 0 & 0 & \cdots & \lambda_m \end{bmatrix} \tag{13}$$

$$\mathbf{PC}_{m \times n} = \mathbf{EOF}_{m \times m}^T \times \mathbf{X}_{m \times n} \tag{14}$$

4.2. Spatial Matrix Empirical Orthogonal Function (EOF)

According to the EOF theory, each spatial mode EOF_k corresponds to an eigenvalue λ_k. The contribution of EOF_k to the total variance is determined by its eigenvalue λ_k; therefore, the significance of EOF_k can be defined by its eigenvalue. The larger the eigenvalue is, the more significant its influence is. Furthermore, if the eigenvalue error ranges $(\lambda_k - \Delta\lambda, \lambda_k + \Delta\lambda)$ between two modes overlap, the mode characteristics will be similar. Here, $\Delta\lambda_k$ is determined according to Equation (15), where N^* represents the effective degrees of freedom of data.

$$\Delta\lambda_k = \lambda_k \sqrt{\frac{2}{N^*}} \tag{15}$$

Figure 11 illustrates the eigenvalues of the first few spatial modes. The first three spatial modes (EOF_1, EOF_2, and EOF_3) contribute the most to the total variance; therefore, only the characteristics of these three spatial modes are analyzed in the following. Compared with λ_2 and λ_3, the λ_1 value for EOF_1 leaps significantly. Evidently, EOF_1 has an overwhelming influence on the snow distribution on the stadium roof. The eigenvalues for EOF_2 and EOF_3 are similar, but the significance of EOF_2 is slightly more pronounced. Furthermore, a relatively large overlap exists in the error ranges between EOF_2 and EOF_3, indicating that their distribution patterns are similar.

The distribution characteristic of EOF_1 is illustrated in Figure 12. The values of EOF_1 are all positive. The distribution pattern is indicated by the strip of local accumulations formed in the concave regions along the radial cables, similar to the unbalanced snowpack formed in the eave region between multi-span gable roofs (Figure 10b). In general, EOF_1 further reflects the influence of the local roof shape on the snow load distribution on the building roof. The distribution characteristics of EOF_2 and EOF_3 are illustrated in Figure 13. The spatial characteristics of EOF_2 and EOF_3 are similar, as discussed previously; that is, the overall surface accumulations along the symmetry axes ($0°$ or $90°$ inflow direction) and skew surface accumulations between the axes ($45°$ inflow direction). These patterns are similar to the fan-shaped depositions on the umbrella-shaped roof (Figure 10a), reflecting the influence of the overall roof shape on the snow load distribution. Moreover, the magnitude of the shape coefficients for EOF_2 and EOF_3 are positive or negative in two perpendicular regions, taking on an opposite distribution pattern with a different inflow direction (the black or gray arrow in Figure 13).

Therefore, EOF_2 and EOF_3 comprehensively reflect the influences of the inflow direction and overall roof shape on the snow load distribution.

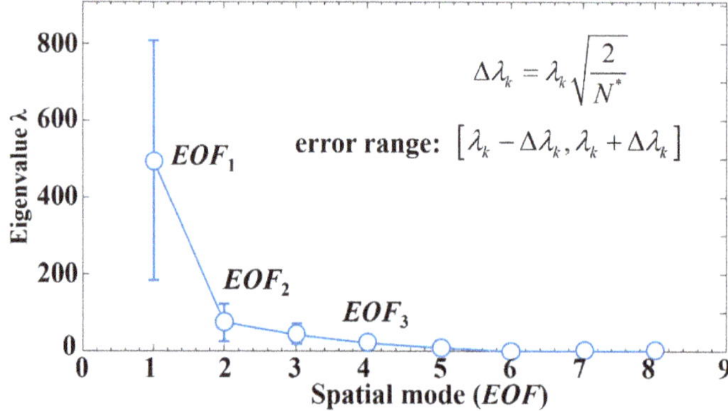

Figure 11. Eigenvalues of each empirical orthogonal function (*EOF*) mode at 95% confidence level.

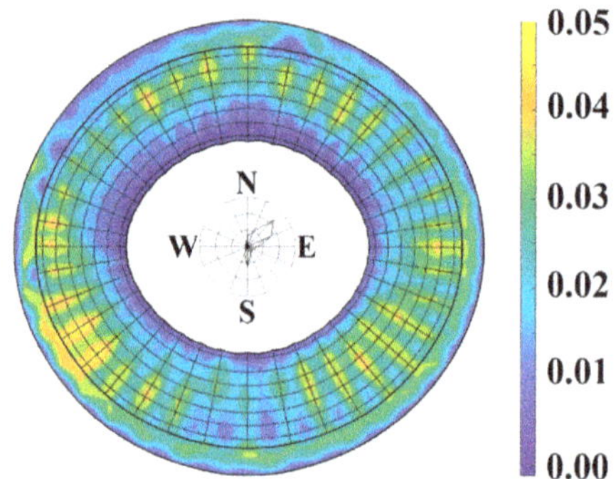

Figure 12. Distribution characteristics of the first spatial mode EOF_1.

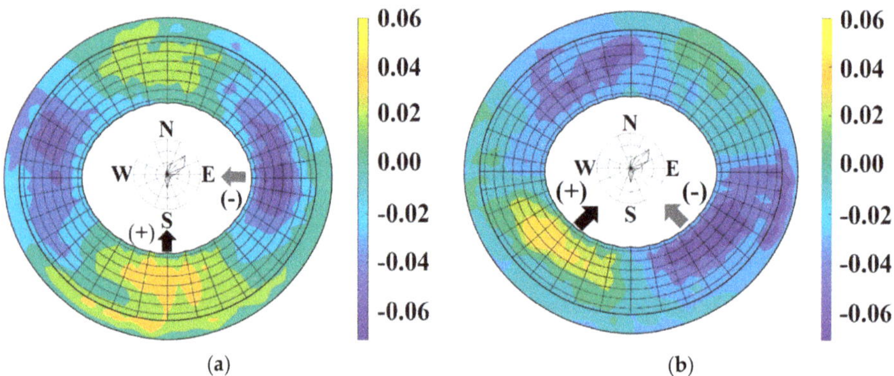

(a) (b)

Figure 13. Distribution characteristics of the second and third spatial modes: (**a**) EOF_2; (**b**) EOF_3.

4.3. Wind Direction Matrix **PC**

As a group of the wind direction-dependent weighted coefficients for the spatial mode, the **PC** coefficient reflects the influence degree and the combination way of the spatial modes. The specific

PC coefficients for EOF_1, EOF_2, and EOF_3 are illustrated in Figure 14. Under the five inflow direction conditions, which range from 0° to 90°, the proportion of the **PC** coefficients for EOF_1 is the largest, indicating an overwhelming influence, as described earlier. The value fluctuates slightly depending on the wind direction and basically remains at 22.5. For EOF_2 and EOF_3, the **PC** coefficients vary sharply with the wind direction, that is, the **PC** coefficients are sensitive to the wind direction. Although the coefficients fluctuate dramatically, the curves of **PC** coefficients are symmetrical along the $PC = 0$ axis with varying inflow directions. Specifically, when the inflow originates from the 0° direction, the inflow direction is consistent with that of the negative state of EOF_2 (gray inflow direction in Figure 13a). The overall snow distribution reflects the characteristics of EOF_2, while the **PC** coefficient for EOF_3 is close to zero. When the inflow originates from the 45° direction, the inflow direction is consistent with that of the negative state of EOF_3 (gray inflow direction in Figure 13b). The overall snow distribution reflects the characteristics of EOF_3, while the **PC** coefficient for EOF_2 is close to zero. Finally, when the inflow moves to 90°, the overall distribution characteristics can be explained only by EOF_2; whereas, the influence of EOF_3 could be neglected. Overall, the local roof shape has the greatest influence on the snowdrift on the complex roof; whereas, the contribution of the whole roof shape to the snowdrift depends on the dominant wind direction. The closer the wind direction represented by the spatial mode is to the dominant wind direction, the greater its contribution to the overall result.

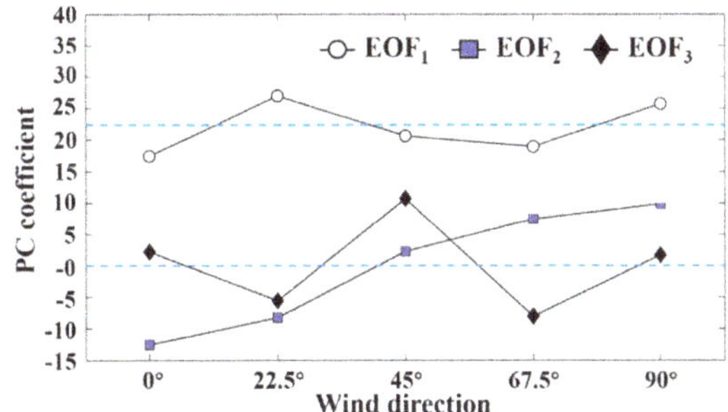

Figure 14. Wind direction coefficients for EOF_1, EOF_2, and EOF_3.

4.4. Combination of **EOF** and **PC**

Since the original field matrix **X** can be decomposed into a spatial matrix **EOF** and a wind direction matrix **PC**, conversely it is possible to generate the snow distribution on this complex membrane roof through the combination of the spatial matrix **EOF** and wind direction matrix **PC**. As the first three spatial modes made the greatest contribution to the snow distribution on this structure and reflected the influences of the local and overall roof shapes, respectively, the three modes were selected for combination. The combination method is shown in Equation (16). Each column of the generated matrix $\mathbf{X}'$ represents the snow distribution under a specific wind direction condition.

$$\mathbf{X}'_{m \times n} = \mathbf{EOF}_{m \times 3} \times \mathbf{PC}_{3 \times n} \tag{16}$$

The generated snowdrifts in different inflow directions are shown in Figure 15. Through comparison with the experimental results as shown in Figure 9, the snowdrift patterns, i.e., the overall fan-shaped deposition occurring on the windward and leeward surfaces along the air stream and the local packing formed in the concave regions along the radial cables and at the center of each membrane piece, show good correspondence with the experimental results. However, it should be noted that a significant underestimation of the shape coefficient values, especially the peak values, is observed. This underestimation was mainly caused by the fact that only the first three modes participated in the combination, and the contribution of the latter modes to the snow distribution was not considered yet. In general, it is preliminarily verified that the snowdrift on a complex roof can be restored by the combination of snow distributions on several corresponding simple roofs.

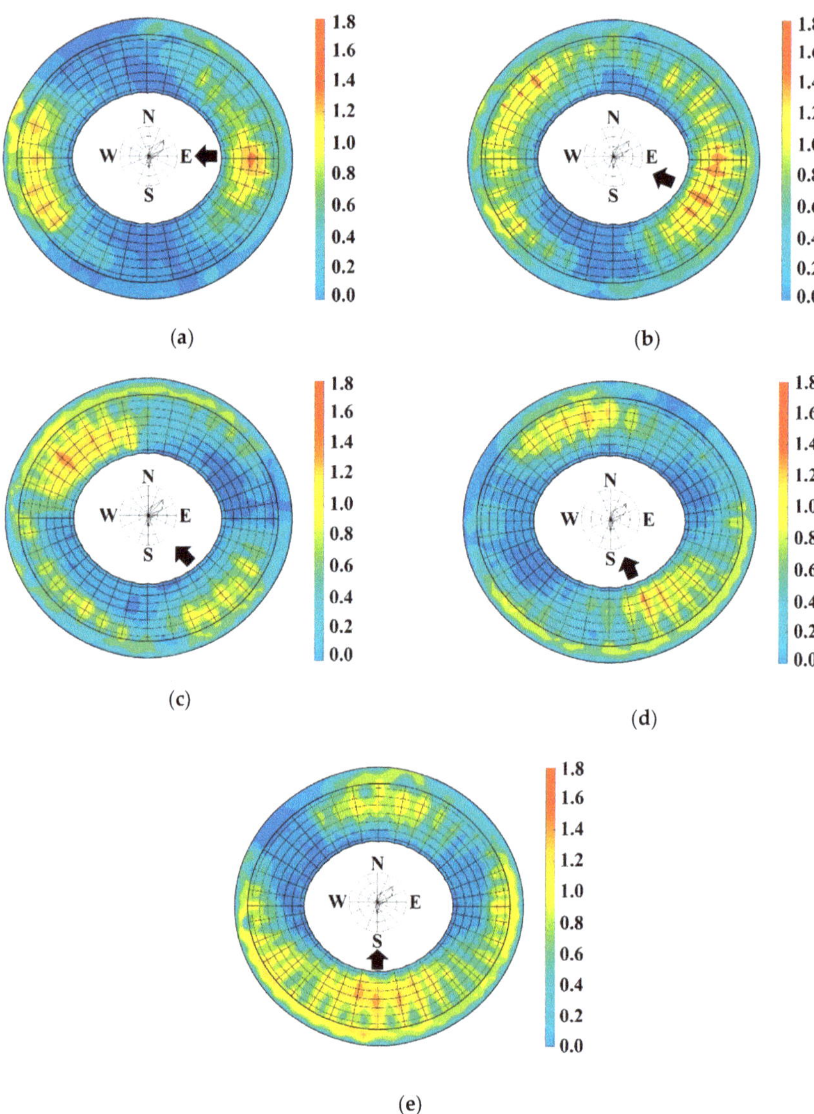

Figure 15. Generated snow distribution shape coefficient μ: (**a**) $0°$; (**b**) $22.5°$; (**c**) $45°$; (**d**) $67.5°$; and (**e**) $90°$.

In comparison with the conventional methods, the existing national load codes only provide the snow load requirements for simple roofs, as mentioned in Section 1. The applications of the CFD and experimental methods in the prediction of snow loads on complex roofs are also limited, due to the characteristics of time and money consumption. However, this study provides a simplified idea of estimating the snow load on a complex roof. Specifically, in the structural design stage, it is possible for the snow load on a complex roof to be estimated preliminarily by the combination of the snow distributions on several simple roofs, which have been provided in the load code (e.g., dome, curved roof, pitch roof, and obstacle, among others) or can be obtained by numerical studies. This will save a lot of time and money. Unfortunately, as an initial exploration, there are still many problems to be solved. Firstly, the correct selection of simple roofs plays a decisive role in determining the accuracy of combinational results, therefore, the selection method of the corresponding simple roofs should be studied emphatically; secondly, the **PC** coefficients are closely related to roof forms, and a large number of in-depth and large-scale studies should be carried out to clarify the **PC** coefficient; finally, reliable measurement data of snowdrifts on actual complex roofs is indispensable for examining prediction results. However, there is no measurement data available now, due to a large number of works and high risks in the process of measurement. These problems should be considered in future investigations.

5. Conclusions

This study has demonstrated the feasibility of snowdrift reproduction based on the air–snow–wind combined experimental system and the experimental approach. A series of scaled tests were carried out to predict the wind–snow flow behavior on a complex membrane roof. The snow distribution characteristics were decomposed and analyzed, and the following results were obtained. (i) The similarity criterion based on the drifting snow flux in the saltation layer was proved to simulate the overall snow distribution effectively. Based on the validated similarity criterion, the snowdrift distributions on a complex long-span membrane roof under different inflow directions were reproduced by using artificial snow particles. The results preliminarily explained the influence of the roof shape and wind direction on the snowdrift, and illustrated the complexity of the snow distribution on the complex roof, compared with that on a simple roof. (ii) Based on the EOF analysis, the snow distribution on the complex roof under different inflow direction conditions was broken down into several spatial distribution modes and wind direction-dependent weighted coefficients. Through the analysis of spatial modes, it was proven that the snow load distribution on a complex roof can be broken down into an integral surface load and local concentrated load. Combined with the roof pattern, it was demonstrated that the snow distribution on a complex roof can be decomposed into the snow distributions on several simple roofs according to the specific roof form. (iii) Through the significance analysis of the main spatial modes, it was demonstrated that the local roof shape has the greatest influence on the snowdrift on the complex roof; whereas, the contribution of the whole roof shape to the snowdrift depends on the dominant wind direction. The closer the wind direction represented by the spatial mode is to the dominant wind direction, the greater its contribution to the overall result. In practice, when no reference is available in the load code, the actual roof snow load can be estimated preliminarily by reasonable combinations of the existing simple roof snow loads in the codes, referring to the specific roof form and the dominant wind direction.

Author Contributions: Conceptualization, F.F. and S.S.; methodology, G.Z.; software, G.Z.; validation, G.Z., Q.Z., and F.F.; formal analysis, G.Z.; investigation, G.Z.; resources, F.F.; data curation, Q.Z.; writing—original draft preparation, G.Z.; writing—review and editing, G.Z.; visualization, Q.Z.; supervision, F.F.; project administration, S.S.; funding acquisition, F.F.

Funding: This was funded by the Chinese National Natural Science Foundation project (51927813, 51978207) and the National Science Fund for Distinguished Young Scholars (51525802).

Acknowledgments: The authors are grateful to the members of the Space Structures Research Center at the Harbin Institute of Technology, for providing invaluable information and advice in this study.

Conflicts of Interest: The authors declare no conflict of interest. The funders had no role in the design of the study; in the collection, analyses, or interpretation of data; in the writing of the manuscript, or in the decision to publish the results.

References

1. Zhou, X.Y.; Li, X.F. Simulation of snow drifting on roof surface of terminal building of an airport. *Disaster Adv.* **2010**, *3*, 42–50.
2. MOHURD. *Load Code for the Design of Building Structures (GB 50009-2012)*, 1st ed.; China Architecture & Building Press: Beijing, China, 2012; pp. 26–29.
3. BSI. *Eurocode 1-Actions on Structures-Part. 1-3: General Actions-Snow Loads (BS EN 1991-1-3)*, 2nd ed.; CEN: London, UK, 2003; pp. 16–30.
4. ASCE. *Minimum Design Loads for Buildings and Other Structures (ASCE/SEI 7-10)*, 1st ed.; ASCE: Reston, VA, USA, 2010; pp. 29–33.
5. AIJ. *AIJ Recommendations for Loads on Buildings*, 2nd ed.; AIJ: Tokyo, Japan, 2004; pp. 14–18.
6. Tominaga, Y.; Mochida, A. CFD prediction of flowfield and snowdrift around a building complex in a snowy region. *J. Wind Eng. Ind. Aerodyn.* **1999**, *81*, 273–282. [CrossRef]
7. Kwok, K.C.S.; Kim, D.H.; Smedley, D.J.; Rohde, H.F. Snowdrift around buildings for Antarctic environment. *J. Wind Eng. Ind. Aerodyn.* **1992**, *41*, 2797–2808. [CrossRef]
8. Thiis, T.K.; Ramberg, J.F. Measurements and numerical simulations of development of snow drifts of curved roofs. In Proceedings of the Snow Engineering VI, Whistler, BC, Canada, 21–27 September 2008.
9. Thiis, T.K.; Potac, J.; Ramberg, J.F. 3D numerical simulations and full scale measurements of snow depositions on a curved roof. In Proceedings of the 5th European & African Conference on Wind Engineering, Florence, Italy, 19–23 July 2009.
10. Beyers, M.; Waechter, B. Modeling transient snowdrift development around complex three-dimensional structures. *J. Wind Eng. Ind. Aerodyn.* **1999**, *81*, 273–282. [CrossRef]
11. Isyumov, N.; Mikitiuk, M. Wind Tunnel Model Tests of Snow Drifting on a Two-Level Flat Roof. *J. Wind Eng. Ind. Aerodyn.* **1990**, *36*, 893–904. [CrossRef]
12. Delpech, P.; Palier, P.; Gandemer, J. Snowdrifting simulation around Antarctic buildings. *J. Wind Eng. Ind. Aerodyn.* **1998**, *74*, 567–576. [CrossRef]
13. Flaga, A.; Flaga, L. Wind tunnel tests and analysis of snow load distribution on three different large size stadium roofs. In Proceedings of the 8th International Conference of Snow Engineering, Nantes, France, 14–17 June 2016; pp. 232–239.
14. Liu, M.; Zhang, Q.; Fan, F.; Shen, S. Experiments on natural snow distribution around simplified building models based on open air snow-wind combined experimental facility. *J. Wind Eng. Ind. Aerodyn.* **2018**, *173*, 1–13. [CrossRef]
15. Kimura, T. Measurements of drifting snow particles. *J. Geogr.* **1991**, *100*, 250–263. [CrossRef]
16. Kind, R.J. A critical examination of the requirements for model simulation of wind-induced erosion/deposition phenomena such as snow drifting. *Atmos. Environ.* **1976**, *10*, 219–227. [CrossRef]
17. Iversen, J.D. Drifting snow similitude. *J. Hydraul. Div.* **1979**, *105*, 737–753.
18. Oikawa, S.; Tomabechi, T.; Ishihara, T. One-day Observations of Snowdrift around a Model Cube. *J. Snow Eng. Jpn.* **1999**, *15*, 3–13. [CrossRef]
19. Kind, R.J. Snowdrifting: A review of modelling methods. *Cold Reg. Sci. Technol.* **1986**, *12*, 217–228. [CrossRef]
20. Beyers, J.H.M.; Harms, T.M. Outdoor modelling of snowdrift at SANAE IV Research Station, Antarctica. *J. Wind Eng. Ind. Aerodyn.* **2003**, *91*, 551–569. [CrossRef]
21. Mo, H.; Fan, F.; Hong, H. Evaluation of input wind speed used in wind tunnel test and numerical simulation to estimate drifting snow profile. *J. Build. Struct.* **2015**, *36*, 75–90.
22. MOHURD. *Technical Specification for Cable Structures (JGJ 257-2012)*, 1st ed.; China Architecture & Building Press: Beijing, China, 2012; pp. 37–38.

Article

Temperature Distribution Characteristics of Concrete during Fire Occurrence in a Tunnel

Seungwon Kim [1,*], Jaewon Shim [2], Ji Young Rhee [2], Daegyun Jung [1] and Cheolwoo Park [1,*]

[1] Department of Civil Engineering, Kangwon National University, 346, Jungang-ro, Samcheok-si 25913, Korea; sso2247@gmail.com
[2] Research Institute, Korea Expressway Corporation, 208-96, Dongbu-daero, 922 beon-gil, Hwaseong-si 18489, Korea; this2pass@ex.co.kr (J.S.); need@ex.co.kr (J.Y.R.)
* Correspondence: inncoms@kangwon.ac.kr (S.K.); tigerpark@kangwon.ac.kr (C.P.); Tel.: +82-33-570-6518 (S.K.); +82-33-570-6515 (C.P.)

Received: 9 October 2019; Accepted: 3 November 2019; Published: 6 November 2019

Abstract: Fire in a tunnel or an underground structure is characterized by a rise in temperature above 1000 °C in 5–10 min, which is due to the characteristics of the closed space. The Permanent International Association of Road Congresses has reported that serious damage can occur in an underground structure as a consequence of high temperatures of up to 1400 °C when a fire accident involving a tank lorry occurs in an underground space. In these circumstances, it is difficult to approach the scene and extinguish the fire, and the result is often casualties and damage to facilities. When a concrete structure is exposed to a high temperature, spalling or dehydration occurs. As a result, the cross section of the structure is lost, and the structural stability declines to a great extent. Furthermore, the mechanical and thermal properties of concrete are degraded by the temperature hysteresis that occurs at high temperatures. Consequently, interest in the fire safety of underground structures, including tunnels, has steadily increased. This study conducted a fire simulation to analyze the effects of a fire caused by dangerous-goods vehicles on the tunnel structure. In addition, a fire exposure test of reinforced-concrete members was conducted using the Richtlinien für die Ausstattung und den Betrieb von Straßentunneln (RABT) fire curve, which is used to simulate a tunnel fire.

Keywords: RABT fire curve; fire simulation; tunnel fire; high temperature; fire safety; fire accident

1. Introduction

With the recent sharp increase in accidents involving transport vehicles carrying hazardous materials (e.g., explosive flammables), damage to highway infrastructure facilities, such as tunnels, has increased substantially [1,2]. Detailed summaries of road and rail tunnel fire events clearly show the importance of considering fire risk in the design of tunnels [3,4]. In particular, due to the extensions and elongations of tunnel structures and increasing passage access to popular town areas, it is urgently necessary to ensure the safety of tunnel structures against unexpected extreme disasters, such as fires [1,2]. For this reason, many developed countries are currently enhancing fire intensity standards and reviewing explosion resistance standards. In most countries, however, the assessment of, and response to, fire risks is still limited, and the maturity of the design goal is relatively low [1,2].

In 2001, two trucks collided in the Gotthard tunnel in Switzerland, resulting in a fire, as well as 11 deaths and many injuries [5]. The scale of the fire was approximately 120–200 MW, and the flame temperature was estimated at over 1000 °C [5]. The fire brigade experienced difficulty in accessing the fire scene for 48 h, causing damage over a length of 700 m inside the tunnel and spalling of up to 350 mm in depth [5]. The damage caused by this accident amounted to approximately $31 billion, and restoration work lasted two months [5]. Even though the authorities were equipped with the latest disaster prevention facilities at the time of the accident, the damage was considerable, and this clearly demonstrates the importance of the response as well as of the preparation of operational facilities [2].

This study analyzed fire intensity and the effects of fire on tunnel structures in terms of depth by simulating fire occurrences resulting from accidents of tank lorries loaded with inflammables in a tunnel environment. Furthermore, in order to verify the validity of the numerical analysis model, the effects of geometrical elements on fire were examined by a fire exposure test on reinforced-concrete members using the Richtlinien für die Ausstattung und den Betrieb von Straßentunneln (RABT) fire curve

2. Characteristics of Fire in a Tunnel

2.1. Material Characteristics of Concrete Exposed to High Temperature

Exposure to high temperatures results in spalling or destruction of the coating on members due to the water vapor pressure created inside the concrete [2,6–8]. At 100–400 °C, Al_2O_3-, Fe_2O_3-, and tobermorite-based hydrates are dehydrated, resulting in the collapse of gel and cement hydrates. In moderate-strength concrete that has been exposed to high temperatures, voids are generated as the vapor inside the concrete evaporates at approximately 200 °C [2,6–8]. The deformation recovery ability is drastically lowered in this temperature range, and the elastic modulus decreases significantly at temperatures of over 600 °C [2,6–8].

Concrete exhibits a tendency toward decreasing compressive strength and elastic modulus when exposed to high temperatures. Concrete exposed to high temperatures causes cross-sectional defects due to surface peeling or scattering [2,6–8]. This phenomenon is called spalling. The main cause of this phenomenon is the water vapor pressure that is generated when the water inside the concrete expands in response to high heat [2,6–8].

It is known that when a concrete structure is exposed to temperatures of approximately 650 °C or higher, it loses 50% of its original strength. When it is exposed to temperatures of approximately 850 °C or higher, it loses its structural performance [2,6–8].

2.2. Temperature Distribution Characteristics Due to Vehicle Fire in a Tunnel

According to a report published by the Permanent International Association of Road Congresses on fire and smoke control in road tunnels [9], in the event of a fire in a tunnel, when air flows in through the tunnel entrance at 6 m/s, the temperature of the ceiling reaches approximately 400 °C from the spot of the fire to a point approximately 100 m from the tunnel exit [9]. According to Dutch regulations, when a fire occurs in a large tank lorry with a loading capacity of 50 m^3 or higher, the temperature rises to approximately 1400 °C.

3. Tunnel Fire Simulation

It is almost impossible to consider all possible fire situations involving dangerous-goods vehicles in experimental assessments and verifications for simulating fires that occur in a road-network system. Various costs and time-consuming limits exist in reality. Therefore, simulation analysis is typically used in such studies, and a real fire experiment is only conducted when necessary to complement the results of the simulation. Fire simulation is generally conducted through computational fluid dynamics (CFD) analysis. CFD analysis is actively applied to fire-modelling research at domestic and international facilities and to establish firefighting design and evacuation parameters [1]. It is the most important tool in fire engineering. Building upon the Field Model developed in the U.K., the National Institute of Standard and Technology (NIST) and the Building and Fire Research Laboratory in the United States have achieved continuous developments in this area since 2000 [1]. CFD analysis, in particular, can be used to examine the thermal-fluid flow phenomenon on a large scale, for which life-sized model experiments are impossible to conduct. It can also quantify the degree of damage, such as the size of the fire, smoke generation, toxic gas generation, and amount of radiation heat [1].

Furthermore, simulation is possible above the normal test-performance limits and can produce results according to specific scenarios, which enables the quantitative assessment of fire risks.

3.1. Modelling

The Fire Dynamic Simulator (FDS, version 6.5.3) [10], which has been developed by the NIST in the U.S., is the numerical analysis model used for the fire analysis in this study. The target space of the analysis was 45 m × 8 m × 5.5 m (length × width × height), and the length of the tunnel was assumed to be 45 m. It was believed that the length of the tunnel would not be affected by the temperature of the fire. As shown in Figure 1, a box-type cross section was applied to mimic the cross-sectional shape of the tunnel. The fire source was located between the center of the tunnel and its side. Tank lorries of 27 m^3 (27,000 L) volume were used with diesel as the fuel. The open-boundary condition was applied at the tunnel entrance and exit. In addition, as in a real situation, the combustion rate method was used to simulate the fire in tank lorries.

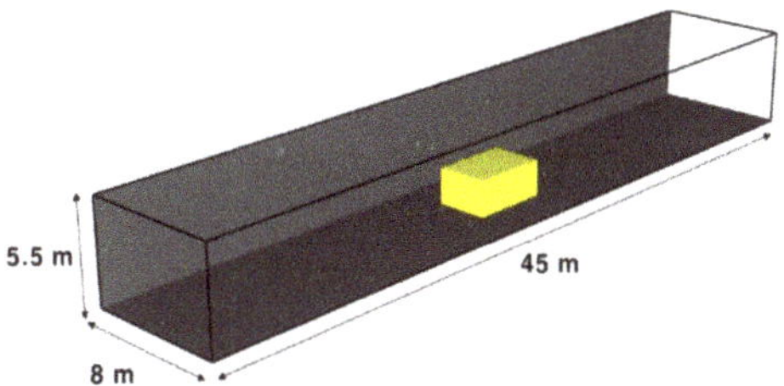

Figure 1. Dimension of tunnel and position of fire in tunnel modelling.

3.2. Analysis of Fire by Combustion Material Type

Comparative analyses were carried out on the changes in the fire characteristics according to the duration of ignition sources that were used to simulate fires for different materials in the FDS. The simulations were conducted by modelling for materials with different properties (e.g., combustion heat, density, and ignition point). Based on the results, the appropriate duration of the ignition source was determined and applied to the fire simulation.

Diesel and octane (C_8H_{18}), which are the commonly used gasoline types, as well as heptane (C_7H_{16}) and ethanol (C_2H_6O), were the hazardous materials applied to the fire simulation analysis. The values of the heat of combustion for these hazardous materials are 44.80 MJ/kg, 47.89 MJ/kg, 48.07 MJ/kg, and 29.65 MJ/kg, respectively. To analyze the changes in fire intensity with respect to the types of hazardous materials, specific analysis conditions were set to verify the fire intensity inside the tunnel when the fire occurred due to a hazardous material with a similar or lower heat of combustion. Table 1 outlines the model and conditions used for the analysis. Table 2 lists the material properties of the combustion materials.

Table 1. Analysis conditions by hazardous material type.

Parameters	Input Data
Space size (m)	45 × 8 × 5.5
Grid size (m)	0.5 × 0.5 × 0.5
Conductivity of concrete (W/m·K)	1.28
Specific heat of concrete (KJ/kg·K)	0.75
Density of concrete (kg/m^3)	2400
Vehicle location	Middle of the tunnel 27,000 L capacity
Combustion material type	Diesel, ethanol, octane, and heptane
Simulation time (s)	3600

Table 2. Properties of combustion materials considered in this study.

Parameters	Diesel	Ethanol	Octane	Heptane
Density (kg/m^3)	840	787	700	684
Specific heat (kJ/kg·K)	1.89	2.45	2.15	2.25
Thermal conductivity (W/m·K)	0.18	0.17	0.13	0.124
Emissivity	0.9	1.0	0.9	0.9
Heat of combustion (kJ/kg)	44,800	29,653	47,898	48,074
Boiling point (°C)	250	76	125	98.5

Figure 2 shows the analysis results for different combustion materials. Except for ethanol, whose heat of combustion is 29.65 MJ/kg, diesel, octane, and heptane have similar values for heat of combustion, i.e., 44.80 MJ/kg, 48.07 MJ/kg, and 47.89 MJ/kg, respectively. These three fuels had a similar fire intensity level of approximately 160 MW. However, the results for diesel showed a longer fire duration by approximately 500 s. In comparison to these three fuels, the heat of combustion of ethanol is approximately 60%, and this fuel also resulted in a lower fire intensity. However, the duration of its fire was longer than 1 h, much longer than the fire durations of the other combustion materials.

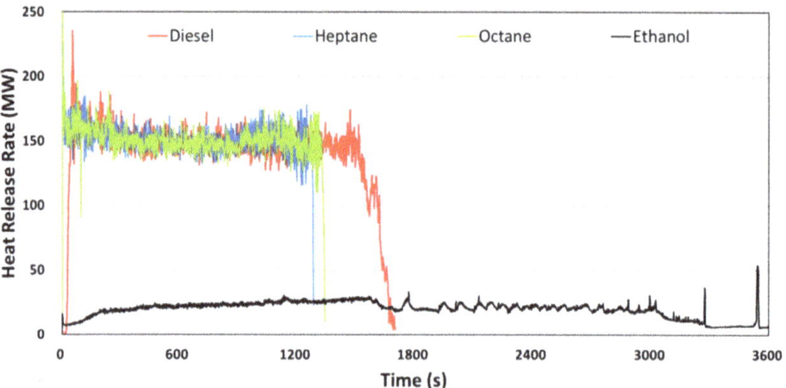

Figure 2. Heat release rates of the tested combustion materials.

The above analysis revealed that different types of hazardous materials on a highway can cause fires with different characteristics (i.e., the fire intensity and duration are likely to differ depending on the material type). The analysis of the simulation results showed that a fire caused by dangerous-goods vehicles carrying diesel results in the most serious conditions of fire intensity and fire duration.

3.3. Analysis of Vehicle Fire in a Tunnel

A simulation was conducted for a fire caused by dangerous-goods vehicles carrying diesel in a tunnel. The tunnel model used for this simulation was the same as that in Figure 2. Table 3 lists the analysis conditions used for this simulation.

Table 3. Vehicle fire analysis conditions.

Parameters	Input Data
Space size (m)	$45 \times 8 \times 5.5$
Grid size (m)	$0.25 \times 0.25 \times 0.25$
Conductivity of concrete (W/m·K)	1.28
Specific heat of concrete (KJ/kg·K)	0.75
Density of concrete (kg/m^3)	2400
Vehicle location	Middle of the tunnel 27,000 L
Measured depth of concrete wall (m)	Surface: 20, 40, 60, and 100
Combustion material type	Diesel
Simulation time (s)	3600

4. Fire Exposure Experiment on Reinforced-Concrete Member

In order to verify the analysis of the fire and heat-transfer characteristics of concrete in the event of a fire in a tunnel structure using the CFD analysis program FDS [10], the heat-transfer characteristics of a concrete member were examined through a fire experiment. The results were then compared with those of the CFD analysis. The fire exposure experiment was conducted using the RABT fire curve, which can simulate a tunnel fire for reinforced-concrete members.

4.1. RABT Fire Curve

The RABT fire curve was developed by the Road Construction Department of the German Ministry of Transportation under the Eureka Project [11]. In a simulated scenario, the temperature sharply rises to 1200 °C within 5 min after the beginning of the fire. The durations of fires involving trains and cars at a temperature of 1200 °C are 55 min and 25 min, respectively. The fire then cools down for 110 min. The RABT fire curve is known to have a shape similar to that of a real tunnel fire [11]. Figure 3 shows the RABT fire curves for railways and highways.

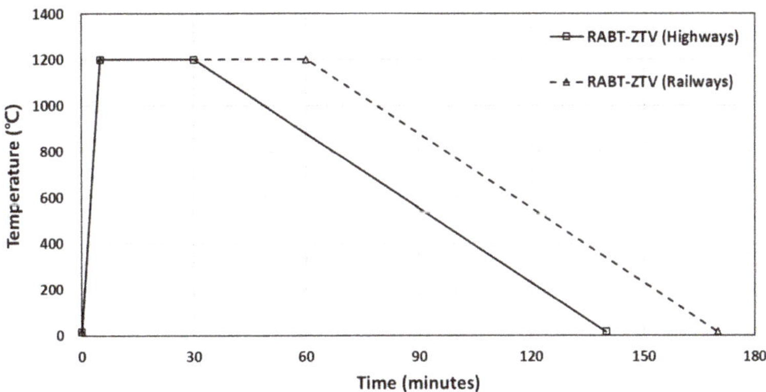

Figure 3. Richtlinien für die Ausstattung und den Betrieb von Straßentunneln (RABT) fire curves of highways and railways [11].

4.2. Dimensions of the Specimen and Experimental Method

The arrangement of reinforcement bars used in the RABT fire curve fire exposure experiment is the same as the arrangement in real road tunnels. Furthermore, Figure 4 shows the horizontal heating furnace for the high-temperature test used in the fire exposure experiment. Only the bottom surface of the specimen was exposed in this experiment. Ceramic fibers that can endure temperatures of up to approximately 1400 °C were installed to provide insulation between the top of the furnace and the specimen. The horizontal heating furnace was designed to allow installation of a rectangular specimen with size of 1400 mm (length) × 1000 mm (width). The actual heating area was 1100 mm (length) × 700 mm (width). The design standard compressive strength of the concrete specimen that was used in this experiment was 27 MPa. Figure 5 shows the experimental setup for the fire exposure test using the RABT fire curve.

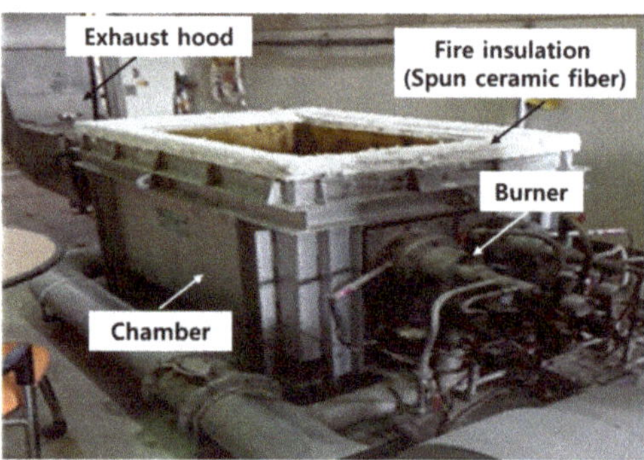

Figure 4. Horizontal heating furnace for the high-temperature test.

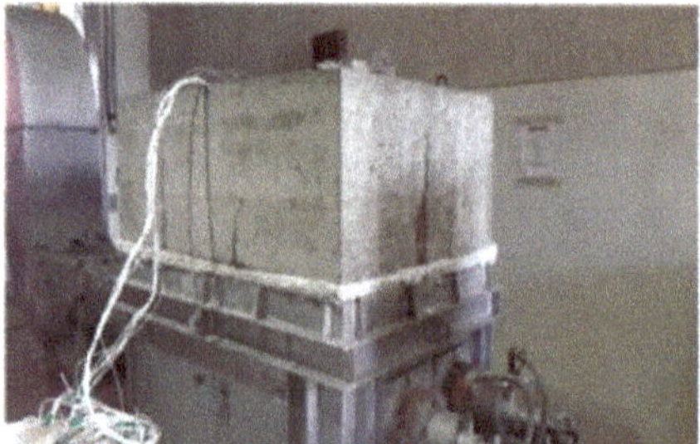

Figure 5. Experimental setup for the fire exposure test using the RABT fire curve.

To examine the heat-transfer characteristics using the RABT fire exposure experiment, thermocouples were installed at the center at 0, 20, 40, 60, 80, and 100 mm from the heating surface. Table 4 lists the positions of the thermocouples.

Table 4. Positions of the thermocouples.

Thermocouples			Depth (mm)
Left	Centre	Right	
-	TC(C)-7	-	0
TC(L)-1	-	TC(R)-1	20
TC(L)-2	-	TC(R)-2	40
TC(L)-3	TC(C)-8	TC(R)-3	60
TC(L)-4	-	TC(R)-4	80
TC(L)-5	-	TC(R)-5	100
TC(L)-6	-	TC(R)-6	120

5. Experimental Results and Analysis

5.1. Analysis of Vehicle Fire in the Tunnel

When a fire source corresponding to 27,000 L of diesel was located at the center of the tunnel, the fire intensity was approximately 150 MW, as shown in Figure 6. Figure 7 shows the results of the temperature measurements on the concrete surface and at depths of 20, 40, 60, and 100 mm. The analysis results for the concrete hydrothermal temperatures at each depth showed that for a fire of such a scale (27,000 L), the surface temperature rose to approximately 1000 °C at approximately 300 s after the occurrence of the fire. This high temperature was maintained for approximately 1800 s.

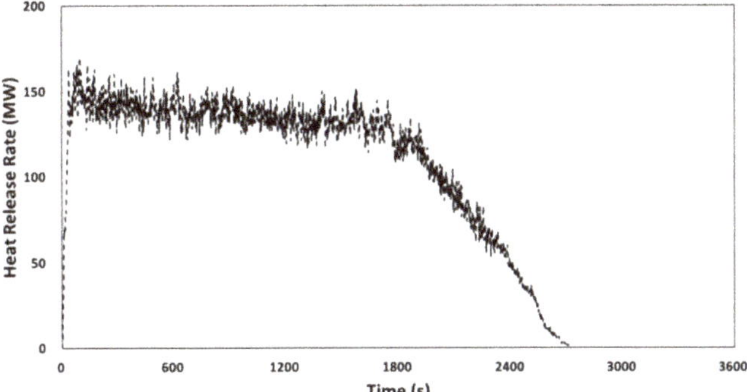

Figure 6. Fire intensity based on vehicle fire analysis.

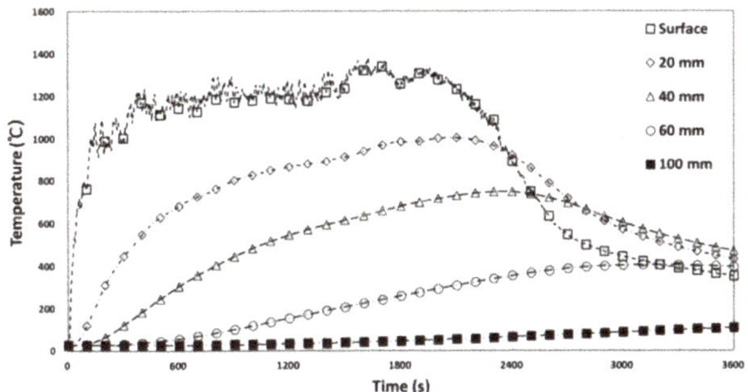

Figure 7. Temperature distribution by depth based on vehicle fire analysis.

The peak temperature of the concrete surface was approximately 1380 °C. The corresponding temperatures at depths of 20, 40, 60, and 100 mm were 1005 °C, 750 °C, 405 °C, and 110 °C, respectively. The International Tunneling Association (ITA) [12] specifies that the maximum critical temperatures for concrete and reinforcement bars should be 380 °C and 250 °C, respectively. On the basis of these results, the maximum critical temperature standard could not be met at concrete depths higher than 60 mm.

Figure 8 shows the temperature distribution inside the tunnel. Furthermore, the fire exposure experiment was conducted for real concrete members by applying the RABT fire curve where the temperature rose sharply to 1200 °C within 5 min. The result was similar to the that of the empirical verification of the tunnel structures' characteristics after exposure to high temperatures [12].

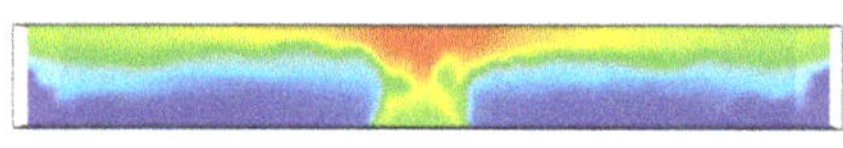
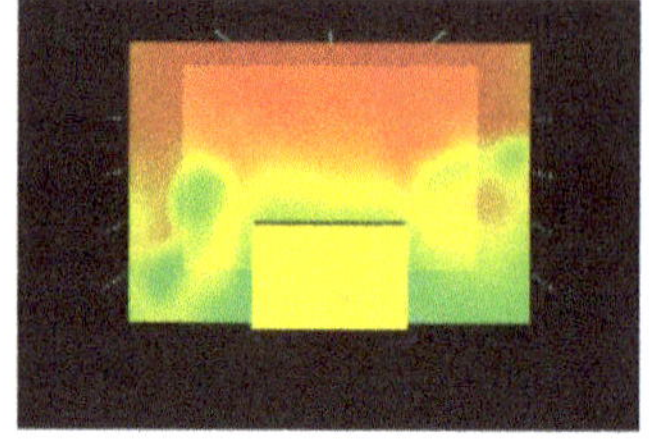

(**a**) Temperature distribution inside the tunnel (floor-plan view)

(**b**) Temperature distribution inside the tunnel (cross-sectional view)

Figure 8. Temperature distribution inside the tunnel.

5.2. Fire Exposure Experiment of Reinforced-Concrete Member

Figures 9 and 10 compare the specimen surface before and after the fire test based on RABT-ZTV (highways) 30 min fire curves and RABT-ZTV (railways) 60 min fire curves. Spalling is a complex process, which occurred in the concrete specimen due to the rapid temperature increase in the furnace. On the basis of the temperature changes in Figures 11 and 12, the result for the left thermocouple shows an abnormal pattern of temperature change with depth. On the basis of the right thermocouple, section loss appeared from the heating surface to 60 mm, and no section loss occurred above 80 mm. The temperature inside the concrete increased sharply depending on the generation of spalling. The temperature measured at approximately 60 mm, where the spalling occurred, was similar to that inside the furnace. At a depth of 80 mm, the temperature was approximately 420 °C during the rising period and approximately 540 °C during the descending period. However, some differences were evident because the spalling by heat varied depending on the condition of the concrete heating surface. The range of section loss can also be seen in Figures 9 and 10, which show the heating surface after the fire test was completed. Considering the arrangement depth of the reinforcement bars inside the specimen, the section loss was found to be 60–80 mm.

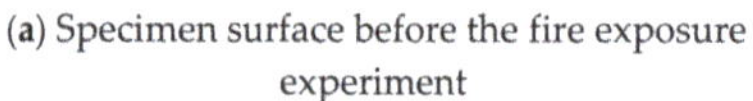

(a) Specimen surface before the fire exposure experiment

(b) Specimen surface after the fire exposure experiment

Figure 9. Specimen surface before and after the RABT-ZTV (highways) fire experiment.

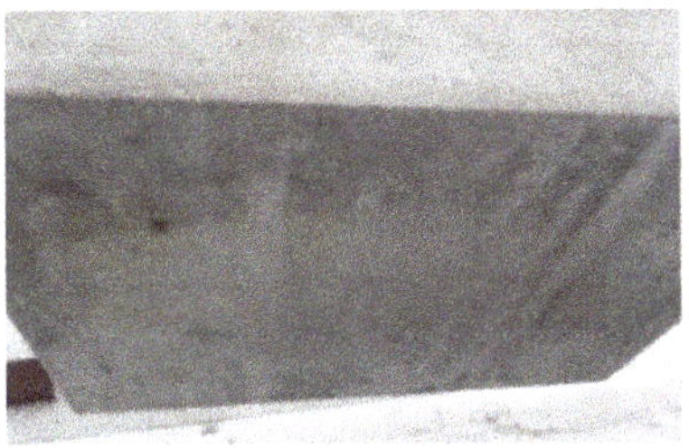
(**a**) Specimen surface before the fire exposure experiment

(**b**) Specimen surface after the fire exposure experiment

Figure 10. Specimen surface before and after the RABT-ZTV (railways) fire experiment.

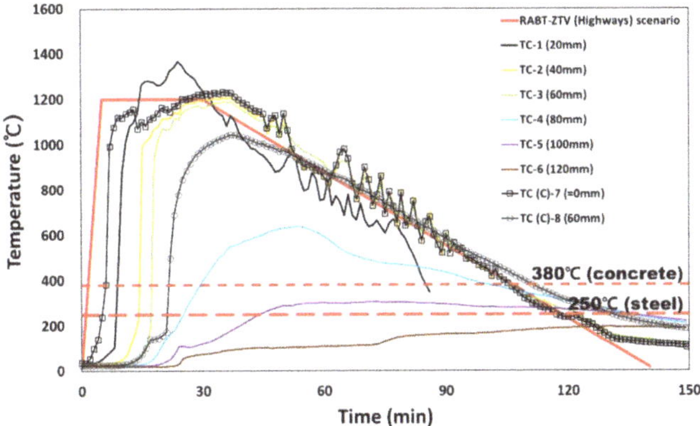

Figure 11. Temperature measurement results for the RABT-ZTV (highways) fire experiment.

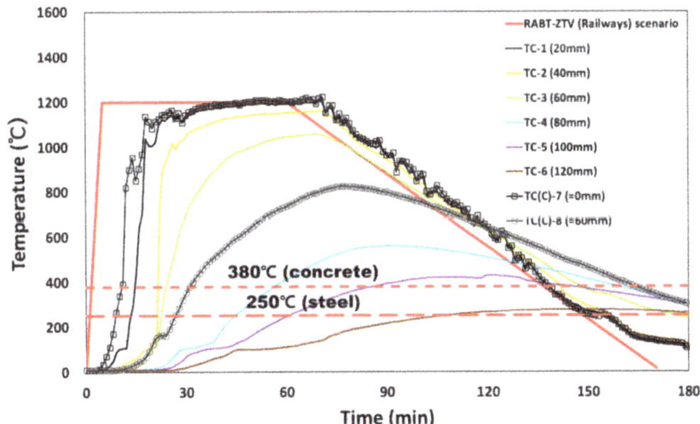

Figure 12. Temperature measurement results for the RABT-ZTV (railways) fire experiment.

On the basis of the results of the fire experiments, the appropriate coating thickness of the tunnel structure should be at least 80 mm in order to protect the concrete and internal reinforcement bars from fire. This contradicts the the simulation result that assumed that there had been no spalling in real concrete.

6. Conclusions

This study analyzed the effects of fire in a tunnel due to an accident involving dangerous-goods vehicles using fire simulation. Furthermore, a fire exposure test of a reinforced-concrete member was conducted using the RABT fire curve, which is employed to simulate a tunnel fire. The conclusions of this study are as follows:

(1) The results of the FDS tunnel fire analysis for each combustion material showed that diesel had a longer fire duration at the same fire intensity in comparison to the other tested fuels. As diesel is also the most commonly used fuel, it was selected as the fire source in the fire analysis that followed.

(2) The results of the fire analysis showed that the peak temperature on the concrete surface was approximately 1380 °C and that at a depth of 60 mm, the temperature was approximately 405 °C. These values do not satisfy the maximum critical temperatures for concrete and reinforcement bars that have been suggested by the ITA. These maximum temperatures are 380 °C and 250 °C, respectively. When a fire corresponding to 27,000 L of diesel occurs, the desirable coating thickness of the tunnel structure is 80 mm, which is greater than 60 mm.

(3) The analysis results for fire caused by dangerous-goods vehicles in the tunnel showed that at approximately 300 s after the beginning of the fire, the surface temperature of the concrete increased to approximately 1000 °C. This high temperature was maintained for approximately 1800 s. Therefore, on the basis of the empirical verification of the high-temperature exposure characteristics of concrete structures, the RABT fire curve was applied, in which the temperature rapidly increased to 1200 °C within 5 min. These results were similar to those of the analysis. Furthermore, since the maximum temperatures for concrete and reinforcement bars are specified as 380 °C and 250 °C, respectively, by the ITA, it is desirable to determine the fire damage that has been caused to the specimen on the basis of the suggested standards when an RABT fire curve with a significant temperature rise is used.

(4) The results of the RABT fire experiment showed serious spalling to a depth of 60 mm, or, in other words, the point at which reinforcement bars are located in concrete. The temperature inside the concrete rose rapidly depending on the spalling. The temperature measured at a depth of approximately 60 mm, where spalling occurred, was similar to that inside the furnace. The temperature at a depth of 80 mm was approximately 420 °C during the rising period and 540 °C during the descending period.

(5) On the basis of the results of the RABT fire curve experiments, 80 mm is the desirable minimum coating thickness of a tunnel concrete member in order to secure the safety of such structures during serious fires.

Author Contributions: Conceptualization, S.K., J.S., J.Y.R., C.P.; methodology, S.K., D.J.; writing and draft preparation, S.K., C.P.

Funding: This research was supported by the Basic Science Research Program through the National Research Foundation of Korea (NRF) and funded by the Ministry of Education (grant no. 2017R1A2B4012678) and the Korea Expressway Corporation Research Institute as part of the research project "Simulation and Experimental Simulation for verification of fire and explosion safety measures for vehicle fires".

Conflicts of Interest: The authors declare no conflicts of interest.

References

1. Shim, J.-W.; Kim, N.-Y.; Rhee, J.-Y.; Jang, T.-S. Development of Fire Protection Design and Blast Resistance Methods of Tunnels and Bridges caused by Fire and Explosions of Vehicles. *Korea Expressway Corp. Res. Inst. Rep.* **2018**, 1–14.
2. Kim, H.-Y.; Kim, H.-J.; Cho, K.-S.; Lee, J.-S.; Kwon, K.-H. An Experimental Study on Thermal Damage and Spalling of Concrete Lining in Tunnel Fire. *J. Korean Inst. Fire Sci. Eng.* **2009**, *23*, 110–120.

3. Maevski, I.Y. *NCHRP Synthesis 415: Design Fires in Road Tunnels—A Synthesis of Highway Practice*; Transportation Research Board: Washington, DC, USA, 2011.

4. Lönnermark, A. On the Characteristics of Fires in Tunnels. Ph.D. Thesis, Department of Fire Safety Engineering, Institute of Technology, Lund University, Lund, Sweden, 2005.

5. Shim, J.-W.; Kim, N.-Y.; You, B.-O.; Park, Y.-H. Fire Cases and Damage Evaluations in Fire of Tunnels and Underground Structures. *Mag. Korea Conc. Inst.* **2011**, *23*, 39–43.

6. Jin, K. A Study on the Improvements of Risk Index through the Analysis of Large Road Tunnel Fire. Master's Thesis, Yonsei University, Seoul, Korea, 2019.

7. Ministry of Land, Infrastructure, and Transport. Assessment of Fire Risk of Bridges on National Routes and Development of Response System. In *Land, Infrastructure and Transport R&D Report*; Ministry of Land, Infrastructure and Transport: Sejong, Korea, 2017.

8. Won, J.-P.; Park, K.-H.; Park, C.-K. Performance of Hybrid Fiber Reinforced Concrete at Evaluated High Temperature. *J. Korea Conc. Inst.* **2008**, *20*, 325–333.

9. WG 6 of the Road Tunnels Committee. Fire and Smoke Control in Road Tunnels. Available online: https://www.piarc.org/en/order-library/3854-en-Fire%20and%20Smoke%20Control%20in%20Road%20Tunnels.htm (accessed on 09 October 2019).

10. McGrattan, K.; Hostikka, S.; McSermott, R.; Floyd, J.; Weinchenk, C.; Overholt, K. Fire Dynamics Simulator Technical Reference Guide. Available online: https://www.thunderheadeng.com/wp-content/uploads/2013/08/FDS_Technical_Reference_Guide.pdf (accessed on 09 October 2019).

11. Experts for Specialised Construction and Concrete Systems (EFNARC). Specification and Guidelines for Testing of Passive Fire Protection for Concrete Tunnels Linings. Available online: http://www.efnarc.org/pdf/Testing%20fire%20protection%20systems%20for%20tunnels.pdf (accessed on 09 October 2019).

12. Working Group No. 6. Maintenance and Repair Guidelines for Structural Fire Resistance for Road Tunnels. Available online: https://about.ita-aites.org/publications/wg-publications/100-guidelines-for-structural-fire-resistance-for-road-tunnels (accessed on 09 October 2019).

 *applied sciences*

Article

Experimental Assessment of the Effect of Vertical Earthquake Motion on Underground Metro Station

Zhiming Zhang [1,2], Emilio Bilotta [2,*], Yong Yuan [3,*], Haitao Yu [4] and Huiling Zhao [5]

[1]   Department of Geotechnical Engineering, Tongji University, Shanghai 200092, China; 1410252@tongji.edu.cn
[2]   Department of Civil, Architectural and Environmental Engineering, University of Naples Federico II, 80125 Naples, Italy
[3]   State Key Laboratory of Disaster Reduction in Civil Engineering, Tongji University, Shanghai 200092, China
[4]   Key Laboratory of Geotechnical and Underground Engineering of Ministry of Education, Tongji University, Shanghai 200092, China; yuhaitao@tongji.edu.cn
[5]   Department of Civil Engineering, Shanghai University, Shanghai 200444, China; hlzhao@shu.edu.cn
*   Correspondence: bilotta@unina.it (E.B.); yuany@tongji.edu.cn (Y.Y.);
    Tel.: +39-81-7683469 (E.B.); +86-21-65980234 (Y.Y.)

Received: 27 September 2019; Accepted: 26 November 2019; Published: 29 November 2019

Featured Application: **This research has proven that the ratio of vertical to horizontal peak ground acceleration (RVH) has a significant influence on dynamic soil-structure interaction. It is believed that under extreme earthquake loading, such as near fault zones, RVH is a parameter of paramount importance and should be accounted for in the seismic analyses and seismic performance assessment of underground structures, especially for those with zero or near-zero buried depth, such as an atrium-style metro station. The conclusions of this research are expected to contribute to the revision of codes for seismic design of underground structures.**

Abstract: This paper presents experimental assessment of the effect of the ratio of vertical to horizontal peak ground acceleration (RVH) on underground metro station. An atrium-style metro station embedded in artificial soil subjected to earthquake loading is examined through shaking table tests. The experimental results for three different RVH, including soil acceleration, soil-structure acceleration difference, dynamic soil normal stress (DSNS), and structural dynamic strain, are presented and the results are compared with the case of horizontal-only excitation. It is found that for an atrium-style metro station, the differences in horizontal acceleration amplitude between the structure and the adjacent soil rise with increasing RVH, which are different at different depths. The most significant differences occur at the depth of the ceiling slab. It is also observed that both the amplitude and distribution of peak DSNS have obvious differences between the left and right side walls at all levels. It is therefore concluded that the RVH has a significant influence on dynamic soil-structure interaction. It is believed that under extreme earthquake loading, such as near fault zones, RVH is a parameter of paramount importance and should be accounted for in the seismic analyses and seismic performance assessments of underground structures, especially for those with zero or near-zero buried depth, such as atrium-style metro stations.

Keywords: vertical earthquake motion; seismic response; atrium-style metro station; shaking table test

---

1. Introduction

The three-dimensional behavior of underground structures during earthquakes is a matter of interest for improving the seismic analysis and design in civil engineering under extreme loads. Although extensive studies were carried out on the effects of the horizontal component of the ground shaking, only limited work has been done in the analysis of the effect of the vertical component.

This topic deserves attention since some field evidence and investigations on the near-field ground motions have demonstrated that vertical components caused destructive consequences, even possible collapse, for some existing structures [1–4]. The latest developments in the underground vibration or deformation monitoring techniques such as the fiber-optic sensors [5,6] are expected to facilitate both the acquisition of the field evidence and the research work concerning the vertical earthquake motion.

In view of the fact that the ratio of vertical to horizontal peak ground acceleration (RVH) is extremely high in the vicinity of the active faults, it seems essential to study the influence of the RVH on the seismic responses of the underground structures, such as metro stations in the seismic active regions. Through numerical analyses based on the finite element-finite difference method, Uenishi and Sakurai [7] discussed the effect of the vertical oscillations and overburden on the damage-concentration mechanism of the columns at several specific underground station sections in the 1995 Hyogo-ken Nanbu (Kobe) earthquake. On the other hand, An et al. [8] discussed the effect of the vertical motion on the failure mode and dynamic response of the metro station, and concluded that in this case the vertical ground motion was not the primary cause of the collapse.

Research had also been carried out to investigate the influencing factor of RVH. Moore and Guan [9] investigated numerically the dynamic interaction of the twin tunnels subjected to the incident seismic waves. They have shown how the dynamic response of the twin tunnels depended on the incident angle of the earthquake wave, together with the spacing between two adjacent tunnels and the modulus ratio of the tunnel to the surrounding soil. Stamos and Beskos [10,11] provided insights into the dynamic responses of the underground infinitely long lined tunnels buried into an elastic or viscoelastic half-space subjected to the obliquely incident waves using the boundary element method (BEM), and proved the influence of the incident directions of the seismic waves on the dynamic responses of the tunnels. Lee and Trifunac [12] proved that the amplitudes of the stresses and deformations near the tunnel were dependent upon the angle of incidence, which determined the overall trends of the amplitudes. Lee and Manoogian [13,14] adopted the weighted residual method to study the scattering and diffraction of the plane SH-waves by an underground cavity of the arbitrary shape in a two-dimensional elastic half-space, and demonstrated same observations as for circular cavities that ground amplifications depended on the orientation of the incident waves. Wong et al. [15] studied the dynamic response of a cylindrical pipe embedded in an elastic semi-infinite medium using hybrid finite element and eigenfunction expansion techniques. They found that dynamic amplification was significantly dependent on the angle of incidence and the depth of embedment. Huang et al. [16] investigated the impact of the incident angles of earthquake shear waves (SV and SH-waves) on the seismic responses of a long lined tunnel, and numerical results indicated that non-linear seismic responses of the long lined tunnels were highly affected by the incident angles of the S-waves, which should be taken into consideration in mapping the seismic risk of the tunnels. Sun and Wang [17] investigated numerically the seismic responses of the underground rectangular tunnel under the vertical seismic excitation. Numerical results show that under vertical seismic excitation the tunnel experienced greater vertical stress from time to time and vertical compressive deformation, and sometimes it completely separated from the above soil layer.

The effect of the vertical shaking during an earthquake has been experimentally investigated previously through just a few shaking table tests. In terms of the experimental study on the tunnels, Xu et al. [18] investigated the mechanism and effect of the seismic measures of the mountain tunnel through the three-dimensional shaking table tests, where the peak ground acceleration (PGA) of the vertical input motion was taken to be 2/3 of the value corresponding to the horizontal input motion. Experimental results show that with 63% probability of exceedance in 50 years, including vertical components of the earthquakes increased the peak dynamic soil pressures on the tunnel sidewall, while the opposite was true for 1% probability of exceedance in 50 years. In both cases there was no significant change in the peak dynamic strain of the inverted arch on two sides of the flexible joints. Wang et al. [19] also adopted the above-mentioned definition of the horizontal and vertical strategies and investigated experimentally two shallow-bias tunnels with a small clear distance subjected to the horizontal–vertical earthquake excitations with different levels. Results show that the acceleration amplification factors of the vertical-direction were generally larger (1.02–3.94 times) that that of the horizontal-direction and the earthquake intensity presented different influence on the acceleration responses in the horizontal and vertical directions. In addition, Zhao et al. [20] investigated experimentally the dynamic responses of a tunnel subjected to near-fault pulse-like earthquake motions, where both vertical excitations and transverse-longitudinal-vertical excitations were included.

With respect to the underground metro stations, for example, Zhao et al. [21] studied the vibration behavior of a framed metro station under horizontal and horizontal–vertical seismic excitation using a shaking table. Experimental results show that for the lower shaking intensity the structural relative displacements under the horizontal seismic excitation were lower than those under the horizontal–vertical seismic excitation, while the opposite was true for the greater shaking intensity. It is believed that in the latter case, the vertical seismic excitation increased soil densification and ground subsidence, which in turn imposed greater constraints on the station. Chen et al. [22] investigated the effect of the vertical ground motion on the dynamic response characteristics of the central columns in a six-story metro station and the test results revealed that with the increasing ratio of the vertical/horizontal acceleration amplitude, the central columns would undertake much more vertical dynamic axial forces compared to side walls. Che et al. [23] experimentally investigated the dynamic behaviors of a single-story subway station embedded in the dry sand excited by the vertical sinusoidal and vertical random waves, respectively. They found that in both cases there was difference in the acceleration amplitude of the ground between the free-field model and soil-station model. Some attempts were also made to explore the relationship between seismic soil pressures and shear strains of the surrounding ground based on the regression analysis of the experimental data.

Numerical and analytical studies of the effect of RVH on underground tunnels dominate most of the previous work. Extremely limited studies have been dedicated to study the influence of the RVH or incident angle of earthquake waves on the seismic responses of the underground metro station, especially for the experimental studies. In terms of the underground metro station, the existing several experimental studies only focus on the effect of the RVH on certain kinds of structural responses, such as structural relative displacement and dynamic internal force. Experimental studies of the effect of RVH on more comprehensive seismic responses also including the acceleration response and dynamic interaction of soil-station system, dynamic soil normal stress (DSNS) along the side wall of station, and structural dynamic tensile strain (DTS), are not available in the literature. Due to the limited number of the existing experiments, further experimental studies of the seismic behavior of the soil and metro station under different vertical earthquake actions are needed to draw more general conclusion. Therefore, in this study the effect of RVH on the seismic responses of an atrium-style metro station including the acceleration, dynamic soil normal stress DSNS, and DTS are the primary foci. Through series of 1 g shaking table tests, this paper presents the key experimental findings, and an attempt is made to evaluate the effect of the RVH both qualitatively and quantitatively. In addition, this paper also provides insights into the seismic behavior of the atrium-style underground structures, which will help researchers and practitioners to better understand the seismic aspects of these kind of structures to improve their seismic design and construction.

2. Shaking Table Tests

2.1. Experimental Facility

The series of shaking table tests were conducted using the MTS Company shaking table facility at the State Key Laboratory for Disaster Reduction in Civil Engineering, Tongji University. The shaking table (Figure 1) has six degrees of freedom: two horizontal (X and Y), one vertical (Z), and three rotational, and the main technical parameters are listed in Table 1. The operating frequency range, maximum input accelerations, and maximum loading capacity all satisfy the experimental demands.

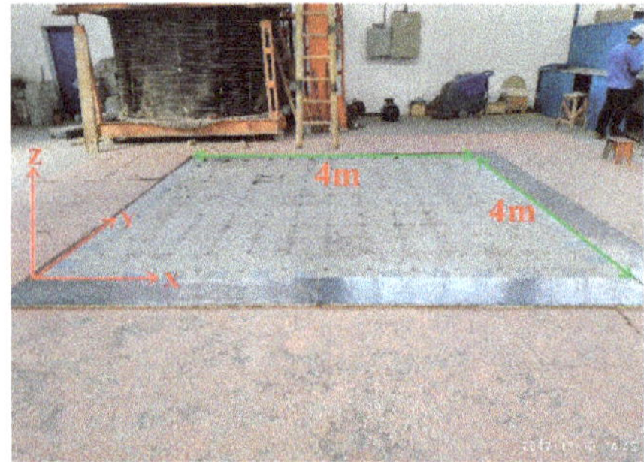

Figure 1. Photos of the shaking table and flexible-wall container.

Table 1. Main technical parameters of the shaking table system.

Table Size/(m × m)	4×4
Maximum specimen mass/ton	25
Operating frequency range/Hz	0.1~50
X direction	±100
Maximum displacement/mm Y direction	±50
Z direction	±50
X direction	±1000
Maximum velocity/(mm/s) Y direction	±600
Z direction	±600
X direction	±4.0
Maximum acceleration (empty table)/g Y direction	±2.0
Z direction	±4.0
X direction	±1.2
Maximum acceleration (with mass 15 ton)/g Y direction	±0.8
Z direction	±0.7

A cylindrical flexible-wall container (Figure 1) with diameter 3.0 m and height 1.8 m was adopted. It was made of 5 mm thick rubber membrane. Reinforced bars with a diameter of 4 mm and a spacing of 60 mm were arrayed circumferentially around the exterior of the rubber membrane. The top-ring, base-ring, and four columns were all manufactured from H-shaped steel members. The top-ring was supported by four columns with universal joints, providing the container with full translational and rotational freedom. Furthermore, the rubber membrane was designed to have similar shear stiffness with model soil, in order to minimize any soil-container interactions. The container had been proven to present good lateral shear-type deformations and to be reliable in terms of the negligible influence of the vertical boundary on the specified experiment [24,25].

2.2. Model

In view that the similitude ratio placed a severe limitation on the choice of suitable model materials, the similitude ratio design would be conducted before choosing the model materials. Based on the capability of the existing facility and equipment, the similitude ratio design of the experiment was accomplished on the basis of the dimensional analysis, which was adopted extensively in the design of the shaking table tests by other researchers. Meymand [26] designed and performed a series of scale model shaking table testing of the piles in clay in a 1 g scale model environment [27], in which he summarized that the method of the governing equations involved the differential equation describing the process and the formation of the similarity variables that related the model to the prototype. The similitude ratio design for this research herein closely follows the aforementioned method.

With respect to dynamics problems, the differential equation of motion expressed through displacement can be stated as (take the equation of x-direction as an example)

$$(\lambda + G)\frac{\partial \varepsilon}{\partial x} + G\nabla^2 u + X = \rho\frac{\partial^2 u}{\partial t^2} \tag{1}$$

where u, t, ρ, and ε are displacement, time, density, and strain, respectively. x and X refer to the coordinate position and the force per unit volume in the x direction, respectively. λ and G are the Lamé constant and shear modulus, respectively. Hence, Equation (1) is also called the Lamé equation. The problem-solving process of general dynamics problems is reduced to solve the Lamé equation under the given boundary conditions. Then, Equation (1) can be expressed as

$$G = \frac{\rho\frac{\partial^2 u}{\partial t^2} - \lambda\frac{\partial \varepsilon}{\partial x} - X}{\nabla^2 u + \frac{\partial \varepsilon}{\partial x}} \tag{2}$$

The most common sets of basic quantities are those of mass M, length L, and time T [28]. In Equation (2), ε is a dimensionless quantity and thus the dimension of $\frac{\partial \varepsilon}{\partial x}$ is L^{-1}. The dimension of ρ is ML^{-3}. The dimension of $\frac{\partial^2 u}{\partial t^2}$, which represents acceleration, a, is LT^{-2}. The three items in the numerator have the same dimensions. The two items in the denominator also have the same dimensions. Since ε is a dimensionless quantity for both prototype and model soil, the similitude ratio for ε must equal unity (1). The similitude ratios for $\nabla^2 u + \frac{\partial \varepsilon}{\partial x}$ and $\rho\frac{\partial^2 u}{\partial t^2} - \lambda\frac{\partial \varepsilon}{\partial x} - X$ are S_l^{-1} and $S_\rho S_a$, respectively. The similitude ratio for G is S_{G_d}, where the subscript G_d represents dynamic shear modulus. Then, the similitude relations between the model soil and the prototype soil can be deduced from Equation (2) as:

$$\frac{G_d^m}{G_d^p} = S_{G_d} = S_\rho S_a S_l \tag{3}$$

where G_d^m and G_d^p are dynamic shear modulus of model and prototype, respectively. S with the subscripts refers to the similitude ratio of the quantities corresponding to those subscripts, i.e., S_{G_d}, S_ρ, S_a, and S_l represent the similitude ratios for soil dynamic shear modulus, soil density, soil acceleration, and soil geometry size, respectively.

The similitude ratios corresponding to other quantities can also be obtained in a similar way. The similitude ratio of the length is set to 1/30 for both the model soil and model station. Table 2 shows the main parameters adopted in defining the similitude relations in this study.

Table 2. Similitude relations.

Types	Properties	Similarity Relations	Similitude Ratios	
			Model Structure	Model Soil
Geometry	Length l	S_l	1/30	1/30
	Linear displacement r	$S_r = S_l$	1/30	1/30
Material	Elastic modulus E	S_E	0.42	0.033
	Equivalent density ρ	$S_\rho = S_E S_l^{-1} S_a^{-1}$	12.6	0.52
	Shear modulus G	S_G	—	0.033
Dynamic	Mass m	$S_m = S_\rho S_l^3$	4.4×10^{-4}	1.2×10^{-6}
	Acceleration a	S_a	1	1
	Duration t	$S_t = S_E^{-\frac{1}{2}} S_l S_\rho^{\frac{1}{2}}$	0.1826	0.1826
	Frequency ω	$S_\omega = 1/S_t$	5.48	5.48
	Dynamic stress σ	$S_\sigma = S_l S_a S_\rho$	0.42	0.033
	Dynamic strain ε	S_ε	1	1

To satisfy the aforementioned similitude relations, an artificial model soil made of sand mixed with sawdust was selected, which had been studied and used by other researchers [29]. By targeting the curves of both the shear modulus and damping ratio decaying with the shear strain for the artificial model soil and prototype soil [30], the optimum mass ratio of sawdust to sand was determined to be 1:2.5 through a series of cyclic triaxial tests performed in the laboratory. The comparison of the curves between artificial model soil and prototype soil can be found in [24].

The prototype of the atrium-style metro station had the dimension of 21.3 × 17.7 m (width × height) and had two stories, including the station hall floor and platform floor. No columns on the station hall floor and thin-walled columns with spacing 7.6 m on the platform floor were adopted. Both the ceiling and middle slabs were mostly replaced with the flat-beams, directly resulting in about 50% opening area of the floorage for both slabs. The prototype station was a cast-in-place reinforced concrete structure and the strength grades of concrete and steel rebar were C35 and HRB400, respectively. Galvanized steel wire and micro-concrete were adopted in the model station to represent two practical materials mentioned above. To satisfy the similitude relations, an optimum mass ratio of micro-concrete was determined to be as cement:sand:lime powder:water = 1:5:0.64:1.18 through a series of material tests, and the compressive yield strength and elastic modulus were measured to be 10.68 MPa and 1.32×10^4 MPa, respectively. Figure 2 shows the details of the model station, including three observation planes. The observation plane-1 was located within the central symmetrical plane of the whole container-soil-structure system, while observation plane-2 and plane-3 were both 50 mm away from it.

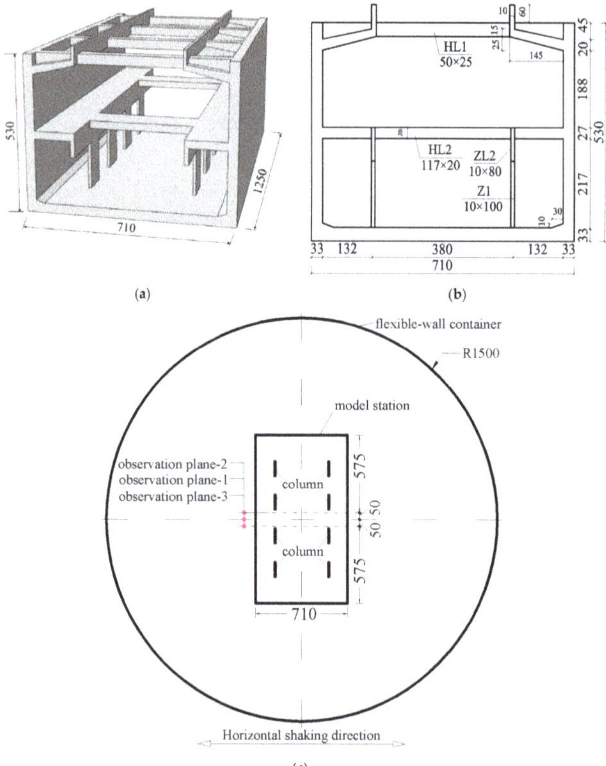

Figure 2. Dimensions of the model station: (**a**) perspective view; (**b**) cross-section; (**c**) locations of three observation planes (unit: mm).

For the whole soil-structure system, various real-time data were collected, including the acceleration and displacement in soil and on structure, the dynamic soil normal stress on both the left and right side walls, and the strain produced on the surface of the structural members.

2.3. Instrumentation

The locations of these instruments were all based on numerous prior numerical analyses using finite element method [31,32], which are shown in Figure 3.

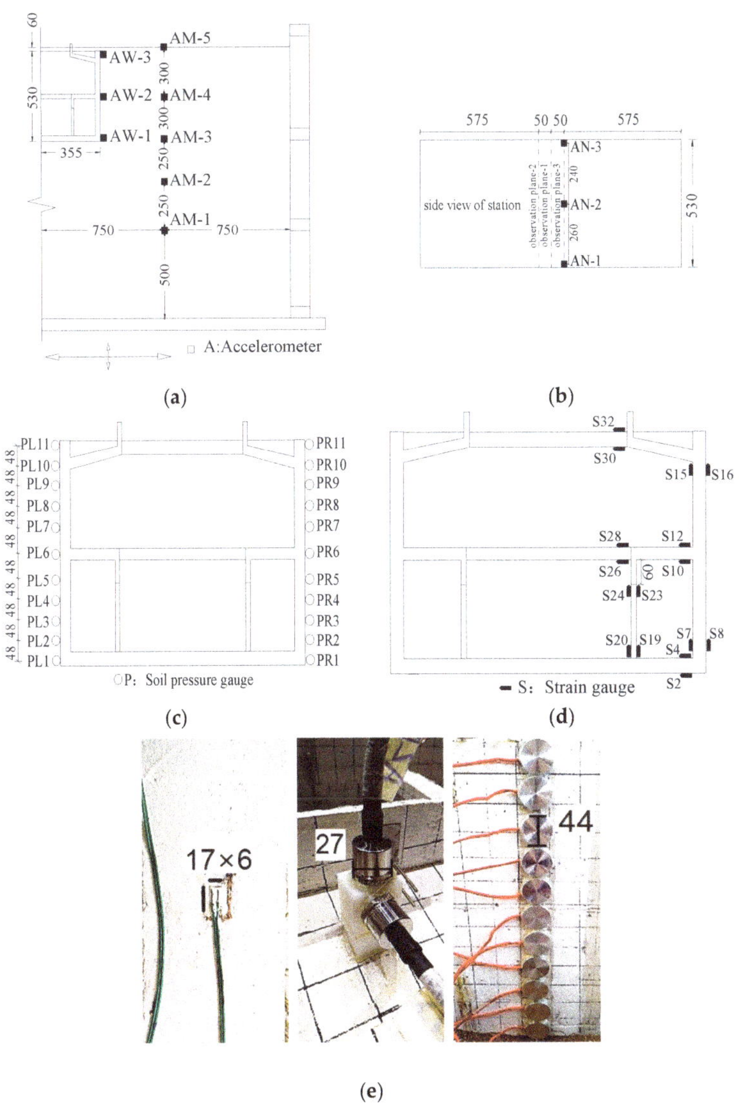

Figure 3. Layouts of sensors for soil-station model: (**a**) accelerometers of side wall and soil across the whole depth in observation plane-1; (**b**) accelerometers of soil adjacent to station in observation plane-2; (**c**) soil pressure cells in observation plane-3; (**d**) strain gauges; (**e**) photos of strain gauge, accelerometers, and soil pressure cells (all dimensions in mm).

Five accelerometers (AM-1, AM-2, AM-3, AM-4, and AM-5), as shown in Figure 3a, were placed in the middle between the center of the model soil and the wall of the flexible-wall container to capture the acceleration responses with soil depth. In considering the symmetry, Figure 3a only shows one-half of the whole container-soil-structure system while the accelerometers are completely displayed.

As displayed in Figure 3a,b three accelerometers (AN-1, AN-2, and AN-3) were arrayed in the model soil adjacent to the side wall, which are set to investigate the effects of the underground metro station on the soil acceleration.

Three more accelerometers (AW-1, AW-2, and AW-3) were installed near the previous ones on the side wall to record its acceleration. The depths of the accelerometers AW-1, AW-2, and AW-3 correspond to the depths of the base slab, middle slab, and ceiling slab, respectively. The comparison of the acceleration responses between the two groups of accelerometers is expected to investigate the effect of RVH on the dynamic interaction, which is a most interesting topic for the soil-structure system.

Soil pressure cells PL1–PL11 on the left side wall and PR1–PR11 on the right side wall were arrayed to investigate the amplitude and distribution of the dynamic soil normal stress on the two walls, Figure 3c. The interval between any two adjacent cells was 48 mm, which allowed to capture the essential data and to represent the complex stress distribution along the side wall as stated by the early finite element analyses.

Strain gauges, plotted in Figure 3d, were arrayed to investigate the values and spatial distribution of the structural strain, especially focusing near the joints between different structural components as the ends of the beam, column, side wall, and slab. In Figure 3d the strain gauges S23 and S24 were located on the top of the column (Z1 in Figure 2b), which was just below the longitudinal beam (ZL2 in Figure 2b).

Figure 3e shows the selected photos of various sensors. The experimental procedure is shown in Figure 4.

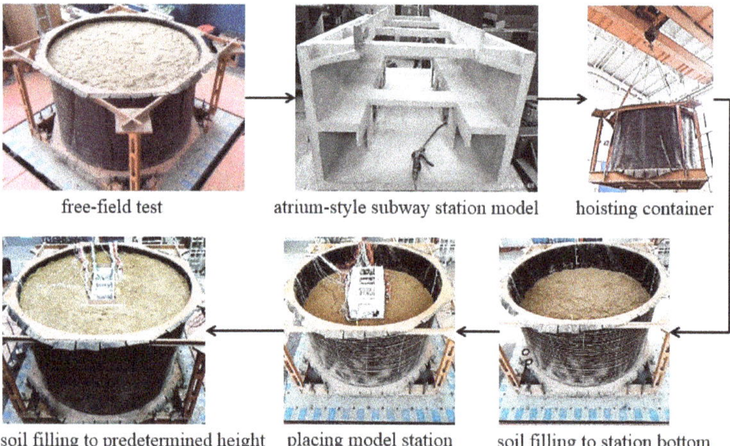

Figure 4. Photos of the experimental procedure.

2.4. RHV Inputs

The shaking table was accelerated incrementally and the earthquake motion in the horizontal and vertical directions were applied synchronously. The acceleration amplitude of all the horizontal input motions was adjusted to 0.1 g and corresponding vertical acceleration amplitudes were adjusted to 0 g, 0.05 g, 0.1 g, and 0.15 g, respectively. Hence, different RVH that equals to 0, 0.5, 1, and 1.5 are obtained to investigate their influence on the seismic behavior of the soil and buried metro station. For convenience, a strong-motion record of the Loma Prieta, California, earthquake of 18 October 1989 from the PEER

Strong Motion Database [33] was selected as the input motion for both types of the loading mentioned above, whose acceleration time history and corresponding Fourier amplitude spectra, together with target design spectrum of prototype ground concerning the experiment, are shown in Figure 5. It is seen that the acceleration response spectra of the selected 1989 Loma Prieta earthquake matched the target design spectrum well. Table 3 shows the test cases in this study.

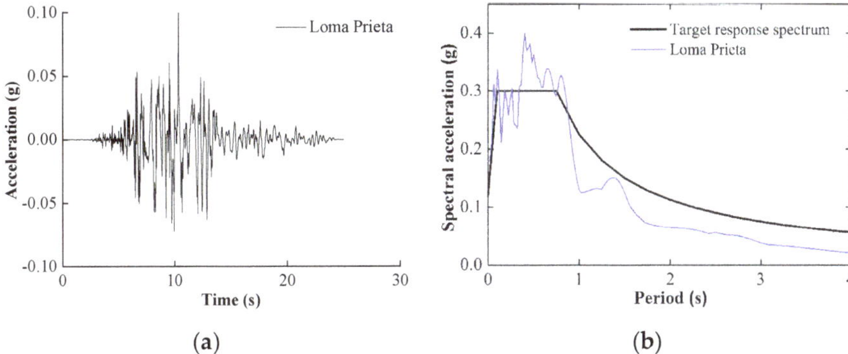

Figure 5. The details of Loma Prieta ground motion: (**a**) accelerogram; (**b**) spectral acceleration compared with target design spectrum [34].

Table 3. Test cases for the shaking table tests.

No	Cases	Waveform	Peak Acceleration (g)	
			Horizontal	Vertical
1	LP-x0.1	Loma Prieta	0.1	0
2	LP-x0.1-z0.05	Loma Prieta	0.1	0.05
3	LP-x0.1-z0.1	Loma Prieta	0.1	0.1
4	LP-x0.1-z0.15	Loma Prieta	0.1	0.15

The targeted base motions were converted from prototype to model scale units. The base accelerations recorded on the shaking table were refered to as achieved motions. Figure 6 shows the comparison between targeted base motions and achieved base motions both in horizontal and vertical components, including all test cases in Table 3. For any test case, the whole acceleration time history was consistent and minor differences in amplitude were observed between achieved and targeted base motions. These minor differences were acceptable, in view of shaking table performance and such large model specimens.

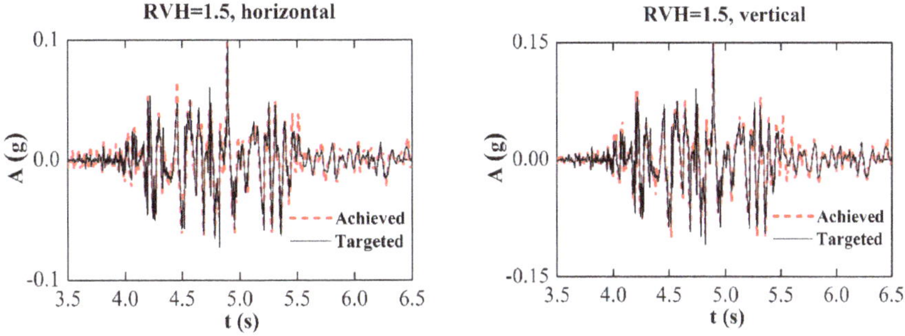

Figure 6. *Cont.*

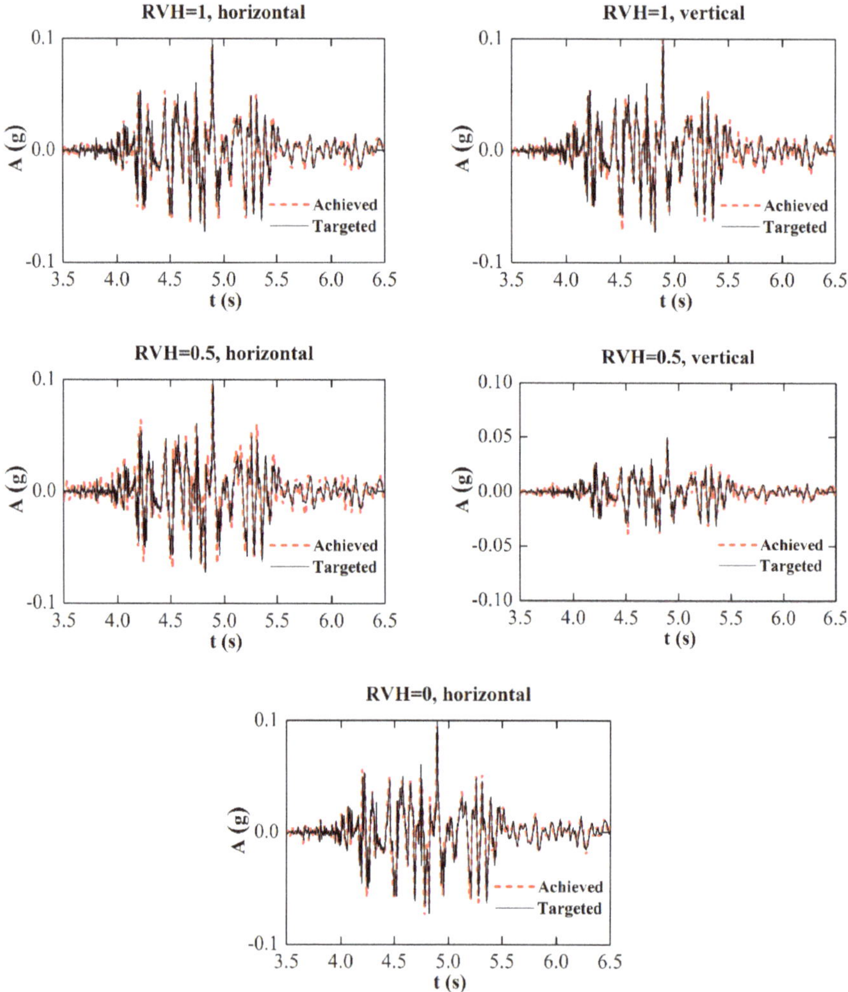

Figure 6. Comparison between targeted base motions and achieved base motions both in the horizontal and vertical components for all test cases.

It is worth mentioning that some details of the experimental program, verification of the free-field test, and interpretation of some preliminary results can be found in prior papers [24,25].

3. Test Results and Discussion

3.1. Influence of RVH on Soil Acceleration

To investigate the influence of the RVH on the vertical propagation of the waves in the ground, as an example, Figure 7 compares the horizontal acceleration time histories of the soil at different measuring points (AM-1, AM-2, AM-3, AM-4, and AM-5) between RVH = 0 and RVH = 1. It is seen that at any measuring point their horizontal acceleration always behaves consistently in the trend, and the differences mainly lie in the amplitude. For both RVH = 0 and RVH = 1, the horizontal acceleration of the soil increases with decreasing depth.

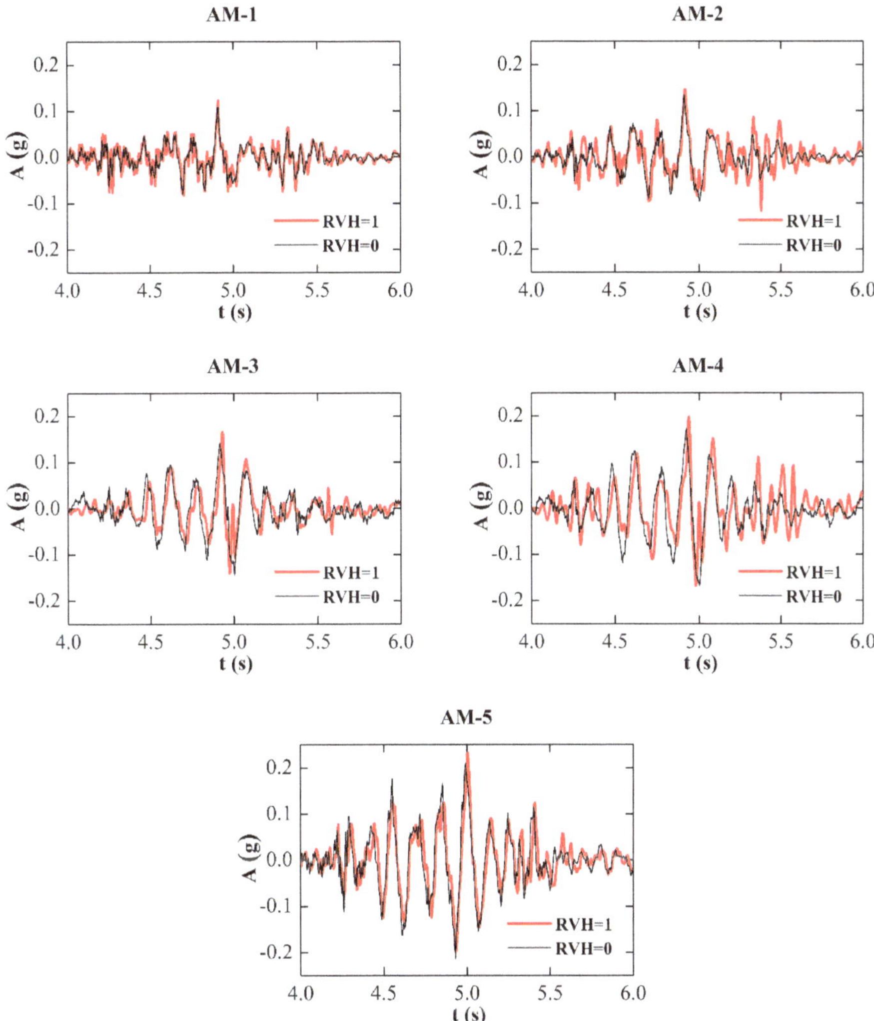

Figure 7. Comparison of horizontal acceleration time histories of soil between RVH = 0 and RVH = 1. RVH: ratio of vertical to horizontal peak ground acceleration.

To better evaluate the trend of the horizontal acceleration amplitude, the acceleration amplification factor (AAF) is obtained by dividing the peak horizontal acceleration at each measuring point by the peak horizontal acceleration of input motion. Therefore, the AAF at the base of the model (at soil depth of 1.6 m in Figure 3a) is equal to unity (1). This definition also applies to the following parts. Figure 8 shows the AAF of the ground along the depth under different RVH. The horizontal acceleration amplitude of the soil tends to increase from the bottom to the ground surface. Under the four conditions the AAFs on the ground surface are about 2.3–3.1 and the soil presents significant amplification. This is reasonable since the amplitude of all the input motions are relatively low.

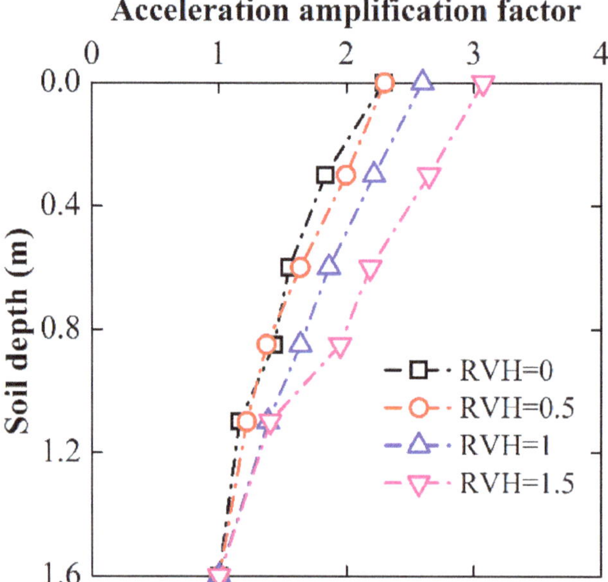

Figure 8. Acceleration amplification factor of ground under different RVH.

In terms of the acceleration levels in this study, with increasing RVH, soil horizontal acceleration increases and the increase in the amplitude grows significantly with increasing RVH on the whole. It demonstrates that soil horizontal acceleration is more sensitive to higher RVH or higher vertical acceleration.

3.2. Acceleration Difference between the Side Wall and Adjacent Soil

To investigate the influence of the RVH on the acceleration response difference between the side wall and adjacent soil, Figure 9 illustrates the comparison of the acceleration time histories between side wall and adjacent soil at three different depths, which are ceiling slab (Figure 9a), middle slab (Figure 9b), and base slab (Figure 9c), respectively.

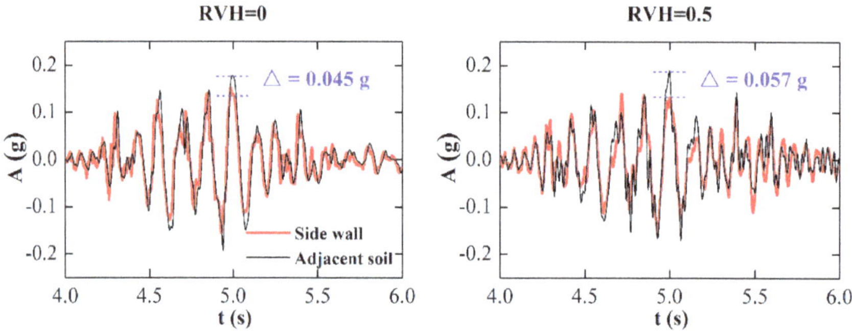

Figure 9. *Cont.*

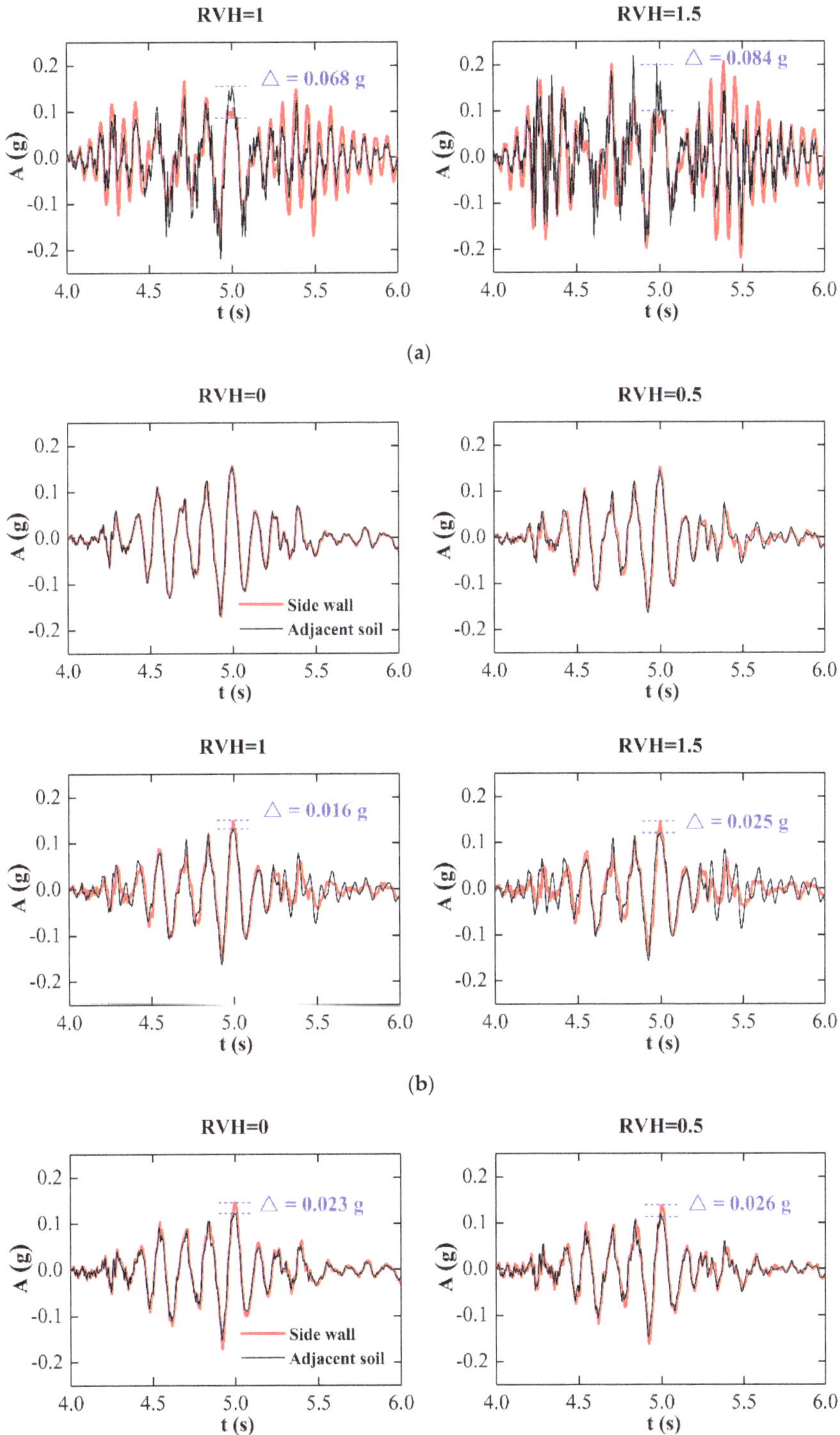

Figure 9. *Cont.*

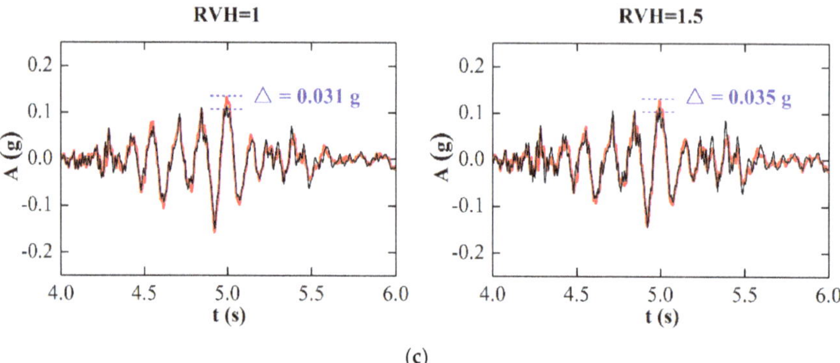

(c)

Figure 9. Comparison of the acceleration time histories between side wall and adjacent soil. at three different depths: (**a**) comparison at depth of ceiling slab; (**b**) comparison at depth of middle slab; (**c**) comparison at depth of base slab.

As can be seen from Figure 9, the phases of the side wall and adjacent soil are always consistent at three depths. The acceleration difference between the side wall and the adjacent soil mainly lies on the amplitude. For the side wall and the adjacent soil, the amplitude of them is almost the same at the depth of the middle slab, where the amplitude differences are about 0–0.025 g at the peak-strain moment. The amplitude of them has minor differences at depth of base slab, where the amplitude differences are about 0.023–0.035 g at the peak-strain moment. It is worth mentioning that some larger differences between the side wall and adjacent soil can be observed at the depth of the ceiling slab, where the amplitude differences reach about 0.045–0.084 g at the peak-strain moment. Under RVH = 1.5 and horizontal input acceleration of 0.1 g, the aforementioned amplitude difference even reaches 0.084 g. It reveals larger differences in the horizontal acceleration between the side wall and the adjacent soil at the depth of the ceiling slab. The differences can be attributed to that the ceiling slab lies too close to the ground surface and there is a significant amplification of the acceleration responses since fewer soil constraints exist there. It reveals that for the underground structures with zero or near-zero buried depth, such as atrium-style metro stations, special attention should be paid to the acceleration response difference between the structure and adjacent soil.

In order to investigate the differences of the horizontal acceleration amplitude between the side wall and the adjacent soil with increasing RVH, the results at three different depths are depicted in Figure 10. It is found that at any depth, the acceleration differences between side wall and adjacent soil appear a linear increase in the amplitude as RVH increases from 0 to 1.5. With regard to the amplitude, the acceleration differences at the depth of the ceiling slab are larger than those at the depth of the base slab. The acceleration differences at the depth of the middle slab are the smallest ones among the three depths. With regard to the increasing rates, the increasing rate at the depth of the ceiling slab is about 2.56%, which is larger than that of 1.65% at the depth of the middle slab. The smallest one is 0.82% at the depth of the base slab.

In conclusion, for an atrium-style metro station, the differences in the horizontal acceleration amplitude between the structure and the adjacent soil rise with increasing RVH, which are different at different depths. The most significant differences occur at the depth of the ceiling slab. The RVH has a significant influence on the dynamic soil-structure interaction, especially for higher RVH.

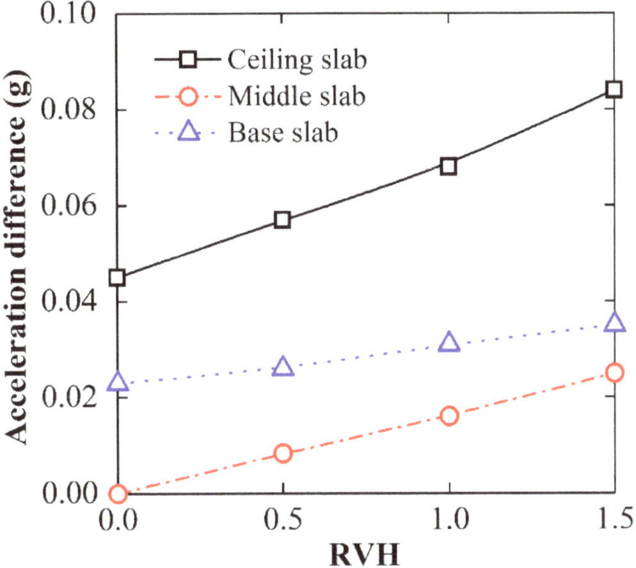

Figure 10. The difference of horizontal acceleration amplitude with increasing RVH between side wall and adjacent soil at three different depths.

3.3. Influence of RVH on Dynamic Soil Normal Stress

Dynamic soil normal stress (DSNS) is defined as the difference between the total stress and initial static stress along the side wall of the model station. As an example, Figure 11 compares the two time histories of the DSNS between RVH = 0 and RVH = 1. It is found that the time histories of the DSNS resemble that of the input motion. The essence of this phenomenon is also revealed by Kramer, who points out that the input seismic excitation in the form of an acceleration can be converted into a stress wave [35]. With different RVH their phases are basically consistent at all measuring points, though the amplitudes of the DSNS behave differently at different depths of the side wall. For example, the amplitudes are almost the same near the bottom of the side wall (e.g., PR1 and PR2) while the amplitudes under RVH = 1 are always larger than those under RVH = 0 at other depths of the side wall (e.g., PR3~PR11).

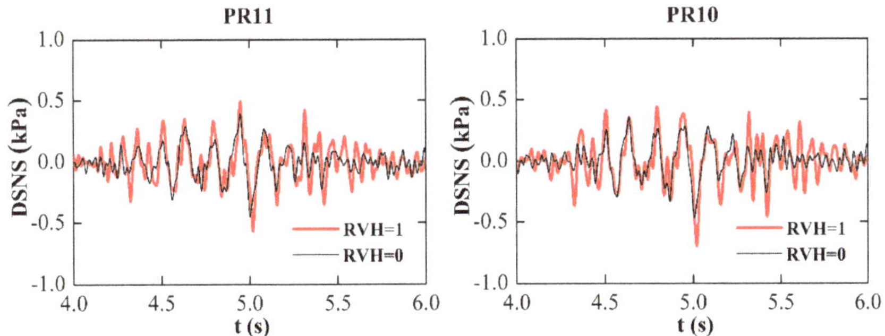

Figure 11. *Cont.*

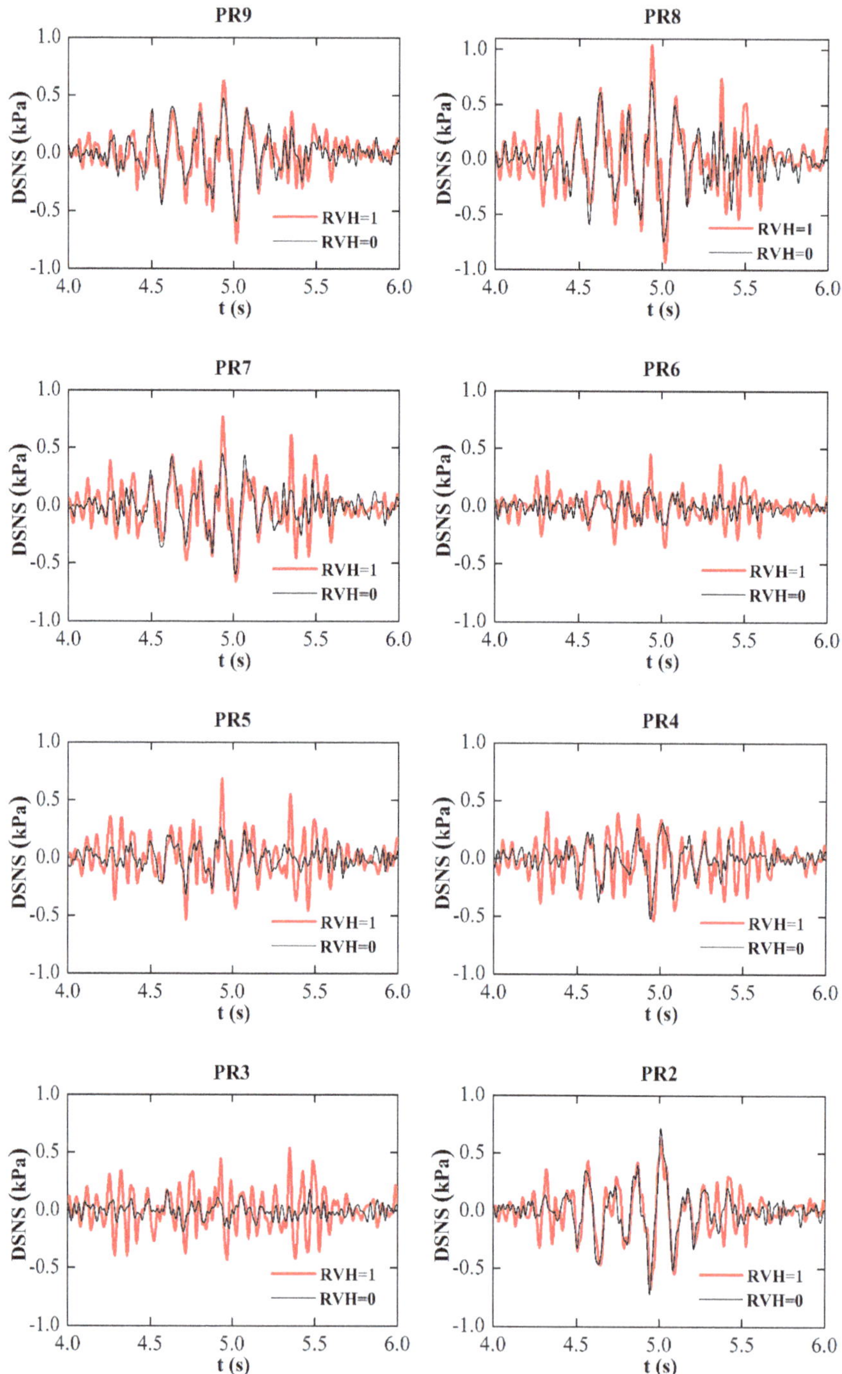

Figure 11. *Cont.*

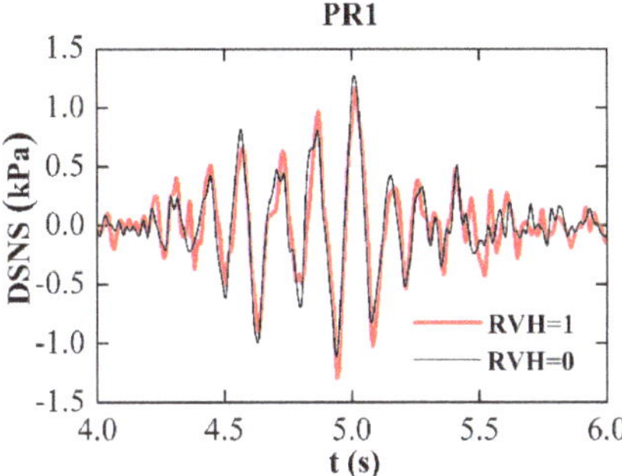

Figure 11. Comparison of time histories of dynamic soil normal stress (DSNS) along right side wall between RVH = 0 and RVH = 1.

To figure out the influence of the RVH on the DSNS along the side wall, Figure 12 displays the peak DSNS along the left and right side walls under different RVH.

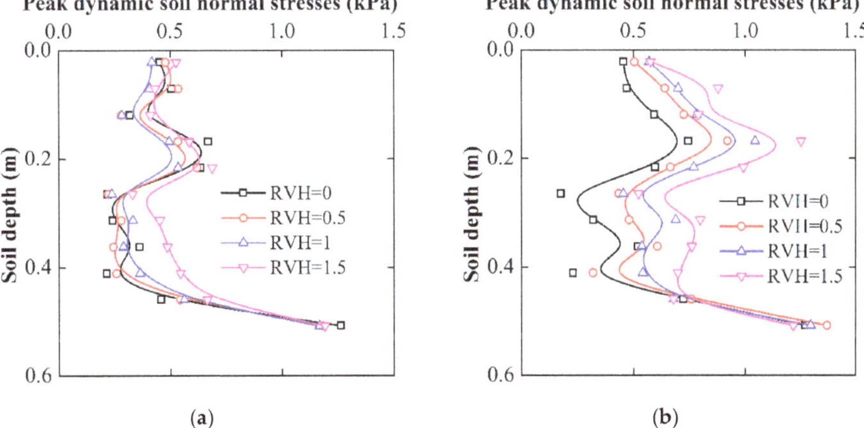

(a)　　　　　　　　(b)

Figure 12. Distribution of peak DSNS under different RVH along (**a**) left side wall, and (**b**) right side wall.

As can be seen from Figure 12a, under the input motion of Loma Prieta, peak DSNS along the left side wall follows an approximate L–shaped distribution. With increasing RVH, on the whole, there is no significant change in the amplitude of the peak DSNS. Compared to other RVH, a relatively large increase at the depth of the platform floor can be observed when RVH equals to 1.5.

With respect to the peak DSNS along the right wall, as shown in Figure 12b, at the depth between the top of the right side wall and the middle of the station hall floor, they all follow an approximately linear distribution. At the depth between the middle of the station hall floor and the bottom of the right side wall, they all follow an approximately Σ–shaped distribution. It is worth mentioning that the maximum stress occurs at the bottom of the side wall basically. With increasing RVH, peak DSNS along the right wall increase on the whole, except for the bottom of the side wall (base slab level). Compared to the results under only horizontal input motion (RVH = 0), the maximum increase in peak DSNS under RVH = 0.5, 1, and 1.5 reach about 148%, 159%, and 199%, respectively. For the bottom of the side wall (base slab level), with increasing RVH, the peak DSNS has no significant change, which is different from the situation at other position of the side wall and has also been concluded above from Figure 11. It might be attributed to the sharp transition in the corner and a possible stress concentration there.

When comparing the distribution of the peak DSNS between the left and right side walls, some obvious differences between them can be found. A similar finding was also drawn from a numerical study of a shallow buried rectangular underground structure under earthquake loading with both the horizontal and vertical components [36]. The differences in stress distribution of the peak DSNS between the left and right side walls can be explained by the shadow effect, which is caused by the entrapment of the waves between the ground and the underground structure [37]. With increasing RVH, the above-mentioned differences between the left and right side walls also increase. From the perspective of dynamic soil normal stress, the experimental results have proven that RVH has significant influence on the dynamic soil-structure interaction.

3.4. Influence of RVH on Structural Dynamic Strain

The dynamic strain is defined as the difference between the total strain during shaking and initial static strain before shaking. As an example, Figure 13 compares the time histories of the dynamic strain between RVH = 0 and RVH = 1 from one side of every cross section (Figure 3d). Just as the characteristics of the DSNS mentioned earlier, the time histories of the dynamic strain also resemble that of the input motion. In terms of the two different RVH, their phases are extremely consistent for all the measuring points and their amplitudes have minor differences for most of the measuring points.

Figure 14 shows the variation of the peak dynamic tensile strain (DTS) with different RVH for different measuring points of the model station. To easily observe the rate of change in the dynamic strain over different RVH, the value of the y–coordinate in Figure 14 represents the ratio of the peak DTS under RVH = i (i = 0, 0.5, 1, and 1.5) to that under RVH = 0. Since two strain gauges are arranged for every cross section (Figure 3d), the strain results on one side are displayed in Figure 14a and the opposite ones are shown in Figure 14b, where the layouts of corresponding strain gauges are plotted again. As can be seen from Figure 14a, with increasing RVH, the DTS will change in an undulating fashion (decrease, increase, and decrease again), which is different from the response characteristics of the soil, in that the soil acceleration at any depth increases rapidly with increasing RVH. It means the influence of the RVH on the structural DTS is rather complex.

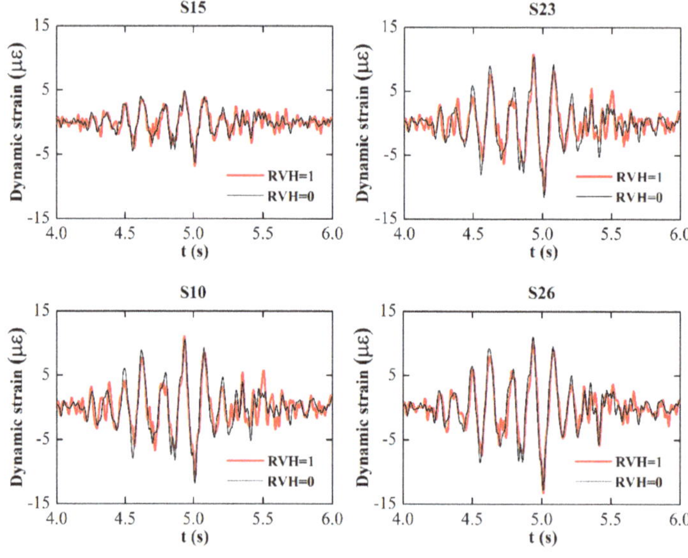

Figure 13. *Cont.*

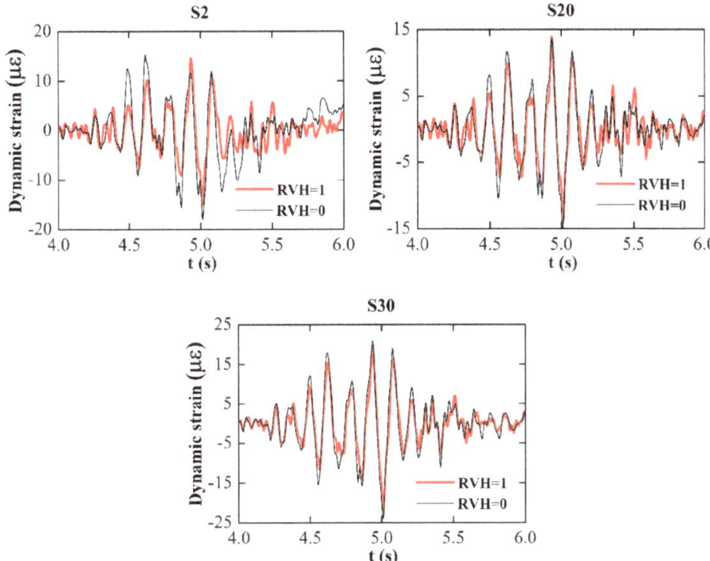

Figure 13. Comparison of time histories of dynamic strain on one side of structural components between RVH = 0 and RVH = 1.

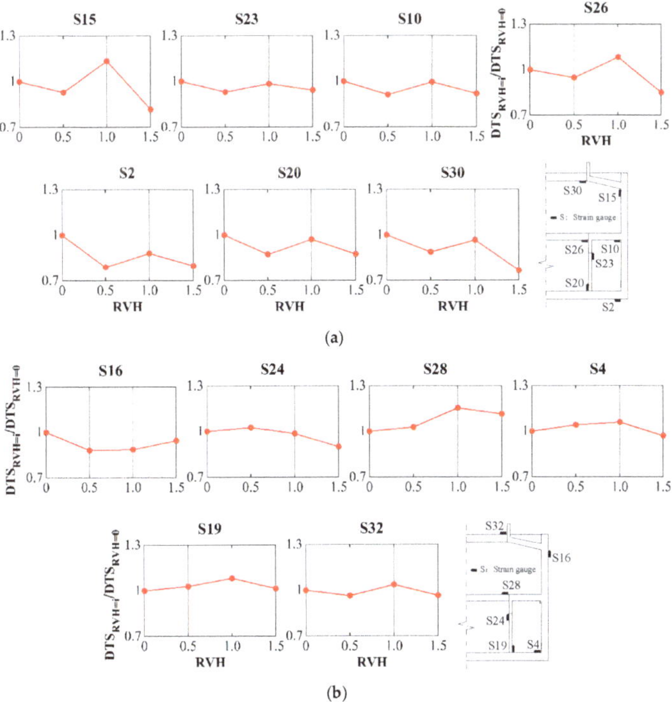

Figure 14. Peak dynamic tensile strain (DTS) of model station under different RVH: (**a**) strains on one side; and (**b**) strains on another side.

The soil and structure behave differently in the seismic responses with increasing RVH. It can also be seen from Figure 14a that the DTS with bidirectional input motion may be less than those with only horizontal input motion (the value of the y–coordinate is less than 1). This conclusion is verified by the numerical simulation conducted on a prototype atrium-style metro station [31], and can also be found in [38]. This phenomenon can be explained as follows: the presence of a cavity can cause, under certain conditions, intense and selective de-amplification of the free-filed motion, which is referred to as "shadow zone" [37]. When comparing the results of the DTS in Figure 14a,b, for every cross section, the DTS on one side is not equal to the opposite one. It reveals that the stresses in the cross section of every structural element are uneven under bidirectional input motion.

4. Conclusions

An atrium-style metro station is a favorite choice owing to its excellence in providing a larger and clearer space for the passengers and commercial clients at the station hall floor. However, concern for the seismic safety of such a station is growing since its hall slabs are replaced with the flat-beams. Tests are needed for the evaluation of its seismic safety. A series of 1 g shaking table tests were conducted to investigate the seismic responses of an atrium-style metro station.

In this study, extensive experimental results were presented to investigate the influence of the RVH (ratio of vertical to horizontal peak ground acceleration) on the seismic responses of both the soil and the underground metro station. Results show that the acceleration amplification factors (AAFs) on the ground surface are about 2.3–3.1, and that the soil presents significant amplification under several considered conditions. Soil horizontal acceleration is more sensitive to the higher RVH or higher vertical acceleration. Significant differences in horizontal acceleration amplitude between the side wall and the adjacent soil at the depth of the ceiling slab are found, and attention should be paid to these kind of differences for the underground structures with zero or near-zero buried depth, such as atrium-style metro stations. With increasing RVH, the differences in the horizontal acceleration amplitude between the side wall and the adjacent soil also rise. The RVH has a significant influence on the dynamic interaction, especially for higher RVH. Under the input motion of Loma Prieta, the distribution of the peak DSNS (dynamic soil normal stress) along the left and right side walls represents significant difference. The peak DSNS along the left side wall follows an approximate L-shaped distribution, while it is different for the right side wall. With increasing RVH, there is no significant change in the amplitude of the peak DSNS along the left side wall on the whole, while it increases a lot (the maximum increase reach about 148%, 159%, and 199% under RVH = 0.5, 1, and 1.5 when comparing with that under RVH = 0) for the right side wall. The soil and structure behave differently in the seismic responses with increasing RVH. The stresses in the cross-section of every structural element are uneven under horizontal–vertical earthquake excitation.

In conclusion, from the perspective of the soil acceleration, structural acceleration, and dynamic soil normal stresses along the side wall of station, experimental results have proven that ratio of vertical to horizontal peak ground acceleration (RVH) has significant influence on the dynamic soil-structure interaction. It is believed that under extreme earthquake loading, such as near fault zones, RVH is a parameter of paramount importance and should be accounted for in the seismic analyses and seismic performance assessments of the underground structures, especially for those with zero or near-zero buried depth, such as atrium-style metro stations.

It is worth mentioning that further investigation of the effect of the vertical earthquake motion is essential through a series of numerical analyses using a minimum number of the earthquake records specified by a certain code. In this way, a cross-check of the findings from the multiple records could be expected to accomplished and this is a work in progress.

Author Contributions: Test scheme, Z.Z., Y.Y., H.Z., and H.Y.; experimental works, Z.Z.; discussion and analysis, E.B. and Y.Y.; writing—original draft preparation, Z.Z.; writing—review and editing, E.B. and Y.Y.; funding acquisition, Y.Y. and H.Z.

Funding: This research was funded by the National Key Research and Development Plan of China (Grants Nos. 2018YFC1504305, 2018YFC0809600, 2018YFC0809602, and 2017YFC1500703), the State Key Laboratory of Disaster Reduction in Civil Engineering, Tongji University (Grant No. SLDRCE15-02), and the China Scholarship Council (Grant No. 201706260143).

Conflicts of Interest: The authors declare no conflict of interest.

References

1. Papazoglou, A.J.; Elnashai, A.S. Analytical and field evidence of the damaging effect of vertical earthquake ground motion. *Earthq. Eng. Struct. Dyn.* **1996**, *25*, 1109–1137. [CrossRef]
2. Building Research Institute. *Final Report of Damage Survey of the 1995 Hyogoken-Nanbu Earthquake*; Building Research Institute: Tsukuba, Japan, 1996.
3. Goltz, J. *The Northridge, California Earthquake of January 17, 1994: General Reconnaissance Report*; Report No. NCEER-94-0005; University at Buffalo: Buffalo, NY, USA; National Centre for Earthquake Engineering Research: Buffalo, NY, USA, 1994.
4. Mitchell, D.; Tinawi, R.; Sexsmith, R.G. Performance of bridges in the 1989 Loma Prieta earthquake—lessons for Canadian designers. *Can. J. Civ. Eng.* **1991**, *18*, 711–734. [CrossRef]
5. Liang, T.C.; Lin, Y.L. Ground vibrations detection with fiber optic sensor. *Opt. Commun.* **2012**, *285*, 2363–2367. [CrossRef]
6. Nedoma, J.; Stolarik, M.; Fajkus, M.; Pinka, M.; Hejduk, S. Use of fiber-optic sensors for the detection of the rail vehicles and monitoring of the rock mass dynamic response due to railway rolling stock for the civil engineering needs. *Appl. Sci.* **2019**, *9*, 134. [CrossRef]
7. Uenishi, K.; Sakurai, S. Characteristic of the vertical seismic waves associated with the 1995 Hyogo-ken Nanbu (Kobe), Japan earthquake estimated from the failure of the Daikai Underground Station. *Earthq. Eng. Struct. Dyn.* **2000**, *29*, 813–821. [CrossRef]
8. An, X.; Shawky, A.A.; Maekawa, K. The collapse mechanism of a subway station during the Great Hanshin earthquake. *Cem. Concr. Compos.* **1997**, *19*, 241–257. [CrossRef]
9. Moore, I.D.; Guan, F. Three-dimensional dynamic response of lined tunnels due to incident seismic waves. *Earthq. Eng. Struct. Dyn.* **1996**, *25*, 357–369. [CrossRef]
10. Stamos, A.A.; Beskos, D.E. Dynamic analysis of large 3-D underground structures by the bem. *Earthq. Eng. Struct. Dyn.* **1995**, *24*, 917–934. [CrossRef]
11. Stamos, A.A.; Beskos, D.E. 3-D seismic response analysis of long lined tunnels in half-space. *Soil Dyn. Earthq. Eng.* **1996**, *15*, 111–118. [CrossRef]
12. Lee, V.W.; Trifunac, M.D. Response of tunnels to incident SH-waves. *J. Eng. Mech. Div. ASCE* **1979**, *105*, 643–659.
13. Lee, V.W.; Manoogian, M.E. Surface motion above an arbitrary shape underground cavity for incident SH waves. *Eur. Earthq. Eng.* **1995**, *1*, 3–11.
14. Manoogian, M.E.; Lee, V.W. Diffraction of SH-Waves by subsurface inclusions of arbitrary shape. *J. Eng. Mech.* **2002**, *122*, 123–129. [CrossRef]
15. Wong, K.C.; Datta, S.K.; Shah, A.H. Three-dimensional motion of buried pipeline. I: Analysis. *J. Eng. Mech.* **1986**, *112*, 1319–1337. [CrossRef]
16. Huang, J.; Du, X.; Zhao, M.; Zhao, X. Impact of incident angles of earthquake shear (S) waves on 3-D non-linear seismic responses of long lined tunnels. *Eng. Geol.* **2017**, *222*, 168–185. [CrossRef]
17. Sun, C.; Wang, Q. Dynamic response of underground structure under vertical earthquake action. *Appl. Mech. Mater.* **2011**, *105–107*, 1480–1483. [CrossRef]
18. Xu, H.; Li, T.; Xia, L.; Zhao, J.X.; Wang, D. Shaking table tests on seismic measures of a model mountain tunnel. *Tunn. Undergr. Sp. Technol.* **2016**, *60*, 197–209. [CrossRef]
19. Wang, F.; Jiang, X.; Niu, J. The large-scale shaking table model test of the shallow-bias tunnel with a small clear distance. *Geotech. Geol. Eng.* **2017**, *35*, 1093–1110. [CrossRef]
20. Zhao, X.; Shuping, J.; Yi, Y. Study on shaking table model test scheme of tunnel subjected to near-fault pulse-like ground motions. *Mod. Appl. Sci.* **2014**, *8*, 126–139. [CrossRef]
21. Zhao, D.; Gong, B.; Yan, C. Experimental analysis of vibration behavior for large-span underground structure. *J. Civ. Archit. Environ. Eng.* **2002**, *24*, 52–57.

22. Chen, W. Static Pushover and Dynamic Response Characteristics of a Multi-Story Subway Station under Near-Fault Ground Motions. Master's Thesis, Tongji University, Shanghai, China, 2016; pp. 64–69.

23. Che, A.L.; Iwatate, T.; Ge, X.R. Dynamic behaviors of subway structure subjected to strong earthquake motions using shaking table tests and dynamic analyses. *Yantu Lixue/Rock Soil Mech.* **2006**, *27*, 1293–1298. [CrossRef]

24. Zhang, Z.; Yuan, Y.; Bilotta, E.; Yu, H.; Zhao, H. Design of shaking table tests on atrium-style subway station models under seismic excitations. In *Tunnels and Underground Cities: Engineering and Innovation Meet Archaeology, Architecture and Art, Proceedings of the World Tunnel Congress, Naples, Italy, 3–9 May 2019*; Peila, D., Viggiani, G., Celestino, T., Eds.; CRC Press: Leiden, The Netherlands, 2019; Paper No. 92982; pp. 1242–1251. [CrossRef]

25. Zhang, Z.; Bilotta, E.; Yuan, Y.; Zhao, H.; Yu, H. Experimental study on seismic ground motion amplification pattern of soil-structure interaction system. In Proceedings of the 7th International Conference on Earthquake Geotechnical Engineering, Rome, Italy, 17–20 June 2019. Paper No. 11063.

26. Meymand, P.J. Shaking Table Scale Model Tests of Nonlinear Soil-Pile-Superstructure Interaction in Soft Clay. Ph.D. Thesis, University of California, Berkeley, CA, USA, 1998.

27. Moss, R.E.S.; Crosariol, V.A. Scale Model shake table testing of an underground tunnel cross section in soft clay. *Earthq. Spectra* **2013**, *29*, 1413–1440. [CrossRef]

28. Moncarz, P.D.; Krawinkler, H. *Theory and Application of Experimental Model. Analysis in Earthquake Engineering*; John A. Blume Earthquake Engineering Center, Stanford University: Stanford, CA, USA, 1981.

29. Chen, Z.; Chen, W.; Li, Y.; Yuan, Y. Shaking table test of a multi-story subway station under pulse-like ground motions. *Soil Dyn. Earthq. Eng.* **2016**, *82*, 111–122. [CrossRef]

30. Bao, Z.; Yuan, Y.; Yu, H. Multi-scale physical model of shield tunnels applied in shaking table test. *Soil Dyn. Earthq. Eng.* **2017**, *100*, 465–479. [CrossRef]

31. Zhang, Z.; Yu, H.; Yuan, Y.; Zhao, H. 3D numerical simulation of seismic characteristics of atrium-style metro station. In Proceedings of the 3rd International Conference on Performance Based Design in Earthquake Geotechnical Engineering, Vancouver, BC, Canada, 16–19 July 2017. Paper No. 145.

32. Zhang, Z.; Yuan, Y.; Bilotta, E.; Yu, H.; Zhao, H. Dynamic soil normal stresses on side wall of a subway station. In *Life Cycle Analysis and Assessment in Civil Engineering: Towards an Integrated Vision, Proceedings of the Sixth International Symposium on Life-Cycle Civil Engineering, Ghent, Belgium, 28–31 October 2018*; Caspeele, R., Taerwe, L., Frangopol, D.M., Eds.; CRC Press: Leiden, The Netherlands, 2018; Paper No. 345; p. 284. [CrossRef]

33. Pacific Earthquake Engineering Research Center (PEER). *User's Manual for the Peer Ground Motions Data Base Web Application*; The Peer Center, University of California: Berkeley, CA, USA, 2010; Available online: https://ngawest2.berkeley.edu (accessed on 27 January 2018).

34. National Standard of the People's Republic of China. *Code for Seismic Design of Buildings (GB 50011–2010)*; China Architecture & Building Press: Beijing, China, 2010; pp. 33–35.

35. Kramer, S.L. *Geotechnical Earthquake Engineering*; Prentice Hall: Upper Saddle River, NJ, USA, 1996.

36. Xu, Z.; Du, X.; Xu, C.; Hao, H.; Bi, K.; Jiang, J. Numerical research on seismic response characteristics of shallow buried rectangular underground structure. *Soil Dyn. Earthq. Eng.* **2019**, *116*, 242–252. [CrossRef]

37. Yiouta-Mitra, P.; Kouretzis, G.; Bouckovalas, G.; Sofianos, A. Effect of underground structures in earthquake resistant design of surface structures. *Geotech. Spec. Publ.* **2007**. [CrossRef]

38. Mwafy, A.; Elnashai, A. Vulnerability of code-compliant RC buildings under multi-axial earthquake loading. In Proceedings of the 4th International Conference on Earthquake Engineering, Taipei, Taiwan, 12–13 October 2006.

Article

An Analytical Framework for the Investigation of Tropical Cyclone Wind Characteristics over Different Measurement Conditions

Lixiao Li [1,2], Yizhuo Zhou [1], Haifeng Wang [3], Haijun Zhou [1,2], Xuhui He [4] and Teng Wu [3,]*

[1] College of Civil and Transportation Engineering, Shenzhen University, Shenzhen 518060, China; lilixiao@szu.edu.cn (L.L.); 1810332047@email.szu.edu.cn (Y.Z.); haijun@szu.edu.cn (H.Z.)

[2] Guangdong Provincial Key Laboratory of Durability for Marine Civil Engineering, Shenzhen University, Shenzhen 518060, China

[3] Department of Civil, Structural and Environmental Engineering, University at Buffalo, State University of New York, Buffalo, NY 14203, USA; hwang48@buffalo.edu

[4] School of Civil Engineering, Central South University, Changsha 410083, China; xuhuihe@csu.edu.cn

[*] Correspondence: tengwu@buffalo.edu

Received: 1 November 2019; Accepted: 6 December 2019; Published: 9 December 2019

Abstract: Wind characteristics (e.g., mean wind speed, gust factor, turbulence intensity and integral scale, etc.) are quite scattered in different measurement conditions, especially during typhoon and/or hurricane processes, which results in the structural engineer ambiguously determining the wind parameters in wind-resistant design of buildings and structures in cyclone-prone regions. In tropical cyclones (including typhoons and hurricanes), the inconsistent wind characteristics may be in part ascribed to the complex flow structure with the coexistence of both mechanical and convective turbulence in the boundary layer of tropical cyclones. Another significant contribution to the scattered wind characteristics is due to various measurement conditions (e.g., terrain exposure and height) and data processing schemes (e.g., averaging time). The removal of the inconsistency in the field-measurement system may offer a more rational comparison of measured wind data from various observation platforms, and hence facilitates a better identification scheme of the wind characteristics to guide the urban planning design and wind-resistant design of buildings and structures. In this study, an analytical framework was firstly proposed to eliminate the potential observation-related effects in wind characteristics and then the wind characteristics of seven field measured tropical cyclones (four typhoons and three hurricanes) were comparatively investigated. Specifically, field measurements of wind characteristics were converted to a standard reference station with a roughness length of 0.03 m, observation duration of 10 min for mean wind and averaging time of 3 s for gusty wind at a 10 m height. The differences of the measured wind characteristics between the typhoons and hurricanes were highlighted. The standardized turbulent wind characteristics under the analytical framework for typhoons and hurricanes were compared with the corresponding recommendations in standard of American Society of Civil Engineers (ASCE 7-10) and Architectural Institute of Japan Recommendations for Loads on Buildings (AIJ-RLB-2004).

Keywords: wind characteristics; boundary layer; typhoon; hurricane; field measurement

---

1. Introduction

Wind characteristics (e.g., mean wind speed, gust factor, turbulence intensity and integral scale, etc.) are the critical factors for wind-resistant design of the wind-sensitive infrastructures and urban planning. Resisting wind effects and reducing wind-induced damage in tropical cyclones is the challenge for the wind sensitive buildings and structures in cyclone-prone regions, as these regions are normally in the economically developed areas with crowded populations and large-scale landmark buildings and structures. Therefore, a rational analytical framework for investigating wind characteristics in tropical cyclones is essential to understand the nature of winds, calibrate codes of practice for wind-resistant design of the large-scale structures and enhance wind tunnel simulations and numerical modeling [1]. Oncoming winds of buildings and bridges are usually simplified as the steady flow part featured by mean wind speed and corresponding vertical profile, and the fluctuating flow part characterized by turbulence intensity, integral scale, gust factor, peak factor, probability distribution, and power spectrum.

Tropical cyclones are characterized by the asymmetric helical flow structure and complex turbulence driven mechanism (both convective and mechanical turbulence). The spatial distribution of the flow structure also varies significantly in the footprint of tropical cyclones. Due to the limited measurements in the lower boundary layer of tropical cyclones, a basic premise of the existing codes and standards is that the turbulent wind characteristics in tropical cyclones are similar to those observed in the boundary layer winds of extratropical storms. However, it is well known that the downward transport of convective cells generated at higher levels together with the boundary layer rolls could modulate the wind structure and turbulence in the lower tropical cyclone boundary layer. These thermodynamics-related activities may lead to the turbulent wind characteristics of the hurricanes/typhoons different from those of the extratropical winds [2–4].

A direct and reliable approach to examine the turbulent wind characteristics is based on the field observations in the paths of landfalling tropical cyclones. Thus, a number of field measurement programs were initiated in the tropical cyclone-prone regions to monitor the hurricane/typhoon winds [5–14]. The field-measured wind characteristics from different observation stations for various tropical cyclones are quite scattered and hard-to-reach unified conclusions to guide the wind-resistance design of buildings and structures in the cyclone-prone regions. The inconsistent wind characteristics of tropical cyclones may be attributed in part to the complexity of turbulence driven mechanisms, e.g., shear (namely roll and streak structures near the surface), convection, rotation, blocking and sheltering effects at the boundary layer, and also the interactive motions of multi-scale eddies in the flow fields of tropical cyclones [15,16]. On the other hand, the underlying surface and the employed schemes to obtain the turbulence parameters may also significantly influence on the variability of wind characteristics. Since the tracks of tropical cyclones are random, most of the field observations were conducted by installing anemometers and accelerometers on structures or observation towers, which were built in the regions frequently attacked by tropical cyclones. The measured turbulent wind characteristics from these observation stations are quite different from one another because of the underlying surrounding terrain conditions and the lack of well-established guidelines for an appropriate documentation of the near surface wind filed in tropical cyclones. In the China wind codes, wind characteristics are specified over standard terrain with roughness length of 0.03 m, averaging time of 10 min for mean wind and duration time of 3 s for gusty wind at 10 m height. Accordingly, it is essential to convert the turbulence characteristics obtained from various stations to the standard terrain and investigate the wind nature in a unified analytical framework. The "standardized" wind characteristics due to their universality could be useful in instructing the structural design in cyclone-prone regions.

This study first presented an analytical framework in which the mean wind speed, turbulence intensity, integral scale, gust factor, and peak factor measured at various terrains, heights and averaging times were properly standardized. Then, the typhoon and hurricane wind data analyzed here were briefly described. Finally, field-observed turbulent wind characteristics of four typhoons and three hurricanes were converted to the standard condition and comparatively investigated. The standardized results were also compared with the corresponding recommendations in ASCE7-10 [17] and AIJ-RLB-2004 [18]. The difference of the wind characteristics in hurricanes and typhoons were also highlighted.

2. Analytical Framework

In this section, an analytical framework will be proposed in which the turbulent characteristics measured in various terrain conditions, heights and averaging times could be converted to a standard station. The standardization of the wind characteristics is based on the assumption of the equilibrium boundary-layer theory [19]. The atmosphere stratification in the boundary layer of tropical cyclones is assumed to be neutrally stable, which implies that the turbulent structure within this region is driven mainly by the local surface roughness effects [20]. In the non-equilibrium boundary-layer, this analytical framework may need further investigations subjecting to specific terrain conditions.

2.1. Mean Wind Speed

To analyze the wind characteristics, an essential step is to convert the wind speeds measured at different station conditions (i.e., various exposures, heights, and averaging times) to the standard condition. The standardization of the mean wind speed in this study follows the three steps: (1) determine the exposure type of the observation station; (2) calculate the gradient wind speeds over the observation exposure; and (3) calculate the mean wind speed at the reference height (10 m high) over standard exposure (open flat terrain) by assuming the gradient wind speeds are equal at the gradient height over different exposures.

2.1.1. Logarithmic Law Wind Profile

Normally, the observations by Global Position System (GPS) dropsonde and Doppler radar show that the variation of mean wind speed with height follows the logarithmic law in the lower part of tropical cyclone boundary layer [21–25]. Thus, the logarithmic law can be used to describe lower boundary layer and the outer-vortex regions of a tropical cyclone:

$$U_s(z_s) = \frac{u_{*s}}{k} \ln\left(\frac{z_s}{z_{0s}}\right), \tag{1}$$

where $U_s(z_s)$ represents the mean wind speed at height z_s over the standard exposure. Specifically, the standard exposure in this study corresponds to the roughness length $z_{0s} = 0.03$ m, the reference height is 10 m, and the time scale for the average value is 10 min. u_{*s} denotes the friction velocity over the standard exposure, and $k \approx 0.40$ is the von Kármán constant.

According to Equation (1), the key procedure to standardize the mean wind speed is to determine the relationship of the friction velocities in various terrains. As all anemometers used in this study are set between 10–60 m height, it is reasonable to assume that the friction velocity in the lower tropical cyclone boundary layer is a height-independent constant [25–27]. Based on the assumption of local equilibrium conditions, the transition model in Engineering Sciences Data Unit (ESDU) [28], which has been applied to convert the 3 s peak speed over open-terrain and the 1-min mean wind speed above open water in hurricane by Simiu et al. [29], is employed:

$$\frac{u_{*s}}{u_{*m}} = \frac{\ln\left(\frac{10^5}{z_{0s}}\right)}{\ln\left(\frac{10^5}{z_{0m}}\right)}, \tag{2}$$

where u_{*m} is the friction velocity over the field measured exposure with roughness length of z_{0m}. Then the relation of mean wind speeds with different terrains can be accordingly expressed as:

$$\frac{U_s(z_s)}{U_m(z_m)} = \frac{\ln\left(\frac{10^5}{z_{0s}}\right) \ln\left(\frac{z_s}{z_{0s}}\right)}{\ln\left(\frac{10^5}{z_{0m}}\right) \ln\left(\frac{z_m}{z_{0m}}\right)}, \tag{3}$$

where $U_m(z_m)$ is the mean wind speed measured at experiment station with height z_m and roughness z_{0m}. In this model, the gradient balance assumption, which has been demonstrated to be valid at a sufficiently high altitude [30], is adopted. The super-gradient flows were observed in the boundary layer of some tropical cyclones, however, it is not systematic in tropical cyclones, especially in overland conditions [31,32]. The case of super-gradient flows in tropical cyclones will discussed in next section.

The relation among mean wind speeds of various averaging times can be expressed as [33]:

$$U_\tau(z) = U_{3600}(z)\left[1 + \frac{\beta^{0.5}c(\tau)}{2.5\ln\left(\frac{z}{z_0}\right)}\right],$$

(4)

where τ denotes the averaging time; $U_\tau(z)$ and $U_{3600}(z)$ are respectively τ-s mean and 1-h mean wind speeds; β represents the ratio of the fluctuating wind speed variance to the square of friction velocity; $c(\tau)$ is an averaging time-related parameter that determined by statistical characteristics of wind speed measurements.

2.1.2. Super-Gradient Wind Profile

The field measurements show the existence of super-gradient wind over ocean surface and the sea land transition regions in tropical cyclones, and the variation of mean wind with height following a logarithmic-quadratic profile [34]. Based on the field measurements in hurricanes over land and ocean surface, Snaiki and Wu [35] proposed a semi-empirical model to depict the mean wind profile. As the empirical model is convenient and accurate, it is adopted here to convert the mean wind speed in landfalling typhoons. The power law-based wind profile is used as follows:

$$U_s(z_s) = U_{10_s}\left[\left(\frac{z_s}{10}\right)^{\alpha_s} + \eta_1\sin\left(\frac{z_s}{\delta_s}\right)\exp\left(-\frac{z_s}{\delta_s}\right)\right],$$

(5)

where $U_s(z_s)$ and U_{10_s} are the mean wind speed at height z_s and 10 m over the standard exposure; α_s is the power law exponent over the standard exposure; δ_s is the height of the wind maximum over the standard exposure; η_{1s} is derived to be:

$$\eta_{1s} = \frac{\left(\frac{\delta_s}{10}\right)^{\alpha_s}\alpha_s e}{\sin 1 - \cos 1},$$

Analogously, the field measured mean wind over the experiment exposure is:

$$U_m(z_m) = U_{10_m}\left[\left(\frac{z_m}{10}\right)^{\alpha_m} + \eta_{1m}\sin\left(\frac{z_m}{\delta_m}\right)\exp\left(-\frac{z_m}{\delta_m}\right)\right],$$

(6)

where $U_m(z_m)$ and U_{10_m} are the mean wind speed at height z_m and 10 m over the measured exposure; α_m is the power law exponent over the measured exposure; δ_m is the height of the wind maximum over the measured exposure; η_{1s} is:

$$\eta_{1m} = \frac{\left(\frac{\delta_m}{10}\right)^{\alpha_m}\alpha_m e}{\sin 1 - \cos 1},$$

By adopting the assumption that the wind speeds at the wind maximum height (δ_s and δ_m) are equal, the following expression can be deduced:

$$\frac{U_s(z)}{U_m(z)} = \frac{\left(\frac{\delta_m}{10}\right)^{\alpha_m}\left[\left(\frac{z_s}{10}\right)^{\alpha_s} + \eta_{1s}\sin\left(\frac{z_s}{\delta_s}\right)\exp\left(-\frac{z_s}{\delta_s}\right)\right]}{\left(\frac{\delta_s}{10}\right)^{\alpha_s}\left[\left(\frac{z_m}{10}\right)^{\alpha_m} + \eta_{1m}\sin\left(\frac{z_m}{\delta_m}\right)\exp\left(-\frac{z_m}{\delta_m}\right)\right]},$$

(7)

Equation (7) could be used to convert the field wind speeds to the standard exposures in the tropical cyclones with super-gradient flow. Actually when the wind speed measured in the lower regions following the logarithmic law, the Equation (7) will merge into the logarithmic law or power law.

2.2. Turbulence Intensity

It is conventional to treat the turbulence ratio (the ratio of the standard deviation of longitudinal wind velocity component σ_u to the friction velocity u_*) as terrain-independent in the equilibrium boundary layer [36,37]. On the other hand, Harris and Deaves [38] proposed an empirical model to consider the variation of turbulence ratio with height as:

$$\frac{\sigma_u}{u_*} = 2.63\eta\left[0.538 + 0.090\ln\left(\frac{z}{z_0}\right)\right]^{\eta^{16}},$$

(8)

where $\eta = 1 - z/h$; $h = u_*/(6f)$; $f = 1.458 \times 10^{-4}\sin\phi$ is the Coriolis parameter; and ϕ denotes the latitude of the observation site. Due to the fact that the derivation of Equation (8) was partly based on non-equilibrium-condition data, the estimation of maximum turbulence ratio, $[\sigma_u/u_*]_{max}$ obtained from Equation (8) is dependent on the terrain roughness length, which is in contradiction to the equilibrium assumption. To correct this issue, the empirical variation of σ_u/u_* with respect to terrain roughness is introduced in ESDU [39] to obtain approximately a constant $[\sigma_u/u_*]_{max}$ for various terrain roughness lengths:

$$\left[\frac{\sigma_u}{u_*}\right](z_0) = 1 + 0.156\ln\left(\frac{u_*}{fz_0}\right),$$

(9)

Since the field measurements give $[\sigma_u/u_*]_{max} = 2.85$ for the terrain with a roughness length of 0.03 m, Equation (8) can be corrected by factoring $2.85/\{1 + 0.156\ln[u_*/(fz_0)]\}$, resulting in an improved model to calculate turbulence ratio as in ESDU [39]:

$$\frac{\sigma_u}{u_*} = \frac{7.496\eta\left[0.538 + 0.090\ln\left(\frac{z}{z_0}\right)\right]^{\eta^{16}}}{1 + 0.156\ln\left(\frac{u_*}{fz_0}\right)}.$$

(10)

As expected, Equation (9) gives a maximum turbulence ratio $[\sigma_u/u_*]_{max}$ of approximately 2.85 for various roughness lengths. However, field measurements show that turbulence ratios in tropical cyclones are usually greater than the values in extratropical storms [12,24,40], making the selection of $[\sigma_u/u_*]_{max} = 2.85$ inapplicable to tropical cyclones. On the other hand, a height-independent relation between turbulence ratio σ_u/u_* and underlying surface roughness length z_0 was proposed by Li et al. [13] based on the analysis of field measurements in typhoons. In this study, this height-independent relation proposed by Li et al. [13] is adopted as:

$$\left[\frac{\sigma_u}{u_*}\right]_{max} = 2.72 - 0.25\log z_0.$$

(11)

As a result, the turbulence ratio in tropical cyclone will be corrected by multiplying Equation (8) by the following factor:

$$\frac{2.72 - 0.25\log z_0}{1 + 0.156\ln[u_*/(fz_0)]}.$$

(12)

Then the turbulence ratio could be expressed as:

$$\frac{\sigma_u}{u_*} = \frac{2.63\eta\left[0.538 + 0.009\ln\left(\frac{z}{z_0}\right)\right]^{\eta^{16}}[2.72 - 0.25\log(z_0)]}{1 + 0.156\ln\left(\frac{u_*}{fz_0}\right)}.$$

(13)

For a standard terrain condition ($z = 10$ m; $z_0 = 0.03$ m; assuming $u_* = 1$ m/s and $\phi = 25°$), the turbulence ratios estimated by Equations (9) and (11) are 2.55 and 2.78, respectively. In this study, Equation (13) is utilized to convert the measured turbulence ratio to the standard condition. Accordingly, the longitudinal turbulence intensity in a standard exposure can be calculated as:

$$[I_u]_s = \frac{2.63\eta_s\left[0.538 + 0.090\ln\left(\frac{z}{z_0}\right)\right]^{\eta_s^{16}}[2.72 - 0.25\ln(z_{0s})][u_*]_s}{\left[1 + 0.156\ln\left(\frac{u_*}{fz_0}\right)\right][U]_s}. \tag{14}$$

2.3. Integral Scale

The approach of integrating correlation function by invoking Taylor's hypothesis is frequently used to estimate the integral scale as it has a clear physical meaning [24,33]:

$$L_u^x = \frac{U}{\sigma_u^2}\int_0^{R_{uu}=0.05\sigma_u} R_{uu}(\tau)\mathrm{d}\tau, \tag{15}$$

where R_{uu} is the autocorrelation function of the longitudinal fluctuating component.

The integration of autocorrelation function, however, usually overestimates the value of integral scale and will result in a deviation of inertial sub-range in the estimated von Kármán-type spectrum [41] compared to that in the field-measured spectra. To improve the accuracy, Harris and Deaves [38] suggested the following model to estimate the longitudinal integral scale:

$$L_u^x = \frac{A^{\frac{3}{2}}\left(\frac{\sigma_u}{u_*}\right)^3 z}{2.5K_z^{\frac{3}{2}}\left(1 - \frac{z}{h}\right)^2\left(1 + 5.75\frac{z}{h}\right)}, \tag{16}$$

where z is the height from the ground.

$$A = 0.115\left(1 + 0.315\eta^6\right)^{\frac{2}{3}} \tag{17}$$

and

$$K_z = 0.19 - (0.19 - K_0)\exp\left[-B\left(\frac{z}{h}\right)^N\right] \tag{18}$$

in which

$$K_0 = \frac{0.39}{R_0^{0.11}}, \tag{19}$$

$$B = 24R_0^{0.155}, \tag{20}$$

$$N = 1.24R_0^{0.008}, \tag{21}$$

$$R_0 = \frac{u_*}{fz_0}. \tag{22}$$

The longitudinal integral scale over standard exposure $[L_u^x]_s$ can be estimated according to Equation (16) by introducing the corresponding values of $[u_*]_s$ and $[z_0]_s$.

2.4. Peak Factor

Peak factor g_u is defined as the ratio of maximum wind speed fluctuation in a duration τ to the standard deviation of the fluctuating wind speed within an observation period of T:

$$g_u(\tau, T) = \frac{\max[u(\tau, T)]}{\sigma_u(\tau, T)}\frac{\sigma_u(\tau, T)}{\sigma_u}. \tag{23}$$

For a stationary stochastic process following Gaussian distribution, the peak factor with $\tau \to 0$ and $T \geq 3600\ s$ could be calculated as [42]:

$$g_u(\tau, T) = \sqrt{2\ln[v(\tau, T)T]} + \frac{0.5772}{\sqrt{2\ln[v(\tau, T)T]}},$$
(24)

where $v(\tau, T)$ is the zero up-crossing rate. It can be calculated by [42,43]:

$$v^2(\tau, T) = \frac{\int_0^\infty n^2 S_u(n)\chi^2(n, \tau, T)\mathrm{d}n}{\int_0^\infty S_u(n)\chi^2(n, \tau, T)\mathrm{d}n},$$
(25)

in which $S_u(n)$ represents the wind velocity spectrum; n denotes the frequency in Hertz, and $\chi^2(n, \tau, T)$ is a filter function used to consider the influence of sampling frequency, averaging time and response characteristics of the anemometer. In this study, the von Kármán-type spectrum is employed as:

$$\frac{nS_u(n)}{\sigma_u^2} = \frac{4\left(\frac{nL_u^x}{U}\right)}{\left[1 + 70.8\left(\frac{nL_u^x}{U}\right)^2\right]^{\frac{5}{6}}},$$
(26)

The filter function is chosen as following for sonic anemometers [44]:

$$\chi^2(n, \tau, T) = \left[\frac{\sin(\pi n\tau)}{\pi n\tau}\right]^2 - \left[\frac{\sin(\pi nT)}{\pi nT}\right]^2,$$
(27)

For propeller anemometers, the following filter function, which takes the mechanical features of propeller anemometers into consideration, is adopted [43]:

$$\chi^2(n, \tau, T) = \left\{\left[\frac{\sin(\pi n\tau)}{\pi n\tau}\right]^2 - \left[\frac{\sin(\pi nT)}{\pi nT}\right]^2\right\}\frac{1}{1 + \left(\frac{2\pi n\lambda}{U}\right)^2},$$
(28)

where λ is the distance constant of the propeller anemometer.

Equation (24) is valid for calculating the average of instantaneous peak factor ($\tau \to 0$) from a long enough wind speed record (e.g., $T \geq 1$ h). With a finite averaging time, τ, and a finite observation period, T, the estimation of standard deviation in Equation (26) could be biased since the measured spectrum is truncated in both high-frequency and low-frequency regions, and might eventually lead to an inaccurate estimation of the peak factor. In the case that these conditions are not satisfied, the following relation is necessary to be introduced to consider the effects of the variance reduction due to the truncation of the velocity spectrum:

$$\frac{\sigma_u(\tau, T)}{\sigma_u} = \frac{\int_0^\infty S_u(n)\chi^2(n, \tau, T)\mathrm{d}n}{\int_0^\infty S_u(n)\mathrm{d}n},$$
(29)

The 3-s peak factor, $[g_u]_s$, in time scale $[T]_s = 600$ s in the standard terrain can be estimated by introducing U_s and L_{us}^x, which were respectively calculated through Equations (3) or (7) and (16).

2.5. Gust Factor

Gust factor, $G_u(\tau, T)$, herein is defined as the ratio of gust speed with gust duration τ to the mean wind speed $U(T)$ with an observation period of T:

$$G_u(\tau, T) - 1 + \frac{\max[u(\tau, T)]}{\sigma_u(\tau, T)}\frac{\sigma_u(\tau, T)}{\sigma_u(\Delta t, T)}\frac{\sigma_u(\Delta t, T)}{U(T)},$$
(30)

where Δt is the sampling interval.

Substituting the peak factor and turbulence intensity into the corresponding terms of Equation (30), the gust factor can be re-expressed as:

$$G_u(\tau, T) = 1 + g_u(\tau, T)I_u \frac{\sigma_u(\tau, T)}{\sigma_u(\Delta t, T)},$$ (31)

where $\sigma_u(\tau, T)/\sigma_u(\Delta t, T)$ can be calculated by:

$$\frac{\sigma_u(\tau, T)}{\sigma_u(\Delta t, T)} = \frac{\int_0^\infty S_u(n)\chi^2(n, \tau, T)dn}{\int_0^\infty S_u(n)\chi^2(n, \Delta t, T)dn},$$ (32)

As a result, the gust factor in the standard exposure, $[G_u(\tau, T)]_s$, can be estimated by:

$$[G_u(\tau, T)]_s = 1 + [g_u(\tau, T)]_s[I_u]_s \left[\frac{\sigma_u(\tau, T)}{\sigma_u(\Delta t, T)}\right]_s,$$ (33)

3. Data Sources

3.1. Tropical Cyclones and Instruments

In this study, the data of four typhoons (0601 typhoon Chanchu, 0606 typhoon Prapiroon, 0812 typhoon Nuri, and 0814 typhoon Hagupit) and three hurricanes (0504 hurricane Katrina, 0510 hurricane Rita, and 0512 hurricane Wilma) were comparatively analyzed. The detailed descriptions of the observation site exposures and the observation tower configurations for the four typhoons and three hurricanes were presented in Li et al. [16] and Masters et al. [12], respectively. The GPS coordinates of the observation stations were listed in Table 1. As the latitudes of all observation stations are around 25°, the latitude of the standard condition is set to be 25° for the convenient of calculation.

Table 1. The GPS coordinates of observation towers. Reproduced with permission from [16], Elsevier, 2019.

Tropical Cyclones	Tower	Latitude	Longitude
Chanchu	RBT	22.7337°	115.5734°
	OT	23.5510°	117.0020°
Prapiroon	BT	21.4519°	111.3149°
Nuri	MFB	22.1810°	113.5630°
	DIT	22.1413°	113.7096°
Hagupit	ZT	21.4509°	111.3745°
	ST	21.2538°	110.6541°
Katrina	T1	29.8253°	−90.0319°
	T2	29.4441°	−90.2628°
	T3	30.4720°	−88.5308°
Rita	T0	29.9512°	−94.0220°
	T3	29.9548°	−93.9542°
	T5	30.0797°	−93.7841°
Wilma	T0	25.9008°	−81.3114°
	T1	26.1458°	−80.5067°
	T2	25.8681°	−80.8997°
	T3	25.7516°	−80.3780°

It is noted that the distance constant λ of the propeller anemometer is an important factor to calculate the peak factor and gust factor as this type of anemometer mechanically filters the amplitudes of gusts with wavelengths less than $2\pi\lambda$ due to the mechanical limitations [10]. In this study, the propeller anemometers of models R.M. Young 05103L and R.M. Young 27106R were respectively used to measure typhoons and hurricanes. The specifications of these two propeller anemometers are listed in Table 2. Based on the parameters in Table 2, the data measured by propeller anemometers were corrected according to Equation (24). In addition to propeller anemometers, the sonic anemometers were also utilized in the field measurement of typhoons. Specifically, two 3-D ultrasonic anemometers (WindMaster™ Pro., Gill Instruments Ltd., Lymington, UK) were installed on tower RBT and one 3-D ultrasonic anemometer (HD2003, Delta Ohm Srl, Selvazzano Dentro, Italy) were setup on tower OT for the measurements of Typhoon Chanchu; one HD2003 anemometer were installed on tower BT to acquire data from Typhoon Prapiroon, and towers DIT and ST were equipped with Gill WindMaster™ Pro. anemometers. The specifications of the utilized sonic anemometers are also listed in Table 2. For the data obtained from sonic anemometers, the filter function presented by Equation (23) were used for the correction.

Table 2. Specifications of anemometers.

Anemometers	Specifications		
05103L	Wind speed	Range	0~100 m/s
		Threshold Sensitivity	1 m/s
		Distance constant	2.7 m for 63% recovery
	Wind direction	Ranges	0 ~ 360°
		Threshold Sensitivity	1.1 m/s at 10° displacement
		Delay Distance	1.3 m for 50% recovery
		Damped Natural Wavelength	7.4 m
27106R	Wind speed	Range	0~25 m/s
		Threshold Sensitivity	0.3 m/s
		Distance constant	2.7 m for 63% recovery
		Damped Natural Wavelength	7.4 m
WindMaster™ Pro	Wind speed	Range	0~ 65 m/s
		Resolution	0.01 m/s
	Wind direction	Ranges	0~359°
		Resolution	0.1°
HD2003	Wind speed	Range	0~60 m/s
		Resolution	0.01 m/s
	Wind direction	Ranges	0~359°
		Resolution	0.1°

3.2. Data Quality Control and Data Source

Tropical cyclones are characterized by strong winds accompanied by torrential rain, ocean waves, and storm surge. The representative wind records are usually located in the eyewall regions of tropical cyclones. In the field measurements of tropical cyclones, however, the anemometers in heavy rain bands usually present some spikes and errors. Therefore, the data quality-control procedure is a necessary step before the analysis of the wind characteristics. In this study, the data quality-control schemes and the criteria for the selection of samples were referred to Li et al. [13]. Specifically, the spikes and errors in the data were first identified and replaced by the five-point weighted averages. Then, the reverse arrangement test [45] and run test [46] with a 95% significance level were employed to test the stationarity of the recorded winds. The datasets that failed to pass both two types of stationary tests were removed from the analysis. The stationarity test ensured that the analyzed data could satisfy the local equilibrium boundary-layer assumption, where the friction velocity is independent of the location in the along-wind direction and the Reynolds number [37].

Table 3 briefly summarized the datasets utilized in this study, together with the observation heights, types of the anemometers used, number of the runs in each group (sample size) and average of the 10 min mean wind speeds. The detailed analysis of the turbulent wind characteristics of the original datasets both in typhoons and hurricanes were presented in Li et al. [16].

Table 3. Datasets analyzed in this study. Reproduced with permission from [16], Elsevier, 2019.

Tropical Cyclones	Sites	Height (m)	Anemometer Types	Number of Runs	Average of the 10 min Mean Wind Speed (m/s)
Chanchu	RBT	10	Sonic	26	19.19
		10	Propeller	15	18.87
		30	Sonic	37	21.96
		60	Propeller	58	22.19
	OT	5	Sonic	21	24.69
		10	Sonic	45	24.28
Prapiroon	BT	10	Sonic	18	20.13
			Propeller	4	22.69
Nuri	MFB	30	Sonic	8	18.25
	DIT	10	Sonic	83	24.26
		60	Sonic	100	25.11
Hagupit	ZT	60	Sonic	38	28.53
	ST	5	Sonic	18	20.34
			Propeller	27	20.34
		10	Sonic	61	21.30
			Propeller	58	20.93
Katrina	T1	5	Propeller	16	21.98
		10	Propeller	43	22.45
	T2	5	Propeller	34	21.50
		10	Propeller	40	24.42
	T3	5	Propeller	5	23.02
		10	Propeller	14	22.46
Rita	T0	5	Propeller	46	21.05
		10	Propeller	60	22.17
	T3	10	Propeller	15	18.85
	T5	5	Propeller	19	18.86
		10	Propeller	32	19.55
Wilma	T0	5	Propeller	13	21.08
		10	Propeller	19	22.41
	T1	5	Propeller	23	23.65
		10	Propeller	37	26.17
	T2	5	Propeller	16	21.34
		10	Propeller	18	23.66
	T3	5	Propeller	9	21.36
		10	Propeller	14	22.63

4. Results and Discussions

The selected datasets from the four typhoons and the three hurricanes summarized in the preceding section were investigated in the analytical framework presented in Section 2. Specifically, both the datasets in typhoons and hurricanes were converted to the standard exposure with a roughness length of 0.03 m at 10 m height and an observation time scale of 10 min. The latitude of the standard terrain is assumed to be 25°.

4.1. Turbulence Intensity

The turbulence intensities were extracted based on the analytical framework and shown in Figures 1a and 2a respectively for investigated typhoons and hurricanes. The corresponding probability density functions (PDFs) are shown in Figures 1b and 2b. For typhoons, the average value of longitudinal turbulence intensities is 0.1952 and the standard deviation is 0.0032. For hurricanes, the average value of longitudinal turbulence intensities is 0.1906 and the standard deviation is 0.0022. The longitudinal turbulence intensity of these four typhoons presents slightly higher values in terms of both mean and standard deviation compared to those three hurricanes. This observation can be further demonstrated by comparing Figures 1b and 2b, where the probability distribution of the longitudinal turbulence intensities in both typhoons and hurricanes follows the normal distribution quite well. One can easily conclude that in these four typhoons the turbulence intensity has a larger value than that in those three hurricanes under the same probability of exceedance.

In ASCE7-10, the longitudinal turbulence intensity is given by:

$$I_u = c\left(\frac{10}{z}\right)^{\frac{1}{6}}, \tag{34}$$

where c equals to $0.30, 0.20$, and 0.15 for category B, C (corresponding to the standard terrain in this study), and D exposures, respectively. In AIJ-RLB-2004 code, the longitudinal turbulence intensity over flat terrain categories is given by:

$$I_u = 0.1\left(\frac{z}{z_G}\right)^{-\alpha-0.05}, \tag{35}$$

in which α and z_G are parameters reflecting the category of exposures. In category II exposure, which is the closest to the standard terrain in this study, α and z_G are respectively 0.15 and 350. The longitudinal turbulence intensities obtained from ASCE7-10 and AIJ-RLB-2004 are 0.2000 and 0.2036, respectively, for the standard exposure. As depicted in Figures 1a and 2a, both ASCE7-10 and AIJ-RLB-2004 present a slightly higher estimation of longitudinal turbulence intensity for hurricanes and typhoons. Generally, the estimation of ASCE7-10 is relatively better compared to that of AIJ-RLB-2004.

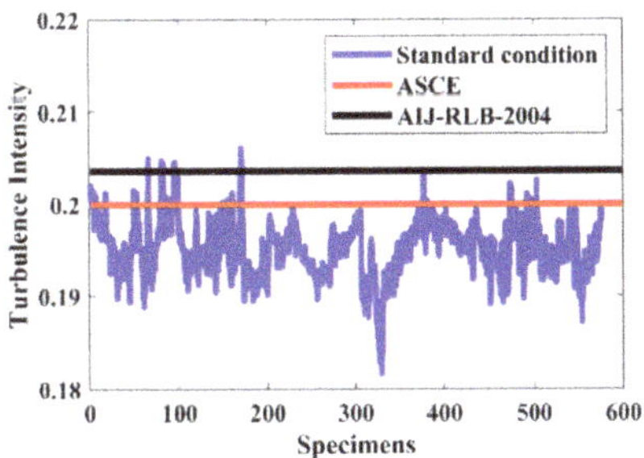

(**a**) Comparison of turbulence intensities from the analytical framework and the recommendation of ASCE and AIJ-RLB-2004

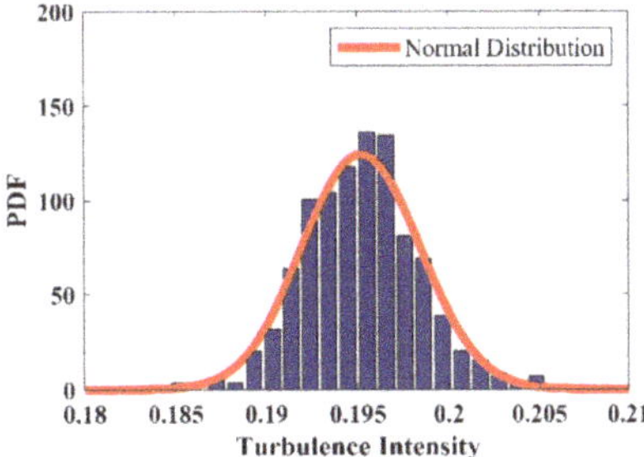

(**b**) PDF of the turbulence intensities from the analytical framework

Figure 1. Turbulence intensities and their probability density functions (PDF) for typhoons.

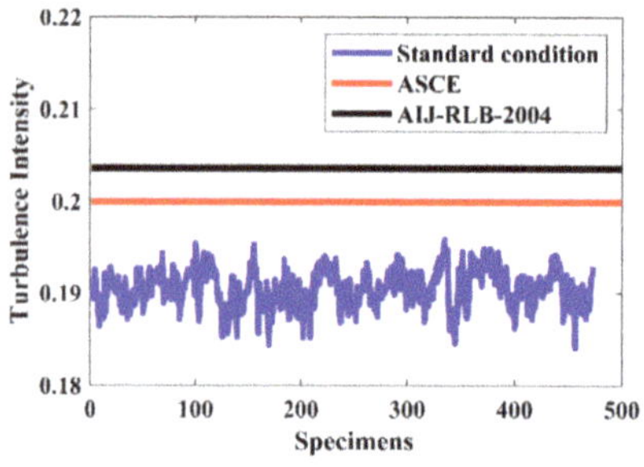

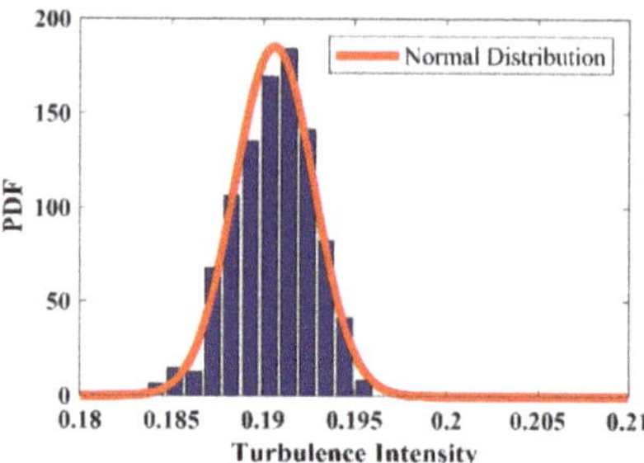

(**a**) Comparison of turbulence intensities from the analytical framework and the recommendation of ASCE and AIJ-RLB-2004

(**b**) PDF of the turbulence intensities from the analytical framework

Figure 2. Turbulence intensities and their PDF for hurricanes.

4.2. Integral Scale

The integral scales and the corresponding probability distributions obtained are presented in Figures 3 and 4 for typhoons and hurricanes, respectively. The longitudinal integral scale in typhoons has an average value of 146.4482 and a standard deviation of 9.2143. The length scales extracted from hurricane measurements have a slightly higher average value of 157.5796 and a significantly lower standard deviation of 3.8106 compared to those of typhoons. As shown in Figures 3a and 4a, the range of integral scales extracted from typhoon measurements is significantly larger than that of hurricanes, which can be better illustrated by the probability distributions of longitudinal integral scales in typhoons and hurricanes. The probability distribution of typhoons follows the Weibull distribution with scale parameter of 150.368 and shape parameter of 20.0454, while the probability distribution of longitudinal integral scales for hurricanes follows the generalized extreme value distribution with scale parameter of −0.3259, shape parameter of 3.8960 and location factor of 156.311. Compared with Figure 4b, where the observed hurricane length scale shows a narrow distribution, the probability distribution of the observed typhoon integral scale of Figure 3b has a significantly wider range. This phenomenon may be in part attributed to the exposures of the observation station for the original datasets. The observation station in those three hurricanes are located in homogeneous open flat terrain as stated in Masters et al. [12]. However, in the observation of those four typhoons, the exposures of the measured stations are a little bit inhomogeneous. Another influence could be ascribed to the differences of turbulent structures of those typhoons and hurricanes. As noted in Li et al. [16], at the same roughness regime, the field measured integral scales in these four typhoons were greater than that in those three hurricanes. The different distributions of the observed hurricane and typhoon length scales might indicate that energy-containing eddies in the observed typhoons have various representative length-scales while those of the observed hurricanes are concentrated around the mean value. The multiple-scale eddy interactions in typhoons and hurricanes need further investigations before fully understanding the observed difference.

In ASCE7-10, the longitudinal integral scale is computed by:

$$L_u^x = l\left(\frac{z}{10}\right)^{\bar{\varepsilon}},\tag{36}$$

where l and $\bar{\varepsilon}$ are respectively 152.4 m and 0.2 for category C exposure (standard exposure in this study). In AIJ-RLB-2004 code, the turbulence integral scale is defined independently of the terrain categories and is given by:

$$L_u^x = \begin{cases} 100\left(\frac{z}{30}\right)^{0.5} & 30\text{ m} < z < z_G \\ 100 & z \leq 30\text{ m} \end{cases},\tag{37}$$

where z_G equals to 350 m for category II exposure, corresponding to the standard exposure in this study. Accordingly, the longitudinal integral scales obtained from ASCE7-10 and AIJ-RLB-2004 are respectively, 152.4 m and 100 m, for the standard exposure. It is noted that ASCE7-10 presents a reasonable estimation of the longitudinal integral scales for both typhoons and hurricanes. However, AIJ-RLB-2004 underestimates the longitudinal integral scales for both typhoons and hurricanes, suggesting that the usage of AIJ-RLB-2004 may lead to an inaccurate estimation of the power spectrum.

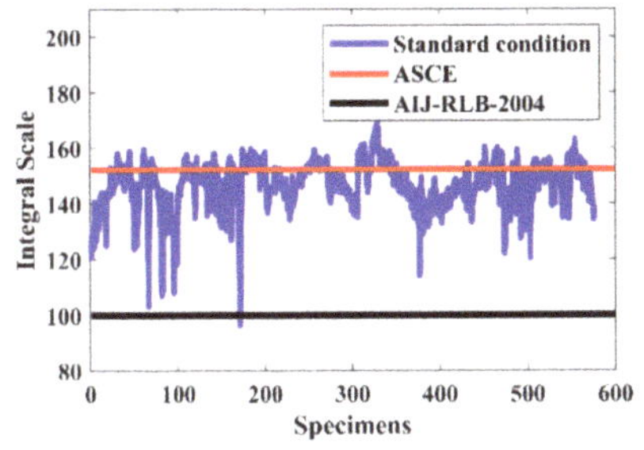

(**a**) Comparison of longitudinal Integral scales from the analytical framework and the recommendation of ASCE and AIJ-RLB-2004

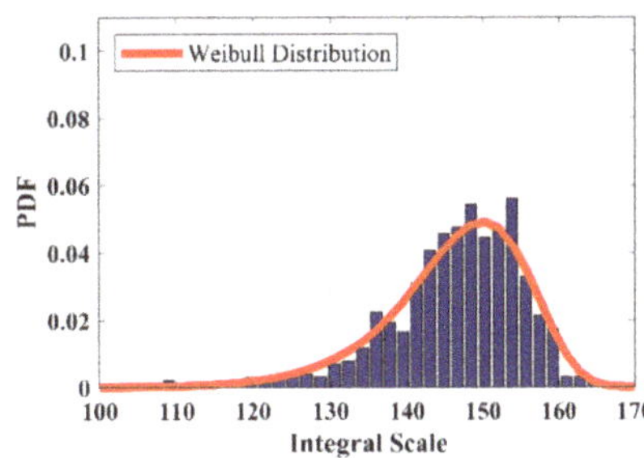

(**b**) PDF of the longitudinal integral scales

Figure 3. Longitudinal integral scales and their PDF for typhoons.

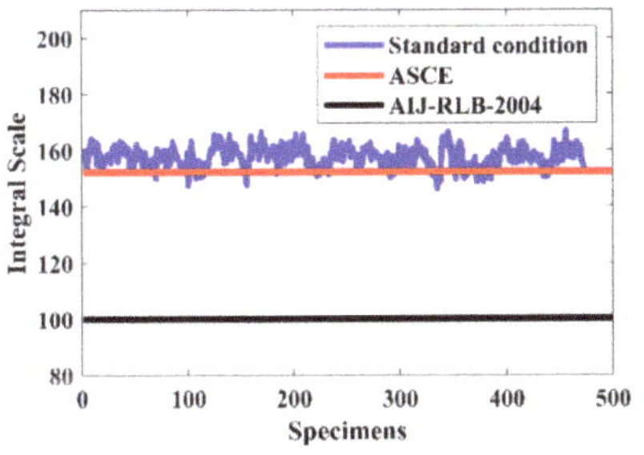

(**a**) Comparison of longitudinal Integral scales from the analytical framework and the recommendation of ASCE and AIJ-RLB-2004

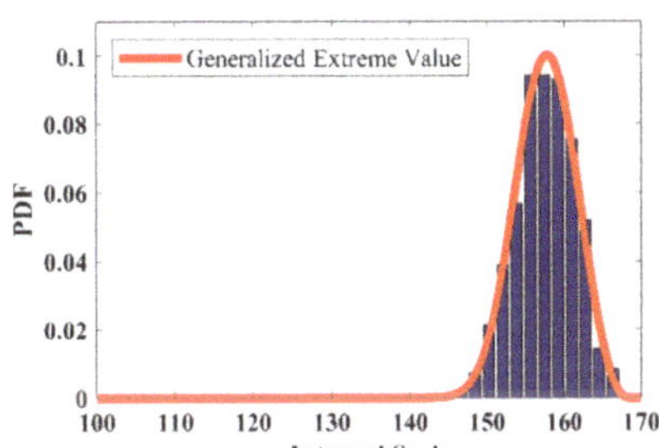

(**b**) PDF of the longitudinal integral scales

Figure 4. Longitudinal integral scales and their PDF for hurricanes.

4.3. Peak Factor

The peak factor is usually utilized for the estimation of gust factor, which plays an important role in determining the wind load on structures [43]. The estimated peak factors of typhoons and hurricanes are respectively presented in Figures 5 and 6, together with the corresponding probability distributions. For typhoons, the average value of the peak factor is 2.5211 and the standard deviation is 0.0198. The fitted PDF is shown in Figure 5b. For hurricanes, the average value of peak factor is 2.5123, slightly smaller than that of typhoons, and the standard deviation of peak factor is 0.0102, significantly smaller than that of typhoons. The probability distribution of peak factors in hurricanes follows t location-scale distribution with location parameter of 2.5174, scale parameter of 0.0022 and shape parameter of 0.9845.

Neither the expression nor the value of peak factor is explicitly prescribed in ASCE7-10. By matching the gust factor over open terrain ($G_u = 1.53$) and the turbulence intensity I_u of

0.2 in Equation (34), a peak factor of 2.65 could be obtained. It should be noted that this calculation is based on an averaging time of 1 h. For a duration of t_g , the gust wind speed can be expressed as:

$$\hat{U}(t_g, T) = \overline{U}(T) + g_u(t_g, T)\sigma_u. \tag{38}$$

Suppose the turbulent wind fluctuations follow the Gaussian distribution, the peak factor will be associated with the exceedance probability of the standard normal distribution. The probability of exceedance of wind gust with a duration of t_g within an observation period of T could be calculated as [47,48]:

$$P\big[U > \hat{U}(t_g, T)\big] = \frac{t_g}{T}. \tag{39}$$

Thus, the peak factor should satisfy:

$$g_u(t_g, T) = \Phi^{-1}\left(1 - \frac{t_g}{T}\right). \tag{40}$$

With the gust duration of 3 s and the averaging time of 10 min, the peak factor is around 2.575, which is slightly higher than the measured values in typhoons and hurricanes. AIJ-RLB-2004 carries out a performance-based wind resistant design procedure. Accordingly, the peak factor is included in the required performance of wind load level and the return period of wind speed. Hence, the comparison of the measurement results with AIJ-RLB-2004 is not discussed here.

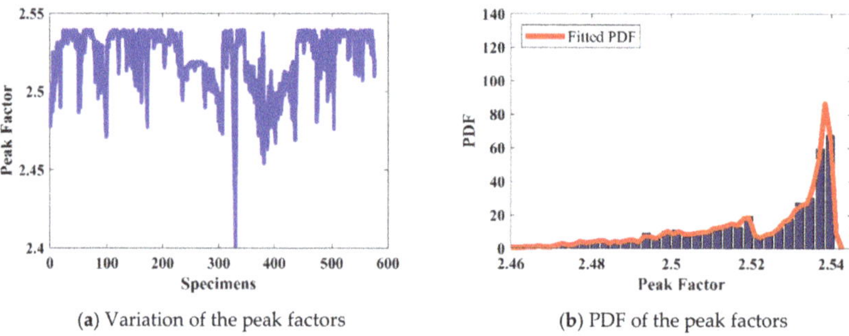

(a) Variation of the peak factors

(b) PDF of the peak factors

Figure 5. Peak factors and their PDF for typhoons.

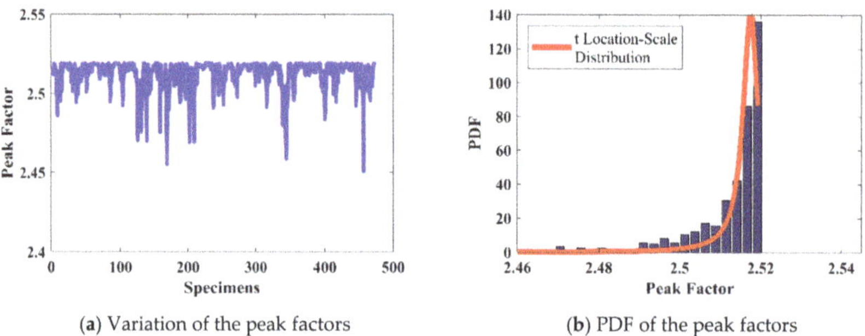

(a) Variation of the peak factors

(b) PDF of the peak factors

Figure 6. Peak factors and their PDF for hurricanes.

4.4. Gust Factor

The gust factor in steady wind conditions depends on several wind characteristics, such as the intensity and integral scale, hence, it is a basic representation of the dynamic properties of wind loads [49]. Figure 7 depicts the gust factors and the corresponding probability distribution of typhoon winds obtained based on the unified analysis framework. For typhoons, the average value of gust factors is 1.4919 and the standard deviation is 0.0069. The gust factors of hurricanes are presented in Figure 8a, where the gust factors have a mean of 1.4787 and a standard deviation of 0.0071. The mean of gust factors for typhoons are slightly higher than that for hurricanes, while the standard deviations of gust factors for typhoons and hurricanes are almost identical with similar probability distribution shapes. The probability distribution of gust factors for typhoons follows the extreme value distribution with a location parameter of 1.4948 and a scale parameter of 0.0051, while the probability distribution of gust factors for hurricanes follows the generalized extreme value distribution with shape parameter of -0.5535, location parameter of 1.4771, and scale parameter of 0.0076.

In ASCE7-10, the calculation of gust factor is referenced to the gust factor curve proposed by Durst [50]. The averaging time of the mean wind speed is 1 h in the Durst gust factor curve, while the gust factor is calculated based on an averaging time of 10 min in this study. Therefore, the conversion scheme for the gust factors with different averaging times presented in Vickery and Skerlj [20] was utilized here. Gust factor with a duration of 3 s and an averaging time of 10 min can be expressed as:

$$G_u = 1 + (SU)[SD(600,\ 3)], \tag{41}$$

where SU is the value of the standard normal deviation associated with the exceedance probability of 0.5% and equals to 2.575. The $SD(600,\ 3)$ could be estimated by the following formula:

$$SD(600,\ 3) = \left[SD^2(3600,\ 3) - SD^2(3600,\ 600) \right]^{1/2}, \tag{42}$$

where $SD(3600,\ 3)$ and $SD(3600,\ 600)$ can be interpolated as indicated in Vickery and Skerlj [20] and equal to 0.1617 and 0.0650, respectively. The gust factor with duration of 3 s and averaging time of 10 min based on Equations (37) and (38) is around 1.3814, which indicates that the gust factors for both typhoons and hurricanes are greater than those for extratropical storms.

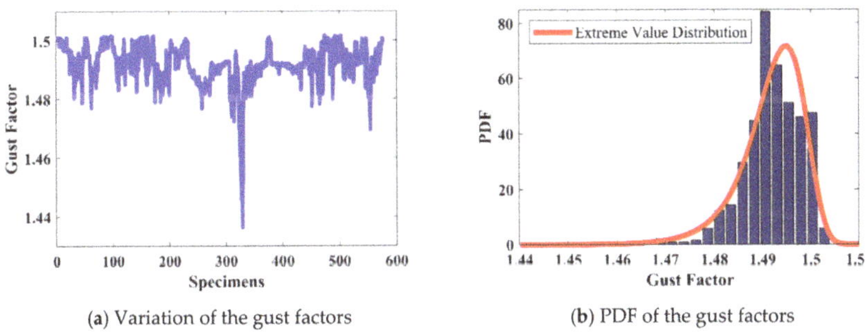

(a) Variation of the gust factors

(b) PDF of the gust factors

Figure 7. Gust factors for typhoons.

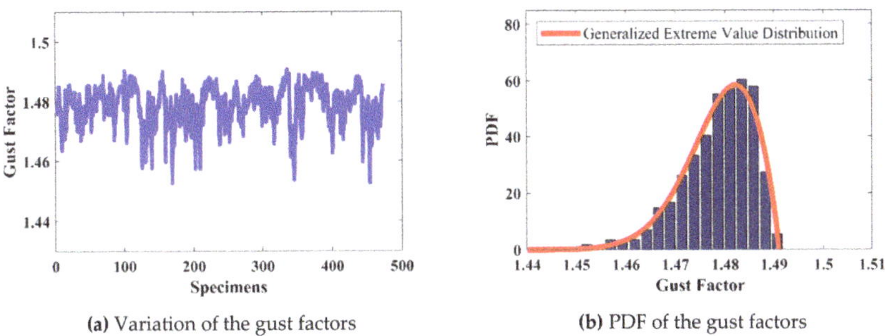

(a) Variation of the gust factors

(b) PDF of the gust factors

Figure 8. Gust factors for hurricanes.

5. Concluding Remarks

An analytical framework was introduced in this study to standardize the turbulent wind characteristics, namely turbulence intensity, integral scale, peak factor and gust factor for various terrain conditions, heights, and averaging times in tropical cyclones. This analytical framework is based on the equilibrium boundary-layer theory and the assumption that the lower tropical cyclone boundary layer is neutrally stable. Field-measured data of the four typhoons and three hurricanes were standardized to the reference exposure with roughness length of 0.03 m, height of 10 m, and averaging time of 10 min, and then utilized for the extraction of wind characteristics under standard exposure. The differences of obtained wind characteristics between typhoons and hurricanes were highlighted, which may be attributed to the basins and latitudes of the genesis of hurricanes and typhoons, the influence of local topography and sea-land transition zone and the differences in turbulent flow structures of typhoons and hurricanes that need further investigations. More specifically, the wind characteristics of these observed typhoons typically present larger values compared to those of observed hurricanes, except for the turbulence integral scale. The turbulence integral lengths of typhoons have a wider distribution compared with those of hurricanes. The obtained turbulent wind characteristics based on the unified analysis framework were comparatively investigated together with the recommendations in ASCE7-10 and AIJ-RLB-2004. The difference between the standardized turbulent characteristics and the corresponding suggested values in the standards (ASCE7-10 and AIJ-RLB-2004) indicates that the tropical cyclone-induced wind loads need be taken into consideration in standards for tropical cyclone-prone regions. It is noted that the ASCE7-10 presents good estimations of the longitudinal turbulence intensity and integral scale for both typhoons and hurricanes, while the peak factor was slightly overestimated and the gust factor was underestimated. The AIJ-RLB-2004 makes a slightly higher estimation of the longitudinal turbulence intensity and a lower estimation of the longitudinal integral scale for both typhoons and hurricanes. The potential reason may be ascribed to the limitation of datasets which used to specify the wind characteristics, although it includes both tropical and extratropical winds. As noted in the AIJ-RLB-2004, the integral scale was treated to be terrain independent. However, the scales of wind eddies are strongly affected by the local roughness.

Author Contributions: Conceptualization, L.L. and T.W.; methodology, L.L. and H.Z.; software, Y.Z. and H.W.; validation, T.W. and H.W.; formal analysis, Y.Z.; investigation, Y.Z. and H.W.; resources, X.H.; data curation, Y.Z.; writing—original draft preparation, L.L. and T.W.; writing—review and editing, L.L., X.H. and T.W.; visualization, Y.Z.; supervision, L.L., H.Z., X.H. and T.W.; project administration, L.L.; funding acquisition, L.L. and T.W.

Funding: This research was funded by National Natural Science Foundation of China (Grant No. 51778373), the Knowledge Innovation Project of Shenzhen (Grant No. JCYJ20170302143625006), Natural Science Foundation of SZU (Grant no. 082017), Natural Science Foundation Grant # CMMI 15-37431 and National Key Research and Development Program of China (2017YFB1201204).

Acknowledgments: The authors gratefully acknowledge K. Gurley of the University of Florida for providing the hurricane data and Lili Song of China Meteorological Administration for providing the typhoon data for this study.

Conflicts of Interest: The authors declare no conflict of interest. The funders had no role in the design of the study; in the collection, analyses, or interpretation of data; in the writing of the manuscript, or in the decision to publish the results.

References

1. Li, Q.S.; Zhi, L.H.; Hu, F. Boundary layer wind structure from observations on a 325 m tower. *J. Wind Eng. Ind. Aerodyn.* **2010**, *98*, 818–832. [CrossRef]
2. Bradbury, W.M.S.; Deaves, D.M.; Hunt, J.C.R.; Kershaw, R.; Nakamura, K.; Hardman, M.E.; Bearman, P.W. The importance of convective gusts. *Meteorol. Appl.* **1994**, *1*, 365–378. [CrossRef]
3. Powell, M.D.; Houston, S.H. Hurricane Andrew's landfall in south Florida. 2. Surface wind fields and potential real-time applications. *Weather Forecast.* **1996**, *11*, 329–349. [CrossRef]
4. Wurman, J.; Winslow, J. Intense sub-kilometer-scale boundary layer rolls observed in Hurricane Fran. *Science* **1998**, *280*, 555–557. [CrossRef] [PubMed]

5.   Melbourne, W.H.; Blackman, D.R. Wind turbulence over seas in tropical cyclones. In Proceedings of the 18th International Conference on Coastal Engineering, Cape Town, South Africa, 14–19 November 1982; pp. 370–383.

6.   Tamura, Y.; Shimada, K.; Hibi, K. Wind response of a tower (typhoon observation at the Nagasaki huis ten bosch domtoren). *J. Wind Eng. Ind. Aerodyn.* **1993**, *50*, 309–318. [CrossRef]

7.   Li, Q.; Fang, J.Q.; Jeary, A.P.; Wong, C.K. Full scale measurements of wind effects on tall buildings. *J. Wind Eng. Ind. Aerodyn.* **1998**, *74–76*, 741–750. [CrossRef]

8.   Sparks, P.R.; Huang, Z. Gust factors and surface-to-gradient wind-speed ratios in tropical cyclones. *J. Wind Eng. Ind. Aerodyn.* **2001**, *89*, 1047–1058. [CrossRef]

9.   Xu, Y.; Zhan, S. Field measurements of Di Wang Tower during Typhoon York. *J. Wind Eng. Ind. Aerodyn.* **2001**, *89*, 73–93. [CrossRef]

10. Schroeder, J.L.; Smith, D.A. Hurricane Bonnie wind flow characteristics as determined from WEMITE. *J. Wind Eng. Ind. Aerodyn.* **2003**, *91*, 767–789. [CrossRef]

11. Cao, S.; Tamura, Y.; Kikuchi, N.; Saito, M.; Nakayama, I.; Matsuzaki, Y. Wind characteristics of a strong typhoon. *J. Wind Eng. Ind. Aerodyn.* **2009**, *97*, 11–21. [CrossRef]

12. Masters, F.J.; Tieleman, H.W.; Balderrama, J.A. Surface wind measurements in three Gulf Coast hurricanes of 2005. *J. Wind Eng. Ind. Aerodyn.* **2010**, *98*, 533–547. [CrossRef]

13. Li, L.; Xiao, Y.; Kareem, A.; Song, L.; Qin, P. Modeling typhoon wind power spectra near sea surface based on measurements in the South China sea. *J. Wind Eng. Ind. Aerodyn.* **2012**, *104–106*, 565–576. [CrossRef]

14. Xiao, Y.; Li, L.; Song, L.; Qin, P. Study on wind characteristics of Typhoon Hagupit based on offshore sea surface measurements. *Acta Aerodyn. Sin.* **2012**, *30*, 380–387.

15. Schroeder, J.L.; Edwards, B.P.; Giammanco, I.M. Observed tropical cyclone wind flow characteristics. *Wind Struct.* **2009**, *12*, 349–381. [CrossRef]

16. Li, L.; Kareem, A.; Xiao, Y.Q.; Song, L.L.; Zhou, C.Y. A comparative study of field measurements of the turbulence characteristics of typhoon and hurricane winds. *J. Wind Eng. Ind. Aerodyn.* **2015**, *140*, 49–66. [CrossRef]

17. ASCE. *Minimum Design Loads for Buildings and Other Structures*; American Society of Civil Engineering: Reston, VA, USA, 2010.

18. AIJ-RLB-2004. *Recommendations for Loads on Buildings*; Architectural Institute of Japan: Tokyo, Japan, 2004.

19. Clauser, F.H. The turbulent boundary layer. *Adv. Appl. Mech.* **1956**, *4*, 1–51.

20. Vickery, P.J.; Skerlj, P.F. Hurricane gust factors revisited. *J. Struct. Eng.* **2005**, *131*, 825–832. [CrossRef]

21. Franklin, J.L.; Black, M.L.; Valde, K. GPS dropwindsonde wind profiles in hurricanes and their operational implications. *Weather Forecast.* **2003**, *18*, 32–44. [CrossRef]

22. Powell, M.D.; Vickery, P.J.; Reinhold, T.A. Reduced drag coefficient for high wind speeds in tropical cyclones. *Nature* **2003**, *422*, 279–283. [CrossRef]

23. Giammanco, I.M.; Schroeder, J.L.; Powell, M.D. Observed characteristics of tropical cyclone vertical wind profiles. *Wind Struct.* **2012**, *15*, 65–86. [CrossRef]

24. Li, L.; Kareem, A.; Xiao, Y.; Song, L.; Qin, P. Wind profile and spectra in typhoon-prone regions in south China. In Proceedings of the ATC & SEI Conference on Advances in Hurricane Engineering, Miami, FL, USA, 24–26 October 2012; pp. 929–940.

25. Tse, K.T.; Li, S.W.; Chan, P.W.; Mok, H.Y.; Weerasuriya, A.U. Wind profile observations in tropical cyclone events using wind-profilers and doppler SODARs. *J. Wind Eng. Ind. Aerodyn.* **2013**, *115*, 93–103. [CrossRef]

26. Baklanov, A.; Grisogono, B. *Atmospheric Boundary Layers: Nature, Theory and Application to Environmental Modelling and Security*; Springer: New York, NY, USA, 2007.

27. Song, L.; Chen, W.; Wang, B.; Zhi, S.; Liu, A. Characteristics of wind profiles in the landfalling typhoon boundary layer. *J. Wind Eng. Ind. Aerodyn.* **2016**, *149*, 77–88. [CrossRef]

28. ESDU. *Strong Winds in the Atmospheric Boundary Layer Part 1: Hourly-Mean Wind Speeds*; Engineering Sciences Data Units: London, UK, 2002.

29. Simiu, E.; Vickery, P.; Kareem, A. Relation between Saffir-Simpson hurricane scale wind speeds and peak 3-s gust speeds over open terrain. *J. Struct. Eng.* **2007**, *133*, 1043–1045. [CrossRef]

30. Willoughby, H.E. Gradient balance in tropical cyclones. *J. Atmos. Sci.* **1990**, *47*, 265–274. [CrossRef]

31. Giammanco, I.M.; Schroeder, J.L.; Powell, M.D. GPS dropwindsonde and WSR-88D observations of tropical cyclone vertical wind profiles and their characteristics. *Weather Forecast.* **2013**, *28*, 77–99. [CrossRef]

32. Tse, K.T.; Li, S.W.; Lin, C.Q.; Chan, P.W. Wind characteristics observed in the vicinity of tropical cyclones: An investigation of the gradient balance and super-gradient flow. *Wind Struct.* **2014**, *19*, 249–270. [CrossRef]

33. Simiu, E.; Scanlan, R.H. *Wind Effects on Structures: Fundamentals and Applications to Design*, 3rd ed.; John Wiley: New York, NY, USA, 1996.

34. Vickery, P.J.; Wadhera, D.; Powell, M.D.; Chen, Y.Z. A hurricane boundary layer and wind field model for use in engineering applications. *J. Appl. Meteorol. Climatol.* **2009**, *48*, 381–405. [CrossRef]

35. Snaiki, R.; Wu, T. A semi-empirical model for mean wind velocity profile of landfalling hurricane boundary layers. *J. Wind Eng. Ind. Aerodyn.* **2018**, *180*, 249–261. [CrossRef]

36. Deaves, D.M. Terrain-dependence of longitudinal rms velocities in the neutral atmosphere. *J. Wind Eng. Ind. Aerodyn.* **1981**, *8*, 259–274. [CrossRef]

37. Kameda, T.; Mochizuki, S.; Osaka, H. Non-equilibrium and equilibrium boundary layers without pressure gradient. In *IUTAM Symposium on Computational Physics and New Perspectives in Turbulence*; Kaneda, Y., Ed.; Springer: Dordrecht, The Netherlands, 2008; pp. 197–202.

38. Harris, R.I.; Deaves, D.M. The structure of strong winds. In Proceedings of the CIRIA Conference on Wind Engineering in the Eightites, London, UK, 12–13 November 1980.

39. ESDU. *Strong Winds in the Atmospheric Boundary Layer Part 2: Discrete Gust Speeds*; Engineering Sciences Data Units: London, UK, 2002.

40. Yu, B.; Chowdhury, A.G.; Masters, F.J. Hurricane wind power spectra, cospectra, and integral length scales. *Bound.-Layer Meteorol.* **2008**, *129*, 411–430. [CrossRef]

41. Von Kármán, T. Progress in the statistical theory of turbulence. *Proc. Natl. Acad. Sci. USA* **1948**, *34*, 530–539. [CrossRef] [PubMed]

42. Davenport, A.G. Note on the Distribution of the Largest Value of a Random Function with Application to Gust Loading. *Proc. Inst. Civil Eng.* **1964**, *28*, 187–196. [CrossRef]

43. Balderrama, J.A.; Masters, F.J.; Gurley, K.R. Peak factor estimation in hurricane surface winds. *J. Wind Eng. Ind. Aerodyn.* **2012**, *102*, 1–13. [CrossRef]

44. Beljaars, A.C.M. The influence of sampling and filtering on measured wind gusts. *J. Atmos. Ocean. Technol.* **1987**, *4*, 613–626. [CrossRef]

45. Bendat, J.S.; Piersol, A.G. *Random Data: Analysis and Measurement Procedures*, 3rd ed.; Wiley: New York, NY, USA, 2000.

46. Wackerly, D.D.; Mendenhall, W.; Scheaffer, R.L. *Mathematical Statistics with Applications*, 7th ed.; Thomson Brooks/Cole: Belmont, CA, USA, 2008.

47. Krayer, W.R.; Marshall, R.D. Gust factors applied to hurricane winds. *Bull. Am. Meteorol. Soc.* **1992**, *73*, 613–617. [CrossRef]

48. Vickery, P.J.; Wadhera, D.; Galsworthy, J.; Peterka, J.A.; Irwin, P.A.; Griffis, L.A. Ultimate wind load design gust wind speeds in the united states for use in ASCE-7. *J. Struct. Eng.* **2010**, *136*, 613–625. [CrossRef]

49. Bardal, L.M.; Setran, L.R. Wind gust factors in a coastal wind climate. In Proceedings of the Eera Deepwind'2016, 13th Deep Sea Offshore Wind R&D Conference, Trondheim, Norway, 20–22 January 2016; Volume 94, pp. 417–424.

50. Durst, C.S. Wind speeds over short periods of time. *Meteor. Mag.* **1960**, *89*, 181–187.

Article

Full-Scale Train Derailment Testing and Analysis of Post-Derailment Behavior of Casting Bogie

Hyun-Ung Bae [1], Jiho Moon [2], Seung-Jae Lim [3], Jong-Chan Park [4] and Nam-Hyoung Lim [4,*]

[1]   R&D Lab., Road Kinematics co., Ltd., Cheonan 31094, Korea; bbnine85@gmail.com
[2]   Department of Civil Engineering, Kangwon National University, Chuncheon 24341, Korea; jmoon1979@kangwon.ac.kr
[3]   Chungnam Railway Research Institute, Chungnam National University, Daejeon 34134, Korea; sngje@cnu.ac.kr
[4]   Department of Civil Engineering, Chungnam National University, Daejeon 34134, Korea; mioso@cnu.ac.kr
*   Correspondence: nhrim@cnu.ac.kr; Tel.: +82-42-821-8867

Received: 10 November 2019; Accepted: 18 December 2019; Published: 19 December 2019

Abstract: In this study, a full-scale train bogie derailment test was conducted. For this, test methodologies to describe the wheel-climbing derailment of the train bogie and to obtain accurate test data were proposed. The derailment test was performed with the casting bogie for a freight train and a Rheda 2000 concrete track. Two different derailment velocities (28.08 km/h and 55.05 km/h) were considered. From the test, it was found that humps in the concrete track affected the post-derailment behavior of the bogie when the derailment velocity was 28.08 km/h. For a higher derailment velocity (55.05 km/h), significant lateral movement of the derailed bogie was observed. This lateral movement was first controlled by wheel–rail contact, followed by contact with the containment wall. Finally, the train was returned to the track center.

Keywords: train derailment; derailment containment provisions; collision testing; post-derailment behavior

---

1. Introduction

In Korea, there were 33 train accident cases in total during the five-year period 2012–2016. Among these, derailment accidents accounted for 78.8% (26 cases) [1], and derailment occurred more often than other types of train accidents. Derailment accidents can cause catastrophic damage to a community. It is hard to prevent 100% of derailment accidents since there are always unexpected factors that can cause derailment, such as human error and natural disasters. Thus, it is necessary to develop technology to reduce damage due to derailment. This technology can be categorized as derailment protection [2–4].

To reduce the damage from derailment accidents, protection facilities can be installed in the railway track. In Korea, guard rails to prevent derailment are used at sharp curves, bridges, and switches, as shown in Figure 1a. Containment walls are also installed on bridges for high-speed railways (where the minimum speed of the line is 200 km/h), as shown in Figure 1b [5–7].

In European countries, three different types of derailment containment provisions (DCPs) are used (DCP types I, II, and III), as shown in Figure 2 [8]. The guard rail is one example of a DCP type I facility, where the DCP is installed inside the track gauge. The wheel of the derailed train comes into direct contact with the facility. DCP type II is similar to DCP type I, but is installed outside of the track gauge. DCP type III facilities are installed outside of the track, similar to DCP type II. However, they are different from DCP type II since the axis of the wheel or bogie of the derailed train impacts this type of DCP.

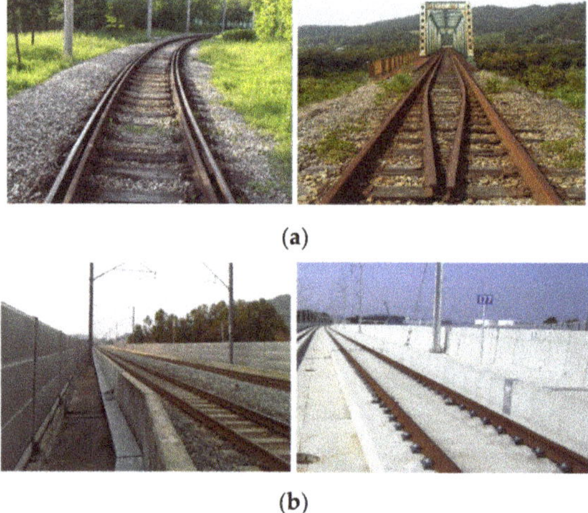

(a)

(b)

Figure 1. Examples of derailment protection facilities: (**a**) guard rail and (**b**) containment wall.

🔵 *DCP Type I*

➤ Installed **Between Running Rails** (Collision at **Wheel Level**)

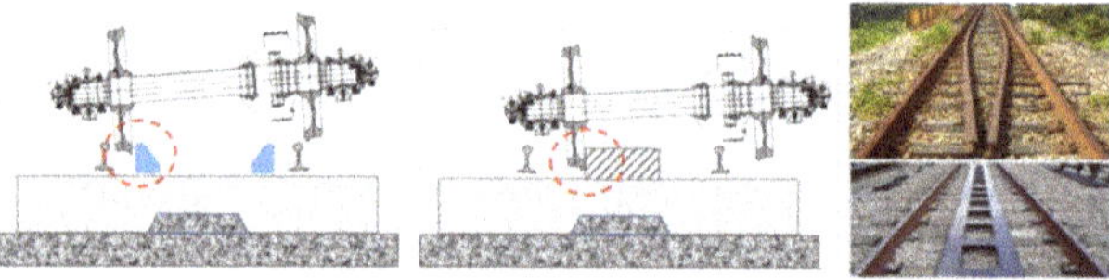

🔵 *DCP Type II*

➤ Installed **Outside of Running Rails** (Collision at **Wheel Level**)

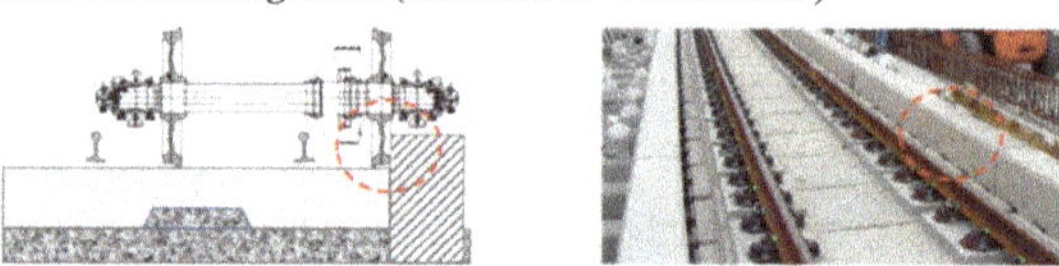

🔵 *DCP Type III*

➤ Installed **Outside of Running Rails** (Collision at **Bogie Level**)

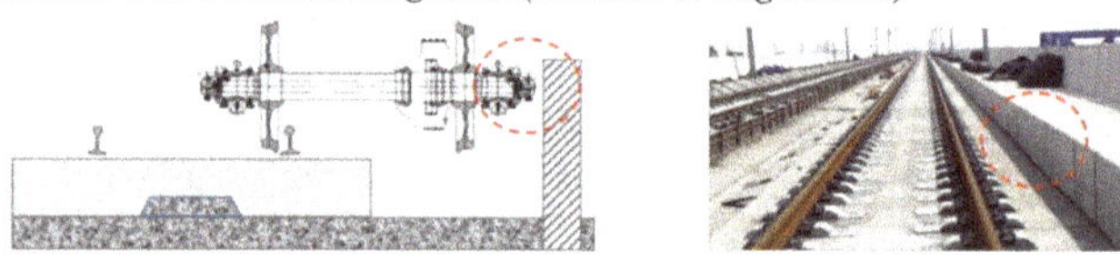

Figure 2. Concept of derailment containment provisions (DCPs).

To verify the performance of such derailment protection facilities, tests and simulations must be conducted, including investigation of the post-derailment behavior of the train. Some researchers have conducted derailment and post-derailment simulations by using 3D finite element analysis. Researchers in Sweden [9,10] analyzed the post-derailment behavior of the wheel of a derailed train colliding with a concrete railroad sleeper. Some researchers in China [11–13] investigated the post-derailment dynamic behavior of a railway vehicle under earthquake excitations. Also, there have been studies simulating collision with a protective facility after the derailment of a high-speed train [2–4].

Full-scale derailment testing is the most reliable method to evaluate the post-derailment behavior of a train and the performance of derailment protection facilities. However, it is very difficult, and only a few such studies can be found in the literature. Wu et al. [14] performed derailment testing, but it was lab-scale testing and the speed was limited to 16 km/h. The test methods relating to train derailment, such as the derailing method and data filtering, are not well established and must be investigated.

In this study, full-scale train derailment testing was conducted. For realistic simulation of the derailment situation and reliable data acquisition from the test, a derailment device and data acquisition system were proposed. Then, from the test, the post-derailment behavior of the train was analyzed.

2. Experimental Methodologies

2.1. Test Site and Track

Full-scale train derailment testing requires a large test site. The test site consisted of acceleration, test, and braking regions, as shown in Figure 3. To increase the speed of the train at the point of derailment, a sufficient acceleration region is needed. Also, adequate test and braking regions should be provided to investigate the post-derailment behavior and to ensure safety, respectively. The lengths of the acceleration, test, and braking regions were 1200 m, 400 m, and 400 m, respectively. In the 400 m test region, a region of 100 m was used to construct a concrete track. The total length of the test line was about 2000 m. Once the target speed of the bogie is reached, a test bogie is released at the end of the acceleration region. The test bogie is derailed at the start of the test region and the post-derailment behavior is observed in the test region. A braking region is provided at the end of the test region. This is a margin region to provide safety after the unexpected behavior of derailed bogie or train. The derailment test site was constructed using a closed railway line to save on costs and replicate actual railway operating conditions.

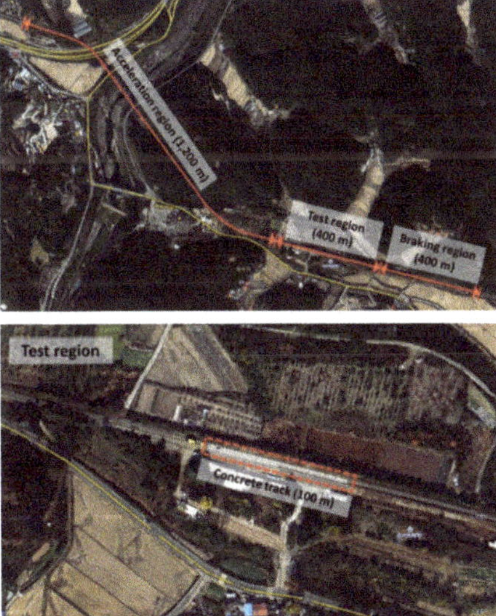

Figure 3. Overview of the test site.

In this study, the focus was on the post-derailment behavior on a concrete track. A Rheda 2000 concrete track, shown in Figure 4, was constructed in the test region where the post-derailment behavior was observed. The Rheda 2000 concrete (ballastless) track was used for the first time in Germany in 2000 as a pilot project on the new rail line between Erfurt and Halle-Leipzig. It was also installed on the high-speed railway in Korea in 2004. Apart for this 100-m concrete track, ballast tracks were used. It should be noted that the derailment containment wall was installed on the left side of the track, as shown in Figure 4. The distance between the wall and the center of the track was 2700 m, considering the geometric condition of the axle of the test vehicle after the derailment.

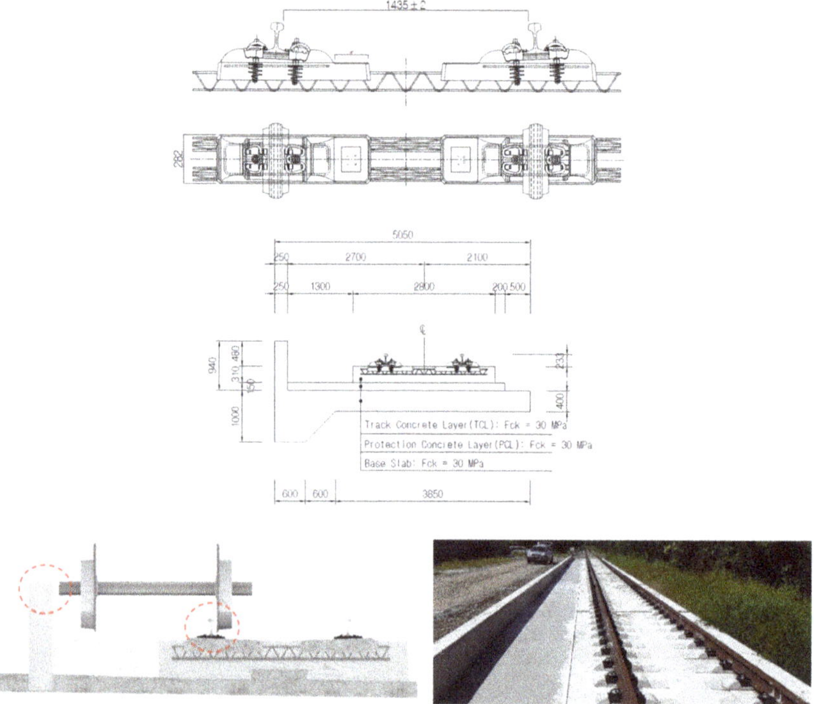

Figure 4. The concrete track used in the test region.

2.2. Test Bogie and Acceleration Method

In this study, a bogie-level test was conducted. The bogie used in this study was a casting bogie for a freight train, as shown in Figure 5. The total weight of the test bogie was 40.91 kN and the specifications of the test bogie are listed in Table 1. The frame structure was made of three pieces of cast steel. The fixed wheel base and wheel diameter were 1676 mm and 860 mm, respectively. A suspension system with coil springs was used between the bolster and side frame.

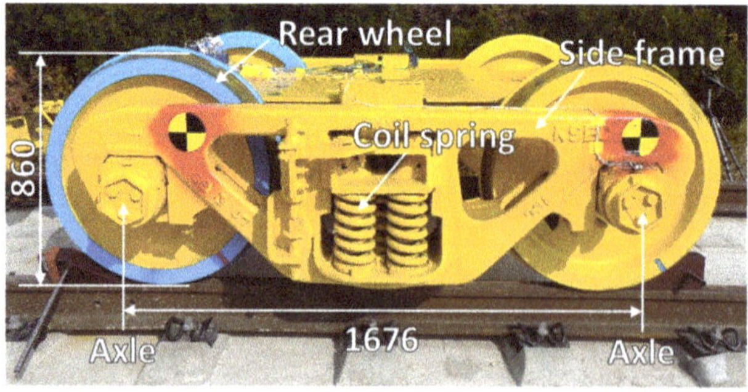

Figure 5. Test bogie (casting bogie for a freight train).

Table 1. Specifications of the test bogie.

Classification	Specification
Frame structure	3-piece cast steel
Fixed wheel base	1676 mm
Wheel diameter	860 mm
Suspension system	Coil spring between bolster and side frame
Friction device	Friction wedge between bolster and side frame
Bearing	AAR D class tapered roller bearings
Test weight	40.91 kN (4.17 tonf)

In order to accelerate the test bogie, several methods can be used, such as a reverse towing system, push system, or remote-controlled system. In this study, the push system shown in Figure 6 was used. The power car was linked to the test bogie with a connector. The test car was accelerated by the power car. Then, the test car was released after reaching the target speed. For this, the releasing system of the connector and braking system of the power car were designed to be controlled by air pressure and a remote controller, as shown in Figure 7. The power car used in this study is shown in Figure 8. The power car had a traction power of 147 kN.

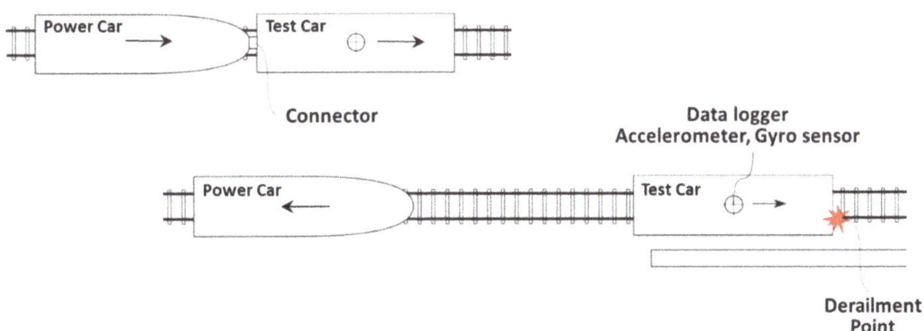

Figure 6. Concept of the push system to accelerate the test bogie.

Figure 7. Release and braking system for the test.

Figure 8. Power car to accelerate the test bogie.

2.3. Derailment Device

A device to induce derailment was needed for the test. Wheel-climbing derailment is one of the most frequent types of derailment. In this study, a derailment device that can induce wheel-climbing derailment was developed and installed at the derailment point, as shown in Figure 9 [15]. The device consisted of a wheel entrance part and a derailment part. In the wheel entrance part, there is an upward slope, and the entered wheel flange reaches the same height as the surface of a rail head. Then, the wheel climbs the surface of a rail head in the derailment part, shown in Figure 9c. Thus, the train or bogie is derailed due to wheel-climbing derailment.

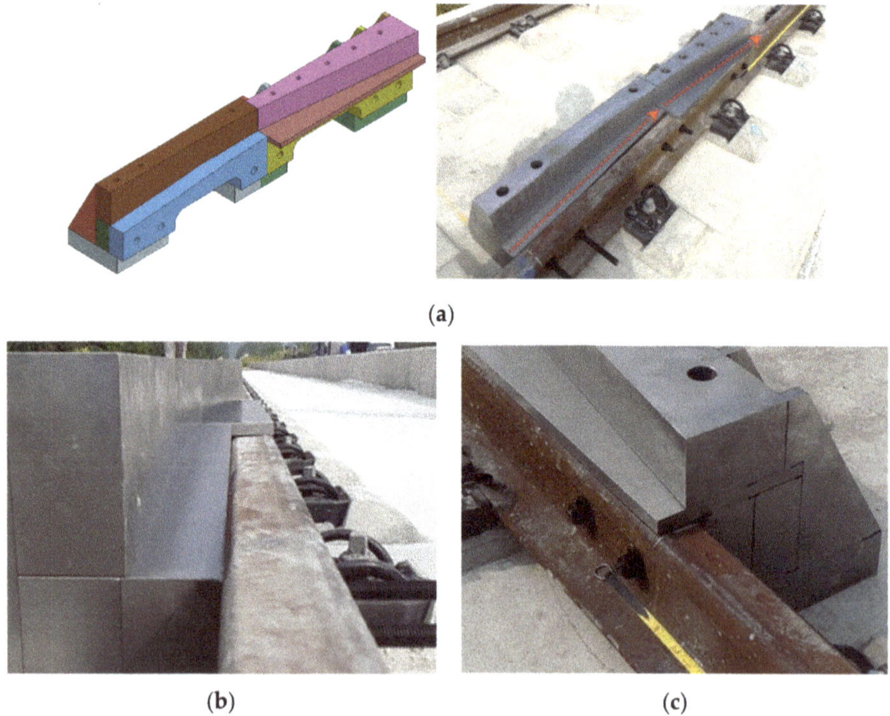

(a)

(b)   (c)

Figure 9. Derailment device: (**a**) overview, (**b**) wheel entrance part, and (**c**) derailment part.

2.4. Data Acquisition System

The position of the derailed train bogie and related data have to be accurately measured during the test. The acceleration and angular velocity of the bogie were measured. For this, an accelerometer and an angular velocity sensor were installed on the geometrical bogie center as shown in Figure 10b, where the capacities of the accelerometer and angular velocity sensor were 2000 g and ±1500 deg/s full-scale range, respectively. Large impact forces are expected after derailment. Thus, the sensors must be securely fixed to the test bogie. To measure the impact force, special accelerometers such as MEMS accelerometers [16] can be used. However, it is hard to determine the exact impact point after derailment. Thus, the impact force to the rail was estimated from the acceleration data of the test bogie. A data logger was installed on the test bogie, as shown in Figure 10c. The data logger used was a shock-resistance data logger (high shock rating of 500 g) for collision testing and it has high sampling capabilities (max. sampling rate of 100 k samples/s/channel). The sampling rate of the data was 1/10,000 s (10,000 Hz) in this test.

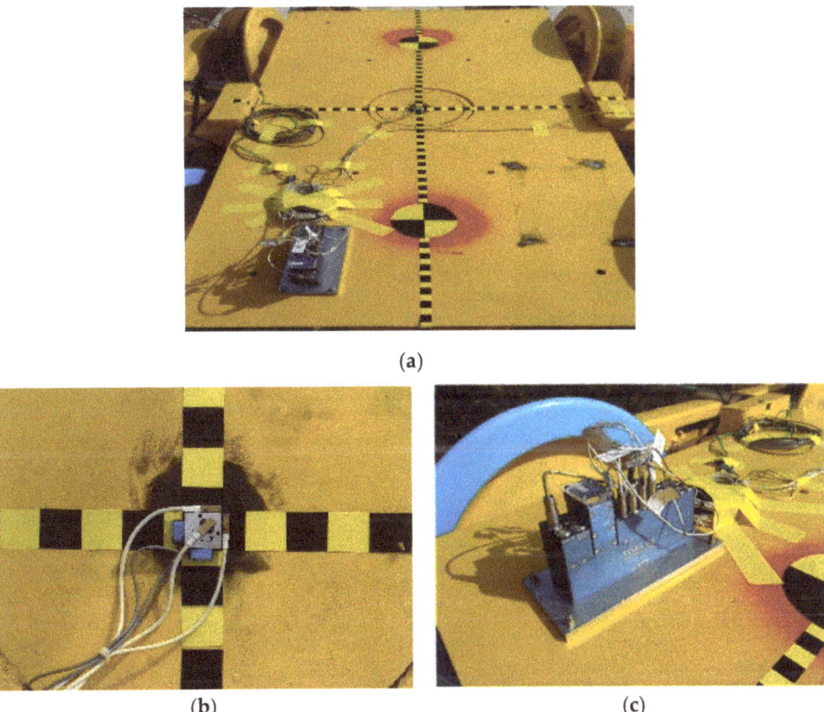

Figure 10. Data acquisition system for the bogie: (**a**) layout, (**b**) accelerometer and angular velocity sensor, and (**c**) a shock-resistance data logger for collision testing.

The velocity of the test bogie at the derailment point is important information for the analysis of post-derailment behavior. Since the test bogie was released before the derailment point, it was necessary to set up an additional velocity measurement system at the derailment point, as shown in Figure 11. In this study, a photoelectric tube speedometer was used. Two lights and two receiver sensors were installed at a specific distance. As the test bogie passes, the passing time between the two lights can be obtained. Then, the initial derailment velocity of the test bogie can be calculated.

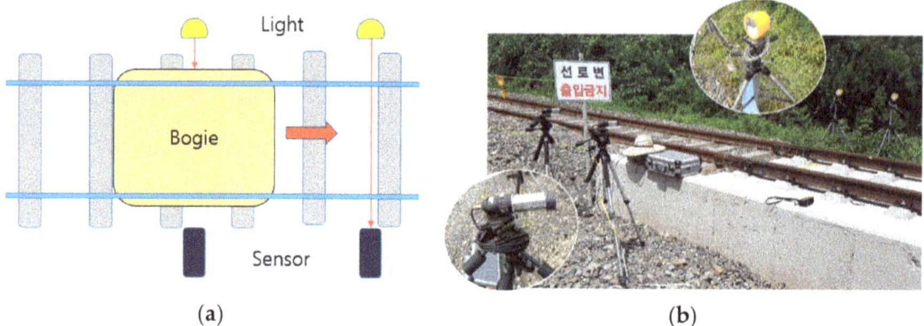

Figure 11. Speed measuring system: (**a**) system conceptual diagram; (**b**) photoelectric tube speedometer.

In this study, a high-speed camera was also used to investigate the post-derailment behavior of the test bogie in depth. Image data are very useful to evaluate the trace of the bogie after derailment. The high-speed camera used in this study is shown in Figure 12. According to its specifications, it can store 3.6 s at 1000 fps (frame/s) when the image format is 1024 × 1024. In total, three high-speed cameras were used. One camera was used to take the top view with a crane, as shown in Figure 12. The others were used to take the side views. For each second, 500 frames were obtained (500 fps). Also, some ordinary digital cameras were installed to obtain image data from various viewpoints.

Figure 12. High-speed cameras: (**a**) side view; (**b**) top view.

2.5. Simultaneous Trigger System

Trigger systems are used to synchronize several different pieces of measurement equipment. By synchronizing the equipment, data analysis is convenient since the initial time of data recording is the same for all equipment. In this study, the three high-speed cameras and data logger were synchronized by a trigger system with a tape switch, as shown in

The tape switch method is simple and free from spurious operation. The trigger for the high-speed cameras was set up on the test track at the derailment point, as shown in Figure 13a. The trigger for the data logger, shown in Figure 13b, was installed on the test bogie. When the test bogie passes the derailment point, the trigger on the test bogie contacts the trigger for the high-speed camera. Then, data recoding is started. It should be noted that the trigger on the bogie was made of flexible materials since the post-derailment behavior could be affected by high stiffness and contact force between the triggers. Figure 13.

(a)   (b)

Figure 13. Simultaneous trigger system by a tape switch: (**a**) trigger for the high-speed cameras; (**b**) trigger for the data logger.

2.6. Running Diagram of the Power Car and Test Bogie

It was necessary to determine the running diagram of the power car and test bogie to achieve the target speed of the test bogie and to ensure the safety of the power car after releasing the test bogie. Figure 14 shows a schematic view of the running diagram of the power car and test bogie. The blue solid line and red dashed line represent the running diagrams of the power car and the test bogie, respectively. Before releasing the test bogie, the power car and the test bogie move together. The velocity at the release time is larger than the target speed. After releasing the test bogie, the power car reduces its speed and stops before the derailment point. The test bogie reaches the target velocity at the derailment point and is derailed.

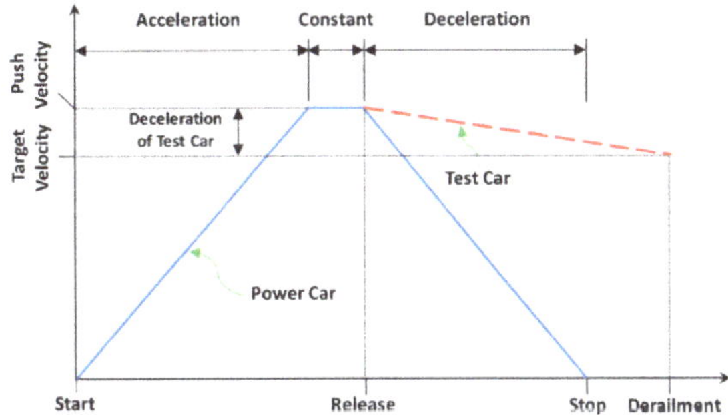

Figure 14. A schematic view of the running diagram of the power car and test bogie.

The distance to the release point depends on the target speed and acceleration capacity of the power car. The running diagram was constructed from the results of several preliminary tests. In this study, the target speeds of the test bogie were 30 km/h and 55 km/h. Figure 15 shows a representative running diagram of the power car through the preliminary tests. An acceleration distance of about 480 m was required for the stationary power car to accelerate to about 56 km/h, and a braking distance of about 150 m was required for subsequent stops. The deceleration to the derailment point after the release of the test bogie was approximately 1–2 km/h. Therefore, the running distance of the power car to ensure safety during the experiment with a target speed of 50 to 55 km/h was determined to be 650 m. In addition, an acceleration distance of about 140 m was required for the power car to accelerate to about 32 km/h, and a braking distance of about 90 m was required for subsequent stops. The deceleration to the derailment point after the release of the test bogie was approximately 2–3 km/h. Thus, the running distance of the power car to ensure safety with a target speed of 25 to 30 km/h was determined to be 250 m.

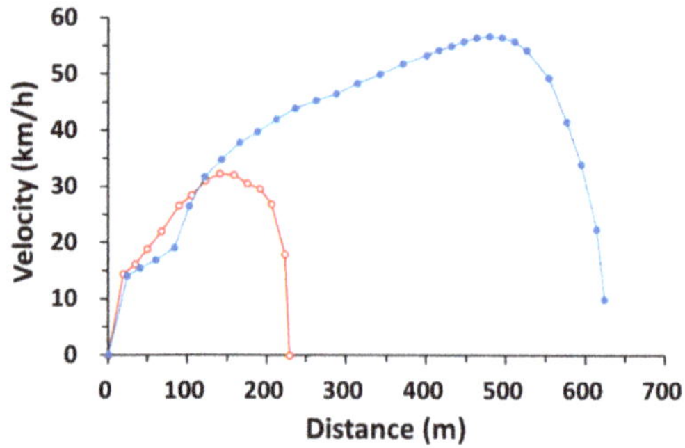

Figure 15. Examples of running diagrams of the power car.

2.7. Post-Processing of Data

It is necessary to conduct a filtering process on raw data since raw data may contain unexpected noise, especially in the case of impact or collision testing. In the case of car collision testing, the data are analyzed after filtering the raw data obtained at a 1000 Hz or 10,000 Hz frequency. Usually, the moving average method is used for the filtering [17,18]. The moving average method is used not only for the smoothing of the data, but also for the evaluation of equivalent static design force. A previous study reported that the moving average method gave a reasonable estimation of the equivalent static design load [19]. Thus, the 50 ms moving average method was applied with 10,000 Hz data sampling in this study.

3. Experimental Results

3.1. Test Cases

After derailment, wheel–rail interaction disappears, and the train runs along the top surface of the track. In this case, various contact and impact conditions, such as gear box–rail, wheel–rail fastener, and wheel–sleeper, arise. These various conditions affect the post-derailment behavior. The post-derailment behavior is an important consideration for the design of derailment protection facilities. In this study, full-scale train bogie derailment tests were conducted. The test bogie and test methodologies are detailed in Section 2. The main test parameter of this study was the derailment velocity. Two different derailment velocities were considered. For Cases #1 and #2, the derailment speeds were 28.08 km/h and 55.05 km/h, respectively.

3.2. Case #1 (Derailment Velocity of 28.08 km/h)

For Case #1, the target speed range was 25–30 km/h. The measured derailment velocity was 28.08 km/h. Figure 16 shows a front view of the post-derailment behavior in Case #1. It can be seen that the bogie ran in the right lateral direction after derailment (0.410–1.810 s), and the left wheel continuously contacted the hump of the concrete track (1.810–3.570 s). Then, the bogie moved to the left direction without contact between the right rail and left wheel (3.570–5.630 s). The direction was changed due to the effect of the hump in this case. In this study, the Rheda 2000 type concrete track was used, and the hump in this track affected the post-derailment behavior.

The acceleration data were analyzed to calculate the velocity and traveling distance of the test bogie. The analysis results were compared with the results obtained in the image data from the high-speed cameras for cross-validation of the data. It is known that the image analysis offers a certain potential for the dynamic investigation [20]. The velocity and traveling distance of the test bogie were also calculated by target mark tracking from the image data, as shown in Figure 17. The results are shown in Figure 18 as a solid line. It can be seen that the velocity obtained from the high-speed camera showed initial fluctuation. However, the overall trend of the velocity was similar to that from the accelerometer. The velocity decreased almost linearly, as shown in Figure 18a. The slope of Figure 18a represents the deceleration, and it was approximately −0.97 m/s². By integrating the acceleration data twice, the traveling distance could be calculated, as shown in Figure 18b. The traveling distances obtained from the accelerometer and high-speed cameras were similar to each other. The discrepancy may come from accumulated error during the integration or distortion of the image by the use of a wide-angle lens.

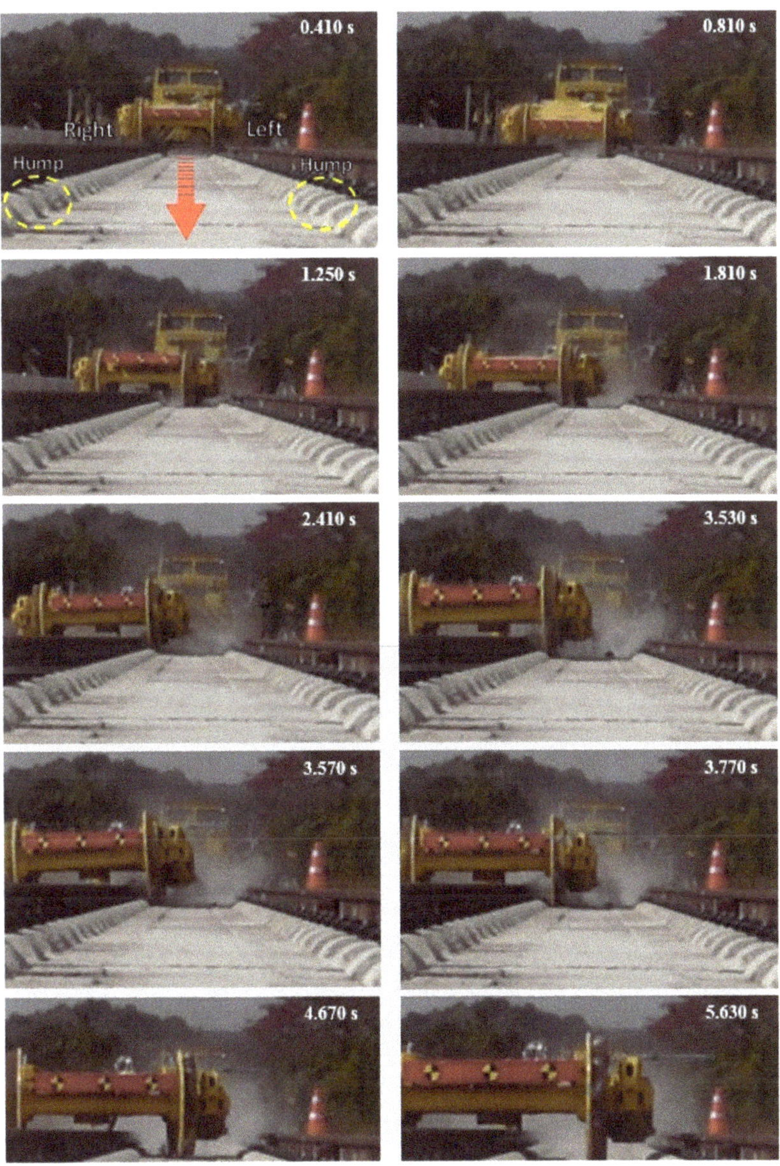

Figure 16. Post-derailment behavior of the bogie for Case #1, front view.

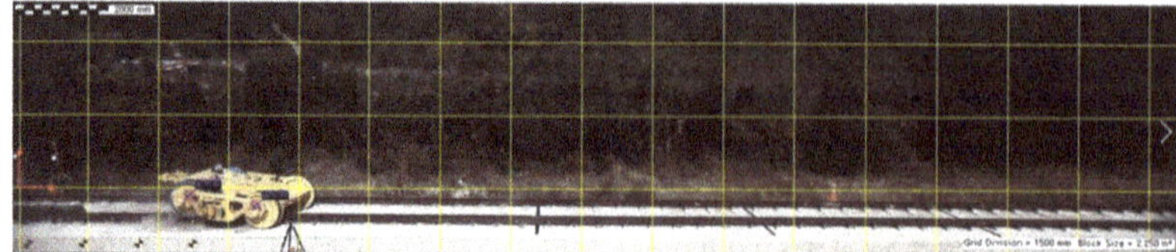

Figure 17. Example of image analysis by target mark tracking.

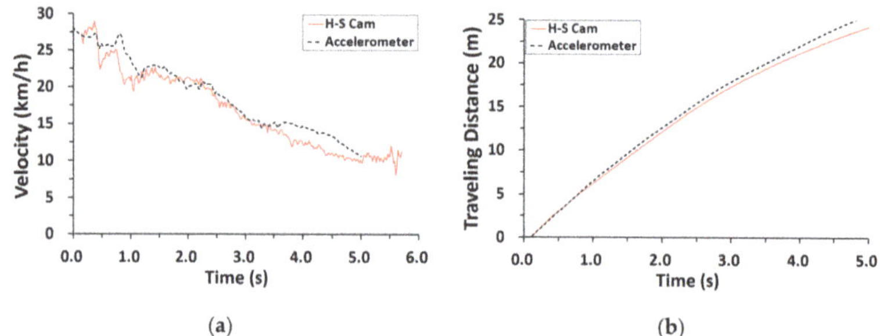

(a)   (b)

Figure 18. Data analysis results, Case #1: (**a**) velocity, (**b**) traveling distance.

The longitudinal, lateral, and vertical acceleration data of the test bogie were analyzed with the video images together. The first impact was observed at 0.410 s after derailment, as shown in Figure 19. It can be seen that there were sudden changes in acceleration at the time of the first impact, as shown in Figure 19a. The maximum acceleration was found in the vertical direction, and it was 5.07 g (49.78 m/s^2). If the total mass of the bogie is included in the calculation of the impact force, it is equivalent to approximately 207 kN of impact force. The first impact occurred between the left wheel and hump, as shown in Figure 19b. Figure 19c shows the damage to the hump and rail fastener due to the first impact.

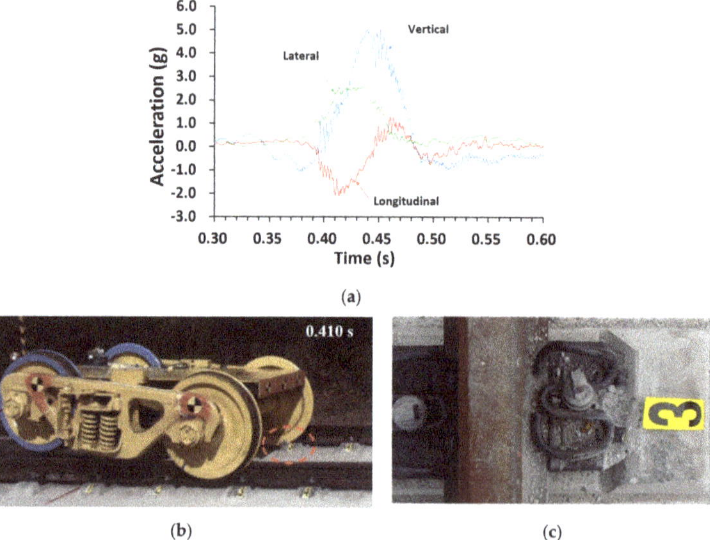

(a)

(b)   (c)

Figure 19. First impact, Case #1: (**a**) acceleration vs. time (0.3–0.6 s); (**b**) test bogie at first impact; (**c**) damage to the hump and rail fastener due to the first impact.

Similar to Figure 19, information about the second and third impacts is presented in Figure 20. The second and third impacts were observed at 0.750 s and 0.810 s, respectively. In the second impact, the right and left rear wheels impacted the fourth hump, as shown in Figure 20b,c. The maximum vertical acceleration was approximately 5.57 g—larger than that from the first impact. This may be attributed to the pitching moment of the bogie after the first impact. In the case of the third impact, the front right wheel contacted the sixth and seventh humps in succession. At the third impact, the peak of the vertical acceleration was markedly decreased.

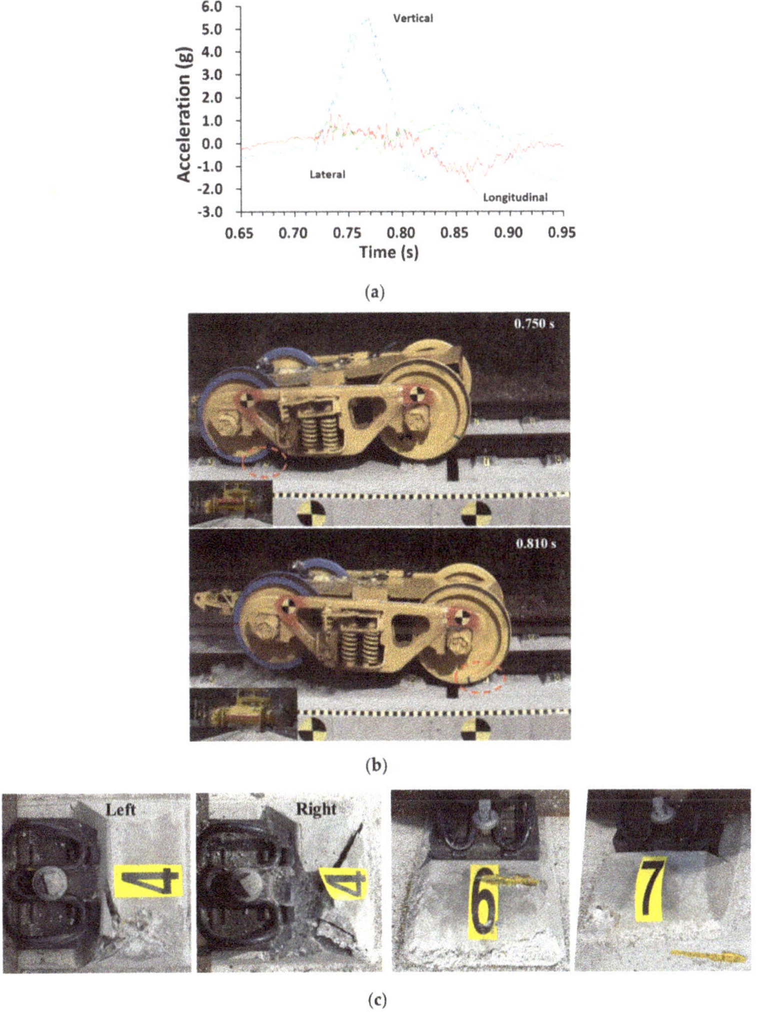

Figure 20. Second and third impacts, Case #1: (**a**) acceleration vs. time (0.65–0.95 s); (**b**) test bogie at the second and third impacts; (**c**) damage to the hump and rail fastener due to the second and third impacts.

After the third impact, the test bogie ran the track without any further significant impact with track components. However, considerable lateral displacement was observed. For example, at 1.170 s, the front left wheel was located at almost the center of the track, as shown in Figure 21a. The lateral displacement continuously increased at 1.170–1.810 s, and the front left wheels contacted the humps in the left rail at 1.810 s. Then, the test bogie returned to the center of the track at 3.750–5.630 s, as shown in Figure 21b.

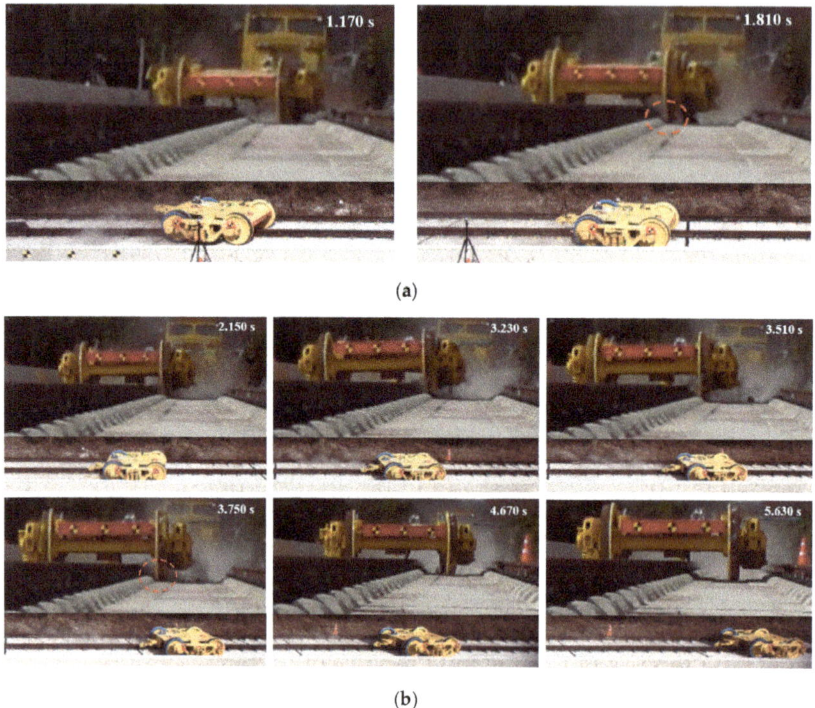

Figure 21. Position of the test bogie: (**a**) 1.170–1.810 s, (**b**) 2.150–5.630 s.

In summary, for Case #1, where the derailment velocity was 28.08 km/h, the derailed test bogie ran the track with considerable lateral movement. The lateral movement was restrained by contact with the humps in the concrete track. Then, the test bogie returned to the center of the track. The whole trace of the test bogie is shown in Figure 22. The maximum vertical acceleration was approximately 5.57 g at the second impact with the humps. The maximum lateral acceleration was approximately 2.62 g at the first impact with the humps. From the results, it can be seen that the effect of the humps on the post-derailment behavior was significant when the Rheda 2000 track system was used. Thus, the type of track must be considered when evaluating the post-derailment behavior of a derailed train.

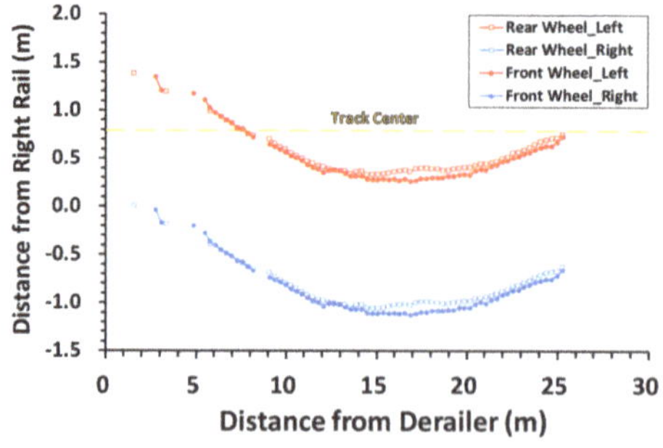

Figure 22. Trace of the left and right wheels for Case #1.

3.3. Case #2 (Derailment Velocity of 55.05 km/h)

In Case #1, the derailed wheel was not guided by the rail or containment wall. However, it was guided by the humps on the track. To identify the effects of the rail and containment wall, the derailment velocity was increased for Case #2. The target derailment velocity range was 50–55 km/h. The measured derailment speed was 55.05 km/h. Figure 23 presents the variation in the velocity and longitudinal traveling distance after derailment for Case #2. The data were obtained by the accelerometer as well as the high-speed cameras. However, the data from the high-speed cameras cannot be interpreted well after 2 s since the test bogie left the camera range due to the increased speed of the test bogie. The deceleration of the test bogie was approximately −1.48 m/s^2. The traveling length was approximately 38 m at 3 s.

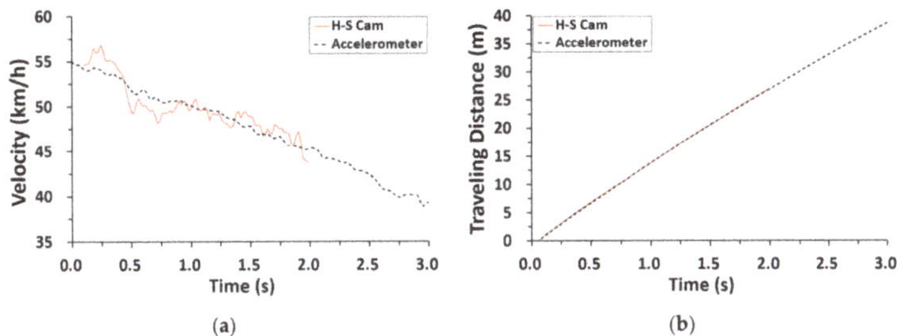

Figure 23. Data analysis results, Case #2: (**a**) velocity, (**b**) traveling distance.

Figure 24 shows the test bogie at derailment. It was found that the test bogie seemed to jump, bypassing the derailment device as shown in Figure 24a. Then, the front left wheel contacted with the third hump and rail fastener at 0.240 s due to the roll of the test bogie, as shown in Figure 24b. When the test bogie passed through the derailment device, the vertical acceleration in Case #2 was much greater than that in Case #1 before the first wheel contact with the track (0.242 s), as shown in Figure 24c. If higher-speed experiments are carried out, the derailment device should be improved because excessive jumping is expected.

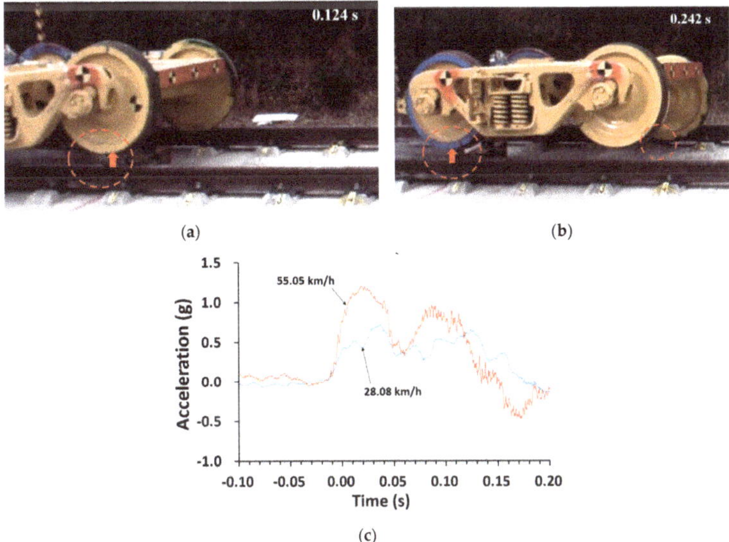

Figure 24. Test bogie at derailment for Case #2: (**a**) 0.124 s; (**b**) 0.242 s (first impact); (**c**) vertical acceleration when derailed (0–0.20 s).

The main impact to the track was observed around from 0.4 s to 0.7 s, as shown in Figure 25. At 0.460 s, the front right wheel came into contact with the eighth rail fastener. Then, the front right wheel impacted the ninth hump and the rear right wheel immediately came into contact with the seventh rail fastener. The maximum vertical acceleration was found at this time, and it was approximately 3.53 g. The maximum acceleration in Case #2 was smaller than that in Case #1. This is because increased velocity is related with the longitudinal direction and vertical impact force is mainly a function of the velocity in the vertical direction.

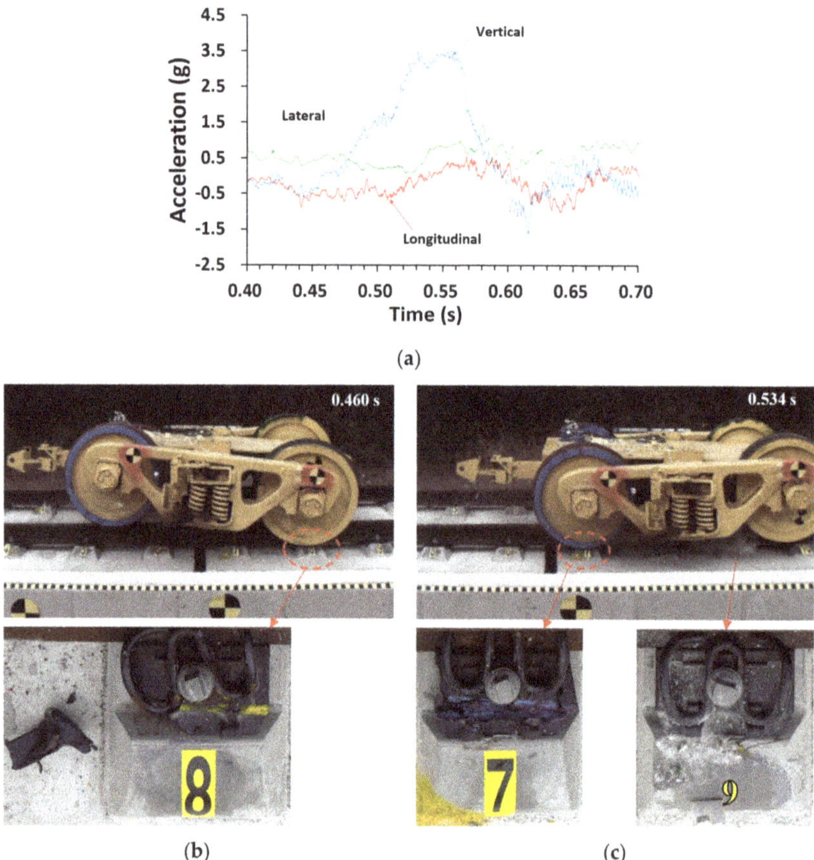

Figure 25. Impact of the test bogie after derailment for Case #2: (**a**) acceleration vs. time (0.4–0.7 s), (**b**) 0.460 s; (**c**) 0.534 s.

After 0.7 s, the test bogie ran the track with considerable lateral movement, similar to Case #1. However, for Case #2, the wheel flange was in contact with the rail as shown in Figure 26b. Then, the test bogie impacted the containment wall at 1.910 s, as shown in Figure 26c. The lateral acceleration was increased due to the impact with the rail and containment wall, as shown in Figure 26a. The lateral acceleration values were approximately 1.17 g (11.51 m/s^2) and 0.75 g (7.36 m/s^2) for the rail and containment wall impact, respectively. Thus, it was expected that the impact force on the containment wall would be reduced approximately 36% by the first guide by the rail for Case #2. Finally, the test bogie returned to the track center direction after containment wall impact, as shown in Figure 27.

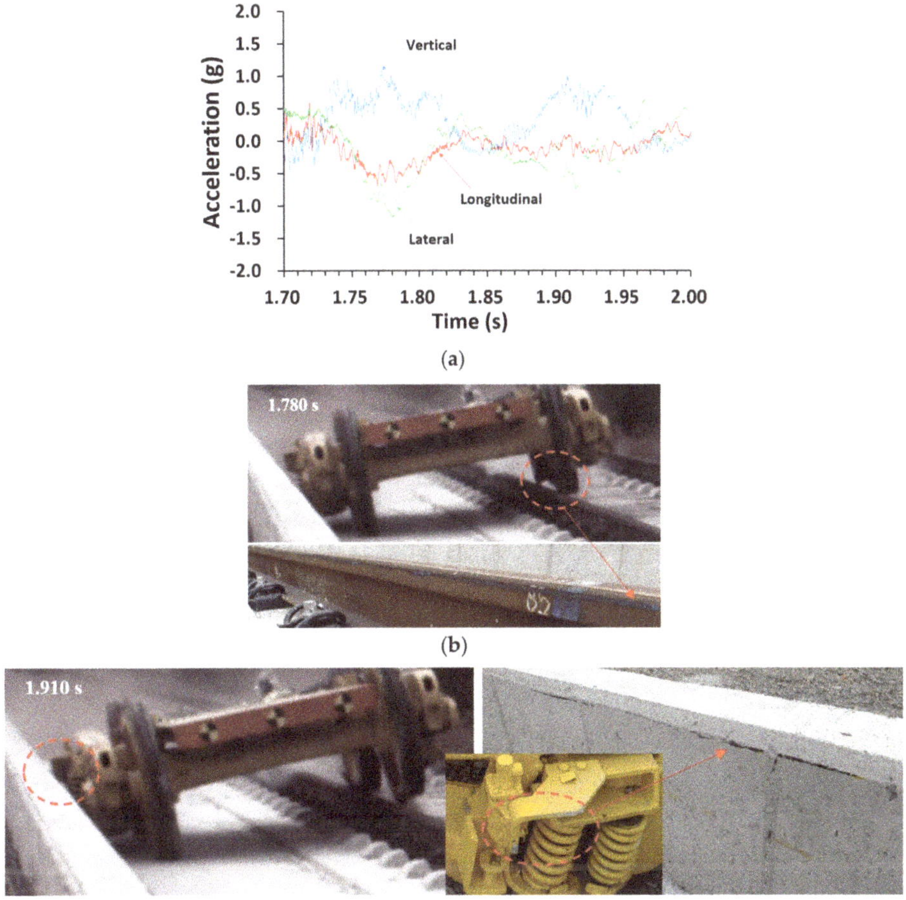

(a)

(b)

(c)

Figure 26. Impact of the test bogie after derailment for Case #2: (**a**) acceleration vs. time (1.7–2 s), (**b**) 1.780 s; (**c**) 1.910 s.

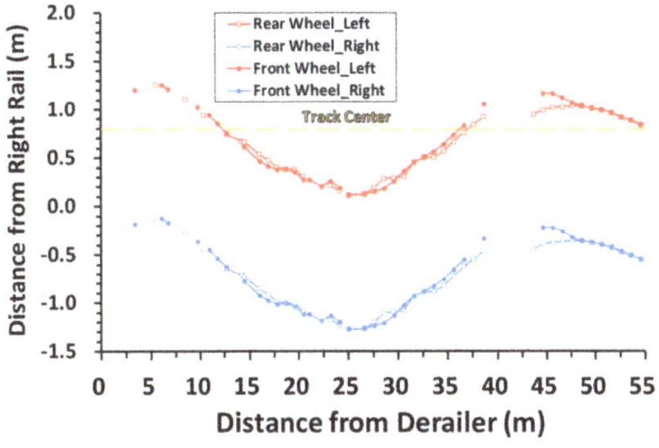

Figure 27. Trace of the left and right wheels for Case #2.

As mentioned in the introduction, the basic concept of DCP type III is that the derailed train is first guided by the rail and the containment wall, preventing excessive lateral movement [8,21,22]. The containment wall installed in this test is DCP type III, and it was seen that the derailed train was effectively guided by DCP type III in test Case #2.

4. Conclusions

This study presented full-scale train bogie derailment test methodologies and post-derailment behavior based on the test results. The derailment tests were conducted at the bogie level, and the test speeds were 28.08 km/h and 55.05 km/h. The major findings of this study are as follows:

1. Train bogie derailment tests can be successfully conducted by using the methodologies proposed in this study. A push system and automatic release device for the test bogie were proposed. Also, a derailment device to induce wheel-climbing derailment and a simultaneous trigger system to synchronize different pieces of equipment were suggested. The proposed methodologies may be applicable to other types of full-scale derailment tests.

2. For the test in Case #1 (derailment velocity of 28.08 km/h), the derailed bogie ran the track with significant right (direction of derailment) lateral movement, and the left (opposite direction to derailment) wheel continuously came into contact with the hump of the concrete track. Then, the bogie moved to the left without contact between the right rail and left wheel. This indicates that the lateral movement for Case #1 was controlled by the humps, and the humps in this track significantly affected the post-derailment behavior of the derailed bogie. In Case #1, the maximum vertical acceleration was approximately 5.57 g at the second impact with the hump. The maximum lateral acceleration was approximately 2.62 g at the first impact with the hump.

3. For the test in Case #2 (derailment velocity of 55.05 km/h), the maximum vertical impact to the track was observed at 0.460–0.560 s, and the maximum vertical acceleration was approximately 3.53 g. The test bogie ran the track with considerable lateral movement, similar to Case #1. However, in Case #2, the wheel flange came into contact with the rail, and then the test bogie impacted the containment wall. The lateral acceleration values were approximately 1.17 g and 0.75 g for the rail and containment wall impacts, respectively. Thus, it was found that the impact force on the containment wall was reduced by approximately 36% by the first guide by the rail. Finally, the test bogie returned to the track center direction after containment wall impact.

Author Contributions: Investigation, Conceptualization, Methodology, Data Curation, Project Administration, Supervision, Writing (original draft, review & editing) from H.-U.B., J.M. and N.-H.L.; Experiment, Data processing from S.-J.L. and J.-C.P. All authors have read and agreed to the published version of the manuscript.

Funding: This research was supported by a grant (19RTRP-B122273-04) from the Railway Technology Research Program, funded by the Ministry of Land, Infrastructure, and Transport of the Korean government.

Conflicts of Interest: The authors declare no conflict of interest.

References

1. Korea Transportation Safety Authority. *Railway Accident and Safety Performance Analysis Report*; Korea Ministry of Land, Infrastructure, and Transport, ASPR (Annual Safety Performance Report); Korea Transportation Safety Authority: Ansan-si, Korea, 2017.

2. Bae, H.U. Advanced Design Concept of Derailment Containment Provisions Using Collision Simulation after Train Derailment. Ph.D. Thesis, Chungnam National University, Daejeon, Korea, 2015.

3. Bae, H.U.; Yun, K.M.; Lim, N.H. Containment capacity and crashworthiness estimation of derailment containment wall on high-speed train. *Proc. Inst. Mech. Eng. Part F* **2018**, *232*, 680–696. [CrossRef]

4. Bae, H.U.; Yun, K.M.; Moon, J.; Lim, N.H. Impact force evaluation of the derailment containment wall for high-speed train through a collision simulation. *Adv. Civ. Eng.* **2018**. [CrossRef]

5. Korea Legal Information Center. *Railroad Safety Act*; Korea Ministry of Government Legislation: Sejong, Korea, 2018; p. 15404.

6. Korea Rail Network Authority. *Track Maintenance Guidelines*; Korea Rail Network Authority: Daejeon, Korea, 2016.

7. Korea Rail Network Authority. *Railway Design Guideline and Handbook—Subsidiary and Safety Facilities for Main Lined*; KR C-02060; Korea Rail Network Authority: Daejeon, Korea, 2017.

8. Booz, A. *Hamilton Report on the Findings of: Current Practice and Effectiveness of Derailment Containment Provisions on High Speed Lines*; HSL-Zuid Organisation: Zoetermeer, The Netherlands, 2004.

9. Brabie, D. Wheel-sleeper impact model in rail vehicles analysis. *J. Syst. Des. Dyn.* **2007**, *1*, 468–480. [CrossRef]

10. Brabie, D.; Andersson, E. High-speed train derailments—Minimizing consequences through Innovative design. In Proceedings of the 8th World Congress of Railway Research (WCRR 2008), Seoul, Korea, 19–21 May 2008; Union Internationale Des Chemins De Fer (UIC): Paris, France, 2008.

11. Wu, X.; Chi, M.; Gao, H. Post-derailment dynamic behavior of a high-speed train under earthquake excitations. *Eng. Fail. Anal.* **2016**, *64*, 97–110. [CrossRef]

12. Wu, X.; Chi, M.; Gao, H.; Ke, X.; Zeng, J.; Wu, P.; Zhu, M. Post-derailment dynamic behavior of railway vehicles travelling on a railway bridge during earthquake. *Proc. Inst. Mech. Eng. Part F* **2016**, *230*, 418–439. [CrossRef]

13. Wu, X.; Chi, M.; Gao, H.; Zhang, D.; Zeng, J.; Wu, P.; Zhu, M. The study of post-derailment measures to limit the extent of a derailment. *Proc. Inst. Mech. Eng. Part F J. Rail Rapid Transit* **2016**, *230*, 64–76. [CrossRef]

14. Wu, X.; Chi, M.; Gao, H. The study of post-derailment dynamic behavior of railway vehicle based on running tests. *Eng. Fail. Anal.* **2014**, *44*, 382–399. [CrossRef]

15. Han, H.H.; Kim, J.W.; Song, I.H.; Bae, H.U.; Koo, J.S.; Lim, N.H. Study on derailment test of bogie using fixed derailer. *J. Korean Soc. Railw.* **2018**, *21*, 957–968. [CrossRef]

16. Chiara, B.; Enrico, B.; Matteo, I.; Salvatore, N. Prototyping and Validation of MEMS Accelerometers for Structural Health Monitoring—The Case Study of the Pietratagliata Cable-Stayed Bridge. *J. Sens. Actuator Netw.* **2018**, *7*, 30.

17. Kim, D.S. Design of Roadside Safety Features Using Vehicle Velocity-Time History. Ph.D. Thesis, Kongju National University, Gongju, Korea, 2012.

18. Kim, K.J. Evaluation of Collision Force by Impact Simulations and Its Application to the Design of Bridge Piers. Ph.D. Thesis, Kongju National University, Gongju, Korea, 2014.

19. Beason, W.L.; Hirsch, T.J.; Campise, W.L. *Measurement of Heavy Vehicle Impact Forces and Inertia Properties*; Final Report, FHWA/RD-89/120; Texas Transportation Institute: College Station, TX, USA, 1989.

20. Chiara, B.; Marco, F.; Claudio, A. Vibration Analysis and Dynamic Characterization of Structural Glass Elements with Different Restraints Based on Operational Modal Analysis. *Buildings* **2019**, *9*, 13.

21. Railtrack PLC (Safety & Standards Directorate). *Recommendations for the Design of Bridges*; Railtrack Approved Code of Practice, GC/RC5110; Railtrack PLC: London, UK, 2000.

22. Network Rail. *Design of Bridges (Design Loading for Accommodation and Occupation Overbridges)*; Network Rail Kings Place, NR/L3/CIV/020; Network Rail: London, UK, 2011.

Article

Prediction of Friction Resistance for Slurry Pipe Jacking

Yichao Ye, Limin Peng, Yang Zhou, Weichao Yang *, Chenghua Shi and Yuexiang Lin

School of Civil Engineering, Central South University, Changsha 410075, China; tunnelye@163.com (Y.Y.);
lmpeng@csu.edu.cn (L.P.); csu.zy@csu.edu.cn (Y.Z.); csusch@163.com (C.S.); csulyx2010@foxmail.com (Y.L.)
* Correspondence: weic_yang@163.com; Tel.: +86-137-8723-2438

Received: 18 November 2019; Accepted: 23 December 2019; Published: 26 December 2019

Featured Application: The new approach established in this paper can provide accuracy prediction of friction resistance for slurry pipe jacking with various soil conditions, which lays a good foundation for better future design and less construction costs.

Abstract: Friction resistance usually constitutes one of the two main components for the calculation of required jacking force. This paper provides a new approach to predict the friction resistance of slurry pipe jacking. First, the existing prediction equations and their establishment methods and essential hypotheses used were carefully summarized and compared, providing good foundations for the establishment of the new model. It was found that the friction resistance can be uniformly calculated by multiplying an effective friction coefficient and the normal force acting on the external surface of the pipe. This effective friction coefficient is introduced to reflect the effect of contact state of pipe-soil-slurry, highly affected by the effect of lubrication and the interaction of pipe-soil-slurry. The critical quantity of pipe-soil contact angle (or width) involved may be calculated by Persson's contact model. Then, the equation of normal force was rederived and determined, in which the vertical soil stress should be calculated by Terzaghi's silo model with parameters proposed by the UK Pipe Jacking Association. Different from the existing prediction models, this new approach has taken into full consideration the effect of lubrication, soil properties (such as internal friction angle, cohesion, and void ratio), and design parameters (such as buried depth, overcut, and pipe diameter). In addition, four field cases and a numerical simulation case with various soils and design parameters were carefully selected to check out the capability of the new model. There was greater satisfaction with the measured data as compared to the existing models and the numerical simulation approach, indicating that the new approach not only has higher accuracy but is also more flexible and has a wider applicability. Finally, the influence of buried depth, overcut, and pipe diameter on the friction resistance and lubrication efficiency were analyzed, and the results can be helpful for the future design.

Keywords: slurry pipe jacking; friction resistance; effective friction coefficient; pipe-soil-slurry interaction; lubrication efficiency

---

1. Introduction

In many parts of the world, the numerous constructions of municipal tunnels are creating unforeseen problems, such as blocking of roads, existing pipelines failure, and buildings subsidence. This has motivated attempts at the development of trenchless construction technology, such as pipe jacking, especially in metropolitan cities [1,2]. Pipe jacking is defined as a trenchless excavation technique, which employs hydraulic jacks to thrust specially made pipes through the ground behind a jacking machine, from a drive shaft to a reception shaft, as illustrated in Figure 1. It has many technical merits, such as a short time limit, high security, low environmental effect, and little traffic disturbance [3–6]. Because of that, pipe jacking has been widely used in the construction of infrastructure for traffic and transportation systems in cities [7,8].

In pipe jacking, the jacking force is a critical factor that determines the thickness of pipe and reaction wall, selection or design of jacking machine and lubricant requirements [9]. The accuracy of prediction of jacking force is directly related to the structural safety and construction cost.

The main component of the jacking force is due to frictional resistance. Application of a lubricant such as bentonite slurry in pipe jacking (so-called 'slurry pipe jacking') is essential to reduce the friction resistance and, therefore, the jacking force [3,10–13]. However, the use of slurry makes it more complex to calculate or predict the friction resistance because of the change in contact conditions between the pipe and soil and lubricant slurry. The new contact state, which is due to the pipe-soil-slurry interaction, is affected by factors such as pipe diameter, soil properties, overcut [3,9], lubrication efficiency [9], pipeline misalignments [10,14–16], and stoppages [9,14,17–19]. The existing prediction models have not fully taken these factors into consideration, leading to an overestimation or underestimation of the friction resistance [9,20–23]. It is therefore obvious that a new prediction approach or model is imperatively needed to be established to solve the problem in slurry pipe jacking [24].

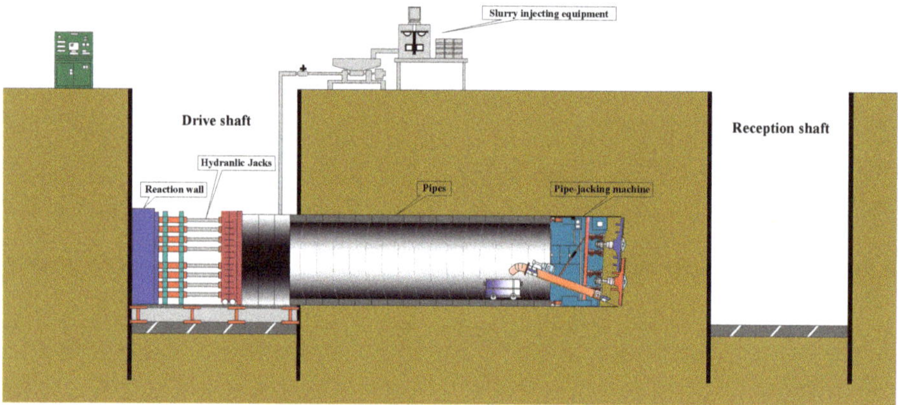

Figure 1. Schematic of slurry pipe jacking.

2. Overview of the Existing Prediction Models of Friction Resistance

Numerous models that calculate the friction resistance of pipe jacking have been proposed by authors from all over the world. The superposition principle is usually used, which holds that the comprehensive outcome of two or more linear factors of a system is equal to the accumulation of the effect of each factor. In pipe jacking, the linear factors to generate friction resistance are due to the weight of pipe (f_W), soil pressure (f_s), slurry pressure (f_m), pipe-soil cohesion (f_{sc}), and pipe-slurry cohesion (f_{mc}). Their equations can be expressed as [5,6,9,10,14,19,21,22,25,26].

$$f_W = \mu_s W \tag{1}$$

$$f_s = \mu_s N \tag{2}$$

$$f_m = \mu_m N \tag{3}$$

$$f_{sc} = c_s B_s \tag{4}$$

$$f_{mc} = c_m B_m \tag{5}$$

where μ_s and μ_m are the kinematic friction coefficient of pipe-soil and pipe-slurry, respectively; W is the weight of pipe per unit length, kN/m; N is the total normal force acting on the pipe, kN/m; c_s and

c_m are the pipe-soil cohesion resistance and pipe-slurry cohesion resistance, respectively, kPa; B_s and B_m are the pipe-soil contact width and pipe-slurry contact width, respectively, m.

Some hypotheses have been made to establish the prediction models, by which one or some of the items listed above should be considered. Typical hypotheses are:

Hypothesis 1. *The excavated tunnel is self-stable, the pipeline simply slides along the bottom of the tunnel due to its own weight (see Figure 2a) [9,10,14].*

Hypothesis 2. *The angular space due to overcut is completely filled with lubricant slurry, and the excavated tunnel is stable under the slurry pressure (see Figure 2b) [22,25].*

Hypothesis 3. *The excavated tunnel is unstable, the surrounding soil collapses and is in full contact with the whole area of the jacking pipes (see Figure 2c) [5,6,9,21,26].*

Hypothesis 4. *The excavated tunnel is stable under the pressure of slurry, and part of the pipe comes in contact with the surrounding soil (see in Figure 2d) [3].*

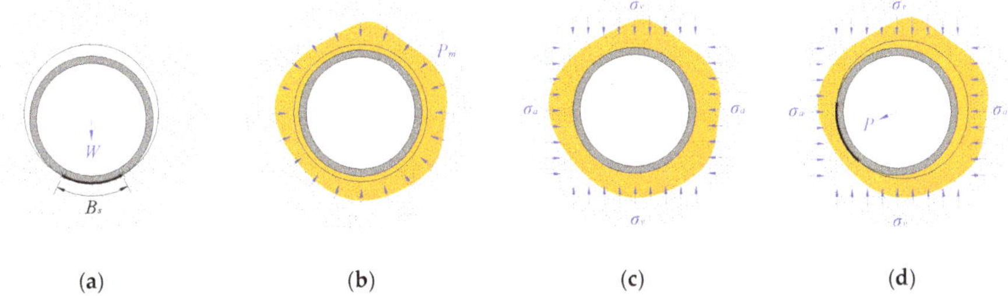

(a)	(b)	(c)	(d)

Figure 2. The models to calculate friction resistance according to different hypotheses: (**a**) Hypothesis 1; (**b**) Hypothesis 2; (**c**) Hypothesis 3; (**d**) Hypothesis 4.

According to Hypothesis 1, there are two kinds of prediction models, in which the item of f_w has to be taken into consideration. The first one assumes that the friction resistance is only due to the weight of pipe [9,21]. It is given as

$$f = \mu_s W \tag{6}$$

$$\mu_s = \tan(\varphi/2) \tag{7}$$

The second one also takes the pipe-soil cohesion resistance (item of f_{sc}) into account [10,14,21], which is given by

$$f = \mu_s W + c_s B_s \tag{8}$$

where the pipe-soil contact width B_s is calculated by Hertzian contact model, as [10,14,21]

$$B_s = 1.6(Pk_d C_e)^{1/2} \tag{9}$$

$$k_d = \frac{D_c D_p}{D_c - D_p} \ , \quad C_e = \frac{1 - v_p^2}{E_p} + \frac{1 - v_s^2}{E_s} \tag{10}$$

where D_c and D_p are the internal diameter of cavity and external diameter of pipe, respectively, m; v_p and v_s are the Poisson's ratio for pipe and soil material, respectively; E_p and E_s are the elasticity modulus of the pipe and soil, respectively, kPa; and P is the effective external force acting on the center of the pipe, kN/m, usually it is considered as equal to the weight of pipe per unit length W.

According to Hypothesis 2, the friction resistance is related only to the properties of lubricant slurry, and the only model found was one that takes both the items of f_m and f_{mc} into account [22,25].

$$f = \mu_m N + c_m B_m \tag{11}$$

$$N = \pi D_p P_m \tag{12}$$

$$B_m = \pi D_p \tag{13}$$

where P_m is the mud slurry pressure.

It is noted that most of the studies completed to date have focused exclusively on the prediction models established by Hypothesis 3. This may be attributed to the assumption of full contact of the pipe and soil that leads to a large prediction value of friction resistance, or in other words, Hypothesis 3 is conservative. Because of that, this kind of model is widely accepted by authorities and standards from all over the world, such as Japan Sewage Association (JSA) [22], UK Pipe Jacking Association (PJA) [27], Chinese Trenchless Technology Association (CTTA) [28], and Germany Standard (AVT A-161) [29]. This kind of model can be summarized and divided into the following four categories.

$$f = \mu_s N \tag{14}$$

$$f = \mu_s N + \mu_s W \tag{15}$$

$$f = \mu_s N + \mu_s W + c_s B_s \tag{16}$$

$$f = \beta \mu_s N + \mu_s W + c_s B_s \tag{17}$$

The fourth kind of model (Equation (17)) introduces an empirical constant β (smaller than 1) on the basis of the third model (Equation (16)) to reflect the effect of lubrication.

For the models summarized above, the item of f_s ($=\mu_s N$) has to be taken into consideration. Thus, the key problem for this kind of model is exactly focused on the calculation of soil pressure N. CTTA suggests N to be calculated by Rankine's formula, which gives that [28]

$$N = 2(1 + K)D_p \sigma_v \tag{18}$$

$$\sigma_v = \gamma h \tag{19}$$

$$K = \tan^2(\pi/4 - \phi/2) \tag{20}$$

where σ_v is the vertical soil stress; K is the coefficient of soil pressure above the pipe; γ is the unit weight of soil; h is the overburden depth of the pipeline; and φ is the internal friction angle of soil.

However, JSA suggests using Terzaghi's silo model, the expression of N is then expressed as [22]

$$N = \pi D_p \sigma_v \tag{21}$$

$$\sigma_v = \frac{b\gamma - 2c_s}{2K \tan(\delta)}\left(1 - e^{-2K \tan(\delta)h/b}\right) \tag{22}$$

where c_s is the cohesion of soil; δ is the friction angle between the pipe and soil; b is the influencing silo width of soil above the pipe, and the other symbols have the same meanings as before.

Although a lot of prediction equations of friction resistance have been proposed and even some of them have been applied to engineering practice, it is obvious that their hypotheses are quite different, and even for the same hypothesis, models and parameters can be different too. Thus, it is bound to make the prediction friction resistances vary greatly. Furthermore, apart from Equations (11) and (17), the other models completely ignore the effect of lubrication, which is very important to slurry pipe jacking.

In the design philosophy of slurry pipe jacking, the angular space due to overcut is expected to be completely filled with lubricant slurry, to reduce the friction resistance with maximum efficiency, creating a 'filter cake' layer around the cavity and is then pressurized to the support pressure required for the soil (see Figure 2b) [3,10]. In this case, the friction resistance should be only related to the slurry pressure and the friction coefficient between slurry and the pipe. From this point of view, the expression Equation (11) seems a convincing explanation here.

However, the more general case is that the excavated tunnel is stable under pressure of slurry and part of the pipe inevitably comes in contact with the soil (see Figure 2d) [3]. The reasons for the occurrence of pipe-soil contact can be complex, such as insufficient design and control of grouting amount of slurry, the pipeline deviates from the intended line and level, irregular deformation of the surrounding soil, and interpenetration between the soil and slurry. Thereby, the state of contact can change from 'pipe-slurry' into 'pipe-soil-slurry' (Figure 2b,d). In such a case, a simple form of Equation (17) seems more suitable to reflect the effect of lubricant slurry. However, it is logical that the greater the contact width between the pipe and the soil, the smaller the effect of lubrication will be, and, therefore, the greater the friction resistance will be. Thus, the value of β should be highly affected by the pipe-soil contact width. For different soils, grouting amount of slurry and design parameters (such as buried depth and overcut) is bound to lead to completely different contact widths of pipe-soil. Thus, β should be in a large range, and it would be rather difficult to pick out a value to use in application.

In fact, a successful prediction model of friction resistance should not only consider the effect of lubrication but also needs to be able to reflect the effect of pipe-soil-slurry interaction. It is based on this understanding that the following model comes into being.

3. Calculation of Friction Resistance for Slurry Pipe Jacking

The general contact state of pipe-soil-slurry due to the interaction of the pipe, surrounding soil, and lubricant slurry is shown in Figure 3. In the picture, the position of pipe-soil contact is arbitrary, with a contact width of B_s and the corresponding contact angle of 2ε.

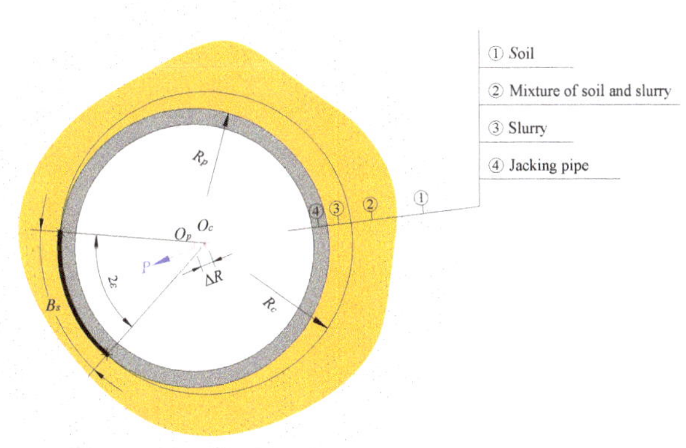

Figure 3. The general contact state of pipe-soil-slurry.

From Figure 3, it is obvious that to calculate friction resistance f, both of the items of f_s and f_m have to be taken into account, which can be expressed as

$$f = \mu N = f_s + f_m = \mu_s N_s + \mu_m N_m \tag{23}$$

where μ is an effective friction coefficient introduced to reflect the effect of lubrication and the influence of pipe-soil contact width. It is generally accepted that $\mu_s = \tan(\varphi/2)$ for the coefficient of kinematic friction between soil and the pipe [21,22]; μ_m for the coefficient of kinematic friction between mud slurry and the pipe can be taken as 0.01, according to the test result reported by Guo [30]. N_s and N_m are the total normal force of pipe-soil and pipe-slurry in contact, respectively.

To calculate N_s and N_m precisely, the location of pipe-soil contact and the magnitude of contact angle (or contact width) and the contact force have to be determined. For various reasons leading to the occurrence of pipe-soil contact, it seems impossible to calculate these quantities in a target section of the pipeline. However, if taking the whole pipeline into consideration, and assuming that the pipe-soil contact can occur at any position of a section of the pipeline with a same probability and the contact force is approximately equal to the soil pressure in the contact area, this problem can be greatly simplified. In this case, we have the following equations:

$$P = N_s = \frac{B_s}{C}N = \frac{\varepsilon}{\pi}N \tag{24}$$

$$N_m = \frac{B_m}{C}N \tag{25}$$

where $C\ (=\pi D_p)$ is the external circumference of pipe and ε is the semi-angle of contact (Figure 3).

By substituting Equations (24) and (25) in Equation (23), after some algebra, giving that

$$\mu = \mu_s \lambda_s + \mu_m \lambda_m$$
$$\lambda_s = \frac{B_s}{C} = \frac{\varepsilon}{\pi}, \quad \lambda_m = \frac{B_m}{C} \tag{26}$$

$$B_m = C - \frac{B_s}{1+e} \tag{27}$$

where e is the void ratio of soil.

By substituting Equation (27) in Equation (26), the expression of μ can be further rewritten as

$$\mu = \mu_s \frac{\varepsilon}{\pi} + \mu_m \left(1 - \frac{\varepsilon}{\pi(1+e)}\right) \tag{28}$$

According to Equation (28), the calculation of ε is essential to calculate μ. Hertzian model provides a simple way for the calculation of the width of contact (or contact angle) as we have mentioned before (see Equations (8) and (10)); however, the Hertzian contact problem is approached only when the applied force is small, or the large radial clearance is large, and the limited angle of contact is smaller than about 30° [31]. Due to the technical limitations, most of the pipe jacking projects encounter clay or sandy soils and with a small overcut, it is therefore important that the applicability of the Hertzian contact model should be extremely limited here. Actually, the Hertzian contact model is just a special case of the Persson's contact model with a small contact width (or angle) [31]. If a large contact angle (larger than 30°) occurs, the more general contact model proposed by Persson should be taken as the first choice. The following singular integro-differential governing equation of contact angle is derived by Persson, as [31,32]

$$B = 4(1-\beta) - 2(1-\alpha)\int_{-\xi}^{+\xi} q(t)\frac{dt}{1+t^2} - \frac{\pi}{2}(1+\alpha)\frac{E_p \Delta R}{(1-v_p^2)P} \tag{29}$$

or

$$\frac{\pi(1+\alpha)E_p \Delta R}{(1-v_p^2)P} = 4(1-\beta) - 2(1-\alpha)\int_{-\xi}^{+\xi} q(t)\frac{dt}{1+t^2} - B \tag{30}$$

The involved auxiliary variables are defined as [31]

$$\Delta R = \frac{D_c - D_p}{2} \; , \quad \xi = \tan\left(\frac{\varepsilon}{2}\right) \, ,$$
$$\alpha = \frac{1-\eta}{1+\eta} \; , \quad \beta = \frac{\lambda}{2(1+\eta)} \; , \tag{31}$$
$$\eta = \frac{E_p}{E_s}\frac{1-v_s^2}{1-v_p^2} \; , \quad \lambda = \frac{1-2v_p}{1-v_p} - \eta\frac{1-2v_s}{1-v_s}$$

After some approximate treatments, the key terms of Equation (30) have been solved by Michele [32], as follows:

$$\int\limits_{-\xi}^{+\xi} q(t)\frac{dt}{1+t^2} = \frac{1}{2\pi}\frac{I_b}{\xi^2(\xi^2+1)} + \frac{\xi^2}{\xi^2+1} \tag{32}$$

$$I_b = \pi\log(\xi^2+1) \tag{33}$$

$$B = \frac{2\xi^4 + 2\xi^2 - 1}{\xi^2(\xi^2+1)} \tag{34}$$

By substituting these equations into Equation (29), Michele obtained an approximate form of general contact angle relation.

$$\frac{\pi(\alpha+1)E_p\Delta R}{(1-v_p^2)P} = \frac{(\alpha-1)\left[\ln(\xi^2+1)+2\xi^4\right]+2}{(\xi^2+1)\xi^2} - 4\beta \tag{35}$$

As compared with Equation (9), Equation (35) is a far more complex nonlinear equation. It can be further simplified with respect to that the elastic modulus of soil E_s is generally much smaller than that of pipe E_p (the difference between the two can be three orders of magnitude). Thus, from Equation (30), the magnitude of auxiliary variable η should be very large, and, therefore, the following approximate relations can be obtained:

$$\frac{\pi(\alpha+1)E_p}{(1-v_p^2)} \approx \frac{2\pi E_s}{(1-v_s^2)},$$
$$\alpha \approx -1, \tag{36}$$
$$\beta \approx \frac{1-2v_s}{2(1-v_s)}$$

Using Equation (36), Equation (35) can be then simplified as

$$\frac{\pi E_s\Delta R}{(1-v_s^2)P} + \frac{1-2v_s}{1-v_s} = \frac{1-\left[\ln(\xi^2+1)+2\xi^4\right]}{(\xi^2+1)\xi^2} \tag{37}$$

From Equation (37), it is essential to calculate P, which requires one to calculate the total normal force N. It can be gained by integrating the normal stress σ_n on an element of the pipe surface and is determined on the basis of vertical and horizontal soil stresses.

$$\sigma_n = \sigma_v \sin\theta + \sigma_a \cos\theta \tag{38}$$

$$N = 4\int_0^{\pi/2} \sigma_n \frac{D_p}{2}d\theta \tag{39}$$

where θ is defined as the angle between the corresponding radius line and the horizontal line at each point of the pipe (Figure 4).

By substituting Equation (38) in Equation (39), it is easy to obtain the equation of N, which has the same form as Equation (18).

To calculate Equation (18), the vertical soil stress σ_v has to be first determined. It is noted that at the present time, by far the most commonly used model for soil pressure calculation is Terzaghi's silo model (Equation (22)) [5,6,9,19,21,26]. According to Equation (22), the calculation of the vertical soil stress requires some physical parameters that may be determined with some accuracy, such as the height of cover h, the cohesion c_s, and the unit weight of soil γ, but also some empirical parameters, such as b, δ, and K. The definition of these empirical parameters varies from one author to another. Here, typical approaches of Terzaghi, Germany Standard ATV-A 161 E-90 [29], Chinese Standard GB 50332-2002 [33], UK Standard BS EN 1594-09 [34], US Standard ASTM F 1962-11 [35], UK PJA [27], Japan JMTA [36], and Japan JSA [22] would be discussed and compared.

For the calculation of silo width b, three kinds of boundary planes of wedge failure assumed by different authors and the corresponding equations are clearly illustrated in Figure 5. The width of the boundary plane is related to the 'vault' effect of soil. Generally, a smaller b means a lower 'vault' effect of soil, leading to a larger vertical soil stress.

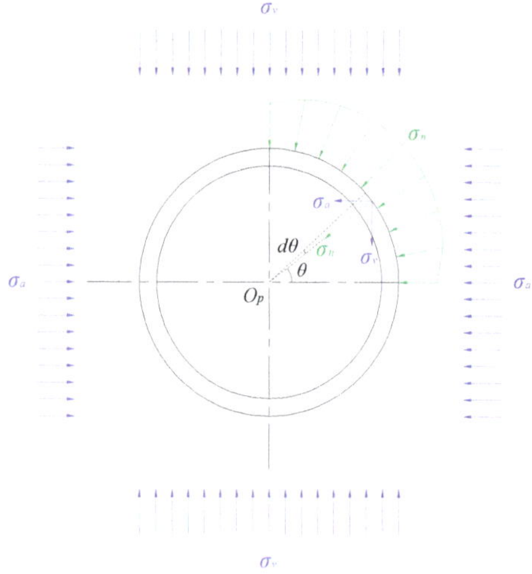

Figure 4. The earth pressure and the normal stress acting on the pipe.

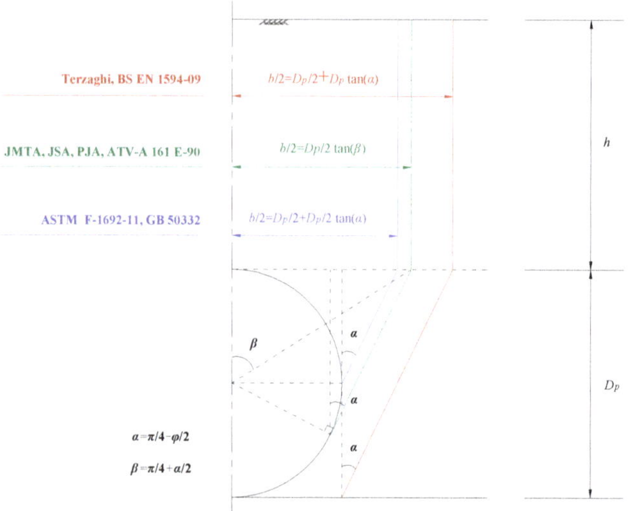

Figure 5. Boundary planes of wedge failure assumed by different authors. JSA: Japan Sewage Association; PJA: UK Pipe Jacking Association.

For the determination of δ, most of the guidelines, such as PJA, JSA, JMTA, BS EN 1594-90, and GB 50332-2002 assume shear planes as perfectly rough and take an angle of friction in the shear planes δ equal to the soil internal friction angle φ. However, ATV-A 161 E-90 and ASTM F 1962-11 make a more cautious assumption and only takes into account half the internal friction angle $\varphi/2$.

For the lateral pressure coefficient K above the tunnel, Terzaghi assumes K coefficient is equal to 1, which corresponds to the range of values encountered in clayey soils. PJA, ASTM F 1962-11, and GB 50332-2002 suggest $K = K_a$ (calculated by Rankine's formula of active soil pressure coefficient), while BS EN 1594 and ATV-A 161 assume $K = K_0$ (calculated by Rankine's formula of soil pressure coefficient at rest). Moreover, according to ATV-A 161, this K coefficient is equal to 0.5, which corresponds to an internal soil friction angle of 30°, a typical value for sandy soils.

Parameters of b, δ, and K chosen by the different authors have been summarized in Table 1.

Table 1. Definition of the empirical parameters in Terzaghi's silo model by different authors [37].

Parameters	b	δ	K	c
Terzaghi (Japan)	$D_p(1 + 2\tan \alpha)$	φ	1	c
JMTA (Japan)	$(D_p + 0.08)\tan \beta$	φ	1	c
JSA (Japan)	$(D_p + 0.1)\tan \beta$	φ	1	c
PJA (UK)	$D_p\tan \beta$	φ	$K_a = \tan^2 \alpha$	c
BS EN 1594 (UK)	$D_p(1 + 2\tan \alpha)$	φ	$K_0 = 1 - \sin \varphi$	c
AVT A-161 (Germany)	$1.732D_p$	$\varphi/2$	$K_0 = 0.5$	None
ASTM F 1962-11 (US)	$1.5D_p$	$\varphi/2$	$K_a = \tan^2 \alpha$	None
GB 50332-2002 (China)	$D_p(1 + \tan \alpha)$	φ	$K_a = \tan^2 \alpha$	None

Note: $\alpha = \pi/4 - \varphi/2$; $\beta = \pi/4 + \alpha/2$.

It is noted that none of the approaches use the same parameters. Consequently, the vertical soil stress calculated by these approaches would be quite different. Thus, it is not convincing to pick out an approach to use without checking the field data. This work will be carried out in the next section.

Thus far, all the equations needed to calculate friction resistance have been determined. If the parameters needed for the prediction equations are quantified, by using Equations (18) and (22) the total normal force N can be determined, then together with Equations (23), (24), (27), (31), and (37), the contact angle 2ε, the effective friction coefficient μ, and the friction force f now can be uniquely identified. The flow chart is shown in Figure 6.

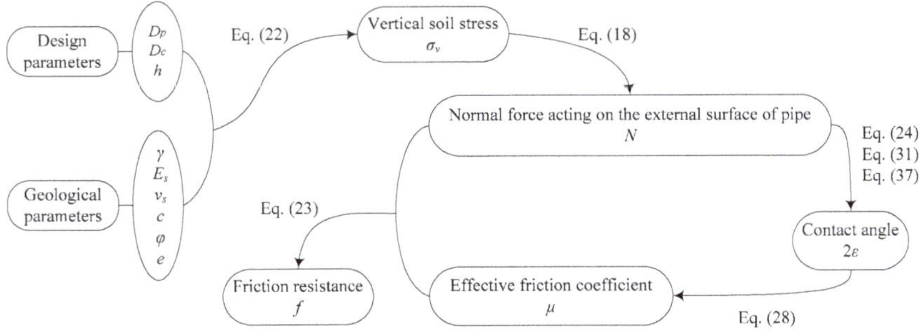

Figure 6. Flow chart of friction resistance prediction.

Apparently, the effective friction coefficient here is not just related to the interfriction angle of soil φ but also the state of pipe-soil contact and the effect of lubrication.

4. The Verification of the Effectiveness of the Proposed New Approach

4.1. Comparison between the Predicted Friction Resistances and the Field Data

Before comparison analysis, the existing prediction equations were numbered from M 1 to M 13 (Table 2). For M 7~M 13, the Equations (7), (14), (18), and (22) should be used and the parameters (b, δ, and K) of Terzaghi's silo model (Equation (22)) in the calculation of earth pressure should be determined by their approaches that are shown in Table 1.

Table 2. The numbering of the existing friction resistance prediction equations.

Number of Models	Equations	Approaches
M 1	Equations (6) and (7)	
M 2	Equations (7)–(10)	
M 3	Equations (7), (11) and (12)	
M 4	Equations (7), (15), (18) and (19)	CTTA (China)
M 5	Equations (7), (16), (21) and (22)	JSA 1 (Japan)
M 6	Equations (7), (17), (21) and (22)	JSA 2 (Japan)
M 7	Equations (7), (14), (18) and (22)	JMTA (Japan)
M 8	Equations (7), (14), (18) and (22)	Terzaghi (Japan)
M 9	Equations (7), (14), (18) and (22)	PJA (UK)
M 10	Equations (7), (14), (18) and (22)	BS EN 1594-09 (UK)
M 11	Equations (7), (14), (18) and (22)	ATV A 161 E-90 (Germany)
M 12	Equations (7), (14), (18) and (22)	ASTM F 1962-11 (US)
M 13	Equations (7), (14), (18) and (22)	GB 50332-2002 (China)

As has been mentioned before, the new approach of this paper introduces an effective friction coefficient μ (Equation (28)) to replace the original pipe-soil friction coefficient μ_s (Equation (7)). The calculation of μ (Equation (28)) should follow the flow chart shown in Figure 2. Equations (18) and (22) are also used to calculate the normal force N and vertical earth pressure σ_v, respectively. The new prediction equations were numbered from M 7 * to M 13 * (Table 3).

Table 3. The numbering of the new friction resistance prediction equations.

Number of Models	Equations	Approaches
M 7 *	Equations (14), (18), (22) and (28)	JMTA (Japan)
M 8 *	Equations (14), (18), (22) and (28)	Terzaghi (Japan)
M 9 *	Equations (14), (18), (22) and (28)	PJA (UK)
M 10 *	Equations (14), (18), (22) and (28)	BS EN 1594-09 (UK)
M 11 *	Equations (14), (18), (22) and (28)	ATV A 161 E-90 (Germany)
M 12 *	Equations (14), (18), (22) and (28)	ASTM F 1962-11 (US)
M 13 *	Equations (14), (18), (22) and (28)	GB 50332-2002 (China)

The superscript "*" is used to distinguish from the number of models in Table 2.

Some parameters required by the predicted equations (such as μ_s, P_m, and W) might not be given in the literature. In principle, during the calculation, the parameters given in the field case should be used, and the missing parameters can be evaluated by the following rules:

1. The pipe-soil friction coefficient μ_s is uniformly calculated by $\mu_s = \tan(\varphi/2)$.
2. The slurry pressure P_m is usually set as equal to the vertical soil stress σ_v plus 30 kPa [28], so $P_m = \sigma_v + 30$ kPa is adopted.
3. The weight of pipe W is evaluated by $W = \gamma_p S$, where $\gamma_p = 24$ kN/m^3 is for reinforced concrete [38], S is the area of pipe cross section.
4. When the overcut ΔR (usually between 0~50 mm) is missed, $\Delta R = 10$ mm is taken.
5. The cohesion of pipe-slurry c_m is usually less than 0.1 kPa [39], $c_m = 0.1$ kPa is used.

6. The average values of soil parameters (such as γ_s, c_s, φ, and e) can be obtained from the Geological Engineering Handbook.

Parameters in each of the four cases were finally determined, which are given in Table 4 [2,9,22]. These cases encountered some representative soils, such as silt, clay, sand, and gravels. Furthermore, they have different overburden depths of 2.72~8.5 m, overcut of 0~20 mm, and pipe diameters of 0.66~4.06 m. All of these characteristics provide good foundations for identifying the capability of the prediction models.

Table 4. Parameters that are needed to calculate the prediction equations in each case.

Cases		C 1 (H City)	C 2 (Shanghai)	C 3 (F City)	C 4 (Neuilly)
Geotechnical Description		**Organic Silt**	**Silty Clay**	**Fine Sand**	**Sand and Gravels**
	h (m)	8.15	8.5	2.72	5
	D_p (m)	0.96	4.06	1.2	0.66
	ΔR (mm)	10	20	10	0
	γ (kN/m3)	18	19.5	20.5	21
	c_s (kPa)	10	30	0	0
	φ (°)	10	25	30	40
Parameters	E_s (MPa)	12	25	30	35
	v_s	0.35	0.3	0.25	0.2
	e	2	1.5	1	0.4
	c_m (kPa)	0.1	0.1	0.1	0.1
	β	0.6	0.5	0.85	0.5
	W (kN/m)	5.5	90.2	8.5	2.2
	μ_m	0.01	0.01	0.01	0.01
	P_m (kPa)	89.97	110.54	62.28	45.94

In Table 5, for each of the drives, measured frictional force values are presented and compared to values calculated by the approaches of the existing models. One can see that, for most cases, the prediction results of the models (M 1 and M 2) established based on Hypothesis 1 are generally too small. This is because it is correct only when the overcut is stable and the pipeline slides on its base inside the annular gap remaining open. The same problem is encountered in M3 (established based on Hypothesis 2), which is probably more due to ignoring the occurrence of pipe-soil contact decreasing the magnitude of the effective friction coefficient. Obviously, for slurry pipe jacking, the prediction friction based on these two assumptions may be insufficient and unsafe.

Table 5. Comparison of friction resistances calculated by the existing models and the measured data.

Cases		C 1		C 2		C 3		C 4	
			Ratio		Ratio		Ratio		Ratio
$f_{mea.}$ **(kN/m)**		1.5		36.16		3.78		9.75	
	M 1	0.48	32%	20.00	55%	2.28	60%	0.80	8%
	M 2	2.68	179%	76.10	210%	2.28	60%	0.80	8%
	M 3	3.02	201%	15.37	43%	2.72	72%	1.16	12%
	M 4	42.47	2832%	439.47	1215%	50.09	1325%	62.22	638%
	M 5	39.72	2648%	575.39	1591%	32.65	864%	12.06	124%
	M 6	38.36	2557%	489.02	1352%	20.50	542%	6.43	66%
$f_{cal.}$ **(kN/m)**	M 7	11.14	743%	218.77	605%	38.35	1014%	13.96	143%
	M 8	14.28	952%	267.04	738%	41.52	1098%	15.32	157%
	M 9	9.42	628%	198.74	550%	37.56	994%	28.88	296%
	M 10	16.68	1112%	247.43	684%	40.31	1066%	26.36	270%
	M 11	30.06	2004%	392.61	1086%	45.37	1200%	37.86	388%
	M 12	30.24	2016%	370.94	1026%	41.90	1108%	42.29	434%
	M 13	24.99	1666%	332.53	920%	36.72	972%	27.63	283%

Note: $f_{mea.}$ is the measured friction resistance; $f_{cal.}$ is the calculated friction resistance.

The calculated results of M4 are much larger than the measured data, which presumably result from ignoring the 'vault effect' of soil, so that the soil pressure has been overestimated. Apart from M4, the other models established based on Hypothesis 3 (M5~13) have shown some applicability in case 4, in which the overcut of C4 is equal to zero, which is exactly in line with the assumption of the full contact of pipe-soil of Hypothesis 3. Except for C4, the prediction results of other cases are much larger, and the amplitude may be up to 30 times the measured data. Therefore, for slurry pipe jacking, Hypothesis 3 is generally over-conservative. Although it can ensure the structural safety of the design, it may also cause a much higher construction cost.

From what we have analyzed above, the existing models have good prediction results only when the field case is consistent with the basic hypothesis of each model. However, it is also these hypotheses that determine their limited applicability.

Table 6 presents the friction resistances calculated by the approach of this paper, by using the parameters of Terzaghi silo model chosen from different authors (Table 1). Better agreements with the field data in each of these cases were obtained, which indicate that compared with the existing prediction models, the approach of this paper that considers the effect of lubrication and interaction of pipe-soil-slurry not only has higher accuracy but is also more flexible and has wider applicability.

Table 6. Comparison of the friction resistances calculated by the new approach and the measured data.

Cases		C 1		C 2		C 3		C 4	
			Ratio		Ratio		Ratio		Ratio
$f_{mea.}$ (kN/m)		1.5		36.16		3.78		9.75	
$f_{cal.}$ (kN/m)	M 7 *	2.44	163%	43.37	120%	3.55	94%	5.19	53%
	M 8 *	3.46	231%	58.53	162%	4.02	106%	5.69	58%
	M 9 *	1.93	129%	41.19	114%	3.43	91%	10.74	110%
	M 10 *	4.30	287%	56.99	158%	3.84	101%	9.80	101%
	M 11 *	2.09	140%	48.22	133%	4.62	122%	14.08	144%
	M 12 *	1.26	84%	36.51	101%	4.08	108%	15.72	161%
	M 13 *	1.89	126%	35.02	97%	3.31	88%	10.27	105%

The superscript "*" is used to distinguish from the number of models in Table 2.

Moreover, the approaches of PJA (M9), ATV A 161 E-90 (M11), ASTM F 1962-11 (M12), and GB 50332-2002 (M13) provide higher prediction accuracy than the other ones (Table 6). However, it is noted that ATV A 161 E-90, ASTM F 1962-11, and GB 50332-2002 suggest the effect of soil cohesion c should be neglected (Table 1), but it is because the cohesion of soil has been considered in the calculation that the good prediction results can be obtained, as shown in Table 6. From this point of view, UK PJA (consider the cohesion of soil) provides the best choice of parameters of Terzaghi' silo model in the calculation of earth pressure. Therefore, the UK PJA approach is supposed to be used in the next analyses.

4.2. Comparison between the Predicted Friction Resistances and the Numerical Simulation Result

Some excellent work of numerical simulation has been conducted for the estimation of the jacking force (or friction resistance) of pipe jacking. For example, Ji et al. [5,40] and Barla and Camusso [41] presented a novel discrete 2D numerical model to evaluate the normal force acting on the pipe. Then, the friction resistance is determined by multiplying the interface friction coefficient by the normal force. Three different drives in a pipe jacking projects were analyzed, where the calculated pattern of jacking force was compared with the measured data, which demonstrated the effectiveness of the proposed approach. Yen and Shou [25] used a model coupling finite element method and a displacement control method to estimate the required jacking force in pipe jacking. The displacement control option in the numerical analysis software ABAQUS (Abaqus Inc., (Palo Alto, CA, USA) 2012) was used to designate the displacement at the end cross section of the pipe in the launch shaft. Accounting for the contact property and the contact range between the pipes and the soil during the jacking process, the stresses exerted on the pipes were used to back-calculate the jacking forces [25].

The numerical simulation strategy of Yen and Shou [25] is quite consistent with the analytical approach of this paper. Therefore, the numerical simulation approach of Yen and Shou will be described in the next section and its estimated result of friction resistance will be discussed and compared with that calculated by the analytical approach of this paper.

One of the analyses focused on a case of slurry pipe jacking in the Taichung Science Park, the basic parameters of this case were summarized in Table 7. In the numerical simulation, the lateral boundaries were fixed by roller, and hinges were used to constrain the bottom boundary. The three dimensional hexahedron element (C3D8R) with eight nodes was used in the simulation. To verify the jacking force obtained from the displacement control simulation, overcut and lubrication were included in the model by setting the contact range (1/3, 1/2, and 1 pipe-soil contact) and the frictional coefficient (Figure 7a–c). The results suggested the 1/3 contact case can estimate the jacking force with a better accuracy towards the middle and the final stage of the pipe jacking process (Figure 8).

Table 7. The details of the slurry pipe jacking in Taichung Science Park.

Geology	h (m)	D_p (m)	ΔR (mm)	γ (kN/m3)	c (kPa)	φ (°)	E_s (MPa)	v_s	e
Gravel	13.0	2.85	10	21	15	37	33	0.20	1

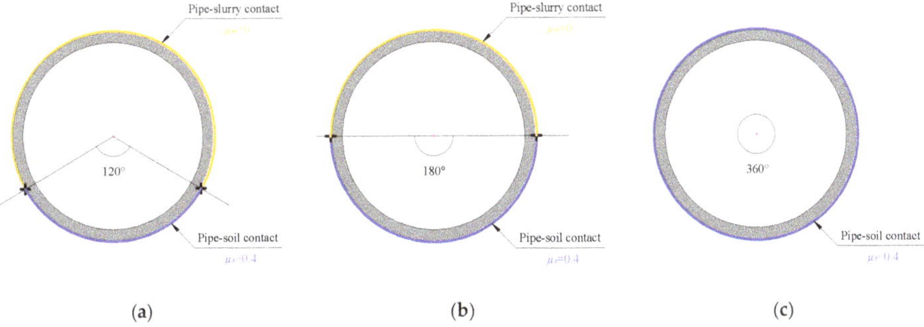

(a)　　　　　　　　　(b)　　　　　　　　　(c)

Figure 7. The setting of pipe-soil contact range and frictional coefficient in the numerical simulation. (**a**) 1/3 of the pipe-soil contact; (**b**) 1/2 of the pipe-soil contact; (**c**) 1 of the pipe-soil contact.

In Figure 8, the slope of the linear regression for the scattered points represents frictional resistance, and the intercept represents face resistance. For the calculated result of this paper, the measured face resistance is used, and the frictional resistance ($f_{cal.}$ = 57.7 kN/m) is calculated by the new approach (Section 3) using parameters of UK PJA in Terzaghi's silo model (Table 1). For the result of numerical simulation, the decrease of jacking force in the initial stage of drives (before 62.5 m) is explained by authors that the weight of the pipe jacking machine (the material of which is steel, featuring a larger unit weight) directly pressing on the soil, causing a larger face resistance. After the drives of 62.5 m, the influence exerted by the machine weight decreased gradually and the increase of jacking force is caused by the accumulation of friction resistance [25]. In other words, the slope of regression from drives of 62.5 to 100 m represents the friction resistance estimated by numerical simulation ($f_{num.}$ = 64.4 kN/m).

It is obvious that both the predicted friction resistances of the analytical equations ($f_{cal.}$ = 57.7 kN/m) and the numerical simulation ($f_{num.}$ = 64.4 kN/m) are acceptable as compared to the measured data ($f_{mea.}$ = 41.5 kN/m). However, better accuracy (especially towards the first 80 m of drives) is obtained by the approach of this paper.

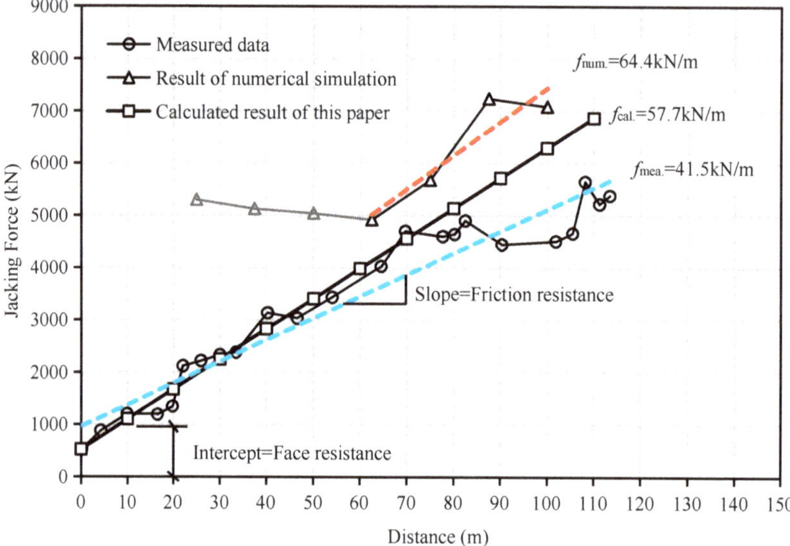

Figure 8. The comparison of jacking forces from monitoring, the numerical simulation, and the predicted equations of this paper.

The predicted accuracy of friction resistance by the numerical simulation approach is highly affected by the setting of pipe-soil contact angle. The pipe-soil contact angle of 71 degrees is that calculated by the analytical equations for this field case. Thus, it seems that the pipe-soil contact angle of 120 degrees (1/3 contact) in the simulation is set too large, resulting in the calculated result slightly larger than the measured data. From this point of view, the numerical simulation approach can accurately predict the friction resistance, but the contact angle of pipe-soil in the simulation needs to be set reasonably to obtain good prediction results. Conversely, in the approach of this paper, the pipe-soil contact angle is theoretically calculated with respect to soil property, overcut, pipe diameter, etc., while human factors or empirical factors can be eliminated as far as possible.

5. Influence of Design Factors on Lubrication Efficiency and Friction Resistance

For a better design of slurry pipe jacking in the future, it is meaningful to study the influence of design parameters (such as buried depth h, pipe diameter D_p, and the overcut ΔR) on lubrication efficiency and friction resistance. To achieve this objective, a set of reference parameters is used, and then, by changing a target parameter according to the design rules, the effect of that parameter on friction resistance and lubrication efficiency can be obtained. The designed cases were shown in Table 8.

Table 8. The reference parameters and cases designed to study the influence of factors on friction resistance.

Reference Parameters	$h = 10$ m, $D_p = 2$ m, $\Delta R = 20$ mm, $\gamma = 20$ kN/m3, $c_s = 15$ kPa, $\varphi = 25°$, $E_s = 25$ Mpa, $v_s = 0.25$, e = 1, $\mu_m = 0.01$	
Designed cases	h (m)	5→8→10→15→20→25→30
	D_p (m)	1→1.5→2→2.5→3→3.5→4
	ΔR (m)	0→5→10→20→30→40→50

5.1. Influence of Design Factors on Friction Resistance

The influences of design factors (h, D_p, and ΔR) on the critical quantities of effective friction coefficient μ, normal force acting on the pipe N, and friction resistance f are respectively shown in Figure 9a–c, Figure 9d–f, Figure 9g–i.

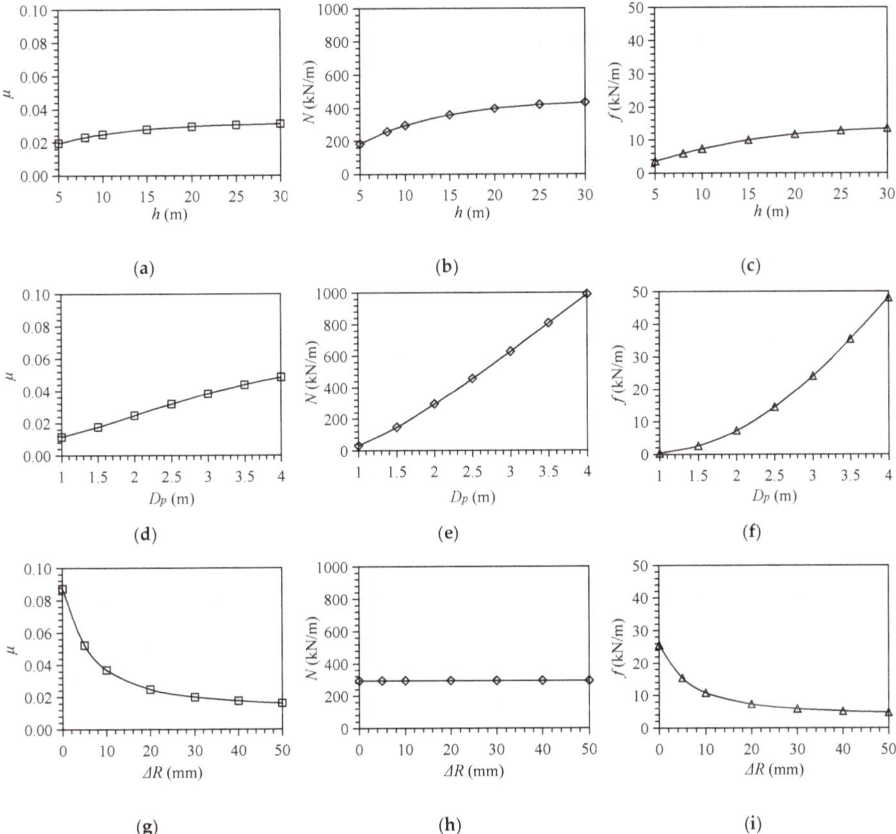

Figure 9. The influence of factors of h, D_p, and ΔR on the quantities of μ (**a–c**), N (**d–f**), and f (**g–i**).

It is evident that increasing buried depth and pipe diameter led to a double action for the increasing of friction resistance. Firstly, this increase then increased the possibility of contact between the pipe and soil, and, therefore, increase the effective friction coefficient μ on the interface. Secondly, the increasing of buried depth increases the vertical soil stress, while the increasing of pipe diameter increases the contact area, both effects of them increase normal force N acting on the pipes. The main difference between the two is that buried depth causes both of the effective friction coefficient and normal force to slightly increase and then gradually stabilize, while they approximately linearly increase with pipe diameter. Especially for the normal force induced by pipe diameter, which is strongly increased from 29.83 to 991.34 kN, this leads to a notable increase of friction resistance from 0.34 to 48.16 kN/m. Thus, the additional friction is strongly affected by the pipe diameter but appears not to be greatly affected by the buried depth.

Different from the buried depth and pipe diameter, the overcut with small values has no effect on the normal force (Figure 9h). However, it has a strongly negative effect on the effective friction coefficient on the interface (Figure 9g). In fact, it does determine the volume of injected lubricant slurry, which has a significant influence on the occurrence of the pipe-soil contact, and, therefore, determine the magnitude of the effective friction coefficient. Thus, it has a strongly negative effect on the friction resistance.

5.2. Influence of Design Factors on Lubrication Efficiency

Except for friction resistance, engineers are also concerned about the lubrication efficiency [9,42,43]. According to Equation (27), the magnitude of μ is between the pipe-slurry friction coefficient μ_m and the pipe-soil friction coefficient μ_s. If there is no contact between the pipe and the soil, the angular space due to overcut is completely filled with lubricant slurry, the effective friction coefficient is equal to μ_m; and if the soil is in full contact with the external surface of the pipe, the effective friction coefficient mainly depends on the pipe-soil nature and approximately equals to μ_s. Thus, the lubrication efficiency can be defined as

$$\chi = \left(1 - \frac{\mu - \mu_m}{\mu_s}\right) \times 100\% \tag{40}$$

By substituting Equation (28) in Equation (31), and considering that μ_m is far smaller than μ_s, the χ can approximately be expressed by Equation (41).

$$\chi = \left(1 - \frac{\varepsilon}{\pi}\right) \times 100\% \tag{41}$$

If pipe-soil contact angle $\varepsilon = 0$, $\chi = 100\%$ for maximum lubrication efficiency, and if $\varepsilon = \pi$, $\chi = 0\%$ for minimum lubrication efficiency. It is noted that $\chi = 0\%$ is not going to happen. According to the Passon model (Equation (37)), the result calculated by the left terms of Equation (37) should be not less than zero, while the right term of Equation (37) is a monotonically decreasing function of pipe-soil contact angle. In other words, when the right term of Equation (37) is equal to zero, the pipe-soil contact angle reaches its maximum value, which is solved by $\varepsilon = \varepsilon_{max} = 72° = 0.4\pi$, corresponding to the minimum lubrication efficiency of 60% (i.e., χ is theoretically between 60% and 100% on the basis of Passon contact model). Although it is not theoretically correct as compared to that counted by Pellet and Kastner as between 45% and 90% [9] and tested by Zhou as between 47.8% and 78.6% [5], it seems practical to estimate the efficiency of lubrication by the approach of this paper.

The pipe-soil contact angle and the corresponding lubrication efficiency calculated by Equation (41) has been shown in Figure 10a–c, Figure 10d–f, respectively. It is found that, as compared to the low effect of overburden depth, special attention should be paid to the effect of pipe diameter and the overcut.

Figure 10f shows that the lubrication efficiency strongly increases from 64% to 91%, while the overcut increases only from 0 to 15 mm, and after that, the effect of overcut is significantly reduced. This observation confirms the importance of overcut, which has to be sufficiently wide so that the decrease of tunnel diameter induced by the elastic ground unloading does not lead to the closure of the annular space. Moreover, Figure 10e shows that the increasing of pipe diameter from 1 to 4 m causes obvious efficiency losses of lubrication from 99% to 82%. Thus, one can conclude that the larger the pipe diameter, the lager the overcut is needed.

The buried depth h is determined by the intended line and level of the tunnel, which is often limited by geological conditions and distributions of the existing buildings and structures, while the pipe diameter D_p highly depends on the practical use or traffic requirements. It seems that there are not many options for the buried depth h and pipe diameter D_p in the design. From this point of view, both for lubrication efficiency and friction reaction, more attention should be paid to the design of overcut.

Parts of the conclusions analytically discussed above have been verified by authors from field observations [9,20], which in turn again confirm the feasibility of the approach of this paper.

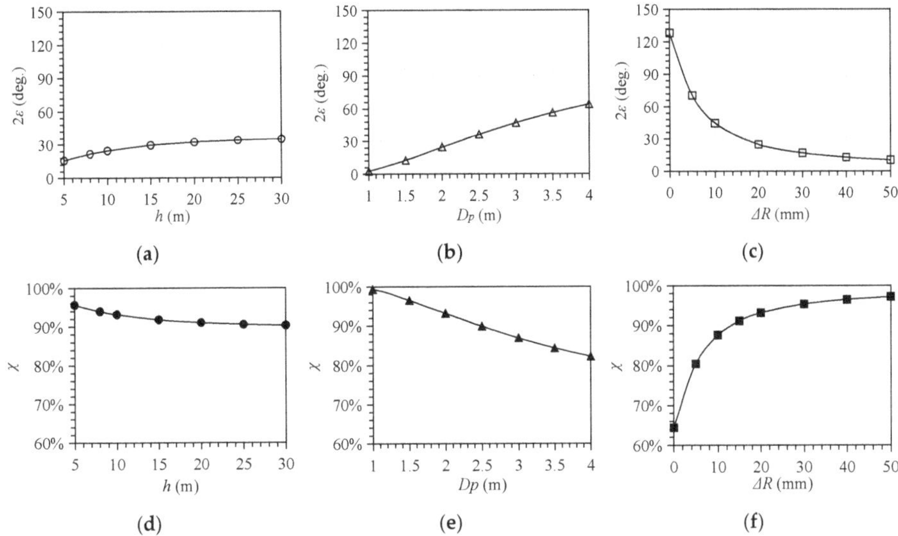

Figure 10. The influences of factors of h, D_p, and ΔR on the pipe-soil contact angle 2ε (**a–c**) and lubrication efficiency χ (**d–f**).

6. Conclusions

Some typical prediction models of friction resistance have been presented and detailed comparisons and analyses have also been made. Then, a new approach considering both the effect of lubrication and the interaction of pipe-soil-slurry, by introducing an effective friction coefficient, has been established. Values of friction resistance calculated using them have been compared with values measured in four field cases and a numerical simulation case with various soils and design parameters. Better agreements are obtained, which indicate a more flexible and wider applicability of the approach in this paper as compared to the existing prediction models and numerical simulation approach. Explanations have also been sought for limited use of the existing models that may be attributed to their hypotheses not that suitable for slurry pipe jacking. The numerical simulation approach can accurately predict the friction resistance, but it is hard to determine the contact angle of pipe-soil reasonably with respect to various soils, overcut, and other conditions to obtain good prediction results.

Using the approach of this paper, for higher prediction accuracy, the cohesion of soil has to be taken into account in the calculation for drives in clayey soils. The Terzaghi's silo model together with parameters determined by the approach of UK PJA is verified as the most well-considered to calculate earth pressure.

For better design in the future, the influences of design factors (buried depth, pipe diameter, and overcut) on friction resistance and lubrication efficiency have been analyzed too. The increase of pipe diameter has a strong influence on the increase of friction resistance; however, the friction amplitude appears not to be greatly affected by the buried depth. As the selection of pipe diameter and buried depth are limited by various objective conditions, special attention should be paid on the design of overcut. The overcut has to be sufficiently wide. When the overcut is small (for example smaller than 15 mm), the decrease of overcut strongly affects the decrease of lubrication efficiency, and, therefore, leads to a notable increase in friction resistance. Moreover, pipe diameter has an obvious influence on the effect of overcut on lubrication efficiency, the larger the pipe diameter the larger the overcut needed.

Author Contributions: L.P. and C.S. put forward the methodology; Y.Z. and Y.L. collected cases' data; Y.Y. derived the formulas and wrote the paper; W.Y. reviewed and edited the paper. All authors have read and agreed to the published version of the manuscript.

Funding: The authors acknowledge the financial support of the National Natural Science Foundation of China (No. 51878670).

Conflicts of Interest: The authors declare no conflict of interest.

Abbreviations

f	friction force per meter length drive
μ	effective friction coefficient for slurry pipe jacking
μ_s	soil-pipe friction coefficient
μ_m	slurry-pipe friction coefficient
δ	pipe-soil friction angle
N	normal force due to soil pressure acting on pipe
P	effective external load applied at the center of the pipes
σ_n	normal soil stress acting on any point of pipes
σ_v	vertical soil stress
σ_a	horizontal soil stress
h	overburden depth
D_c	internal diameter of cavity
D_p	external diameter of pipe
R_c	internal radius of cavity
R_p	external radius of pipe
ΔR	radial clearance (or overcut)
B_s	width of contact between the pipe and soil
B_m	width of contact between the pipe and mud slurry
b	influencing width of soil above the pipe
ε	semi-angle of contact area
c	soil cohesion
φ	inner friction angle of soil
γ	unit weight of soil
e	void ratio of soil
K	lateral coefficient of soil pressure above the pipe
E_p	elasticity modulus of pipe
E_s	elasticity modulus of soil
v_p	Poisson's ratio of pipe
v_s	Poisson's ratio of soil

References

1. Ji, X.; Zhao, W.; Jia, P.; Qiao, L.; Barla, M.; Ni, P.; Wang, L. Pipe Jacking in Sandy Soil under a River in Shenyang, China. *Indian Geotech. J.* **2017**, *47*, 246–260. [CrossRef]
2. Wang, J.F.; Wang, K.; Zhang, T.; Wang, S. Key aspects of a DN4000 steel pipe jacking project in China: A case study of a water pipeline in the Shanghai Huangpu River. *Tunn. Undergr. Space Technol.* **2018**, *72*, 323–332. [CrossRef]
3. Khazaei, S.; Shimada, H.; Kawai, T.; Yotsumoto, J.; Matsui, K. Monitoring of over cutting area and lubrication distribution in a large slurry pipe jacking operation. *Geotech. Geol. Eng.* **2006**, *24*, 735–755. [CrossRef]
4. Ji, X.; Ni, P.; Barla, M.; Zhao, W.; Mei, G.X. Earth pressure on shield excavation face for pipe jacking considering arching effect. *Tunn. Undergr. Space Technol.* **2018**, *72*, 17–27. [CrossRef]
5. Ji, X.; Zhao, W.; Ni, P.; Barla, M.; Han, J.; Jia, P.; Chen, Y.; Zhang, C. A method to estimate the jacking force for pipe jacking in sandy soils. *Tunn. Undergr. Space Technol.* **2019**, *90*, 119–130. [CrossRef]
6. Ong, D.E.L.; Choo, C.S. Assessment of non-linear rock strength parameters for the estimation of pipe-jacking forces. Part 1. Direct shear testing and backanalysis. *Eng. Geol.* **2019**, *244*, 159–172. [CrossRef]

7.   Ren, D.J.; Xu, Y.S.; Shen, J.S.; Zhou, A.; Arulrajah, A. Prediction of ground deformation during pipe-jacking considering multiple factors. *Appl. Sci.* **2018**, *8*, 1051. [CrossRef]

8.   Zhang, Y.; Yan, Z.G.; Zhu, H.H. A full-scale experimental study on the performance of jacking prestressed concrete cylinder pipe with misalignment angle. In *Proceedings of Geo. Shanghai 2018 International Conference: Multi-physics Processes in Soil Mechanics and Advances in Geotechnical Testing*; Springer: Singapore, 2018; pp. 345–354.

9.   Pellet-Beaucour, A.L.; Kastner, R. Experimental and analytical study of friction forces during microtunneling operations. *Tunn. Undergr. Space Technol.* **2002**, *17*, 83–97. [CrossRef]

10. Milligan, G.W.E.; Norris, P. Pipe-soil interaction during pipe jacking. *Proc. Inst. Civ. Eng. Geotech. Eng.* **1999**, *137*, 27–44. [CrossRef]

11. Shou, K.J.; Yen, J.; Liu, M. On the frictional property of lubricants and its impact on jacking force and soil-pipe interaction of pipe-jacking. *Tunn. Undergr. Space Technol.* **2010**, *25*, 469–477. [CrossRef]

12. Yang, X.; Liu, Y.; Yang, C. Research on the slurry for long-distance large-diameter pipe jacking in expansive soil. *Adv. Civ. Eng.* **2018**, *2018*, 9040471. [CrossRef]

13. Sterling, R.L. Developments and research directions in pipe jacking and microtunneling. *Undergr. Space* **2018**, in press. [CrossRef]

14. Milligan, G.W.E.; Norris, P. Site-based research in pipe jacking-objectives, procedures and a case history. *Tunn. Undergr. Space Technol.* **1996**, *11*, 3–24. [CrossRef]

15. Cui, Q.L.; Xu, Y.S.; Shen, S.L.; Yin, Z.-Y.; Horpibulsuk, S. Field performance of concrete pipes during jacking in cemented sandy silt. *Tunn. Undergr. Space Technol.* **2015**, *49*, 336–344. [CrossRef]

16. Cheng, W.C.; Ni, J.C.; Arulrajah, A.; Huang, H.W. A simple approach for characterising tunnel bore conditions based upon pipe-jacking data. *Tunn. Undergr. Space Technol.* **2018**, *71*, 494–504. [CrossRef]

17. Cheng, W.C.; Ni, J.C.; Shen, S.L.; Huang, H.W. Investigation into factors affecting jacking force: A case study. *Proc. Inst. Civ. Eng. Geotech. Eng.* **2017**, *170*, 322–334. [CrossRef]

18. Zhang, P.; Behbahani, S.S.; Ma, B.; Iseley, T.; Tan, L. A jacking force study of curved steel pipe roof in Gongbei tunnel: Calculation review and monitoring data analysis. *Tunn. Undergr. Space Technol.* **2018**, *72*, 305–322. [CrossRef]

19. O'Dwyer, K.G.; McCabe, B.A.; Sheil, B.B. Interpretation of pipe-jacking and lubrication records for drives in silty soil. *Undergr. Space* **2019**, in press.

20. Chapman, D.N.; Ichioka, Y. Prediction of jacking forces for microtunnelling operations. *Trenchless Technol. Res.* **1999**, *14*, 31–41. [CrossRef]

21. Sofianos, A.I.; Loukas, P.; Chantzakos, C. Pipe jacking a sewer under Athens. *Tunn. Undergr. Space Technol.* **2004**, *19*, 193–203. [CrossRef]

22. Shimada, H.; Khazaei, S.; Matsui, K. Small diameter tunnel excavation method using slurry pipe-jacking. *Geotech. Geol. Eng.* **2004**, *22*, 161–186. [CrossRef]

23. Barla, M.; Camusso, M.; Aiassa, S. Analysis of jacking forces during microtunnelling in limestone. *Tunn. Undergr. Space Technol.* **2006**, *21*, 668–683. [CrossRef]

24. Chapman, D.N.; Rogers, C.D.F.; Burd, H.J. Research needs for new construction using trenchless technologies. *Tunn. Undergr. Space Technol.* **2007**, *22*, 491–502. [CrossRef]

25. Yen, J.; Shou, K. Numerical simulation for the estimation the jacking force of pipe jacking. *Tunn. Undergr. Space Technol.* **2015**, *49*, 218–229. [CrossRef]

26. Zhang, H.; Zhang, P.; Zhou, W.; Dong, S.; Ma, B. A new model to predict soil pressure acting on deep burial jacked pipes. *Tunn. Undergr. Space Technol.* **2016**, *60*, 183–196. [CrossRef]

27. PJA. *Guide to Best Practice for the Installation of Pipe Jacks and Microtunnels*; Pipe Jacking Association: London, UK, 1995.

28. CTTA. *Specifications for Construction and Acceptance of Pipe Jacking*; China Trenchless Technology Association (CTTA): Wuhan, China, 2006; pp. 38–39.

29. ATV-A 161 E-90. *Structural Calculation of Driven Pipes*; German ATV Rules and Standards: Hennef, Germany, 1990; pp. 18–20.

30. Guo, W.; Xie, H.; Wu, R.; Zhou, B. Experimental study on bentonite lubrication during pipe jacking construction. *J. Henan Sci. Technol.* **2015**, *555*, 115–118. (In Chinese)

31. Michele, C.; Paolo, D. The state of stress induced by the plane frictionless cylindrical contact. I. The case of elastic similarity. *Int. J. Solids Struct.* **2001**, *38*, 4507–4523.

32. Michele, C.; Paolo, D. The state of stress induced by the plane frictionless cylindrical contact. II. The general case (elastic dissimilarity). *Int. J. Solids Struct.* **2001**, *38*, 4523–4533.

33. GB 50332-2002. *Structural Design Code for Pipeline of Water Supply and Waste Water Engineering*; The Ministry of Construction of China & General Administration of Quality Supervision, Inspection and Quarantine of the People's Republic of China: Beijing, China, 2002; pp. 23–24.

34. BS EN:1594-09. *Gas Supply System-Pipelines for Maximum Operating Pressure over 16 Bar-Functional Requirements*; British Standards Institution: Brussels, UK, 2009; pp. 76–78.

35. ASTM F 1962-11. *Standard Guide for Use of Maxi-Horizontal Directional Drilling for Placement of Polyethylene Pipe or Conduit under Obstacles Including River Crossings*; American Society for Testing and Materials: West Conshohocken, PA, USA, 2011.

36. JMTA. *Mictotunnelling Methods Serious II, Design, Construction Management and Rudiments*; Japan Microtunnelling Association (JMTA): Tokyo, Japan, 2013; pp. 69–72.

37. Cheng, W.C.; Wang, L.; Xue, Z.F.; Ni, J.C.; Rahman, M.M.; Arulraja, A. Lubrication performance of pipejacking in soft alluvial deposits. *Tunn. Undergr. Space Technol.* **2019**, *91*, 102991. [CrossRef]

38. ASCE/CI 27-17. *Standard Practice for Direct Design of Precast Concrete Pipe for Jacking in Trenchless Construction*; American Society of Civil Engineering: Reston, VA, USA, 2001; p. 22.

39. GB-50268-2008. *Code for Construction and Acceptance of Water and Sewerage Pipeline Words*; The Ministry of Construction of China & General Administration of Quality Supervision, Inspection and Quarantine of the People's Republic of China: Beijing, China, 2008.

40. Ji, X. Estimation of Jacking Force during Jacking Pipes in Shenyang Sandy Stratum. Ph.D. Thesis, Northeastern University, Shenyang, China, 2017.

41. Barla, M.; Camusso, M. A method to design microtunnelling installations in randomly cemented Torino alluvial soil. *Tunn. Undergr. Space Technol.* **2013**, *33*, 73–81. [CrossRef]

42. Khazaei, S.; Wu, W.; Shimada, H.; Matsui, K. Effect of lubrication strength on efficiency of slurry pipe jacking. In *Underground Construction and Ground Movement, Proceedings of the Geo. Shanghai International Conference 2006, Shanghai, China, 6–8 June 2006*; American Society of Civil Engineers: Reston, VA, USA, 2006.

43. Zhou, S.; Wang, Y.; Huang, X. Experimental study on the effect of injecting slurry inside a jacking pipe tunnel in silt stratum. *Tunn. Undergr. Space Technol.* **2009**, *24*, 466–471. [CrossRef]

Article

Fragility Curves for RC Structure under Blast Load Considering the Influence of Seismic Demand

Flavio Stochino *, Alessandro Attoli and Giovanna Concu

Department of Civil Environmental Engineering and Architecture, University of Cagliari, 09123 Cagliari, Italy; ale-attoli@tiscali.it (A.A.); gconcu@unica.it (G.C.)
* Correspondence: fstochino@unica.it; Tel.: +39-070-675-5115

Received: 10 December 2019; Accepted: 4 January 2020; Published: 8 January 2020

Featured Application: The fragility curves can be useful for the early design of strategic RC buildings under blast load.

Abstract: The complex characteristics of explosion load as well as its increasingly high frequency in the civil environment highlight the need to develop models representing the behavior of structures under blast load. This work presents a probabilistic study of the performance of framed reinforced concrete buildings designed according to the current Italian NTC18 and European EC8 technical standards. First, a simplified single degree of freedom model representing the structural system under blast load has been developed. Then, a probabilistic approach based on Monte Carlo simulation analysis highlighted the influence of seismic demand on the behavior of Reinforced Concrete RC buildings subjected to blast load.

Keywords: concrete; blast load; Monte Carlo analysis; seismic demand; pushover

---

1. Introduction

In the last years, structural safety under blast load has become a dramatic problem. Extreme events, such as impacts, explosions, etc., can occur in everyday life with unexpectedly high frequency [1,2]. In fact, the problem of terrorist attacks, important for strategic and military building design [2,3], can be put side by side with civil building explosion accidents [4,5].

Recently, many studies were aimed at assessing the performance of new and advanced materials under blast load: glass [6], fiber reinforced polymer [7,8], layered composite materials [9], and foam [10]. On the other hand, the structural design itself is evolving to a more general framework in which structural elements are designed and assembled to obtain general properties like robustness. The latter is the ability of a structure to withstand extreme loads without being damaged to an extent disproportionate to the cause. When an extreme load is concerned, structural damages are common and robustness is of paramount relevance. See [11] for a current state of the art review and [12] for a detailed analysis of the problem and of the available quantitative indexes.

RC structures designed and built in seismic zones should be robust in order to withstand the extreme earthquake load and many studies on this topic have been developed in the last years: [13–17]. Fewer studies deal with the interaction between earthquake and blast load. Abdollahzadeh and Faghihmaleki [18] evaluated the robustness of a seismic designed RC structure under blast load with deterministic, probabilistic and risk-based methods but did not investigated the influence of seismic demand. The latter risk-based approach has been developed in [19] considering a multi-hazard analysis for seismic and blast critical events.

The uncertainties due to blast load imply the need of a probabilistic approach in order to have an accurate estimation of the structural behavior and integrity [20,21]. Performance based fragility estimates have been adopted to assess the reliability of structures under impact in [22] showing how it is possible to formulate a Bayesian physical model for these kinds of problems.

This paper reports on a probabilistic analysis of the effect of a seismic demand on the structural fragility in case of blast load. Starting from the capacity curves of framed structures designed for different seismic loads in the Italian territory, an equivalent single degree of freedom model is developed in order to perform a probabilistic analysis based on Monte Carlo approach. Fragility curves and performance analysis are obtained with a general methodology that can be extended to many other structures. After this introduction, in Section 2 the blast load model is presented, while Section 3 describes the selected structure. Section 4 depicts the structural model while Section 5 presents the probabilistic framework. Results are in Section 6, while some concluding remarks and prospective developments are stated in Section 7.

2. Load Model

In this work, the case of hemispheric explosion load was considered. The stand-off pressure P_{so} in MPa was estimated using the Mills' approach [23]:

$$P_{sO} = 1.772\left(\frac{1}{Z^3}\right) - 0.114\left(\frac{1}{Z^2}\right) + 0.108\left(\frac{1}{Z}\right) \tag{1}$$

where z is the scaled distance representing the ratio between the distance from the explosive charge to the building and the cubic root of the explosive charge, it is expressed by:

$$z = \frac{R}{W^{\frac{1}{3}}} \tag{2}$$

where R is the stand-off distance and W is the mass of explosive in kg of equivalent TNT [24,25]. The incident impulse is represented by Held's [24] equation:

$$I_{sO} = B\,\frac{W^{2/3}}{R} \tag{3}$$

where B is a numerical coefficient that has been considered equal to 4.5×10^5 for $R > 10$ m and 3.5×10^5 for $R \le 10$ m following the indications reported in [24,25]. Instead, the reflected pressure peak can be expressed as [21]:

$$Pr = 2 \cdot P_{sO}\left(\frac{7P_{atm} + 4P_{sO}}{7P_{atm} + P_{sO}}\right) \tag{4}$$

where $P_{atm} = 0.1$ MPa. The positive phase duration t_d can be expressed assuming a triangular impulse:

$$t_d = \frac{2\,I_{sO}}{P_{sO}} \tag{5}$$

The blast load time history is usually expressed with an exponential function of time t as proposed by Friedlander [26] considering $\beta = 1.8$:

$$P_r(t) = P_r\left(1 - \frac{t}{t_d}\right)^{\frac{-\beta t}{t_d}} \tag{6}$$

In this work, in order to reduce the computational cost, the nonlinear Equation (6) can be simplified with an equivalent triangular time-history, as shown in Figure 1:

$$P_r(t) = P_r\left(1 - \frac{t}{t_d}\right) \tag{7}$$

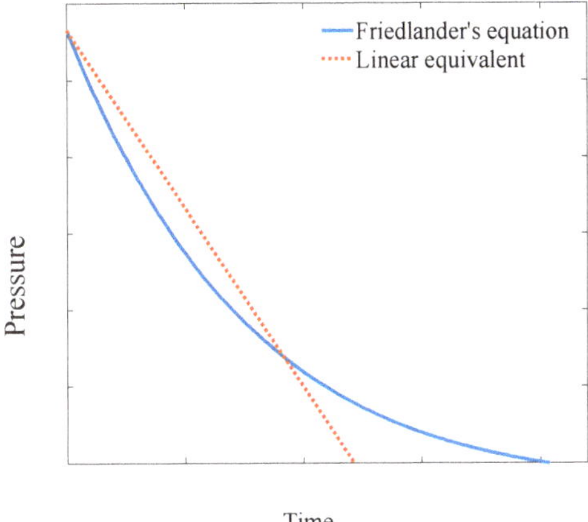

Figure 1. Blast load time-histories: exponential and triangular.

In this linear case, the positive phase duration is obtained by equating the area underneath the two curves in order to have an equivalent impulse for the two models.

3. Case Study

A framed RC structure with squared cross section has been considered as a case study, see Figure 2a for the geometrical sizes. Beam and column characteristics are detailed in Figure 2b. This kind of structure can serve as watchtower in a military environment.

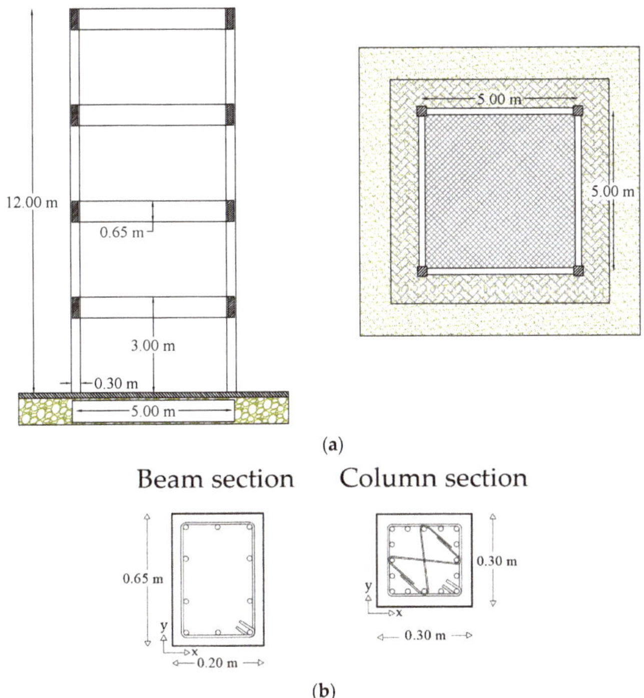

Figure 2. Structure lateral view and cross section (**a**), beam and column cross section (**b**).

In order to study the influence of the seismic demand in the structural design, the same structure has been designed considering four different locations in Italy: L'Aquila, Catania, Bari and Cagliari characterized by different seismic load, from the highest to the lower, see Table 1 and Figure 3. In this way, the reinforcements distributions will be different for each construction site. Clearly, the city with the highest seismic demand is L'Aquila and the highest reinforcement ratio is obtained when the structure is located in this city. At the same time, the lowest reinforcement ratio is obtained for Cagliari, that is the location with the lowest seismic demand.

Table 1. Location of the considered construction sites for the structure.

Location	Seismic Zone	Max PGA
L'Aquila	1	PGA > 0.25 g
Catania	2	0.15 g < PGA < 0.25 g
Bari	3	0.05 g < PGA < 0.15 g
Cagliari	4	PGA < 0.05 g

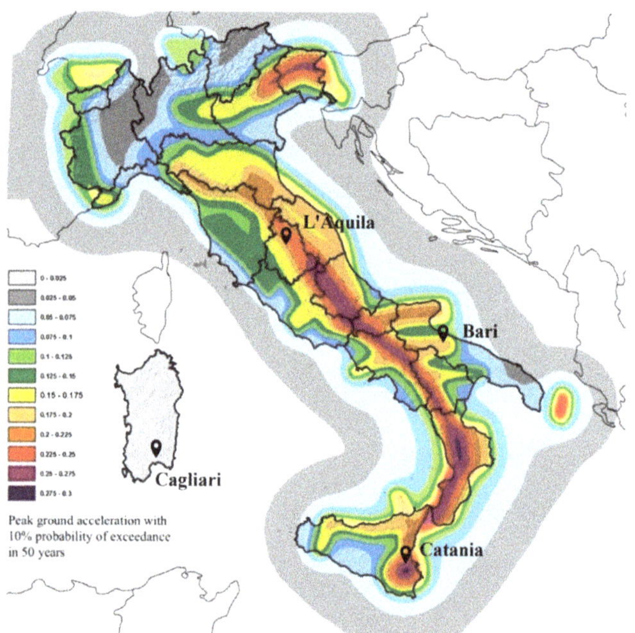

Figure 3. Italian map of seismic PGA (peak ground acceleration), taken from [27] and geographical locations of the four design sites.

For the sake of synthesis, the same structure was designed in order to fulfill four different seismic demands corresponding to different Italian locations, see Tables 1 and 2. Consequently, four different structural models are considered in the next sections in order to investigate how the seismic demand can influence also the blast resistance.

Table 2. Materials characteristics.

f_{ck} (MPa)	ε_{c3}‰	ε_{cu}‰	f_{yd} (MPa)	ε_{sy}‰
28	1.75	3.5	450	2.9

4. Structural Model

In case of blast load the structural behaviour of a mechanical system can be represented by a single degree of freedom (SDOF) model characterized by a spring denoting the stiffness and a mass expressing the inertia, see Figure 4. Indeed, in this kind of problem damping can be disregarded because the maximum displacement is obtained in the first cycle of loading, see [28]. In fact, the aim of this structural model is to evaluate the maximum displacement of the structure at collapse.

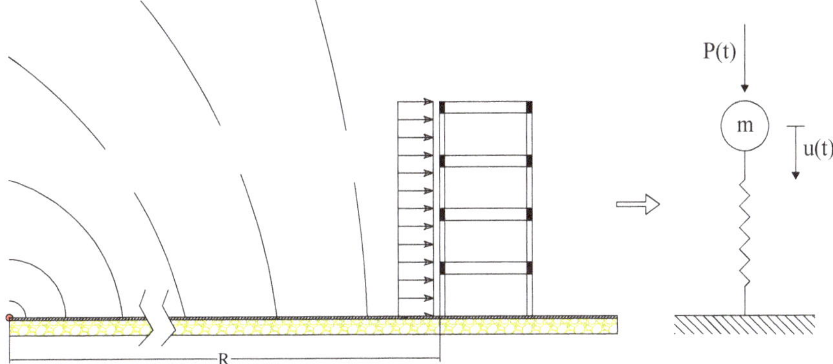

Figure 4. Hemispherical aboveground blast (**left**) and structural model (**right**).

If the mechanical non-linearities are taken into account the SDOF constitutive law can be simplified with load-displacement bilinear diagram, as shown in Figure 5.

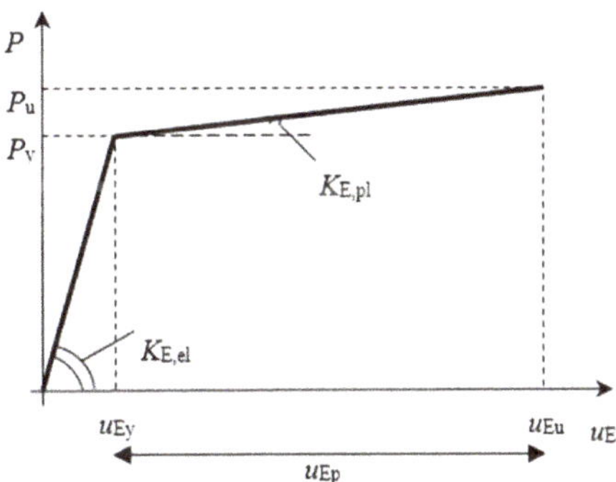

Figure 5. SDOF constitutive law, P_y and u_{Ey} respectively are the yielding load and displacement, while P_u and u_{Eu} are the corresponding ultimate ones. $K_{E,el}$ represents the elastic stiffness while $K_{E,pl}$ the plastic one.

For simple structures like beams or columns, it is possible to obtain the bilinear force-displacement diagram quite easily, just by identifying the collapse mechanism and, consequently, yielding and ultimate displacement values. In the case of a complex structure, this process becomes difficult, and in general cases it is not always possible to represent the structural behavior with an equivalent SDOF system. However, in the present case, the framed structure is quite slender and a simple modal analysis (the modal analysis was performed with the numerical model presented in Section 4.2) showed that the first eigenmode is characterized by the 85% of participant mass. For this reason, it is possible to assume that the dynamic behavior of the structure under a uniform blast load pressure can be represented by an equivalent SDOF system. Push-over analysis [29–32] can produce the force-displacement diagram known as a capacity curve. From this capacity curve it is possible to obtain an equivalent bilinear force–displacement diagram that represents the SDOF constitutive behavior [33–35] as shown in Section 4.2.

4.1. Materials and Strain Rate Effects

The materials constitutive laws and characteristics are shown in Figure 6 and Table 2.

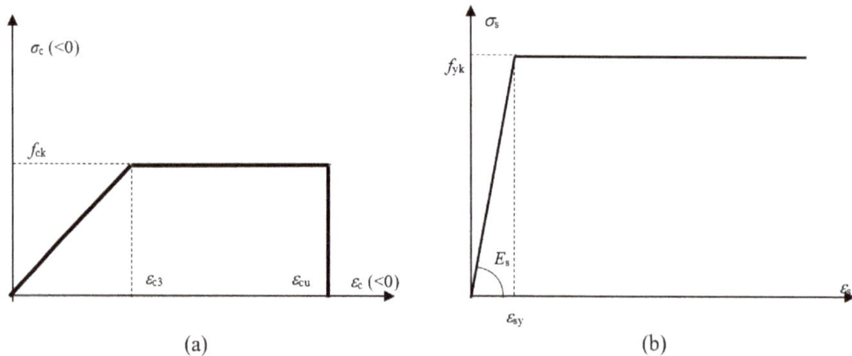

Figure 6. Materials constitutive law: (**a**) concrete, (**b**) steel.

The time dependency of the mechanical characteristics of concrete and steel on strain rate is already known. Indeed, in case of blast or impulsive load the characteristics of materials can be strongly influenced by strain rate, see [36–39]. In the literature, it is possible to find quite advanced analytical models for the strain rate effects [40], but obviously they would increase the computational cost and the complexity of the model. In order to simplify the problem and reduce the computational cost in this paper the approach proposed in [25] has been applied. A set of Dynamic Increase Factors (DIF) equal to the ratio between a dynamic mechanical characteristic f_d and the equivalent static one f has been defined as reported in Table 3.

Table 3. Dynamic increase factor (DIF) for RC elements, extracted from [25].

Type of Stress	Reinforcing Bars		Concrete
	f_{dy}/f_y	f_{du}/f_u	f_{dc}/f_c
Bending	1.17	1.05	1.19
Diagonal Tension	1.00	-	1.00
Direct Shear	1.10	1.00	1.10
Bond	1.17	1.05	1.00
Compression	1.10	-	1.12

Thus, given the critical internal force for each structural component, the appropriate DIF has been chosen from Table 3 in order to modify the mechanical characteristics of the structural model.

4.2. Capacity Curves

The structure presented in Section 3 has been designed for permanent, service and earthquake load following the Italian [41] and European Standard [42] in each location, see Table 4. Then a finite element (FE) model of the framed structure has been developed using the commercial software JASP 6.5 [43]. This FE model was used to perform a static non-linear analysis with a uniform horizontal load. The lumped plastic hinge behavior has been modelled by standard approach [42–44], as illustrated in Figure 7.

Table 4. Reinforcements details for columns in each structure, *Ac* represents the concrete area, *As* the tensile reinforcement, *A's* the compressive reinforcement, *r* is the reinforcement ratio (*r = As/Ac*) while the letter *x* or *y* denotes the reinforcements for the bending moment around x axis or y axis respectively, see Figure 2.

	Ac	*As,x*	*As,y*	*r,x*	*r,y*	*A's,x*	*A's,y*	*r',x*	*r',y*
Location	[mm^2]	[mm^2]	[mm^2]	[-]	[-]	[mm^2]	[mm^2]	[-]	[-]
L'Aquila	90,000	770	770	0.0086	0.0086	770	770	0.0086	0.0086
Catania	90,000	616	616	0.0068	0.0068	616	616	0.0068	0.0068
Bari	90,000	462	616	0.0051	0.0068	462	616	0.0051	0.0068
Cagliari	90,000	308	462	0.0034	0.0051	308	462	0.0034	0.0051

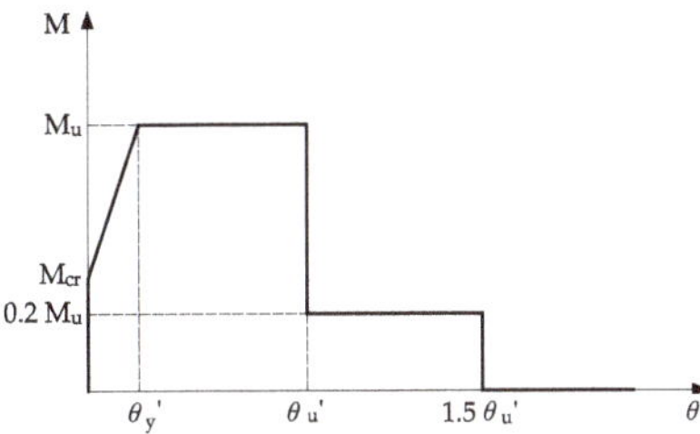

Figure 7. Moment-rotation constitutive behavior for a frame member-end considered in JASP 6.5, where M_u is the ultimate bending moment, M_{cr} is the first cracking moment, θ_y' is the yielding rotation and θ_u' is the ultimate rotation.

The results in terms of horizontal force and top horizontal displacement (capacity curves) are reported in Figure 8 for the structures designed in the selected locations. The equivalent bilinear SDOF constitutive curves have been calculated equating the area underneath bilinear and capacity curves assuming that the ending point and the first elastic slope should be the same for the two curves.

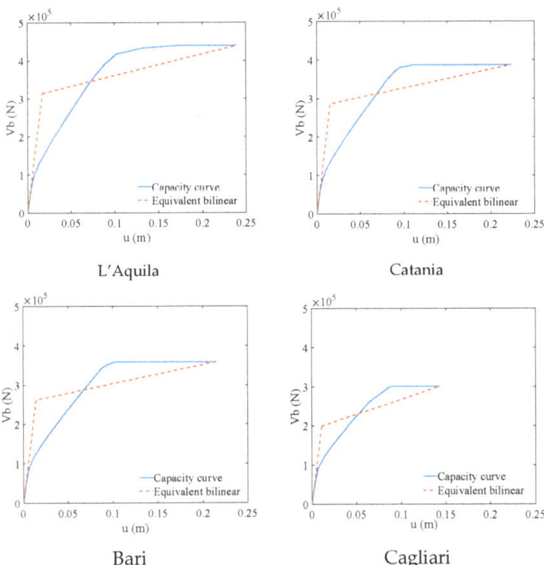

Figure 8. Capacity curves and SDOF constitutive law for each structure location. The total horizontal load Vb is plotted versus the horizontal displacement of the top floor.

4.3. Analytical Model

The equations of motion of the equivalent SDOF oscillator in the elastic and plastic regimes have the following forms:

$$M_{E,el}\frac{d^2v_E(t)}{dt^2} + K_{E,el}u_E(t) = P_E(t) \quad for\ 0 \leq u_E \leq u_{Ey} \tag{8a}$$

$$M_{E,pl}\frac{d^2v_E(t)}{dt^2} + K_{E,pl}u_E(t) + \left[K_{E,el} - K_{E,pl}\right]u_{Ey} = P_E(t) \quad for\ u_{Ey} < u_E \leq u_{Eu} \tag{8b}$$

where $u_E(t)$ is the model displacement, $P_E(t)$ is the total load on the structure. $M_{E,el}$ and $M_{E,pl}$ denote the equivalent mass of the oscillator respectively for the elastic and plastic field. In general: $M_E = K_{LM}M$ where M is the structural mass and K_{LM} is a coefficient which accounts for the boundary conditions of the structural element, the type of load and the regime considered (elastic or plastic). The structure presented in Section 3 can be considered similar to a vertical cantilever fixed in the bottom part. Thus, in case of a uniformly distributed load for a cantilever structure $M_{E,el} = 0.65 \cdot M$ and $M_{E,pl} = 0.66 \cdot M$, see [45]. v_{Ey} and v_{Eu} are the equivalent yieding and ultimate model displacement, while $K_{E,el}$ and $K_{E,pl}$ respectively are the equivalent stiffnesses in the elastic and plastic range.

Given the analytical expression of the load time history it is possible to find the close form solution of Equation (8), see [46]. In this work, $P_E(t)$ is approximated by the linear expression presented in Equation (7).

4.4. Load Scenario

In this paper, only external explosion produced by a terrorist attack has been taken into account. Stewart et al. [47] described some of the possible scenarios that can generate an external hemispheric explosion. It is interesting to distinguish theme by the ways in which a mass of explosives could be transported near the object of the attack: 5 kg body explosive; 25 kg suitcase explosive; 200 kg car explosive.

In this work the load scenario obtained with 200, 300, 400, and 500 kg of TNT has been considered. These situations can be easily obtained considering a car or a truck containing the explosives. Various stand-off distances have been investigated studying the effects of the explosives for the structure described in Section 3.

4.5. Damage Thresolds

In order to measure the structural performance under blast load the drift values proposed by [48] have been adopted, see Table 5. It is important to point out that this approach considers the whole structural response given that the stand-off distance is sufficiently large to obtain a planar blast wave acting on the building, see Figure 4. Thus, for the sake of simplicity, localized column or beam collapse has not been considered, in addition, also the harmful damages on secondary elements which can lead to a loss of life are neglected. Instead, the top floor maximum displacement u_{MAX} related to the building height $h=12$ m has been considered to define the relative drift:

$$X = \frac{u_{MAX}}{h} \tag{9}$$

Table 5. Assumed drift x_0 thresolds for performance levels, extracted from [48].

Slight Damage	Moderate Damage	Severe Damage
0.0012	0.0080	0.011

This simplified approach is clearly limited to its assumptions but can be useful in case of preliminary or early design because it can easily provide a synthetic parameter describing the damage condition of a building after the blast load.

5. Probabilistic Analysis

Fragility curves describe the conditional probability of exceedance ($P(X > x0|Z)$) of the response parameter X (drift in this case) given a demand intensity measure (scaled distance Z in this case). Thus, the structural fragility can be expressed as the cumulative distribution of the probability that a damage threshold x_0 is exceeded [21,49]:

$$P(X > x_0) = \int_{-\infty}^{+\infty} P(X > x_0|Z)\, p(Z)dz \cong \sum_{i=0}^{\infty} P(X > x_0|Z)_i\, p(Z)_i \Delta Z_i \tag{10}$$

where the discretization of the integral calculation is represented by a discrete sum of conditions in which the scaled distance is varied with a given step ΔZ.

In this paper, the structural characteristics have been considered deterministic while the uncertainties of the load have been modelled considering the explosive mass and stand-off distance as stochastic variables characterized by lognormal distributions whose characteristics are shown in Table 6.

Table 6. Probabilistic Analysis Input Data.

Symbol	Description	COV	Distribution
R	Stand-off distance	0.05	Lognormal
W	Explosive mass	0.15	Lognormal

A Monte Carlo analysis has been developed with the above described SDOF model in order to obtain fragility curves presented in Section 6.2 and the probability of thresholds exceedance shown in Section 6.3. The coefficient of variation (COV) of the maximum drift has been checked in order to define the convergence condition of the analysis.

6. Results and Discussion

6.1. Maximum Drift

In the first load scenario, a 500 kg TNT bomb was blown up at various distance from the structure. The maximum drift values have been plotted as a function of the scaled distance in Figure 9. In the same picture the above-mentioned damage thresholds have been plotted in order to easily find the safety scaled distance for each structure. As expected, the structure designed in L'Aquila with the highest seismic load produced the best performance reaching the damage thresholds with smallest scaled distance in comparison with the other structures. This can be explained given the higher reinforcement ratio (see Table 4) and robustness of this structure in comparison with the others. At the same time, it is clear that 500 kg of TNT represents a huge amount and it is necessary to reach very high scaled distance (higher than 10 $/\text{kg}^{1/3}$) to avoid any damage.

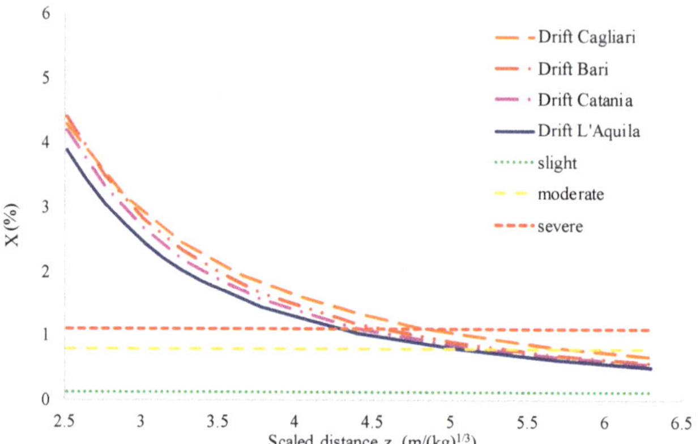

Figure 9. Maximum drift for the structure under blast load for each location, 500 kg TNT.

6.2. Fragility Curves

The fragility curves for the structures designed in the four locations are presented in Figures 10–12. In this case the TNT mass has been varied considering a Lognormal distribution with different mean values: 200, 300, 400, and 500 kg. The three damage thresholds have been considered: slight (Figure 10), moderate (Figure 11) and severe (Figure 12). Instead, the stand-off distance has been varied in a deterministic way in order to analyze all the scaled distance values.

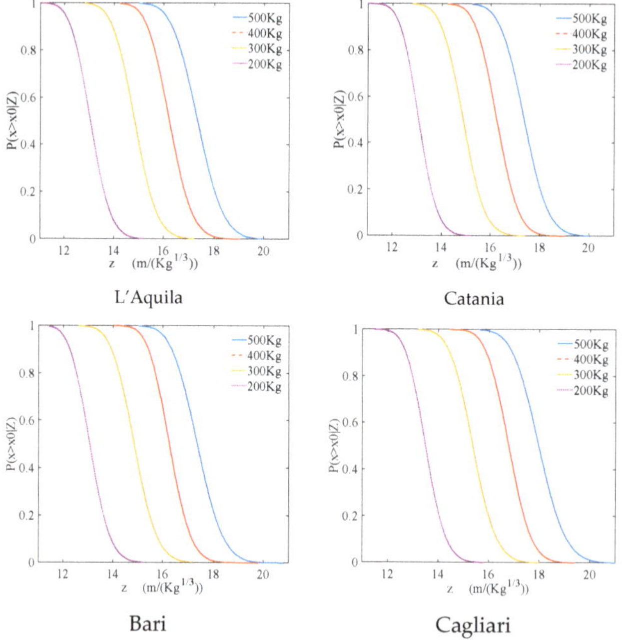

Figure 10. Fragility curves for slight damage condition for each structure location.

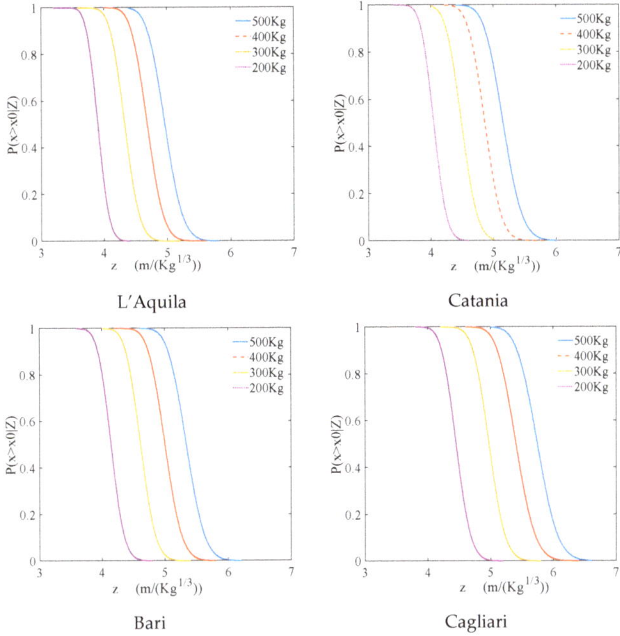

Figure 11. Fragility curves for moderate damage condition for each structure location.

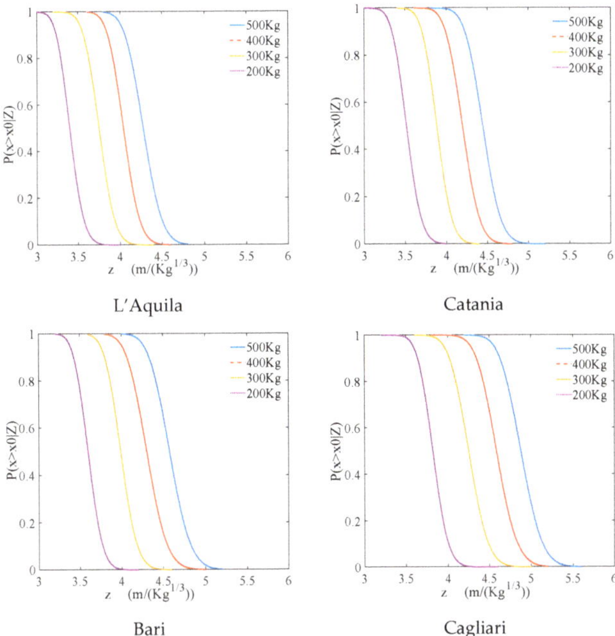

Figure 12. Fragility curves for severe damage condition for each structure location.

As a representative case, the COV of the maximum displacement for the severe damage threshold in case of 500 kg bomb for each structure location has been shown in Figure 13. Also, the other cases present similar trends.

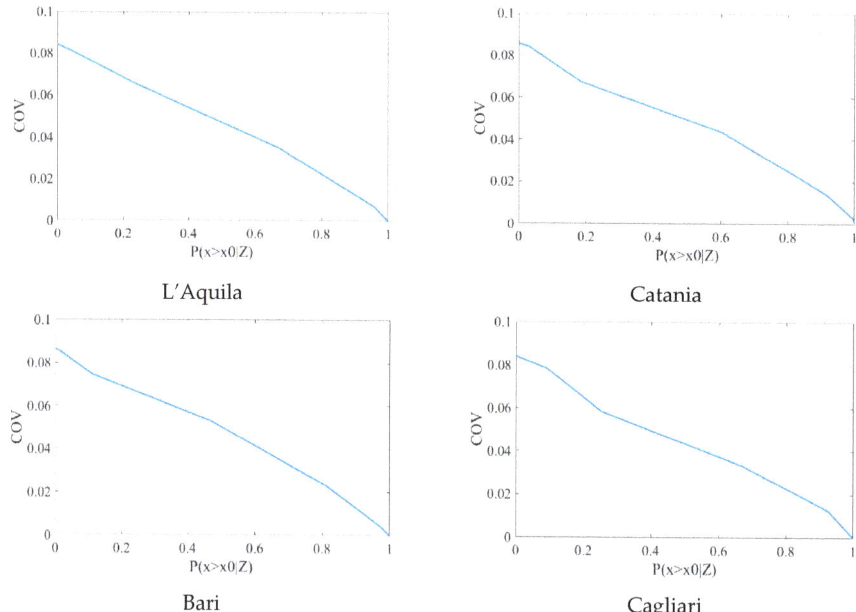

Figure 13. Maximum drift COV variation for severe damage condition in case of 500 kg bomb for each structure location.

In Figures 10–13, the curves corresponding to 500 kg are the ones placed on the far right of each figure highlighting that this is the most severe condition in each case. Instead, if the structure designed in the site with less seismic demand (Cagliari) is compared to the one with the highest (L'Aquila) it is possible to note how the curves are shifted to the left in the latter case and to the right in the former one. This is another proof of the better structural performance of the structure designed in L'Aquila.

6.3. Probability of Thresholds Exceedance

The Monte Carlo analysis has been developed considering both stand-off distance R and mass weight W as stochastic variables following a Lognormal probability distribution as described in Section 5. Tables 7–10 present the probability of exceedance for each different threshold and for each structure designed in the different location.

Also in this case, it is possible to highlight the better structural performance of the structure designed in L'Aquila in comparison with the others. In addition, it is possible to note how, as the stand-off distance increases, the probability of exceedance for each threshold is reduced. In most cases the safe stand-off distance for a 500 kg TNT bomb is about 40–45 m. Instead, in the case of 200 kg, it is 25 m for L'Aquila, 30 m for Catania and Bari, and 35 m for Cagliari.

Table 7. Percentage probability of thresholds exceedance for L'Aquila location.

	200 Kg	300 Kg	400 Kg	500 Kg
R = 20 m				
slight	100	100	100	100
moderate	93.88	100	100	100
severe	64.92	99.8	100	100
R = 25 m				
slight	100	100	100	100
moderate	10.72	98.68	100	100
severe	0.02	42.96	98.78	100
R = 30 m				
slight	100	100	100	100
moderate	0.02	41.82	96.52	100
severe	0	0.76	48.38	96.78
R = 35 m				
slight	100	100	100	100
moderate	0	0.42	37.62	94.18
severe	0	0	2.9	44.6
R = 40 m				
slight	100	100	100	100
moderate	0	0	1.88	44.42
severe	0	0	0	1.82
R = 45 m				
slight	100	100	100	100
moderate	0	0	0.06	4.74
severe	0	0	0	0.02

Table 8. Percentage probability of thresholds exceedance for Catania location.

	200 Kg	300 Kg	400 Kg	500 Kg
R = 20 m				
slight	100	100	100	100
moderate	98.54	100	100	100
severe	65.48	99.96	100	100
R = 25 m				
slight	100	100	100	100
moderate	25.68	99.74	100	100
severe	0.14	66.32	99.74	100
R = 30 m				
slight	100	100	100	100
moderate	0.14	52.68	98.58	100
severe	0	2.48	69.86	99.4
R = 35 m				
slight	100	100	100	100
moderate	0	1.48	59.22	98.5
severe	0	0	7.62	61.02
R = 40 m				
slight	100	100	100	100
moderate	0	0	6.82	66.26
severe	0	0	0.1	3.86
R = 45 m				
slight	100	100	100	100
moderate	0	0	0.18	13.7
severe	0	0	0	0.08

Table 9. Percentage probability of thresholds exceedance for Bari location.

	200 Kg	300 Kg	400 Kg	500 Kg
R = 20 m				
slight	100	100	100	100
moderate	99.7	100	100	100
severe	68.18	99.98	100	100
R = 25				
slight	100	100	100	100
moderate	42.04	99.92	100	100
severe	0.46	81.68	99.96	100
R = 30 m				
slight	100	100	100	100
moderate	0.38	62.24	99.62	100
severe	0	7.36	83.88	99.92
R = 35 m				
slight	100	100	100	100
moderate	0	4.14	76.38	99.48
severe	0	0.06	14.02	66.04
R = 40 m				
slight	100	100	100	100
moderate	0	0.02	16.48	81.3
severe	0	0	0.02	7.5
R = 45 m				
slight	100	100	100	100
moderate	0	0	0.46	23.44
severe	0	0	0	0.16

Table 10. Percentage probability of thresholds exceedance for Cagliari location.

	200 Kg	300 Kg	400 Kg	500 Kg
R = 20 m				
slight	100	100	100	100
moderate	99.96	100	100	100
severe	89.54	100	100	100
R = 25 m				
slight	100	100	100	100
moderate	78.12	100	100	100
severe	5.86	97.1	100	100
R = 30 m				
slight	100	100	100	100
moderate	2.14	91.96	100	100
severe	0	32.28	95.5	99.86
R = 35 m				
slight	100	100	100	100
moderate	0	28	96.18	100
severe	0	0.48	25.64	88.8
R = 40 m				
slight	100	100	100	100
moderate	0	1.08	52.24	95.58
severe	0	0	0.92	31.92
R = 45 m				
slight	100	100	100	100
moderate	0	0	4.32	55.14
severe	0	0	0	2.02

7. Conclusions

This paper presented a simplified procedure to evaluate the safety of a framed RC structure under blast load highlighting the seismic design influence. The considered framed building has been designed considering four different seismic demand corresponding to specific locations in Italy characterized by different PGA. Then, a simplified SDOF system has been obtained from a pushover analysis in order to perform several Monte Carlo analyses aimed at highlighting what is the performance of the considered building under blast load in both a deterministic and probabilistic framework.

The structure designed to withstand the highest seismic load (L'Aquila) has been proven to have better performance also in the case of blast load. Thus, the influence of the seismic demand in the building design is evident. Indeed, the structure designed with the lowest seismic demand (Cagliari) presents the worst structural behavior under the blast load. This can be clearly seen from the fragility curves Figures 10–12, probability Tables 7–10, and also from the deterministic maximum drift presented in Figure 9.

The obtained results can be useful for safety evaluation in the case of a terrorist attack. Indeed, the fragility curves and the probability of threshold exceedance can help the designer in evaluating what should be a "safe" distance for the given structure that can be obtained with fence system or bollards.

Further development of this work is expected, applying this method to existing structures, like those described in [50,51]. It is also interesting to merge the proposed method with other assessment approaches and retrofitting techniques considering other kind of structures [52–54].

Author Contributions: F.S. conceived the illustrated strategy and the theoretical formulation contributing also to the numerical analysis; A.A. developed the numerical and analytical analysis; G.C. analyzed the numerical results and contributed to the theoretical formulation. All the authors wrote the paper. All authors have read and agreed to the published version of the manuscript.

Funding: The financial support of the Autonomous Region of Sardinia under grant PO-FSE 2014–2020, CCI: 2014-IT05SFOP021, through the project "Retrofitting, rehabilitation and requalification of the historical cultural architectural heritage (R3-PAS)" is acknowledged by Flavio Stochino.

Conflicts of Interest: The authors declare no conflict of interest.

References

1. Hadianfard, M.A.; Malekpour, S.; Momeni, M. Reliability analysis of H-section steel columns under blast loading. *Struct. Saf.* **2018**, *75*, 45–56. [CrossRef]
2. Sławiński, G.; Malesa, P.; Świerczewski, M. Analysis Regarding the Risk of Injuries of Soldiers Inside a Vehicle during Accidents Caused by Improvised Explosive Devices. *Appl. Sci.* **2019**, *9*, 4077. [CrossRef]
3. Stewart, M.G. Reliability-based load factor design model for explosive blast loading. *Struct. Saf.* **2018**, *71*, 13–23. [CrossRef]
4. Xu, J.; Wu, C.; Xiang, H.; Su, Y.; Li, Z.X.; Fang, Q.; Hao, H.; Liu, Z.; Zhang, Y.; Li, J. Behaviour of ultra high performance fibre reinforced concrete columns subjected to blast loading. *Eng. Struct.* **2016**, *118*, 97–107. [CrossRef]
5. Faber, M.H.; Kübler, O.; Fontana, M. Modeling consequences due to failure of extraordinary structures. In *Probabilistic Safety Assessment and Management*; Springer: London, UK, 2004; pp. 488–493.
6. Zhang, X.; Bendon, C. Vulnerability and protection of glass windows under blast: Experiments, methods and current trends. *Int. J. Struct. Glass Adv. Mater. Res.* **2017**, *1*, 10–23. [CrossRef]
7. Razaqpur, A.G.; Tolba, A.; Contestabile, E. Blast loading response of reinforced concrete panels reinforced with externally bonded GFRP laminates. *Compos. Part B Eng.* **2007**, *38*, 535–546. [CrossRef]
8. Gribniak, V.; Misiūnaitė, I.; Rimkus, A.; Sokolov, A.; Šapalas, A. Deformations of FRP–Concrete Composite Beam: Experiment and Numerical Analysis. *Appl. Sci.* **2019**, *9*, 5164. [CrossRef]
9. Tekalur, S.A.; Shukla, A.; Shivakumar, K. Blast resistance of polyurea based layered composite materials. *Compos. Struct.* **2008**, *84*, 271–281. [CrossRef]
10. Ma, G.W.; Ye, Z.Q. Analysis of foam claddings for blast alleviation. *Int. J. Impact Eng.* **2007**, *34*, 60–70. [CrossRef]

11. Stochino, F.; Bedon, C.; Sagaseta, J.; Honfi, D. Robustness and Resilience of Structures under Extreme Loads. *Adv. Civ. Eng.* **2019**, *2019*, 4291703. [CrossRef]

12. Starossek, U.; Haberland, M. Approaches to measures of structural robustness. *Struct. Infrastruct. Eng.* **2011**, *7*, 625–631. [CrossRef]

13. Qu, Z.; Sakata, H.; Midorikawa, S.; Wada, A. Lessons from the behavior of a monitored 11-story building during the 2011 Tohoku earthquake for robustness against design uncertainties. *Earthq. Spectra* **2015**, *31*, 1471–1492. [CrossRef]

14. Chen, Y.L.; Huang, L.; Lu, Y.Q.; Deng, L.; Tan, H.Z. Assessment of structural robustness under different events according to vulnerability. *J. Perform. Constr. Facil.* **2016**, *30*, 04016004. [CrossRef]

15. Asprone, D.; De Risi, R.; Manfredi, G. Defining structural robustness under seismic and simultaneous actions: An application to precast RC buildings. *Bull. Earthq. Eng.* **2016**, *14*, 485–499. [CrossRef]

16. Puppio, M.; Giresini, L.; Doveri, F.; Sassu, M. Structural irregularity: The analysis of two reinforced concrete (rc) buildings. *Eng. Solid Mech.* **2019**, *7*, 13–34. [CrossRef]

17. Puppio, M.L.; Ferrini, M. Parametric analysis on external dissipative link system for seismic protection of low rise RC buildings. *Fratt. Integr. Strut.* **2019**, *13*, 706–739. [CrossRef]

18. Abdollahzadeh, G.; Faghihmaleki, H. A method to evaluate the risk-based robustness index in blast-influenced structures. *Earthq. Struct.* **2017**, *12*, 47–54. [CrossRef]

19. Abdollahzadeh, G.; Faghihmaleki, H. Seismic-explosion risk-based robustness index of structures. *Int. J. Damage Mech.* **2017**, *26*, 523–540. [CrossRef]

20. Stewart, M.G.; Netherton, M.D. Security risks and probabilistic risk assessment of glazing subject to explosive blast loading. *Reliab. Eng. Syst. Saf.* **2008**, *93*, 627–638. [CrossRef]

21. Olmati, P.; Petrini, F.; Gkoumas, K. Fragility analysis for the Performance-Based Design of cladding wall panels subjected to blast load. *Eng. Struct.* **2014**, *78*, 112–120. [CrossRef]

22. Sharma, H.; Gardoni, P.; Hurlebaus, S. Probabilistic demand model and performance-based fragility estimates for RC column subject to vehicle collision. *Eng. Struct.* **2014**, *74*, 86–95. [CrossRef]

23. Mills, C.A. The design of concrete structures to resist explosions and weapon effects. In Proceedings of the 1st International Conference for Hazard Protection, Edinburgh, UK, 27–30 September 1987.

24. Held, M. Blast waves in free air. *Propellants Explos. Pyrotech.* **1983**, *8*, 1–7. [CrossRef]

25. UFC 3-340-02. *Structures to Resist the Effects of Accidental Explosions*; Department of Defense: Virginia, VA, USA, 2008.

26. Freidlander, F.G. The diffraction of sound pulses. I. Diffraction by a semi-infinite plate. *Proc. R. Soc. Lond. A* **1946**, *186*, 322–344.

27. Italian Institute of Geophysics and Volcanology: Map of Seismic Hazard 2004. Available online: http://zonesismiche.mi.ingv.it/ (accessed on 7 January 2020).

28. Riedel, W.; Fischer, K.; Kranzer, C.; Erskine, J.; Cleave, R.; Hadden, D.; Romani, M. Modeling and validation of wall–window retrofit system under blast loading. *Eng. Struct.* **2012**, *37*, 235–245. [CrossRef]

29. Xiao, T.L.; Qiu, H.X.; Li, J.L. Seismic Behaviors of Concrete Beams Reinforced with Steel-FRP Composite Bars under Quasi-Static Loading. *Appl. Sci.* **2018**, *8*, 1913. [CrossRef]

30. Yang, W.; Bao, C.; Ma, X.; Zhang, S. Study on structural robustness of isolated structure based on seismic response. *Appl. Sci.* **2018**, *8*, 1686. [CrossRef]

31. Sassu, M.; Giresini, L.; Bonannini, E.; Puppio, M. On the use of vibro-compressed units with bio-natural aggregate. *Buildings* **2016**, *6*, 40. [CrossRef]

32. Puppio, M.; Pellegrino, M.; Giresini, L.; Sassu, M. Effect of material variability and mechanical eccentricity on the seismic vulnerability assessment of reinforced concrete buildings. *Buildings* **2017**, *7*, 66. [CrossRef]

33. Attard, T.; Fafitis, A. Modeling of higher-mode effects using an optimal multi-modal pushover analysis. *WIT Trans. Built Environ.* **2005**, *81*, 405–414.

34. Graziotti, F.; Penna, A.; Bossi, E.; Magenes, G. Evaluation of displacement demand for unreinforced masonry buildings by equivalent SDOF systems. In Proceedings of the IX International Conference on Structural Dynamics (EURODYN2014), Porto, Portugal, 30 June–2 July 2014.

35. Moghadam, A.S.; Tso, W.K. Damage assessment of eccentric multistory buildings using 3-D pushover analysis. In Proceedings of the Eleventh World Conference on Earthquake Engineering, Acapulco, Mexico, 23–28 June 1996.

36. Stochino, F. RC beams under blast load: Reliability and sensitivity analysis. *Eng. Fail. Anal.* **2016**, *66*, 544–565. [CrossRef]

37. Stochino, F.; Carta, G. SDOF models for reinforced concrete beams under impulsive loads accounting for strain rate effects. *Nucl. Eng. Des.* **2014**, *276*, 74–86. [CrossRef]

38. Bai, Y.L.; Yan, Z.W.; Ozbakkaloglu, T.; Dai, J.G.; Jia, J.F.; Jia, J.B. Dynamic Behavior of PET FRP and Its Preliminary Application in Impact Strengthening of Concrete Columns. *Appl. Sci.* **2019**, *9*, 4987. [CrossRef]

39. Xie, Z.; Duan, Z.; Guo, Y.; Li, X.; Zeng, J. Behavior of Fiber-Reinforced Polymer-Confined High-Strength Concrete under Split-Hopkinson Pressure Bar (SHPB) Impact Compression. *Appl. Sci.* **2019**, *9*, 2830. [CrossRef]

40. Comité Euro-International duBéton. *Concrete Structures under Impact and Impulsive Loading*; CEB Bulletin n. 187; Comité Euro-International duBéton: Lausanne, Switzerland, 1988.

41. NTC18. *NormeTecniche Per le Costruzioni*; D.M. 17.01.2018; Italian Ministry of Infrastructures and Transportation: Rome, Italy, 2018.

42. Code, P. *Eurocode 8: Design of Structures for Earthquake Resistance-Part 1: General Rules, Seismic Actions and Rules for Buildings*; European Committee for Standardization: Brussels, Belgium, 2005.

43. Java Structural Program. Available online: http://www.ingegnerianet.it/software-calcolo-strutturale.php (accessed on 5 December 2019).

44. ASCE. *Seismic Rehabilitation of Existing Buildings*; ASCE/SEI 41-06; ASCE: Reston, VA, USA, 2007.

45. Biggs, J.M. *Introduction to Structural Dynamics*; McGraw-Hill Book Company: New York, NY, USA, 1964.

46. Mays, G.; Smith, P.D.; Smith, P.D. *Blast Effects on Buildings: Design of Buildings to Optimize Resistance to Blast Loading*; Thomas Telford: London, UK, 1995.

47. Stewart, M.G.; Netherton, M.D.; Rosowsky, D.V. Terrorism risks and blast damage to built infrastructure. *Nat. Hazards Rev.* **2006**, *7*, 114–122. [CrossRef]

48. Akkar, S.; Sucuoğlu, H.; Yakut, A. Displacement-based fragility functions for low-and mid-rise ordinary concrete buildings. *Earthq. Spectra* **2005**, *21*, 901–927. [CrossRef]

49. Giresini, L.; Casapulla, C.; Denysiuk, R.; Matos, J.; Sassu, M. Fragility curves for free and restrained rocking masonry façades in one-sided motion. *Eng. Struct.* **2018**, *164*, 195–213. [CrossRef]

50. Mistretta, F.; Piras, M.V.; Fadda, M.L. A reliable visual inspection method for the assessment of RC structures through fuzzy logic analysis. In Proceedings of the International Symposium on Life-Cycle Civil Engineering (IALCCE2014), Tokyo, Japan, 16–19 November 2014; pp. 1154–1160.

51. Fadda, M.L.; Mistretta, F.; Piras, M.V. Vulnerability assessment of concrete bridges using different methods of visual inspection. *Civ. Comp Proc.* **2014**, *105*, 13.

52. Mistretta, F.; Sanna, G.; Stochino, F.; Vacca, G. Structure from Motion Point Clouds for Structural Monitoring. *Remote Sens.* **2019**, *11*, 1940. [CrossRef]

53. Mistretta, F.; Stochino, F.; Sassu, M. Structural and thermal retrofitting of masonry walls: An integrated cost-analysis approach for the Italian context. *Build. Environ.* **2019**, *155*, 127–136. [CrossRef]

54. Sassu, M.; Stochino, F.; Mistretta, F. Assessment method for combined structural and energy retrofitting in masonry buildings. *Buildings* **2017**, *7*, 71. [CrossRef]

Article

The Seepage and Soil Plug Formation in Suction Caissons in Sand Using Visual Tests

Liquan Xie, Shili Ma * and Tiantian Lin

College of Civil Engineering, Tongji University, Shanghai 200092, China; xie_liquan@tongji.edu.cn (L.X.); gotobest8@163.com (T.L.)
* Correspondence: ma_shili@163.com; Tel.: +86-191-2176-3067

Received: 4 December 2019; Accepted: 8 January 2020; Published: 13 January 2020

Abstract: The rapid development of offshore wind energy in China is becoming increasingly relevant for movement toward green development. This paper presents the results of visual tests of a suction caisson used as foundation for offshore wind turbines. The distribution of hydraulic gradients of sand at the mudline in the caisson was obtained to find out the relationship with the heights of soil plugs. The relationship equation was proposed and obtained by using quadratic regression, guiding project designs, and construction. It was found that there was no soil plug in the caisson when small suction was applied during the suction penetration. The relationship between the heights of the soil plugs and the hydraulic gradient of the soil was proposed and obtained by using quadratic regression to predict (roughly) the height of soil plugs in suction caissons in sand during suction penetration. The influence of settlement outside caissons on the soil plug was found to decrease as the buried depth rose.

Keywords: suction caisson; suction penetration; soil plug; hydraulic gradient; visual tests

---

1. Introduction

In the past few decades, it has become increasingly important to rapidly develop the offshore wind industry, which provides practical sources of energy with a low carbon footprint [1,2]. Foundations play an important role in guaranteeing the safe operation of offshore wind turbines [3]. The suction caisson, being installed economically and efficiently into soil deposits, has been increasingly used as a competitive foundation for offshore wind turbines in deep water [4,5]. A suction caisson is a large cylindrical structure that is typically made of steel, open at the bottom and closed at the top [6]. To set up a perfect offshore wind turbine, two aspects need to be considered for the engineering design of this foundation: suction installation and in-service performance [7]. The capacity of suction caissons as the foundation for offshore wind turbines is enhanced by means of peripheral embedded thin walls, which confine the internal soil [8]. A caisson is installed by penetrating the seabed under its own weight, and then by pumping water out of the caisson to create suction that forces the foundation into the seabed [9].

During suction penetration, the induced seepage flow through highly permeable sand into the caisson interior can create some negative effects. To investigate these effects, a number of studies have been completed, and they show that the seepage facilitates the installation process at the caisson tip and along the inner wall [10–13]. Erbrich and Tjelta [14] presented a series of finite element analyses and found that the suction forced water to migrate through the soil from outside to the inner caisson. Previous research [15,16] still has certain issues, in that the seepage around the suction was assumed to follow Darcy's law, which requires that we carry on further discussion. Experimental investigations in sand have revealed that soil plugs are likely to occur during suction-assisted installation [15,17,18]. A series of centrifuge tests on the installation of suction caissons were carried out by Tran et al. [19,20], who found that the soil plugs heave up to 20% of the caisson penetration as excessive suction, and they developed a void ratio–permeability relationship to check the sand heave against the plugs. There is no previous study on the relationship between soil plugs and the seepage velocity of sand around suction caissons.

This paper presents the results of visual tests of a suction caisson used as a foundation for offshore wind turbines. The process of suction installation of the caisson foundations with axisymmetric geometry is simplified to a plane problem. In order to study their composition, the heights of the final soil plugs in the caisson were measured. The distribution of the hydraulic gradient of sand at the mudline in the caisson was obtained to find out its relationship with the heights of the soil plugs. The relationship between the heights of soil plugs and the hydraulic gradient of soil was proposed and obtained by using quadratic regression, guiding project designs, and construction.

2. Experimental Program

The process of suction installation of the caisson foundations (shown in Figure 1a) with axisymmetric geometry is simplified to a plane problem in this paper. Due to the fact that the soil was replaced by the walls of the soil tank in tests, it is assumed that the forces of water adhesion to the walls are equal to that of the sand around the suction caissons. A soil tank (shown in Figure 1), as used in tests, has dimensions of 20×0.5 cm in its plan view, and a depth of 28 cm. Transparent fiberglass was used to make the soil tank, with a thickness of 1.0 cm. The two pumping outlets in the soil tank were connected to the suction loading system by a thin pipe and a drainpipe, respectively. Figure 1b shows the diagram of the test device, with a constant suction S in the caisson caused by the suction loading system creating the different head H. The value of the applied constant suction can be expressed as $S = \gamma_w H$, where γ_w is the unit weight of water (10 kN/m^3).

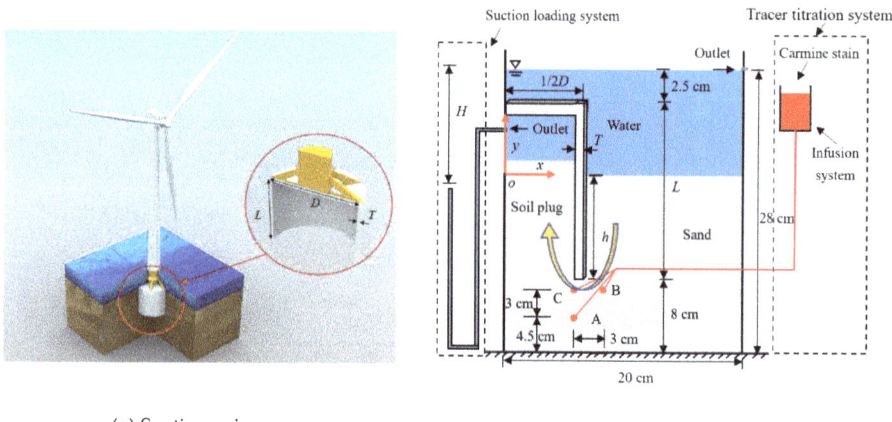

<table>
<tr><td>(**a**) Suction caisson.</td><td>(**b**) Diagram of test devices.</td></tr>
</table>

Figure 1. Test devices.

The suction caisson with an external diameter D of 12 cm had a height of 17.5 cm L and a thickness T of 0.5 cm. In order to create an enclosed compartment between the soil tank, the suction caisson, and the soil, UV glue was used to stick the soil tank and the suction caisson together. The suction caisson model was located at a distance 7.5 cm from the bottom of the soil tank.

The siliceous sand shown in Figure 2 was used in tests because it is commercially available and shows a deep contrast with the carmine stain. The void ratio e of the soil was determined according to the standard soil testing methods, and can be written as $e = \rho_w G_s / \rho_d - 1$, where ρ_d is the dry density of the sand and G_s is the specific gravity of the sand. Table 1 shows the properties of the siliceous sand in tests. The siliceous sand particles had an average radius of 0.748 mm, and they obeyed uniform distribution, with a ratio of maximum to minimum radii of 2.0. The pluviation method and the compaction method were used to prepare uniform sand specimens in layers in the soil tank. The permeability coefficient of the siliceous sand was obtained by using an empirical equation, $k = 2d_{10}^2 e^2$ [21], where d_{10} is the effective size of the sand. Due to the uniform particle size of the sand, it is suggested that d_{10} is equal to the average particle size (0.718 mm). Before carrying out every test, the sand specimen was left to stand for 24 h.

Figure 2. The siliceous sand in tests.

Table 1. Properties of siliceous sand.

Property	Specific Gravity G_s	Void Ratio e	Saturated Weight γ_{sat} (kN/m^3)	Permeability Coefficient k (cm/s)	Cohesion c (kPa)	Angle of Internal Friction φ (°)
Value	2.66	0.75	19.5	0.28	0	31

The seepage field around the suction caisson model was visualized by the tracer titration system, which consisted of an infusion system and carmine stain. The infusion system was made of a medical syringe with a thin pipe (radius of 2 mm), shown in Figure 1b. Three holes were drilled in the soil tank at points A, B, and C, connecting to the infusion system with the thin pipes. At points A, B, and C, the carmine stain was released with pinpoint accuracy by the infusion system. To reduce the effect of turbulence, the carmine stain was injected into the soil used in tests as slowly as possible.

The suction caisson model had a buried depth h (5, 10, and 15 cm) before each test, to simulate the process of suction installation. The influence of the temperatures of the environment and the water were not considered in the tests. A video camera was used to collect images of the experimental phenomena of seepage and the soil plug in the suction caisson. At the end of the trial, the height of the soil plug did not seem to change. The testing programs are listed in Table 2 in tests. The pre-test results show that seepage failure took place in the sand when the suction in the caisson with buried depths of 5, 10, and 15 cm were greater than 2.0, 3.5, and 4.0 kPa, respectively.

Table 2. List of visual tests on suction caissons.

Buried Depth h (cm)	Suction S (kPa)						
5	0.5	1.0	1.5	-	-	-	-
10	0.5	1.0	1.5	2.0	2.5	3.0	-
15	0.5	1.0	1.5	2.0	2.5	3.0	3.5

3. Test Results and Discussion

3.1. Visual Seepage Paths

The seepage field can be visualized where the carmine stain flows along streamlines in the soil around the suction caisson. Figure 3 shows the visualization of the seepage flow at point B as $h = 15$ cm and $S = 0.5$ kPa. It was observed that the visual tests achieved good results to study the seepage of sand around a suction caisson during suction penetration. It took 500 s for the carmine stain to move from point B to the mudline, and the length of the seepage path was 3.21 cm. Tests revealed that the seepage paths of soil with different S in suction caissons had the same motion path. The results indicated that the seepage path of soil is unrelated to the S applied in the caisson and is affected by the penetration depth during the installation of the foundation. The seepage paths of the carmine stain in sand are plotted with a plane coordinate system, shown in Figure 4. It can be seen that there is an obvious trend: the streamline dyed by carmine stain moves toward the wall of the caisson model. Additionally, the larger the penetration depth, the more obvious was the streamline–adherent trend. It was observed that the seepage path lengths were 1.95 and 1.34 h when the penetration depths were equal to 5 and 15 cm, respectively. The distance between the carmine stain in the mudline and the inner wall of the caisson model was 0.375 times the diameter of the suction caisson, for $h = 0.42$, 0.83, and 1.25 L, respectively.

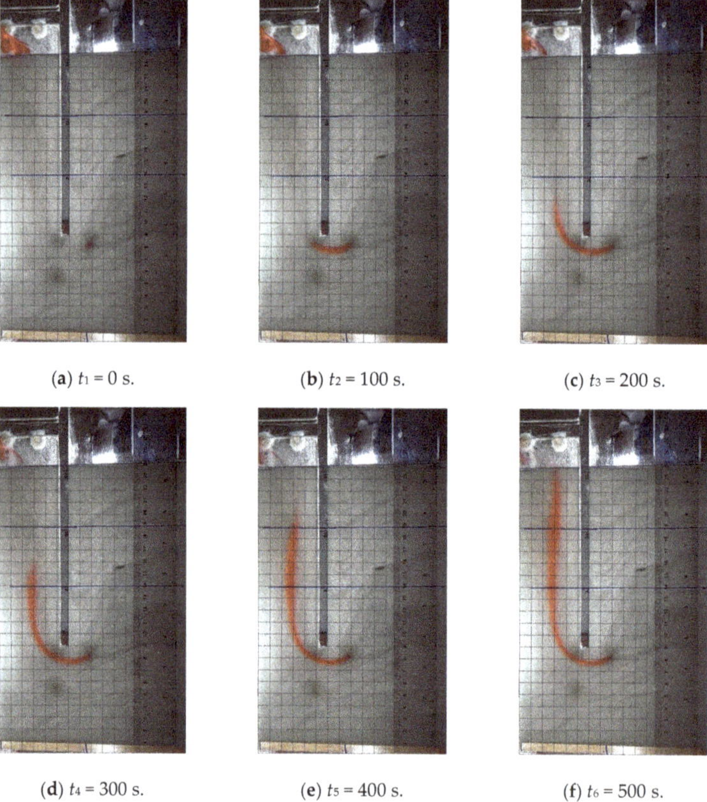

(**a**) $t_1 = 0$ s.   (**b**) $t_2 = 100$ s.   (**c**) $t_3 = 200$ s.

(**d**) $t_4 = 300$ s.   (**e**) $t_5 = 400$ s.   (**f**) $t_6 = 500$ s.

Figure 3. The visualization of seepage flow at point B with $h = 15$ cm and $S = 0.5$ kPa.

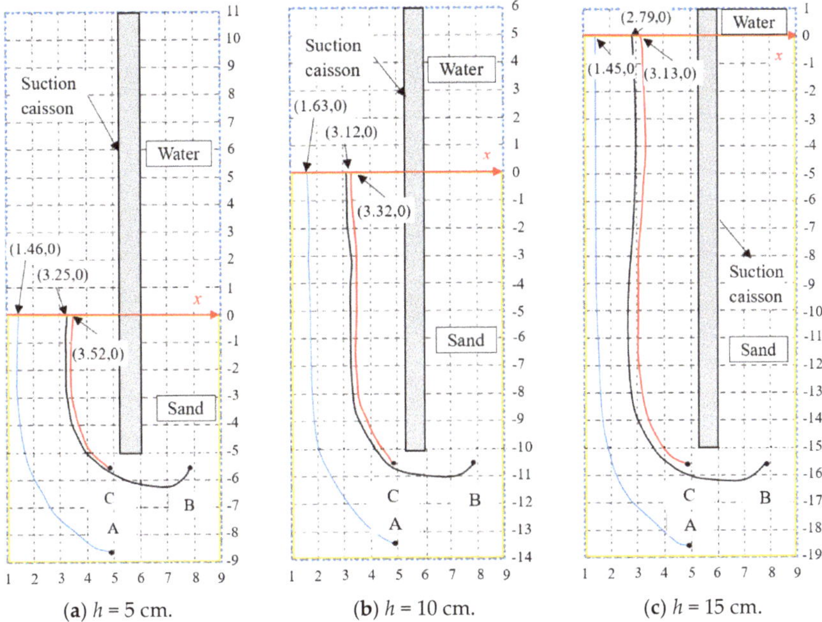

Figure 4. The seepage path of the carmine stain in sand.

3.2. Hydraulic Gradient Analysis

In this paper, the test results prove that Darcy's law is not applicable, and this phenomenon was also observed from the large test pressure difference found in the literature [22]. For the seepage velocity v, it was assumed that the streamlines around the caissons were stable and not affected by each other. The path lengths of the carmine stain were measured from test images. All results are presented in terms of dimensionless forms. The seepage velocity v of the sand was assumed to be affected by x/D and $S/\gamma'h$. The hydraulic gradient v/k of the sand in tests at the mudline was proposed by using the regression function, and can be expressed as follows:

$$\frac{v}{k} = 0.163\left[\frac{0.94DS}{(0.5D-x)\gamma'h}\right]^{0.76} \tag{1}$$

where x is the abscissa value in Figure 1 and γ' is the effective unit weight of soil.

The comparison of actual values and calculated values is shown in Figure 5. It can be seen that the fitted values agree well with the actual values. Compared with v/k obtained from tests, the fitted values have a residual sum of squares of about 1.467.

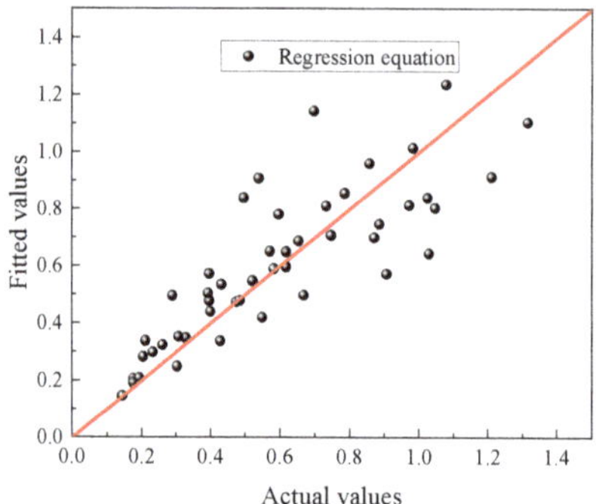

Figure 5. The comparison of actual values and fitted values for v/k.

Figure 6 shows the hydraulic gradient of sand around a suction caisson with different x/D when the suction in the caisson is equal to 1.5 kPa. It can be seen that the hydraulic gradient v/k increases with the increase of x/D. The results indicated that the seepage velocity is larger as the streamline moves toward the wall of the suction caissons. The hydraulic gradient v/k, with $x/D = 0.2675$ and $h/D = 1.25$, is shown in Figure 7. It can be observed that the v/k increases with the increase of $S/\gamma'h$ and the fitted values of the v/k agree well with the test results. The v/k when $S/\gamma'h = 2.451$ is 5.85 times greater than when $S/\gamma'h = 0.351$ in tests. The minimum relative error is 5.05% for Equation (1) when $S/\gamma'h = 5.478$. When $h/D = 0.42$, the streamline around the suction becomes disordered as a result of the larger applied suction ($S = 1.5$ kPa). Due to the assumption that the streamlines are stable and not affected by each other, the seepage velocity obtained is smaller than the actual situation, leading to large deviations as compared with Equation (1).

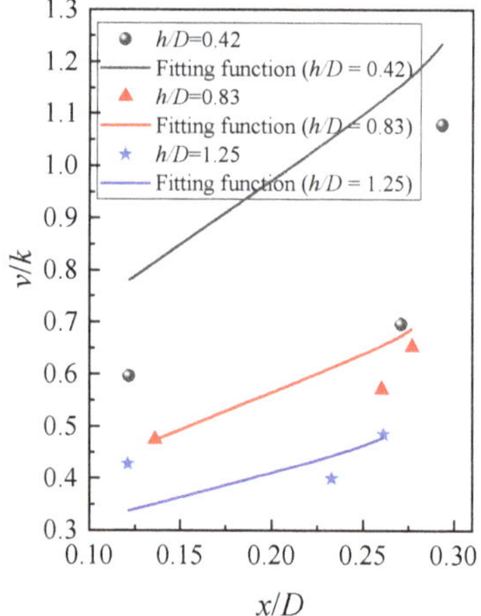

Figure 6. The hydraulic gradient with $S = 1.5$ kPa.

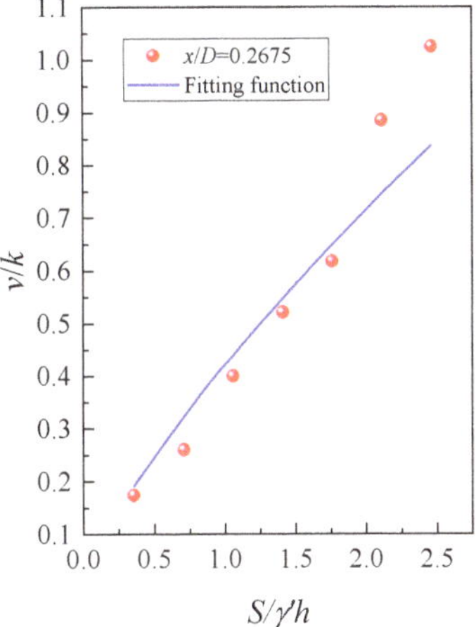

Figure 7. The hydraulic gradient with x/D = 0.2675 and h/D = 1.25.

3.3. Soil Plug and Settlement Formation

During suction penetration, a part of the soil in the caissons gets into the open-ended hollow caisson cavity, forming a soil plug [23]. The final soil plugs in the caisson at the buried depth of 5 cm (caused by the caisson) are shown in Figure 8. The test results show that there is no soil plug in the caissons with small suction applied during suction penetration. The maximum heights of soil plugs in the suction caisson models with buried depths of 5, 10, and 15 cm are 0.169, 0.085, and 0.087 times the buried depths, respectively. It was observed that the suction caused settlement of the surrounding soils outside the caisson. The maximum soil settlement appeared a certain distance from the outer wall of a suction caisson due to caisson–soil friction.

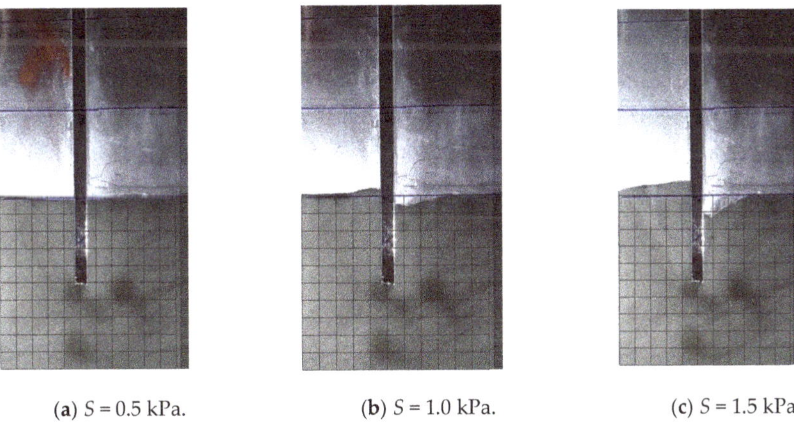

(**a**) S = 0.5 kPa.   (**b**) S = 1.0 kPa.   (**c**) S = 1.5 kPa.

Figure 8. The soil plug in the caisson.

Figure 9 shows the curves of the heights of soil plug h_{sp} and of soil settlement $h_{settlement}$ outside the caisson. Other test conditions not drawn in Figure 9 did not have the obvious soil plug and soil settlement around the suction caisson model. There is no obvious soil settlement outside the suction caisson with a buried depth of 15 cm. The height of soil plug h_{sp} increases with the increase of x as a result of the distribution of seepage velocity, which is greater as it moves toward the wall of suction caissons.

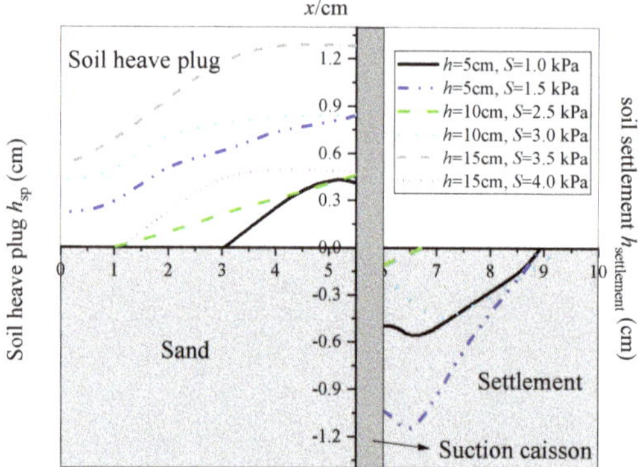

Figure 9. The height of soil plug and soil settlement outside the caisson.

The volume of soil plug V_{sp} and that of soil settlement $V_{settlement}$ were measured by using the volume formula after obtaining the size of the soil plug and settlement from the images. The comparison between the volume of soil plug V_{sp} and that of soil settlement $V_{settlement}$ in tests is shown in Table 3. The soil plug consists of the soil heaved by seepage force into the caisson, and the soil from outside into the caisson, leading to soil settlement around the foundation. It can be seen that the influence of settlement outside the caisson on the soil plug decreases as the buried depth rises. The soil plug is mainly influenced by the soil from outside into the caisson, and $V_{settlement}/V_{sp} = 1.581$ due to the decrease in the void ratio of soil around the caisson caused by the seepage force during suction penetration when $h = 5$ cm and $S = 1.0$ kPa. The influence of the soil heaved by the soil plug increases with the increase of h. The volume of soil heaved begins to rise as the velocity of seepage reaches a certain value.

Table 3. The volume of soil plug and soil settlement in tests.

	$h = 5$ cm		$h = 10$ cm		$h = 15$ cm	
	$S = 0.1$ kPa	$S = 1.5$ kPa	$S = 2.5$ kPa	$S = 3.0$ kPa	$S = 3.5$ kPa	$S = 4.0$ kPa
$V_{settlement}$ (cm^3)	0.544	0.970	0.020	0.473	0.000	0.000
V_{sp} (cm^3)	0.344	1.520	0.519	1.954	0.849	2.827
$V_{settlement}/V_{sp}$	1.581	0.638	0.039	0.242	0.000	0.000

3.4. Prediction of Soil Plug Height

The relationship between the height of the soil plug h_{sp} and the hydraulic gradient v/k was proposed by using quadratic regression, and can be expressed as follows:

$$\frac{h_{sp}}{h} = A + B\frac{S}{\gamma' h} + C\left(\frac{S}{\gamma' h}\right)^2 + D\frac{v}{k} + E\left(\frac{v}{k}\right)^2 + F\frac{Sv}{\gamma' hk}, \tag{2}$$

where A–F are constant coefficients (given in Table 4). Figure 10 shows the comparison of actual values and fitted values for the dimensionless soil plug h_{sp}/h. Fitted values are evenly distributed around the actual values. Compared with the actual values, the fitted values have an average relative error of about 28.74%.

Table 4. Optimal value of constant coefficients for h_{sp}/h.

A	B	C	D	E	F
0.57495	−0.49370	0.09759	0.05075	−0.02168	0.02252

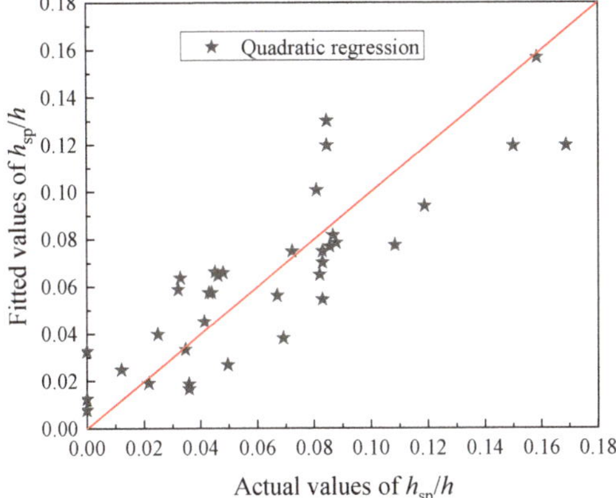

Figure 10. The comparison of actual values and fitted values for the dimensionless soil plug h_{sp}/h.

Figure 11 shows the relationship between h_{sp}/h and v/k when h = 15 cm. It is shown that the dimensionless soil plug h_{sp}/h first increases and then decreases with the increase of v/k. The reason behind the decreased trend is that the soil plug is subjected to downward friction applied by the inner wall of the suction caisson. The combination of Equations (1) and (2) is used to predict roughly the height of soil plugs in suction caissons in sand during suction penetration, guiding project designs, and construction.

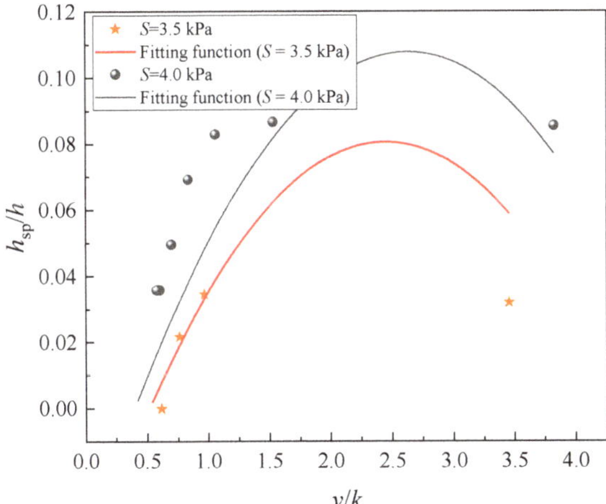

Figure 11. The relationship between h_{sp}/h and v/k when h = 15 cm.

4. Conclusions

A series of model tests were conducted in this study to investigate the visualization of suction caisson penetration in sand. The following conclusions can be drawn:

(1) The seepage field can be visualized while the carmine stain flows along a streamline in the soil around the suction caisson. The seepage velocity is larger as the streamline moves toward the wall of the suction caissons. The pre-test results show that the seepage failure took place in the sand when the suction in the caisson with buried depths of 5, 10, and 15 cm was greater than 2.0, 3.5, and 4.0 kPa, respectively. The results indicated that the seepage path of the soil is unrelated

to S applied in the caisson and is affected by the penetration depth during the installation of the foundation.

(2) The hydraulic gradient v/k of the sand in tests at the mudline was proposed by using the regression function. It can be seen that the fitted values agree well with the actual values. Darcy's law is not applicable, and the hydraulic gradient v/k increases with the increase of x/D. The results indicated that the seepage velocity is larger as the streamline moves toward the wall of the suction caissons. The v/k when $S/\gamma'h = 2.451$ is 5.85 times greater than when $S/\gamma'h = 0.351$ in tests. The minimum relative error is 5.05% for Equation (1) when $S/\gamma'h = 5.478$.

(3) There is no soil plug in the caissons with small suction applied during suction penetration. The maximum heights of the soil plugs in the suction caisson models with buried depths of 5, 10, and 15 cm are 0.169, 0.085, and 0.087 times the buried depths, respectively. The height of soil plug h_{sp} increases with the increase of x as a result of the distribution of seepage velocity, which is greater as it moves toward the wall of suction caissons. The influence of the soil heaved by the soil plug increases with the increase of h. The volume of the soil heaved begins to rise as the velocity of the seepage reaches a certain value. The dimensionless soil plug h_{sp}/h first increases and then decreases with the increase of v/k.

Author Contributions: Conceptualization, S.M. and L.X.; methodology, L.X. and S.M.; validation, T.L.; formal analysis, T.L. and S.M.; investigation, T.L. and S.M.; resources, data curation, T.L. and S.M.; writing original draft preparation, T.L. and S.M.; visualization, T.L. and S.M; supervision L.X.; project administration, L.X.; funding acquisition, L.X. All authors have read and agreed to the published version of the manuscript.

Funding: This work was financially supported by the Chinese National Natural Science Foundation (No. 51479137).

Conflicts of Interest: The authors declare no conflict of interest.

Notation

S	Applied suction in the caisson
H	Water head
γ_w	Unit weight of water (10 kN/m^3)
D	External diameter of the suction caisson
L	Height of the suction caisson
T	Thickness of the wall of the suction caisson
ρ_d	Dry density of the sand
G_s	Specific gravity of the sand
d_{10}	Effective size of the sand
e	Void ratio of the sand
γ_{sat}	Saturated weight of the sand
k	Permeability coefficient
h	Buried depth of the suction caisson
x	Abscissa value in Figure 1
γ'	Effective unit weight of soil
c	Cohesion of soil
φ	Angle of internal friction
h_{sp}	Height of soil plug
$h_{settlement}$	Soil settlement outside the caisson

References

1. Jalbi, S.; Arany, L.; Salem, A.; Cui, L.; Bhattacharya, S. A method to predict the cyclic loading profiles (one-way or two-way) for monopile supported offshore wind turbines. *Mar. Struct.* **2019**, *63*, 65–83. [CrossRef]

2. Chen, J.; Hu, Z.; Liu, G.; Wan, D. Coupled aero-hydro-servo-elastic methods for floating wind turbines. *Renew. Energy* **2019**, *130*, 139–153. [CrossRef]

3. Li, D.; Ma, S.; Zhang, Y. Undrained pullout capacity of modified suction caisson in clay by finite element limit analysis. *Mar. Georesour. Geotechnol.* **2019**, *37*, 291–300. [CrossRef]

4. Guo, Z.; Jeng, D.; Guo, W.; Wang, L. Failure mode and capacity of suction caisson under inclined short-term static and one-way cyclic loadings. *Mar. Georesour. Geotechnol.* **2018**, *36*, 52–63. [CrossRef]

5. Lee, J.; Tran, N.X.; Kim, S. Development and field application of GFRP suction pile. *Ocean Eng.* **2019**, *173*, 308–318. [CrossRef]

6. Li, D.; Ma, S.; Zhang, Y.; Chen, F. Lateral Bearing Capacity of Modified Suction Caissons Determined by Using the Limit Equilibrium Method. *China Ocean Eng.* **2018**, *32*, 461–466. [CrossRef]

7. Byrne, B.W.; Houlsby, G.T. Experimental Investigations of the Response of Suction Caissons to Transient Combined Loading. *J. Geotech. Geoenviron.* **2004**, *130*, 240–253. [CrossRef]

8. Kourkoulis, R.S.; Lekkakis, P.C.; Gelagoti, F.M.; Kaynia, A.M. Suction caisson foundations for offshore wind turbines subjected to wave and earthquake loading: Effect of soil-foundation interface. *Géotechnique* **2014**, *64*, 171. [CrossRef]

9. Ukritchon, B.; Wongtoythong, P.; Keawsawasvong, S. New design equation for undrained pullout capacity of suction caissons considering combined effects of caisson aspect ratio, adhesion factor at interface, and linearly increasing strength. *Appl. Ocean Res.* **2018**, *75*, 1–14. [CrossRef]

10. Chen, F.; Lian, J.; Wang, H.; Liu, F.; Wang, H.; Zhao, Y. Large-scale experimental investigation of the installation of suction caissons in silt sand. *Appl. Ocean Res.* **2016**, *60*, 109–120. [CrossRef]

11. Houlsby, G.T.; Byrne, B.W. Design procedures for installation of suction caissons in sand. *Proc. Inst. Civil Eng. Geotechnol.* **2005**, *158*, 135–144. [CrossRef]

12. Senders, M.; Randolph, M.F. CPT-Based Method for the Installation of Suction Caissons in Sand. *J. Geotech. Geoenviron.* **2009**, *135*, 14–25. [CrossRef]

13. Andersen, K.H.; Jostad, H.P.; Dyvik, R. Penetration Resistance of Offshore Skirted Foundations and Anchors in Dense Sand. *J. Geotech. Geoenviron.* **2008**, *134*, 106–116. [CrossRef]

14. Erbrich, C.T.; Tjelta, T.I. Installation of bucket foundations and suction caissons in sand-geotechnical performance. In Proceedings of the Offshore Technology Conference 1999, Houston, TX, USA, 3–6 May 1999.

15. Harireche, O.; Mehravar, M.; Alani, A.M. Suction caisson installation in sand with isotropic permeability varying with depth. *Appl. Ocean Res.* **2013**, *43*, 256–263. [CrossRef]

16. Tran, M.N.; Randolph, M.F.; Airey, D.W. Study of Sand Heave Formation in Suction Caissons Using Particle Image Velocimetry (PIV). In *Frontiers in Offshore Geotechnics, Proceedings of the 1st International Symposium on Frontiers in Offshore Geotechnics, Perth, Australia, 19–21 September 2005*; Francis: London, UK, 2005; pp. 259–265.

17. Harireche, O.; Mehravar, M.; Alani, A.M. Soil conditions and bounds to suction during the installation of caisson foundations in sand. *Ocean Eng.* **2014**, *88*, 164–173. [CrossRef]

18. Allersma, H. Centrifuge Research on Suction Piles: Installation and Bearing Capacity. In *BGA International Conference on Foundations: Innovations, Observations, Design and Practice, Proceedings of the International Conference Organised by British Geotechnical Association and Held in Dundee, Dundee, UK, 2–5 September 2003*; Thomas Telford Publishing: London, UK, 2003; pp. 91–98.

19. Tran, M.N.; Randolph, M.F.; Airey, D.W. Installation of Suction Caissons in Sand with Silt Layers. *J. Geotech. Geoenviron.* **2007**, *133*, 1183–1191. [CrossRef]

20. Tran, M.N.; Airey, D.W.; Randolph, M.F. Study of Seepage Flow and Sand Plug Loosening in Installation of Suction Caissons in Sand. In Proceedings of the Fifteenth International Offshore and Polar Engineering Conference, Seoul, Korea, 19–24 June 2005; International Society of Offshore and Polar Engineers: Mountain View, CA, USA, 2005.

21. Lu, T.; Liu, Z. *Advanced Soil Mechanics*; China Machine Press: Beijing, China, 2006.

22. Wang, L.; Li, Y.; Zhao, G.; Chen, N.; Xu, Y. Experimental Investigation of Flow Characteristics in Porous Media at Low Reynolds Numbers (Re→0) under Different Constant Hydraulic Heads. *Water* **2019**, *11*, 2317. [CrossRef]

23. Guo, W.; Chu, J.; Kou, H. Model tests of soil heave plug formation in suction caisson. *Proc. Inst. Civil Eng. Geotechnol.* **2016**, *169*, 214–223. [CrossRef]

Article

Experimental and Numerical Investigation of Wind Characteristics over Mountainous Valley Bridge Site Considering Improved Boundary Transition Sections

Xiangyan Chen [1], Zhiwen Liu [1,2,*], Xinguo Wang [3], Zhengqing Chen [1,2], Han Xiao [1] and Ji Zhou [3]

[1]   Hunan Provincial Key Lab for Wind & Bridge Engineering, Hunan University, Changsha 410082, China; xyanchen@hnu.edu.cn (X.C.); zqchen@hnu.edu.cn (Z.C.); hnuxh@hnu.edu.cn (H.X.)
[2]   College of Civil Engineering, Hunan University, Changsha 410082, China
[3]   China Railway Siyuan Survey and Design Group Co., Ltd., Wuhan 430063, China; wxg13971581277@163.com (X.W.); zhouji5085@126.com (J.Z.)
*   Correspondence: zhiwenliu@hnu.edu.cn; Tel.: +86-1397-588-0715

Received: 17 December 2019; Accepted: 16 January 2020; Published: 21 January 2020

Featured Application: The results of research in this paper can provide a certain design basis for the correction of an inlet boundary for both wind tunnel test and CFD simulation over complex terrain. By adopting the improved transition sections recommended in this work, engineers will have access to more precise assessment of wind characteristics over mountainous terrain, which contributes to wind-related issues such as wind actions of long-span bridges, transmission lines, wind farm site selection, and the prediction of pollutant dispersion.

Abstract: To study wind characteristics over mountainous terrain, the Xiangjiang Bridge site was employed in this paper. The improved boundary transition sections (BTS) were adopted to reduce the influence of "artificial cliffs" of the terrain model on the wind characteristics at the bridge site over the mountainous terrain. Numerical simulation and experimental investigations on wind characteristics over mountainous terrain with/without BTS were conducted for different cases, respectively. The research results show that the cross-bridge wind speed ratios and wind attack angles at the main deck level vary greatly along the bridge axis, which can be roughly divided into three parts, namely the mountain (I, III) and central canyon areas (II). The cross-bridge wind speed ratios at the main deck level with BTS is generally larger than that without BTS in the central canyon area (II) for most cases, while the opposite trend can be found in wind attack angles. The longitudinal wind speed ratios of the terrain model with BTS at L/4, L/2, and 3L/4 of the bridge length are larger than that of the terrain model without BTS for most cases. In general, the maximum relative error between numerical results and experimental results is about 30% for most cases.

Keywords: mountainous valley; bridge site; boundary transition section (BTS); wind characteristics; numerical simulation; wind tunnel test

---

1. Introduction

Wind characteristics over mountainous valleys are critical to many wind-related issues, such as wind actions of long-span bridges and transmission lines, wind farm site selection, prediction of pollutant dispersion, and so on. In particular, wind characteristics at mountainous valley bridge sites such as design wind speed, wind yaw angles, wind attack angles, and turbulence spectra play a critical role in wind-resistant design of long-span bridges. The wind characteristics over complex mountainous terrain are significantly different from those of open areas. Therefore, reasonable determination of wind parameters at the bridge sites over mountainous terrain is important to the balance between wind-resistance safety and the economy of long-span bridges. However, the current wind-resistant design codes for bridges generally make simple corrections for mountain wind fields based on wind characteristics of flat terrain (Chock et al., [1]). Therefore, it is especially necessary to investigate wind characteristics over complex mountainous terrain precisely.

The main methods for studying the wind characteristics over complex mountainous terrain bridge sites include theoretical study, field measurement, experimental study, and numerical study. Theoretical studies were used to predict the wind speed-up effects over simple topographic features with slopes low enough to avoid flow separation (Jackson and Hunt [2]; Hunt et al. [3]), which cannot be adopted to study the wind characteristics over complex mountainous terrains. Field measurement is regarded as the most reliable approach to investigate the wind characteristics in the ABL (Li, et al. [4]), which can accurately measure the wind speed and direction data at different measurement points. During the past two decades, numerous field measurements of wind characteristics over mountainous valley were conducted (Harstveit [5]; Hannesen et al. [6]; Lubitz et al. [7]; Sharples et al. [8]; Abiven et al. [9]; Risan et al. [10]; Lystad et al. [11]; Peng et al. [12]; Zhang et al. [13]; Yu et al. [14]; Jing et al. [15]). However, field measurements of wind characteristics over mountainous bridge site are usually expensive, time-consuming, and can only provide wind data at a limited number of measurement points, which are easily affected by the terrain.

Experimental studies of wind characteristics over complex mountainous terrain were conducted by many researchers over the last three decades (Cermak [16], Chock et al. [1], and Xu et al. [17]; Kozmar et al. [18]; Yan et al. [19]; Mattuella et al. [20]; Muhammad et al. [21]; Kozmar et al. [22]; Chen et al. [23]; Flay et al. [24]). Bowen [25] pointed out that the geometric scales of complex terrain model should not be less than 1:2500~5000, and the Reynolds number should be larger than $Re_h = 10^5$ (h is the model hill height) for wind tunnel tests. However, for engineering structures located at complex mountainous terrain, such as long-span bridges, power transmission towers, and wind turbines, it is especially important, and an urgent matter, to carry out numerical simulation of wind characteristics since complex terrain makes it difficult for scholars to propose a common wind characteristic model. When the wind tunnel test model or numerical simulation model are adopted to study wind characteristics of mountainous terrain, it is inevitable to take a certain range of areas from mountainous terrain, causing "artificial cliffs" (shown in Figure 1) on the edge of terrain models. These artificial cliffs can deviate practical wind characteristics at the inlet boundary from these of theoretical ones, such as wind speed, wind attack angles, and turbulence characteristics. To cope with the negative influence of an artificial cliff on wind tunnel tests, some scholars have put forward with different solutions. Maurizi et al. [26] and Pang et al. [27] used an inclined boundary transition section (BTS) to connect the ground and the top of the terrain model. Hu et al. [28] proposed a curved transition section for complex terrain and concluded that this kind of transition section has a better flow transition performance compared with the traditional inclined one. Li et al. [29] combined a terrain model with three-dimensional transition sections in the wind tunnel. Huang et al. [30] evaluated speed-up ratios, mean vertical attack angles, and mean exceeding turbulence intensities of different transition curves to determine the optimal transition section for terrain models. Hu et al. [31] established two different BTS in the computational domain for comparison purposes and the results show that the updated curved BTS has a better flow transition efficiency than those reported previously. Liu et al. [32] proposed an improved BTS to modify the inlet boundary by combining the Witozinsky curve and straight line.

With the development of computational fluid dynamics (CFD), many scholars (Uchida and Ohya [33]; Tong et al. [34]; Deleon et al. [35]; Risan et al. [10]; Tamura et al. [36] and Cassiani et al. [37]) investigated wind characteristics over complex terrain with CFD method. Maurizi et al. [26], Kim et al. [38] and Castellani et al. [39] confirmed the availability of the $k - \varepsilon$ model and RNG $k - \varepsilon$ model. Although there is some research focusing on wind characteristics over complex terrain by CFD simulation, few studies on transition sections on the edge of terrain models have been found recently. Due to diversity and complexity of complex terrain, it is necessary to conduct further research on the BTS of mountainous terrain models.

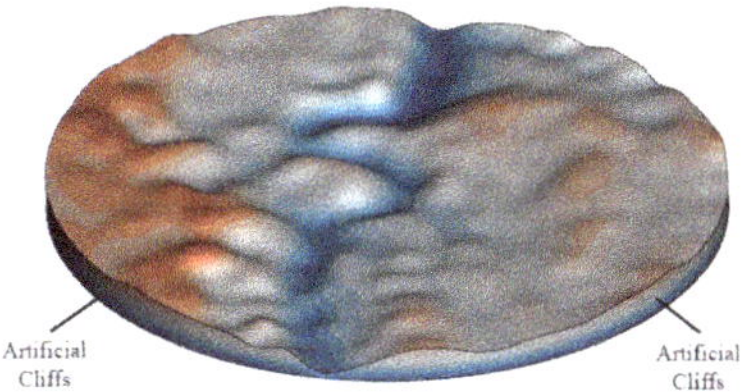

Figure 1. Schematic diagram of Artificial Cliffs.

The Xiangjiang Bridge in Guizhou, China, was employed as an engineering example to study the wind characteristics at bridge site over complex terrain and effects of the improved BTS. The mid-span of the cable-stayed bridge was regarded as the origin. The terrain model with a geometric scale of 1:1500 (as suggested by Bowen [25]) was fabricated to simulate a region with a diameter of 8 km. Wind tunnel tests were conducted with and without the proposed BTS in order to investigate the BTS effects on wind characteristics (Liu et al. [32]). To verify the feasibility of the improved BTS numerically, wind characteristics over the scaled mountainous terrain model with and without the proposed BTS were predicted by CFD simulation. The wind characteristics, such as mean cross-bridge wind speed and wind attack angles at the main deck level, longitudinal wind speed profiles at L/4, L/2 and 3L/4 of the bridge length were studied in detail to verify the accuracy of CFD simulation and effects of improved BTS.

This paper is organized as follows: Section 2 introduces the numerical method, including governing equations of fluids, turbulence model, optimization of terrain model with BTS, the and numerical model. Section 3 describes the details of wind tunnel test setup. Section 4 investigates the cross-bridge wind speed ratios and wind attack angles at main deck level, longitudinal wind speed ratios profiles at L/4, L/2 and 3L/4 of bridge length of the terrain model with/without BTS under different cases. Section 5 presents conclusions of the study.

2. Numerical Method

2.1. Governing Equations of Fluids

Air flow passing through mountainous terrain can be approximately considered to be incompressible viscous fluid, which can be expressed based on the Reynolds-averaged Navier-Stokes (RANS) equations as follows:

$$\frac{\partial u_i}{\partial x_i} = 0 \tag{1}$$

$$\frac{\partial u_i}{\partial t} + u_j \frac{\partial u_i}{\partial x_j} = -\frac{1}{\rho}\frac{\partial p}{\partial x_i} + \frac{\partial}{\partial x_j}\left[v\left(\frac{\partial u_i}{\partial x_j} + \frac{\partial u_j}{\partial x_i}\right) - \overline{u_i' u_j'}\right] \tag{2}$$

where t and x_i are time and Cartesian coordinates, respectively; u_i and u_i' are the time-averaged and fluctuating flow velocity components, respectively; p is the time-averaged pressure; ρ and v are the fluid density and kinematic viscosity; $\overline{u_i' u_j'}$ is the Reynolds stress tensor.

2.2. Turbulence Model

The Large eddy simulation (LES) has inherent advantages over the steady-state and unsteady RANS turbulence models in physical simulation, and is well suited for the simulation of turbulence and nonlinear features of complex terrain wind. The use of LES for complex mountainous terrain requires large computing resources. However, for wind field modeling over complex terrain, RANS remains the main turbulence model adopted in CFD simulation up to present, where it is often being applied with a satisfactory degree of success. Many researchers (Bitsuamlak et al. [40]; Kim et al. [38]; Mohamed et al. [21]; Yassin, et al. [41] and Yan et al. [42]) found that results obtained by RNG $k - \varepsilon$ agree well with these of experiments. Therefore, the RNG $k - \varepsilon$ model is adopted to numerical simulation in this paper.

The transport equations for RNG $k - \varepsilon$ model are given as follows,

$$\frac{\partial}{\partial t}(\rho k) + \frac{\partial}{\partial x_i}(\rho k u_i) = \frac{\partial}{\partial x_j}\left(\alpha_k \mu_{eff} \frac{\partial k}{\partial x_j}\right) + G_k - \rho\varepsilon + S_k \tag{3}$$

$$\frac{\partial}{\partial t}(\rho\varepsilon) + \frac{\partial}{\partial x_i}(\rho\varepsilon u_i) = \frac{\partial}{\partial x_j}\left(\alpha_\varepsilon \mu_{eff} \frac{\partial \varepsilon}{\partial x_j}\right) + C_{1\varepsilon}\frac{\varepsilon}{k}G_k - C_{2\varepsilon}\rho\frac{\varepsilon^2}{k} - R_\varepsilon + S_\varepsilon \tag{4}$$

where G_k is the turbulent kinetic energy caused by mean speed gradient whose definition is $G_k = -\rho\overline{u_i'u_j'}\partial u_j/\partial x_i$ and the equation is $G_k = \mu_t S^2$ using Boussinesq hypothesis where S is the strain rate tensor of mean speed, i.e., $S = \sqrt{2S_{ij}S_{ij}}$. α_k and α_ε are reciprocals of turbulent kinetic energy, k and dissipation rate, ε respectively. S_k and S_ε are source terms defined by users. R_ε is the addition item of RNG $k - \varepsilon$ model compared with $k - \varepsilon$ model and its expression is

$$R_\varepsilon = \frac{C_\mu \rho \eta^3 (1 - \eta/\eta_0)}{1 + \beta\eta^3}\frac{\varepsilon^2}{k} \tag{5}$$

where $\eta = Sk/\varepsilon$ is the ratio of turbulence and mean flow time scale. η_0 is the typical value of η in uniform shear flow. C_μ and β are constants of the turbulence model. ρ is the air density.

2.3. Optimization of Terrain Model BTS

2.3.1. Form of Boundary Transition Section Curve

The Witozinsky curve was obtained in the case of an ideal axis of uncompressed axisymmetric flow. Wind tunnel operation results show that this kind of contraction curve can achieve good flow quality in test sections. Many low-speed wind tunnels constructed in 1960s by Chinese scholars were designed based on the Witozinsky formula (Wang et al. [43]). Evaluating the results of different contraction curves in terms of average wind speed reduction factors, vertical wind attack angles, and turbulence intensity, Huang et al. [44] drew a conclusion that the performance of the Witozinsky curve is the best and proposed an optimal curve expression of the Witozinsky curve at the same time. In his research, the terrain model boundary is directly connected to the BTS as shown in Figure 2a, which may lead to insufficient development of the inflow. In order to make the inflow develop fully, BTS in the present paper is improved by combining the Witozinsky curve with a horizontal line as shown in Figure 2b. The formula of the Witozinsky curve is shown as follows:

$$y = H\left\{1 - \frac{\left[1 - (x/L)^2\right]^2}{\left[1 + A(x/L)^2\right]^3}\right\} \tag{6}$$

where $A = 1/3$, H is the vertical distance from the highest point of the curve to the ground, L is the horizontal projection length of the whole curve and the corresponding coordinates of any point on the curve is (x, y).

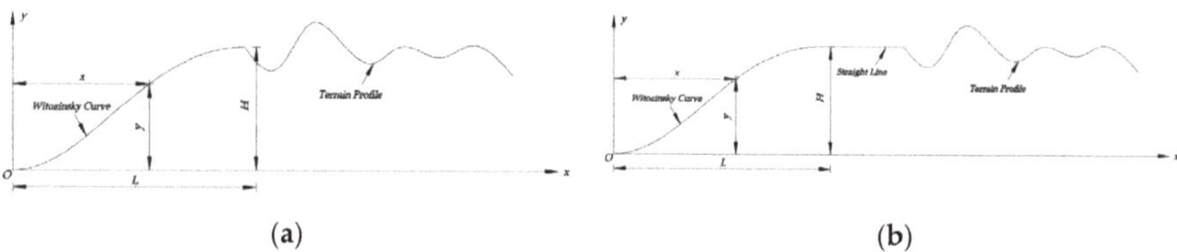

(a)   (b)

Figure 2. Terrain model BTS with/without horizontal straight line (Liu, et al., 2019). (**a**) BTS without horizontal straight line; (**b**) BTS with horizontal straight line.

2.3.2. Weight Allocation of Evaluation Indexes

Because the evaluation of BTS at the entrance of the terrain model mainly focuses on speed-up effect, the equivalent wind attack angle and the increase ratio of turbulence intensity, the specific evaluation indexes of effects of the terrain model BTS are defined as follows:

$$V_x = \sum_{k=1}^{N} \frac{v_k - v_{0k}}{v_{0k}} \cdot \frac{\Delta h_k}{H_w} \tag{7}$$

$$A_x = \sum_{k=1}^{N} \frac{a_k \cdot \Delta h_k}{H_w} \tag{8}$$

$$I_x = \sum_{k=1}^{N} \frac{I_k - I_{0k}}{I_{0k}} \cdot \frac{\Delta h_k}{H_w} \tag{9}$$

where N is the total amount of monitoring points. v_k is the downward wind speed of the kth monitoring point from the ground and v_{0k} is the wind speed at the kth monitoring point of the wind speed entrance. Δh_k is the vertical distance between the kth monitoring point and the $(k-1)^{th}$ monitoring point. H_w is the total height of the monitoring points. a_k is the wind attack angle of the kth monitoring point from the ground. I_k is the turbulence intensity at the kth monitoring point and I_{0k} is the turbulent intensity at the kth monitoring point from the ground at the wind speed entrance.

The two-dimensional CFD simulation was used to optimize the parameters of the Witozinsky curve and horizontal line. For the indexes such as the speed-up ratio of wind speed, equivalent wind attack angle, and turbulence intensity increase ratio at the boundary of the terrain model, the weight allocation based on the correlation degree proposed by Tang (Tang et al. [45]) was adopted to determine the BTS curve parameters for a given total length of the transition section (Liu, et al. [32]).

There are $m = 13$ schemes and $n = 4$ indexes that have an influence on the comprehensive evaluation of each scheme. They can compose the matrix $X = (x_{ij})_{m \times n}$. In this paper, the evaluation indexes are all cost indexes, which means that the smaller the index is, the better the result is. Standardize values in each column of $X = (x_{ij})_{m \times n}$ using min-max standardization (shown in formula (10) to map each index of the matrix into [0,1]. After conversion, the dimensionless matrix is $X' = (x_{ij})_{m \times n}$. Then, select the maximum in each column of $X' = (x_{ij})_{8 \times 6}$ to obtain the optimal objects S_0^+ and select the minimum in each column to get the worst object S_0^-. The correlation coefficient of each scheme with the optimal object and the worst object in each index can be calculated by Equations (11) and (12),

$$x' = \frac{x - \min}{\max - \min} \tag{10}$$

$$\zeta_{0+i}(j) = \frac{1}{2\Delta_{0+ij} + 1} \tag{11}$$

$$\zeta_{0-i}(j) = \frac{1}{2\Delta_{0-ij} + 1} \tag{12}$$

where $\Delta_{0+ij} = \left| x_{0+j} - x_{ij} \right|$ and $\Delta_{0-ij} = \left| x_{0-j} - x_{ij} \right|$. The larger $\zeta_{0+i}(j)$ is, the larger the correlation coefficient between scheme to be evaluated and the optimal scheme is in terms of index j. On the other hand, the performance of $\zeta_{0-i}(j)$ is absolutely opposite.

Assume the weight vectors of n indexes are $\omega = (\omega_1, \omega_2, \ldots\ldots, \omega_{n-1}, \omega_n)$ respectively and

$$f(\omega) = \sum_{j=1}^{n} \omega_j^2 [1 - \zeta_{0+i}(j)]^2 + \sum_{j=1}^{n} \omega_j^2 \zeta_{0-i}(j)^2$$ represents the square sum of the weighted distance between

the optimal solution and the worst one. Therefore, the smaller $f(\omega)$ is, the better scheme i is. Establish the multi-objective programming model as follows:

$$\min f(\omega) = (f_1(\omega), f_2(\omega), \ldots \ldots, f_{m-1}(\omega), f_m(\omega))^T$$

$$s.t. \begin{cases} \sum\limits_{i=1}^{n} \omega_j = 1 \\ \omega_j \geq 0, j = 1, 2, \ldots n-1, n \end{cases} \qquad s.t. \begin{cases} \min \sum\limits_{i=1}^{m} f_i(\omega) \\ \sum\limits_{i=1}^{n} \omega_j = 1 \\ \omega_j \geq 0, j = 1, 2, \ldots, n-1, n \end{cases} \qquad (13)$$

Since $f_i(\omega) \geq 0 (i = 1, 2, \ldots, m-1, m)$, the multi-objective programming model is transformed into single-objective programming model. Establish Lagrange Function,

$$F(\omega, \lambda) = \sum_{i=1}^{m} \sum_{j=1}^{n} \omega_j^2 \{[1 - \zeta_{0+i}(j)]^2 + \zeta_{0-i}(j)^2\} - \lambda(\sum_{j=1}^{n} \omega_j - 1) \qquad (14)$$

$$\begin{cases} \frac{\partial F}{\partial \omega_j} = 2\omega_j \sum\limits_{i=1}^{m} \{[1 - \zeta_{0+i}(j)]^2 + \zeta_{0-i}(j)^2\} - \lambda = 0 \\ \frac{\partial F}{\partial \lambda} = \sum\limits_{j=1}^{n} \omega_j - 1 = 0 \end{cases} \qquad (15)$$

Solve Equation (15), the weight of n indexes can be calculated by

$$\omega_j = \frac{1}{u_j \sum\limits_{i=1}^{n} \frac{1}{u_j}} \qquad (16)$$

where u_j is an intermediate variable and can be calculated by the following equation,

$$u_j = \sum_{i=1}^{m} \{[1 - \zeta_{0+i}(j)]^2 + \zeta_{0-i}^2(j)\} \qquad (17)$$

The final calculation results of all the wind profile positions are shown in Table 1. The virtual angle is defined as the angle between the horizontal line and the line connecting the ground and the end of the Witozinsky curve. The horizontal length refers to the horizontal straight line length after the Witozinsky curve. Since the improved BTS proposed in this paper aims to provide an effective corrections for terrain models of wind tunnel test and CFD simulation, the total length of the improved BTS need to be sufficiently short to reduce blockage ratio while ensuring the accuracy of the wind characteristics over complex terrain; therefore, the total length of the improved BTS is regarded as the major consideration. If the total length of the horizontal projection is smaller than 1.0 m, the minimal final result is obtained at the combination of a virtual angle of 57° and a horizontal length of 0.8 m. Different optimal solutions can be obtained by changing the total horizontal projection length.

Table 1. Final results concerning weights at different wind profiles.

Horizontal Length (m)	Virtual Angle of BTS												
	27°	30°	33°	36°	39°	42°	45°	48°	51°	54°	57°	60°	63°
0.0	1.06	0.89	1.46	1.51	1.56	1.62	1.98	1.7	1.75	1.79	1.89	1.98	2.17
0.2	0.88	0.92	0.96	0.99	1.01	1.05	1.74	1.1	1.12	1.15	1.22	1.3	1.51
0.4	0.6	0.64	0.66	0.68	0.7	0.73	1.14	0.76	0.78	0.8	0.83	0.88	1.03
0.6	0.36	0.38	0.4	0.41	0.43	0.45	0.7	0.46	0.47	0.49	0.51	0.53	0.6
0.8	0.2	0.22	0.23	0.24	0.25	0.26	0.44	0.27	0.28	0.29	0.3	0.31	0.34
1.0	−0.3	0.05	0.06	0.07	0.08	0.09	0.21	0.1	0.1	0.11	0.11	0.11	0.12

2.4. Numerical Terrain Model

2.4.1. Survey of the Xiangjiang Bridge

The Xiangjiang Bridge is a four-span three-pylon cable-stayed bridge with span of 120 m + 235 m + 235 m + 120 m = 710 m. The main deck is 252.2 m above the bottom of the valley, as shown in Figure 3a. The balanced cantilever construction method are adopted to build the bridge, and the wind loads of the bridge with maximum double cantilever is the focus of wind-resistant design. Note that the bridge site is located in the mountainous terrain, as shown in Figure 3b. It can be seen from Figure 3b that the river approximately runs southwest to northeast, and the bridge is located in the center of the terrain. In order to determine the wind characteristics parameters for wind-resistant design of the bridge, numerical simulations and wind tunnel tests on wind characteristics over the mountainous terrain with/without BTS were carried out, respectively.

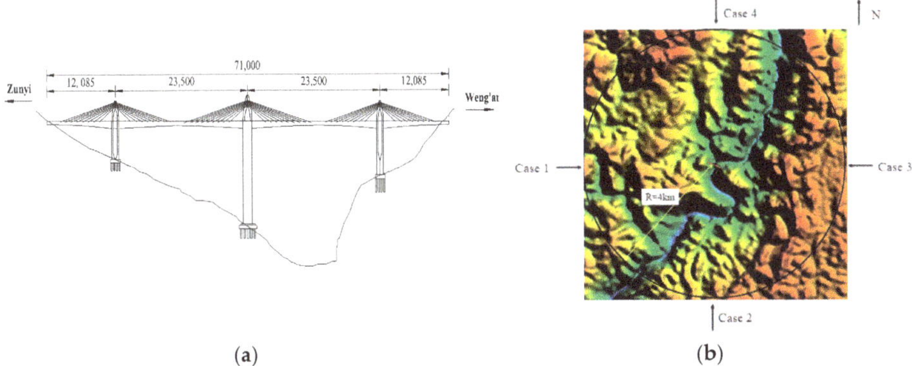

(a)   (b)

Figure 3. Elevation of the Xiangjiang Bridge and the mountainous terrain (Unit: cm). (**a**) Elevation of the Xiangjiang Bridge; (**b**) Complex terrain at the bridge site.

2.4.2. Terrain Model Construction

With reference to the existing literatures on mountainous terrain wind characteristics, as shown in Table 2, the bridge site is taken as the center, and an area with a diameter of 8.0 km was selected to establish the mountainous terrain model. The computational domain with 32.34 m long, 8.0 m wide, 2.0 m high, and the geometry scale of 1:1500 were adopted according to the value of blockage ratio of the terrain model. The lowest elevation of the terrain model was taken as the bottom of the computational domain. The left side of the computational domain is defined as velocity-inlet boundary condition. The right side of the computational domain is defined as pressure-outlet boundary condition. The front, back, and top sides of the computational domain are defined as symmetry boundary conditions. The bottom of the computational domain is defined as non-slip wall boundary condition. The boundary conditions and computational domain of the terrain model without BTS are given in Figure 4. As can be seen in Figure 4, the distance between the center of the terrain model and the front and back walls of the computational domain is 4.0 m, respectively. The center of the terrain model is 9.67 m and 22.67 m away from the left and right sides of the computational domain.

Table 2. Summary of main parameters of mountainous terrain models.

Number	Terrain Region	Geometry Scale Ratio	Methods and Domain	Authors
1	9.5 km × 7.3 km	-	CFD	Risan A., et al. [10]
2	D = 15 km	1:1000	Wind tunnel test: 36 m (L) × 22.5 m (B) × 4.5 m (H)	Li Y. et al. [29]
3	D = 12 km	1:1000	Wind tunnel test: 25 m(L) × 12.0 m(B) × 16.0 m (H)	Xu H. et al. [17]
4	D = 5.0 km	1:1500	Wind tunnel test	Pang J. et al. [27]
5	D = 9.0 km	1:1000	CFD: 15 m (L) × 15 m (B) × 4.5 m (H)	Hu P., et al. [31]
6	15 km × 14 km	-	CFD	Maurizi A., et al. [26]
7	D = 27.2 km	1:4000	Wind tunnel test: 14 m (L) × 15 m (B) ×2.0 m (H)	Chen F., et al. [23]

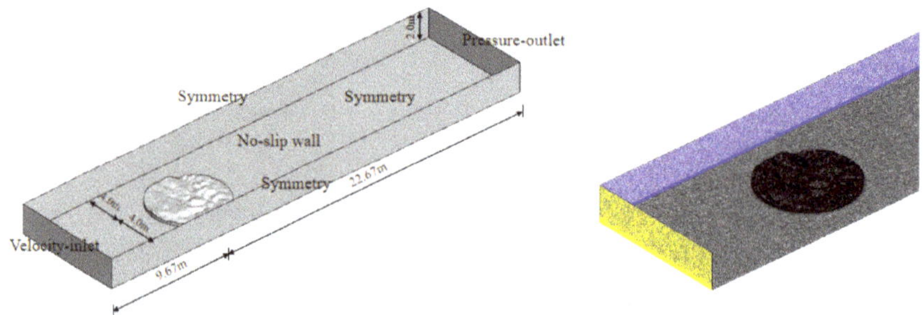

Figure 4. Computational domain, boundary conditions, and grid of terrain model without BTS.

The Power Law Profile Model is currently the only model of boundary layer wind profile [46]. The wind speed at the velocity inlet boundary after scaling of the computational domain are defined as follows:

$$V(z) = V_{0.007}\left(\frac{z}{0.007}\right)^{a}, 0 \leq z \leq 0.233\text{m}$$
$$V(z) = V_{ref}, z > 0.233\text{m} \tag{18}$$

where z is the distance between the monitoring point and the ground; a is the wind profile index and $a = 0.16$ when the ground surface belongs to class-B wind field; $V(z)$ is the wind velocity of the monitoring point which is z away from the ground; $V_{0.007}$ is the basic wind speed at the level of $z_0 = 0.007$m in the power law profile. $V_{ref} = V_{0.233}$ is the wind velocity when $z = 0.233$m.

In order to investigate the effects of the improved BTS on wind characteristics over mountainous terrain model, numerical simulation of the terrain model with/without BTS were conducted, respectively. With reference to the results of two-dimensional numerical simulation of the terrain model BTS, it is determined that the virtual inclination of the transition section is $\theta = 57°$ and the length of the horizontal line is $L = 0.80$m (Liu et al. [32]).

2.4.3. Grids of the Terrain Model

To ensure the numerical simulation accuracy of the terrain model grid, the grid independence checking was performed firstly. The mesh is divided into different parts. Unstructured prism was used to find grids near the ground, and the tetrahedral grid was adopted to reflect the undulating changes of the terrain. The near-wall grids of the terrain model are given in Figure 4. The detailed parameters of the three grids for independence checking are shown in Table 3. The computational domain and grid arrangement of the terrain model with BTS are shown in Figure 5.

Table 3. Detailed setups of three grids of terrain model without BTS.

Grid	Mesh Size in Terrain Surface (m)	Mesh Size in Computation Domain (m)	Thickness of the First Boundary Layer Mesh (m)	Growth Rate of the Boundary Layer Mesh	Amount of Boundary Layer	Total Mesh Amount
G1	0.08	0.25	5	1.1	20	1,546,916
G2	0.07	0.20	5	1.1	20	3,881,449
G3	0.05	0.18	5	1.1	20	10,509,942

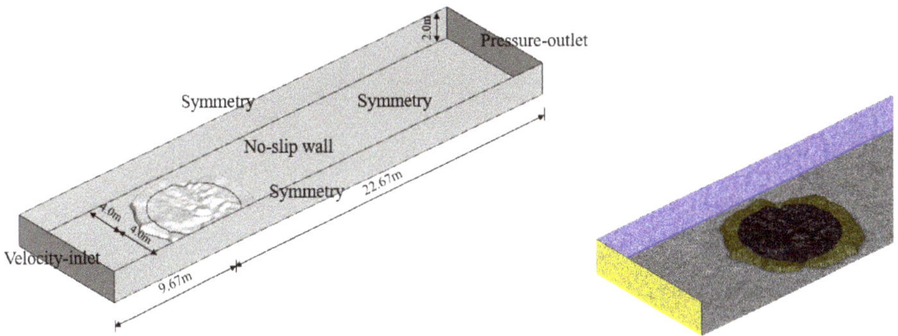

Figure 5. Computational domain, boundary conditions and grid of terrain model with BTS.

The flow solver used in this paper to simulate the wind characteristics over complex terrain is based on the finite-volume discretization in the space and the SIMPLEC (Semi implicit method for pressure linked equation consistent) pressure-correlation algorithm on a non-staggered grid arrangement (Kim, et al. [38]). The second-order cell-centered is adopted to deal with spatial discrete. The second-order upwind is used for time discrete. The time step is 0.015 s and the total amount of time steps is 10,000. According to the main parameters of wind-resistant design of the bridge, the mean cross-bridge wind speed and wind attack angles at the main deck level, longitudinal wind speed profiles at L/4, L/2 and 3L/4 of the bridge length were monitored to evaluate the calculation accuracy of different grids. The wind speed monitoring points are given in Figure 6.

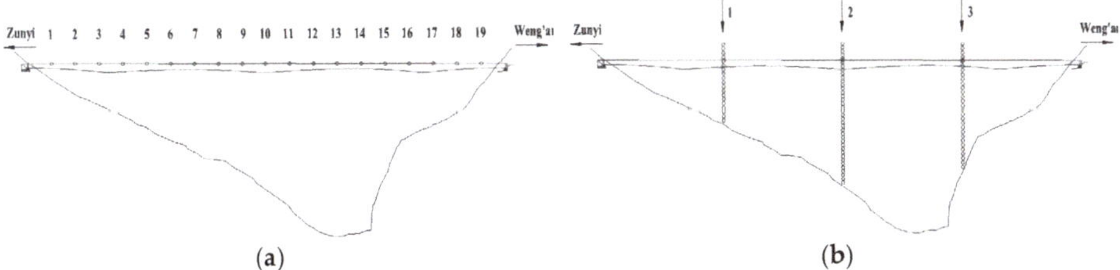

Figure 6. Schematic diagram of monitoring points. (**a**) Monitoring points at the main deck level; (**b**) Monitoring points profiles at L/4, L/2 and 3L/4 of the bridge length.

The grid sensibility was conducted by taking case 1 without BTS as an example. Numerical results of mean cross-bridge wind speed and wind attack angles of grid G1, G2, and G3 at the main deck level over the terrain model without BTS are given in Figure 7a,b, respectively. The longitudinal wind speed profiles at L/4, L/2 and 3L/4 of the bridge length of grid G1, G2, and G3 over the terrain model without BTS are shown in Figure 7c. It can be found in Figure 7a that the mean cross-bridge wind speed of G2 agrees well with that of G3 with a relative error of 23% while the result of G1 departs from the results of the other two grid generations to a great extent with relative error 116%. Note that the same tendency can be found in Figure 7b (relative error 41% for terrain model with BTS while 160% without

BTS) and (c), which means that relative sparse grid generation in the terrain model and computational domain will cause significant discrepancies in terms of average wind characteristics at bridge site. Therefore, the grid generation, G2, is considered to be the optimal grid since its amount of grids is less compared to G3 and the calculation accuracy can be guaranteed.

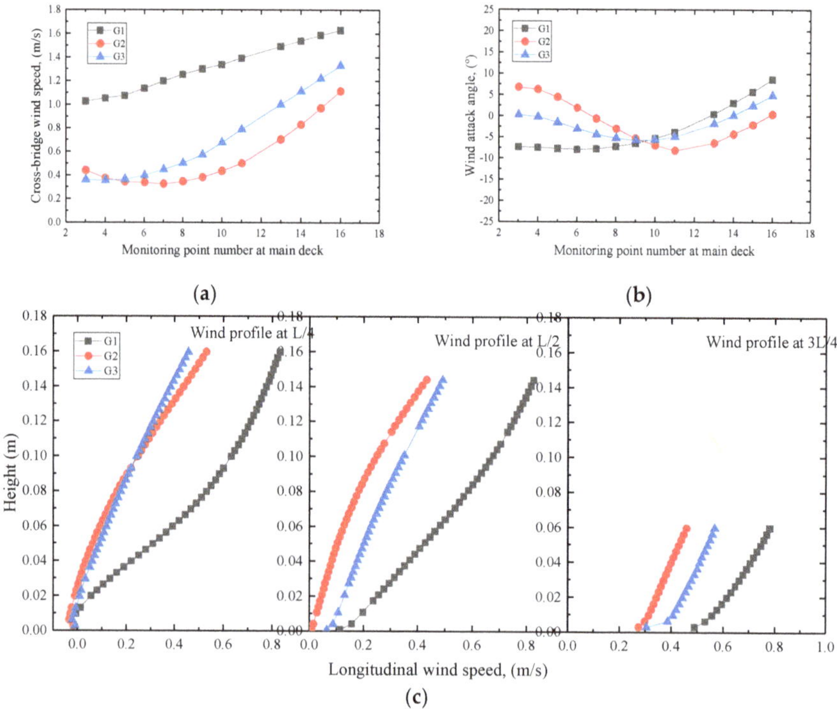

Figure 7. Comparisons for average wind characteristics with different grid generations. (**a**) Mean cross-bridge wind speed at main deck level; (**b**) Wind attack angle at main deck level; (**c**) Longitudinal wind speed profiles at L/4, L/2 and 3L/4 of the bridge length.

3. Wind Tunnel Test Set-Up

The wind tunnel tests were conducted in the third test section of HD-2 wind tunnel of Hunan University, Changsha, Hunan, China. HD-2 wind tunnel of Hunan University is 53 m long and 18 m wide, including three test sections. The first test section is 17 m long, 3 m wide, and 2.5 m high, and the wind speed of the test section was 0~58 m/s, which can be continuously adjusted. The second test section is 15 m long, 5.5 m wide and 4.4 m high, and the maximum wind speed was 18 m/s. The third test section was 15 m long, 8.5 m wide and 2 m high, and the maximum wind speed was 15 m/s. Considering the size of the third section of HD-2 wind tunnel, the terrain model centered on the mid-span of the bridge with a geometric scale of 1:1500 was designed and fabricated to simulate a region with a diameter of 8 km. The average height of the terrain model was roughly 0.2 m. The blockage ratio is defined as the ratio of cross-sectional area of the terrain model and the wind tunnel test section. If the blockage ratio of the terrain model is too large, the inflow cannot develop fully, leading to inaccurate prediction of the wind field over complex terrain. The blockage ratio of the terrain model is about 6.27%, which is within the range of 5.0% and 10% required in Design Rules for Aerodynamic Effects on Bridges published by British.

To investigate the BTS effects on wind characteristics over mountainous terrain model, the wind tunnel tests of terrain model with/without BTS were conducted, respectively. The geometric parameters of the BTS are consistent with these of the CFD simulation model. Figure 8 shows the terrain model with/without BTS.

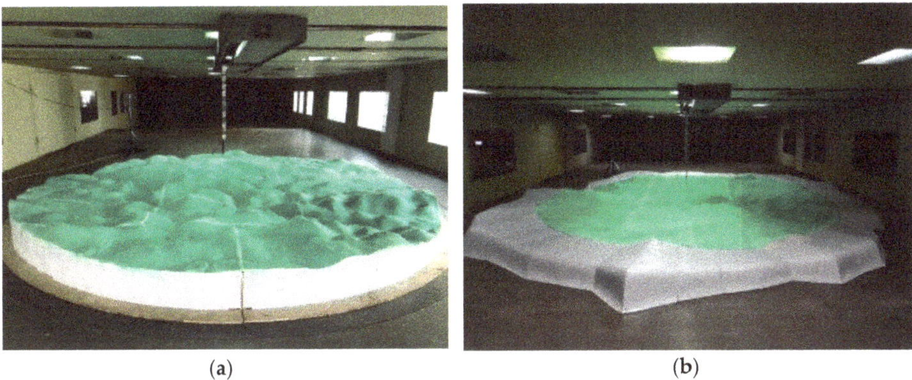

(a)   (b)

Figure 8. Mountainous terrain model with/without BTS in wind tunnel. (**a**) Terrain model without BTS; (**b**) Terrain model with BTS.

Before the wind tunnel test of terrain model was carried out, the inflow wind speed and turbulence intensity profiles in the empty wind tunnel section was conducted first. The total height of the monitoring points was 0.6 m, and the interval between two adjacent monitoring points was 0.03 m. The first monitoring point was 0.03 m away from the bottom of the wind tunnel section. The wind speed was measured by Cobra probes from TFI (Turbulent Flow Instrumentation Pty Ltd., Victoria, Australia). Figure 9 shows the comparison of the wind profile of the incoming wind speed in the empty wind tunnel section and the wind profile of terrain category B suggested in wind-resistant design specification for highway bridges (JTG/T 3360-01-2018). As shown in Figure 9a, the measured inflow wind speed profile in the empty wind tunnel section was closed to the corresponding wind speed profile of category B. The main reason is that the geometric scale of the terrain model is 1:1500, the height of the boundary layer of the terrain category B was 0.233 m, which was approximately close to the boundary layer height in the wind tunnel test section under a uniform inflow condition. It can be seen from Figure 9b that the longitudinal turbulence intensity was about 3.0% when the height of the monitoring points is larger than 0.2 m.

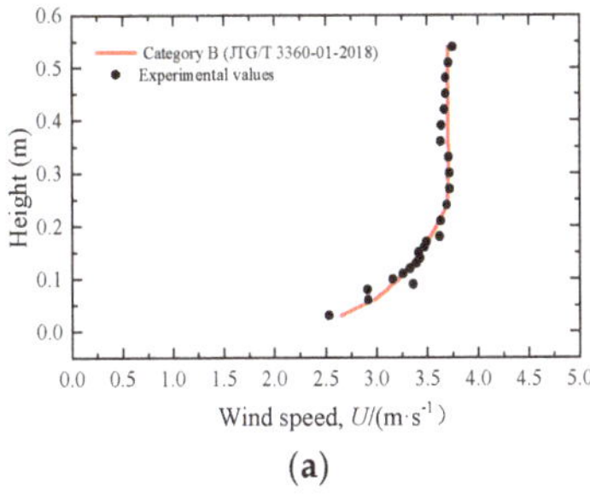

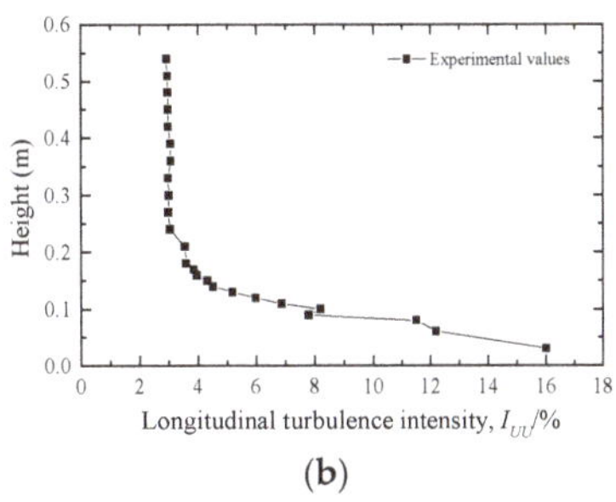

(a)   (b)

Figure 9. Wind speed and turbulence intensity profiles at different height. (**a**) Wind speed profile; (**b**) Longitudinal turbulence intensity profile.

To find out the influence of wind yaw angles on wind characteristics over mountainous terrain, four cases of wind from the west, south, east, and north were investigated by numerical simulations and wind tunnel tests, respectively. The wind yaw angle was 0° when the inflow enters the terrain model along the west. Counterclockwise rotation was positive for the wind yaw angle and the detailed test cases are shown in Figure 3b.

4. Verification of Improved BTS

Due to space limitations, only wind speed, turbulence kinetic energy, and wind attack angle of case 1 as well as 2, without and with improved BTS are compared to verify its effect.

It can be seen from Figure 10 that when there is improved BTS, wind speed distribution is more uniform and closer to that of the inflow wind speed profile. Besides, there is no flow separation near the boundary of the terrain model.

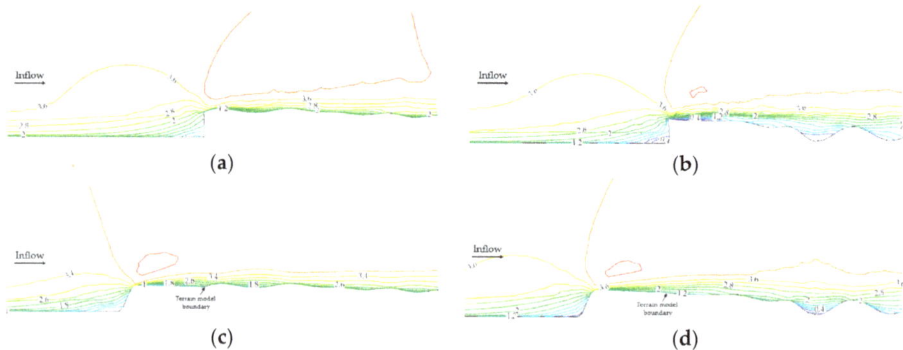

Figure 10. Wind speed distribution along the inflow direction near the inlet of the terrain model without or with improved BTS from CFD simulation (unit: m/s). (**a**) Case 1 without BTS; (**b**) Case 2 without BTS; (**c**) Case 1 with BTS; (**d**) Case 2 with BTS.

Turbulence kinetic energy (TKE) is a measure of the intensity of turbulence and determines the ability of the flow to maintain turbulence or become turbulence, thus indicating flow stability. From Figure 11, it can be concluded that at the inlet of the terrain model without improved BTS, the TKE with BTS is smaller than that of the terrain model without BTS, which indicates that the improved BTS is able to reduce turbulence intensity induced by artificial cliffs and uplift the inflow gradually.

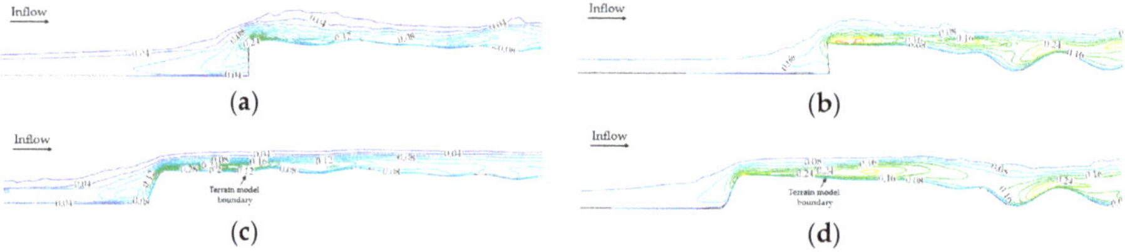

Figure 11. Turbulence kinetic energy distribution along the inflow direction near the inlet of the terrain model without or with improved BTS from CFD simulation (unit: m²s⁻²). (**a**) Case 1 without BTS; (**b**) Case 2 without BTS; (**c**) Case 1 with BTS; (**d**) Case 2 with BTS.

Wind attack angles induced by artificial cliffs will have a significant influence on the wind attack angle at the bridge site, thus reducing wind attack angles at the inlet boundary hold a decisive place in wind characteristic prediction over complex terrain. As shown in Figure 12, the wind attack angle at the inlet boundary with the improved BTS is much smaller than that without BTS and is closer to zero, indicating that the improved BTS can reduce the wind attack angle caused by artificial cliffs to a great extent.

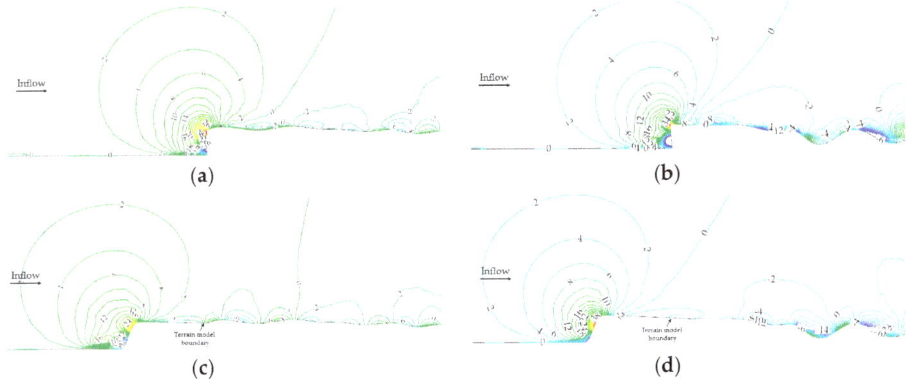

Figure 12. Wind attack angle distribution along the inflow direction near the inlet of the terrain model without or with improved BTS from CFD simulation (unit: °). (**a**) Case 1 without BTS; (**b**) Case 2 without BTS; (**c**) Case 1 with BTS; (**d**) Case 2 with BTS.

In conclusion, by comparing wind speed, turbulence kinetic energy, and wind attack angle near the inlet of the terrain model without and with improved BTS, the BTS can effectively reduce the impact of artificial cliffs on wind characteristics after the inlet boundary.

5. Numerical and Experimental Results

According to the wind-resistance design requirements of the bridge, the cross-bridge wind speed, wind attack angles at main deck level, and longitudinal wind speed profiles at L/4, L/2, and 3L/4 of the bridge length were analyzed for numerical and experimental cases. In this paper, the positive direction of cross-bridge wind is from southwest to northeast, and the positive of attack angle is from underside to upside of the main deck. Here, cross-bridge wind speeds are all normalized by the horizontal component of the gradient wind speed of inflow wind, namely

$$\overline{V}_H = \frac{V_H}{V_{ref} \cdot \cos\beta} \tag{19}$$

where $\overline{V}_H$ is the non-dimensional cross-bridge wind speed, V_H is the cross-bridge wind speed, V_{ref} is the gradient wind speed of inflow above which the wind speed remains unchanged with the increase of altitude, $V_{ref} = 3.70\text{m/s}$, and β is the angle between inflow wind direction and perpendicular of the bridge axis.

Wind attack angle at the main deck level is defined as follows:

$$\alpha = \arctan\left(\frac{V_V}{V_H}\right) \tag{20}$$

where α is the wind attack angle at main deck level, V_V is the vertical wind speed at main deck level, and V_H is the cross-bridge wind speed at main deck level.

5.1. Cross-Bridge Wind Speed at Main Deck Level

The numerical and experimental results of the non-dimensional cross-bridge wind speed at main deck level under different cases are given in Figure 13. Considering the span layout characteristics large-span bridges, it can be roughly divided into three parts along bridge axis; namely, mountain areas (I, III) and the central canyon area (II).

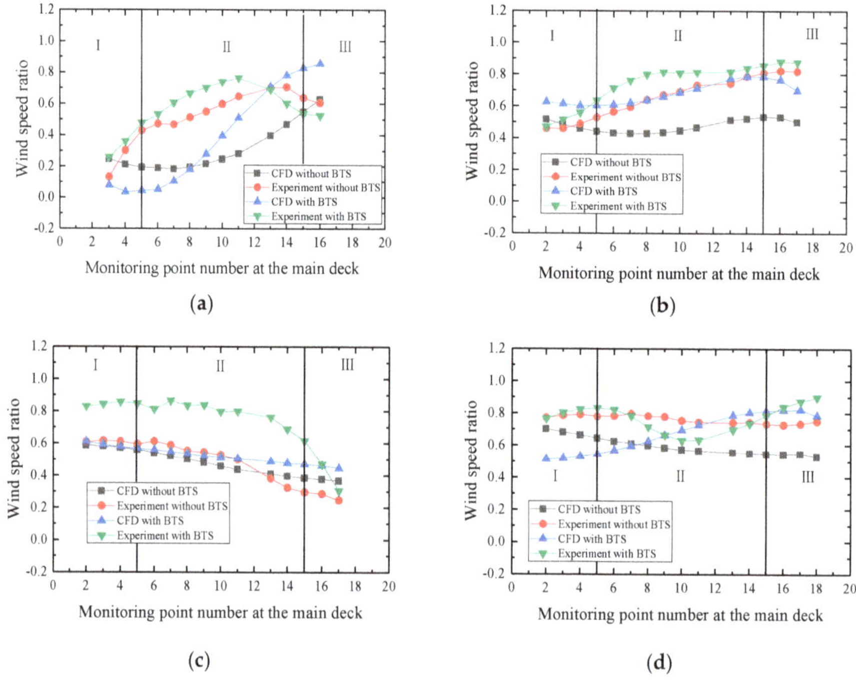

Figure 13. Non-dimensional cross-bridge wind speed at main deck level under different cases. (**a**) Case 1; (**b**) Case 2; (**c**) Case 3; (**d**) Case 4.

From Figure 13a, it can be seen that for case 1 (wind blows from the west), the non-dimensional cross-bridge wind speed at the main deck level varies greatly along the bridge axis. The non-dimensional cross-bridge wind speed in part III was relatively large, while that in part I was relative small. For experimental results, the cross-bridge wind speed of the terrain model with BTS was generally larger than that of the terrain model without BTS. For case 1, due to the topographic effects, there was a certain deviation between the numerical simulation results of the cross-bridge wind speed at the main deck level and the experimental results.

From Figure 13b, it can be seen that for case 2 (wind blows from the south), the non-dimensional cross-bridge wind speed at the main deck level varies along the bridge axis. The non-dimensional cross-bridge wind speed in part III is relatively large, while the cross-bridge wind speed in part I was relative small. For experimental results, the cross-bridge wind speed of the terrain model with BTS was generally larger than that of the terrain model without BTS. The numerical simulation results also show the same trend.

From Figure 13c, it can be seen that for case 3 (wind blows from the east), the non-dimensional cross-bridge wind speed at the main deck level varies little along the bridge axis. The non-dimensional cross-bridge wind speed in part III was relatively large, while the non-dimensional cross-bridge wind speed in part I is relatively small. For both the experimental results and numerical results, the cross-bridge wind speed of the terrain model with BTS was generally larger than that of the terrain model without BTS. Besides, it can be seen from Figure 13c that little difference of the wind characteristics with and without improved BTS was found, which indicated that the effect of improved BTS depends on the inflow direction. The altitude in the east over the complex terrain was high. When the height difference between the inlet boundary and the bridge site is large, the flow separation is severe and the effect of the improved BTS on flow separation is limited. Thus, the relation between the height difference and the improved BTS can be evaluated in future research.

From Figure 13d, it can be seen that for case 4 (wind blows from the north), the non-dimensional cross-bridge wind speed at the main deck level varies little along the bridge axis. For experimental results, the cross-bridge wind speed of the terrain model with BTS is generally smaller than that of the terrain model without BTS in part II. For numerical simulation results, the cross-bridge wind speed of the terrain model with BTS was generally larger than that of the terrain model without BTS in part II.

In conclusion, in cases 1 to 3, the wind tunnel test results of the terrain model show that the cross-bridge wind speed ratio at the main deck level of the terrain model with the improved BTS was relatively larger than that of the terrain model without BTS. A similar tendency can be found from numerical results of the terrain model with/without BTS. Thus, if the there is no improved BTS in front of the terrain model, the cross-bridge design wind speed of the bridge deck may be underestimated. It is recommended to use the improved BTS for terrain model boundary correction.

The statistical values of numerical results and experimental results of non-dimensional cross-bridge wind speed at main deck level under different cases are shown in Figure 14. It can be seen from Figure 14 that though there are some discrepancies between the wind tunnel tests results and numerical simulation results, the majority of the data is located within relative errors of 30%. This kind of error is indeed unavoidable because the scale of terrain model was small which made it a challenge to place the model in the exact center of the test section and find the accurate monitoring point positions as CFD simulation did. Thus, the wind tunnel test results can be regarded as accurate ones, and the results from CFD simulations can be used for qualitative analysis.

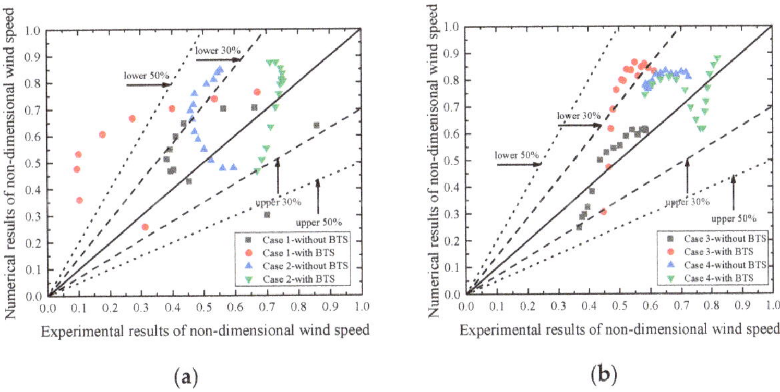

Figure 14. Numerical results vs. experimental results of non-dimensional cross-bridge wind speed at main deck level under different cases. (**a**) Case 1 and 2; (**b**) Case 3 and 4.

5.2. Wind Attack Angles at Main Deck Level

The numerical and experimental results of the wind attack angle at main deck level of the bridge for different cases are given in Figure 15. From Figure 15a, it can be seen that for case 1 (wind blows from the west), the wind attack angle at the main deck level of the bridge varied greatly along the bridge axis, which can be roughly divided into three parts; namely, mountain areas (I, III) and the central canyon area (II). For experimental results, the wind attack angles at the main deck level of the terrain model with BTS is about $-15°\sim-10°$ in part II, while the wind attack angles at the main deck level of the terrain model without BTS is $-20°\sim-5°$ in part II. For numerical results, the wind attack angles at the main deck level of the terrain model with BTS was about $-10°\sim0°$, while the wind attack angles at the main deck level of the terrain model without BTS was $-7.5°\sim5°$.

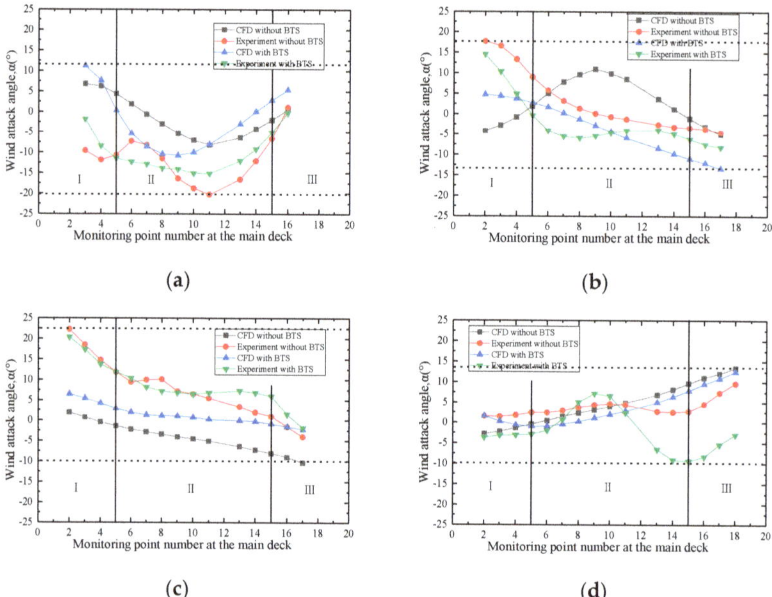

Figure 15. Wind attack angle at main deck level of wind tunnel tests and CFD simulation. (**a**) Case 1; (**b**) Case 2; (**c**) Case 3; (**d**) Case 4.

From Figure 15b, it can be seen that for case 2 (wind blows from the south), the wind attack angle at the main deck level of the bridge changes along the bridge axis. For experimental results, the wind attack angles at the main deck level of the terrain model with BTS was about −5°~0° in part II, while the wind attack angles at the main deck level of the terrain model without BTS was −5°~10° in part II. For numerical results, the wind attack angles at the main deck level of the terrain model with BTS was about −12°~2°, while the wind attack angles at the main deck level of the terrain model without BTS was about 0°~10°.

From Figure 15c, it can be seen that for case 3 (wind blows from the east), for experimental results, the wind attack angles at the main deck level of the terrain model with BTS were about 5°~12° in part II, while the wind attack angles at the main deck level of the terrain model without BTS were 1°~12° in part II. For numerical results, the wind attack angles at the main deck level of the terrain model with BTS are about 0°~3°, while the wind attack angles at the main deck level of the terrain model without BTS were about −7°~−2°.

From Figure 15d, it can be seen that for case 4 (wind blows from the north), the experimental results of the wind attack angles at the main deck level of the terrain model with BTS were about −10°~7°, while the wind attack angles at the main deck level of the terrain model without BTS were about 3°~ 5°. For numerical simulation results, the wind attack angles at the main deck level of the terrain model with BTS were about −0.5°~7.5°, while the wind attack angles at the main deck level of the terrain model without BTS were about −0.5°~10°.

In general, both numerical simulation and experimental results under different cases show that the range of the wind attack angles at the main deck level in part II with BTS is smaller than that of the terrain model without BTS. The reason is that the existence of the BTS effectively reduces the influence of artificial cliffs. Thus, the turbulence development of wind field with BTS is more sufficient than that without BTS; the wind attack angles with BTS are smaller than that without BTS.

The statistical values of numerical and experimental results of the wind attack angles at main deck level under different cases are shown in Figure 16. It can be seen from the Figure 16 that though there are some discrepancies between the wind tunnel tests and numerical simulation, the majority of the deviation is located within relative errors of 30%.

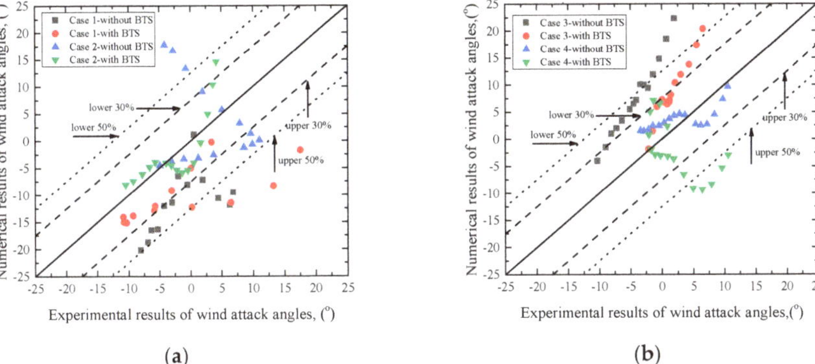

Figure 16. Numerical results vs. experimental results of wind attack angles at main deck level under different cases. (**a**) Case 1 and 2; (**b**) Case 3 and 4.

5.3. Wind Profiles at L/4, L/2 and 3L/4 of Bridge Length

The longitudinal wind speed ratio is defined as the measured longitudinal wind speed above the terrain model divided by longitudinal wind speed of the corresponding inflow boundary layer. The numerical and experimental results of longitudinal wind speed ratio profiles at L/4, L/2 and 3L/4 of the bridge length are given in Figures 17–19. As shown in Figures 17–19, the longitudinal wind speed ratio at different positions, namely L/4, L/2 and 3L/4 of the bridge length, generally increase with height.

As shown in Figure 17, the experimental results of longitudinal wind speed ratio were larger than the numerical results at L/4 of bridge length for case 1 (wind blows from the west) and case 4 (wind blows from the north), respectively. The experimental results of longitudinal wind speed ratio were slightly larger than the numerical results at L/4 of bridge length for case 2 (wind blows from the south) and case 3 (wind blows from the east), respectively. The experimental results of the longitudinal wind speed ratio at L/4 of the bridge length shows that the longitudinal wind speed ratio of terrain model with BTS were larger than that of the terrain model without BTS for different cases. The numerical results at L/4 of the bridge length shows that the longitudinal wind speed ratio of terrain model with BTS were slightly larger than that of terrain model without BTS under cases 2 to 4, except for case 1.

As shown in Figure 15, the experimental results of longitudinal wind speed ratio at L/2 of bridge length were larger than the numerical results for case 1 (wind blows from the west) and case 2 (wind blows from the north), respectively. The experimental results of longitudinal wind speed ratio were close to the numerical results for case 3 (wind blows from the east) and case 4 (wind blows from the north), respectively. The experimental results of the longitudinal wind speed ratio at L/2 of the bridge length shows that the longitudinal wind speed ratio of terrain model with BTS were larger than that of the terrain model without BTS for cases 1 to 3, except for case 4. The numerical results of the longitudinal wind speed ratio at L/2 of the bridge length shows that the longitudinal wind speed ratio of terrain model with BTS were slightly larger than that of terrain model without BTS under cases 1, 2, and 4, except for case 3.

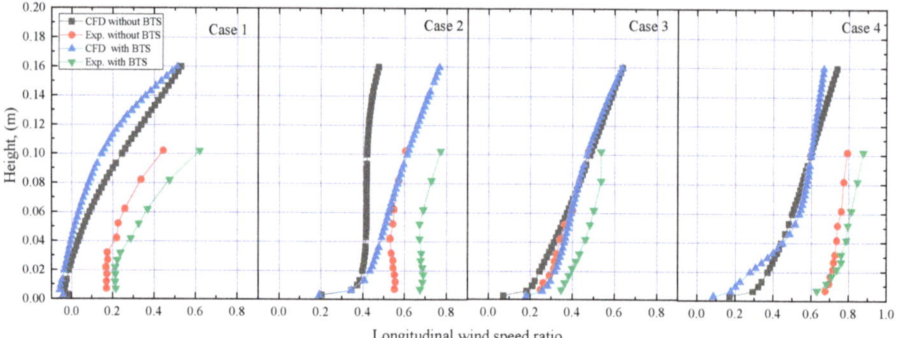

Figure 17. Numerical and experimental results of longitudinal wind speed ratio profiles at L/4.

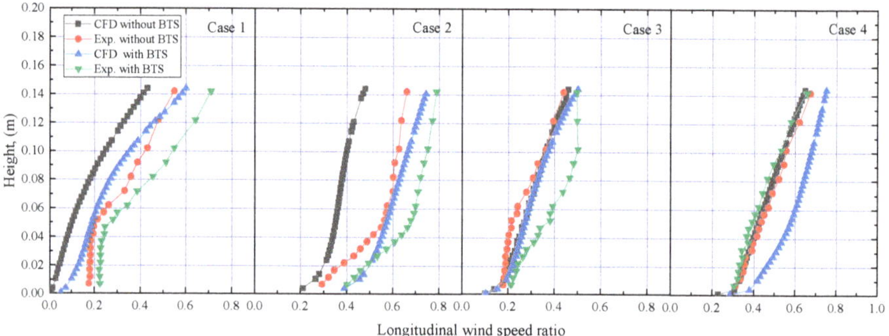

Figure 18. Numerical and experimental results of longitudinal wind speed ratio profiles at L/2.

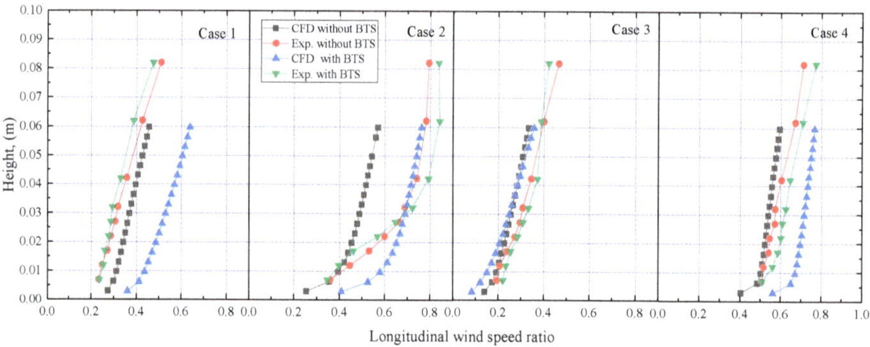

Figure 19. Numerical and experimental results of longitudinal wind speed ratio profiles at 3L/4.

As shown in Figure 16, the experimental results of longitudinal wind speed ratio at 3L/4 of bridge length are close to the numerical results for cases 2 to 4. The experimental results of longitudinal wind speed ratio at 3L/4 of bridge length were smaller than the numerical results for case 1. The experimental results shows that the longitudinal wind speed ratio of terrain model with BTS were slightly smaller than that of the terrain model without BTS for case 1 while the longitudinal wind speed ratios of the terrain model with BTS were slightly larger than that of the terrain model without BTS for cases 2 to 4.

The statistical values of the numerical and experimental results of longitudinal wind speed ratio at L/4, L/2 and 3L/4 under different cases are given in Figures 20–22. From Figure 20a, it can be seen that the relative error between numerical and experimental results of longitudinal wind speed ratios at L/4 of the bridge length under case 1 with/without BTS were about 50%, while the relative error between numerical and experimental results of longitudinal wind speed ratios at L/4 of the bridge length under case 2 with/without BTS were about 30%. From Figure 20b, it can be seen that the relative error between numerical and experimental results of longitudinal wind speed ratio at L/4 of the bridge length under case 3 and 4 with/without BTS were about 30%. As shown in Figure 18, the relative error between numerical and experimental results of longitudinal wind speed ratios at L/2 of the bridge length under different cases with/without BTS were less about 30% except for case 3 with BTS. From Figure 22, the relative error between numerical and experimental results of longitudinal wind speed ratios at 3L/4 of the bridge length under different cases with/without BTS were less than 30%. In general, it can be seen from the Figures 20–22 that though there were some discrepancies between the wind tunnel tests and numerical simulation, the majority of the data are located within relative errors of 30%.

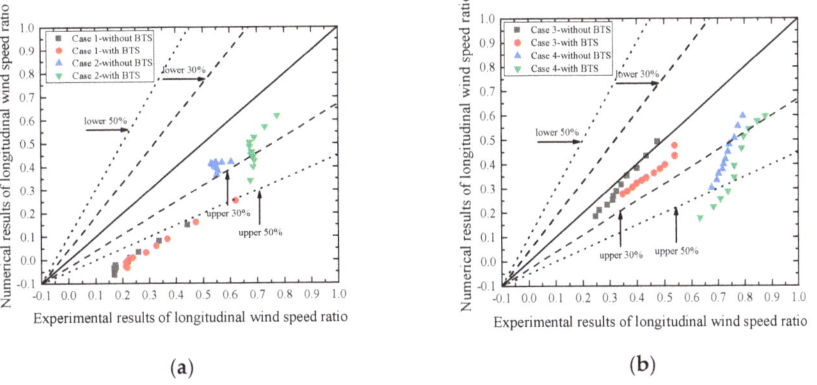

Figure 20. Numerical results vs. experimental results of longitudinal wind speed ratio at L/4 under different cases. (**a**) Case 1 and 2; (**b**) Case 3 and 4.

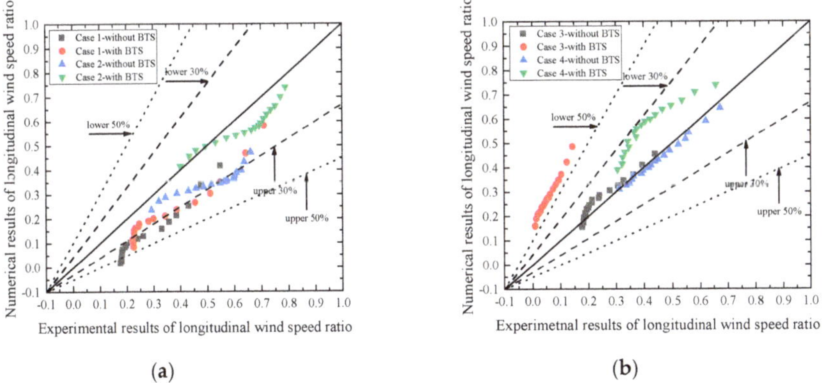

Figure 21. Numerical results vs. experimental results of longitudinal wind speed ratio at L/2 under different cases. (**a**) Case 1 and 2; (**b**) Case 3 and 4.

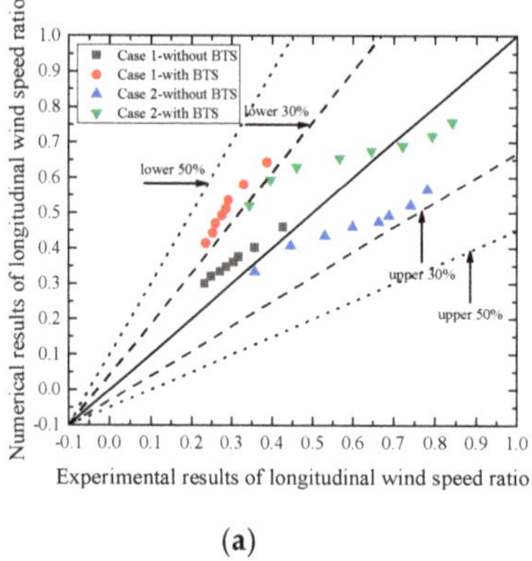
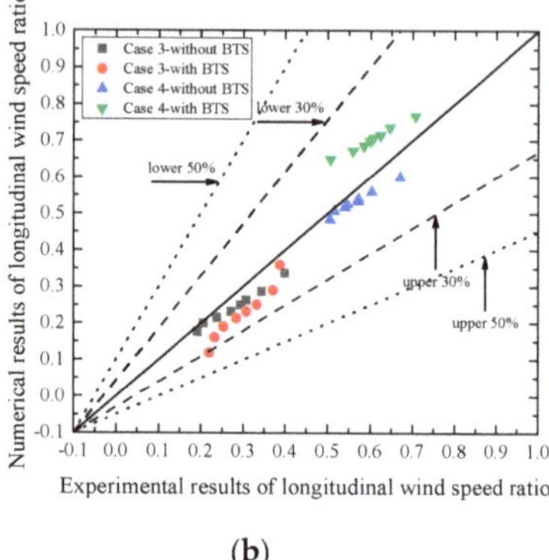

(a)   (b)

Figure 22. Numerical results vs. experimental results of longitudinal wind speed ratio at 3L/4 under different cases. (**a**) Case 1 and 2; (**b**) Case 3 and 4.

6. Conclusions

To study the wind characteristics over mountainous terrain, the Xiangjiang Bridge site was employed in this paper. The improved boundary transition sections were adopted to reduce the influence of the artificial cliffs on the edge of the terrain model on the wind characteristics at the bridge site over the mountainous terrain. The mean cross-bridge wind speed and wind attack angles at the main deck level, longitudinal wind speed profiles at L/4, L/2, and 3L/4 of the bridge length were investigated in detail with the assistance of numerical simulations and wind tunnel tests. The main conclusions are summarized as follows:

1. The cross-bridge wind speeds and wind attack angles at the main deck level of the bridge at mountainous terrain site vary greatly along the bridge axis. Considering the span layout characteristics large-span bridges, it can be roughly divided into three parts along bridge axis, namely, mountain areas (I, III) and central canyon area (II). The changes in wind characteristics near the mountain areas (I, III) were relatively large, while the changes in wind speed and wind attack angles in the central canyon area (III) were relatively small.

2. Experimental and numerical results show that the cross-bridge wind speed at the main deck level of the mountainous terrain model with BTS was generally larger than that of the mountainous terrain model without BTS in the central canyon area (II) for most cases, which needs to be paid special attention in wind-resistance design of the bridge. If there is no improved BTS in front of the terrain model, the cross-bridge design wind speed of the bridge deck may be underestimated.

3. In general, the range of wind attack angles at the main deck level of the mountainous terrain model with BTS within 1/4 to 3/4 of the bridge length; namely, in the central canyon area (II), are smaller than the range of wind attack angles of the mountainous terrain model without BTS. It is recommended to use the improved BTS for terrain model boundary correction.

4. The longitudinal wind speed ratio at different positions; namely, L/4, L/2 and 3L/4 of the bridge length, generally increase with height. The longitudinal wind speed ratio of the terrain model with BTS at L/4, L/2 and 3L/4 of the bridge length are larger than that of the terrain model without BTS for most cases.

5. In general, there are some discrepancies between the numerical results and wind tunnel tests results of wind characteristics, namely cross-bridge wind speed ratios, wind attack angles, and longitudinal wind speed ratios at L/4, L/2, and 3L/4 of the bridge length, but the maximum relative error between numerical and experimental results for most cases is about 30%.

Author Contributions: Conceptualization, Z.C., Z.L. and X.C.; methodology, Z.L. and X.C.; software, X.C.; validation, Z.L., X.W. and J.Z.; formal analysis, H.X.; investigation, X.C.; resources, Z.L., X.W. and J.Z.; data curation, H.X.; writing—original draft preparation, X.C.; writing—review and editing, Z.L.; visualization, H.X.; supervision, Z.C.; project administration, X.W. and J.Z.; funding acquisition, Z.L. All authors have read and agree to the published version of the manuscript.

Funding: This research was funded by the National Natural Science Foundation of China, grant number 51478180 and 51778225 and China Railway Siyuan Survey and Design Group Co., Ltd.

Acknowledgments: We deeply appreciate the assistance provided by Shuqiong Li, Ruilin Zhang, Yuefei Chen, Beisong Sun, Yafeng Li, Jianzhong Wang and Jingsi Shen in the process of wind tunnel test and the administrative and technical support by the Hunan Provincial Key Lab of Wind Engineering and Bridge Engineering.

Conflicts of Interest: The authors declare no conflict of interest. The funders of China Railway Siyuan Survey and Design Group Co., Ltd. provided authors with the access to the engineering example, Xiangjiang Bridge site and collection, analyses or interpretation of data.

References

1. Chock, G.Y.K.; Cochran, L. Modeling of topographic wind speed effects in Hawaii. *J. Wind Eng. Ind. Aerodyn.* **2005**, *93*, 623–638. [CrossRef]
2. Jackson, P.S.; Hunt, J.C.R. Turbulent wind flow over a low hill. *Q. J. R. Meteorol. Soc.* **1975**, *101*, 929–955. [CrossRef]
3. Hunt, J.C.R.; Leibovich, S.; Richards, K.J. Turbulent shear flows over low hills. *Q. J. R. Meteorol. Soc.* **1988**, *114*, 1435–1470. [CrossRef]
4. Li, Q.S.; Li, X.; He, Y.; Yi, J. Observation of wind fields over different terrains and wind effects on a super-tall building during a severe typhoon and verification of wind tunnel predictions. *J. Wind Eng. Ind. Aerodyn.* **2017**, *162*, 73–84. [CrossRef]
5. Harstveit, K. Full scale measurements of gust factors and turbulence intensity, and their relations in hilly terrain. *J. Wind Eng. Ind. Aerodyn.* **1996**, *61*, 195–205. [CrossRef]
6. Hannesen, R.; Dotzek, N.; Handwerker, J. Radar analysis of a tornado over hilly terrain on 23 July 1996. *Phys. Chem. Earth Part B Hydrol. Ocean. Atmos.* **2013**, *25*, 1079–2084. [CrossRef]
7. Lubitz, W.D.; White, B.R. Wind-tunnel and field investigation of the effect of local wind direction on speed-up over hills. *J. Wind Eng. Ind. Aerodyn.* **2007**, *95*, 639–661. [CrossRef]
8. Sharples, J.J.; Mcrae, R.H.D.; Weber, R.O. Wind characteristics over complex terrain with implications for bushfire risk management. *Environ. Model. Softw.* **2010**, *25*, 1099–1120. [CrossRef]
9. Abiven, C.; Palma, J.M.L.M.; Brady, O. High-frequency field measurements and time-dependent computational modelling for wind turbine siting. *J. Wind Eng. Ind. Aerodyn.* **2011**, *99*, 123–129. [CrossRef]
10. Risan, A.; Lund, J.A.; Chang, C.Y.; Sætran, L. Wind in complex terrain—Lidar measurements for evaluation of CFD simulations. *Remote Sens.* **2018**, *10*, 59. [CrossRef]
11. Lystad, T.M.; Fenerci, A.; Øiseth, O. Evaluation of mast measurements and wind tunnel terrain models to describe spatially variable wind field characteristics for long-span bridge design. *J. Wind Eng. Ind. Aerodyn.* **2018**, *179*, 558–573. [CrossRef]
12. Peng, Y.; Wang, S.; Li, J. Field measurement and investigation of spatial coherence for near-surface strong winds in Southeast China. *J. Wind Eng. Ind. Aerodyn.* **2018**, *172*, 423–440. [CrossRef]
13. Zhang, J.W.; Li, Q.S. Field measurements of wind pressures on a 600m high skyscraper during a landfall typhoon and comparison with wind tunnel test. *J. Wind Eng. Ind. Aerodyn.* **2018**, *175*, 391–407. [CrossRef]
14. Yu, C.; Li, Y.; Zhang, M.; Zhang, Y.; Zhai, G. Wind characteristics along a bridge catwalk in a deep-cutting gorge from field measurements. *J. Wind Eng. Ind. Aerodyn.* **2019**, *186*, 94–104. [CrossRef]
15. Jing, H.; Liao, H.; Ma, C.; Tao, Q.; Jiang, J. Field measurement study of wind characteristics at different measuring positions in a mountainous valley. *Exp. Therm. Fluid Sci.* **2020**, *112*, 1–18. [CrossRef]
16. Cermak, J.E. Physical modelling of flow and dispersion over complex terrain. *Bound. Layer Meteorol.* **1984**, *30*, 261–292. [CrossRef]
17. Xu, H.; He, Y.; Liao, H.; Ma, C.; Xian, R. Experiment of wind field in long-span bridge site located in mountainous valley terrain. *J. Highw. Transp. Res. Dev.* **2011**, *7*, 44–50.
18. Kozmar, H.; Allori, D.; Bartoli, G.; Borri, C. Complex terrain effects on wake characteristics of a parked wind turbine. *Eng. Struct.* **2016**, *110*, 363–374. [CrossRef]

19. Yan, B.W.; Li, Q.S. Coupled on-site measurement/CFD based approach for high-resolution wind resource assessment over complex terrains. *Energy Convers. Manag.* **2016**, *117*, 351–366. [CrossRef]
20. Mattuella, M.L.J.; Loredo-Souza, M.A.; Oliveira, G.K.M.; Petry, P.A. Wind tunnel experimental analysis of a complex terrain micrositing. *Renewable Sustain. Energy Rev.* **2016**, *54*, 110–119. [CrossRef]
21. Muhammad, J.C.; Horia, H. A hybrid approach for evaluating wind flow over a complex terrain. *J. Wind Eng. Ind. Aerodyn.* **2018**, *175*, 65–76. [CrossRef]
22. Kozmar, H.; Allori D.; Bartoli, G.; Borri, G. Wind characteristics in wind farms situated on a hilly terrain. *J. Wind Eng. Ind. Aerodyn.* **2018**, *174*, 404–410. [CrossRef]
23. Chen, F.; Peng, H.; Chan, P.; Zeng, X. Low-level wind effects on the glide paths of the North Runway of HKIA: A wind tunnel study. *Build. Environ.* **2019**, *164*, 1–8. [CrossRef]
24. Flay, G.J.R.; King, A.B.; Revell, M.; Carpenter, P.; Turner, R.; Cenek, P.; Pirooz, A.A.S. Wind speed measurements and predictions over Belmont Hill, Wellington, New Zealand. *J. Wind Eng. Ind. Aerodyn.* **2019**, *195*, 1–18. [CrossRef]
25. Bowen, A.J. Modelling of strong wind flows over complex terrain at small geometric scales. *J. Wind Eng. Ind. Aerodyn.* **2003**, *91*, 1859–1871. [CrossRef]
26. Maurizi, A.; Palma, J.M.L.M.; Castro, F.A. Numerical simulation of the atmospheric flow in a mountainous region of the north of Portugal. *J. Wind Eng. Ind. Aerodyn.* **1998**, *74–76*, 219–228. [CrossRef]
27. Pang, J.; Song, J.; Lin, Z. Determination of design wind speed on bridge site over mountainous areas. *China J. Highw. Transp.* **2008**, *21*, 39–44.
28. Hu, P.; Li, Y.; Huang, G.; Kang, R.; Liao, H. The appropriate shape of the boundary transition section for a mountain gorge terrain model in a wind tunnel test. *Wind Struct.* **2015**, *20*, 15–36. [CrossRef]
29. Li, Y.; Hu, P.; Xu, X.; Qiu, J. Wind characteristics at bridge site in a deep-cutting gorge by wind tunnel test. *J. Wind Eng. Ind. Aerodyn.* **2017**, *160*, 30–46. [CrossRef]
30. Huang, G.; Cheng, X.; Peng, L.; Li, M. Aerodynamic shape of transition curve for truncated mountainous terrain model in wind field simulation. *J. Wind Eng. Ind. Aerodyn.* **2018**, *178*, 80–90. [CrossRef]
31. Hu, P.; Han, Y.; Xu, G.; Asce, M.A.; Li, Y.; Xue, F. Numerical Simulation of Wind Fields at the Bridge Site in Mountain-Gorge Terrain Considering an Updated Curved Boundary Transition Section. *J. Aerosp. Eng.* **2018**, *31*, 1–14. [CrossRef]
32. Liu, Z.; Chen, X.; Chen, Z. Optimization of Transition sections around terrain model at mountain canyon bridge site. *China J. Highw. Transp.* **2019**, *32*, 266–278.
33. Uchida, T.; Ohya, Y. Large-eddy simulation of turbulent airflow over complex terrain. *J. Wind Eng. Ind. Aerodyn.* **2003**, *91*, 219–229. [CrossRef]
34. Tong, H.; Walton, A.; Sang, J.; Chan, J.C.L. Numerical simulation of the urban boundary layer over the complex terrain of Hong Kong. *Atmos. Environ.* **2005**, *39*, 3549–3563. [CrossRef]
35. Deleon, R.; Sandusky, M.; Senocak, I. Simulations of turbulent flow over complex terrain using an immersed-boundary method. *Bound. Layer Meteorol.* **2018**, *167*, 399–420. [CrossRef]
36. Tamura, T.; Okuno, A.; Sugio, Y. LES analysis of turbulent boundary layer over 3D steep hill covered with vegetation. *J. Wind Eng. Ind. Aerodyn.* **2007**, *95*, 1463–1475. [CrossRef]
37. Cassiani, M.; Katul, G.G.; Albertson, J.D. The effects of canopy leaf area index on airflow across forest edges: Large-eddy simulation and analytical results. *Bound. Layer Meteorol.* **2008**, *126*, 433–460. [CrossRef]
38. Kim, H.G.; Patel, V.C.; Lee, C.M. Numerical simulation of wind flow over hilly terrain. *J. Wind Eng. Ind. Aerodyn.* **2000**, *87*, 45–60. [CrossRef]
39. Castellani, F.; Astolfi, D.; Burlando, M.; Terzi, L. Numerical modelling for wind farm operational assessment in complex terrain. *J. Wind Eng. Ind. Aerodyn.* **2015**, *147*, 320–329. [CrossRef]
40. Bitsuamlak, G.T.; Stathopoulos, T.; Bédard, C. Numerical evaluation of wind flow over complex terrain: Review. *J. Aerosp. Eng.* **2004**, *17*, 135–145. [CrossRef]
41. Yassin, M.F.; Al-Harbi, M.; Kassem, M.A. Computational fluid dynamics (CFD) simulations on the effect of rough surface on atmospheric turbulence flow above hilly terrain shapes. *Environ. Forensics* **2014**, *15*, 159–174. [CrossRef]
42. Yan, B.W.; Li, Q.S.; He, Y.C.; Chan, P.W. RANS simulation of neutral atmospheric boundary layer flows over complex terrain by proper imposition of boundary conditions and modification on the k-ε model. *Environ. Fluid Mech.* **2016**, *16*, 1–23. [CrossRef]

43. Wang, S.; Liu, X.; Lu, L. Simulation analysis of low-speed blow down wind tunnel with contraction curves. *Mach. Tool Hydraul.* **2012**, *40*, 100–104.

44. Huang, G.; Jiang, Y.; Peng, L.; Solari, G.; Liao, H.; Li, M. Characteristics of intense winds in mountain area based on field measurement: Focusing on thunderstorm winds. *J. Wind Eng. Ind. Aerodyn.* **2019**, *190*, 166–182. [CrossRef]

45. Tang, W. Quantitative techniques and weight allocation of evaluation indexes for complex systems. *Sci. Technol. Process Policy* **2009**, *26*, 116–118.

46. Ge, Y. Aerodynamic Design of Lupu Bridge in Shanghai. In Proceedings of the 3rd International Conference on Current and Future Trends in Bridge Design, Construction and Maintenance, Shanghai, China, 28–30 April 2003; pp. 69–80.

Article

Study on the Horizontal Axis Deviation of a Small Radius TBM Tunnel Based on Winkler Foundation Model

Shifan Qiao [1], Ping Xu [1], Ritong Liu [2] and Gang Wang [3],*

[1] Department of Civil Engineering, Central South University, Changsha 410075, China; qiaosf@csu.edu.cn (S.Q.); xplt0615@csu.edu.cn (P.X.)

[2] China Energy Engineering Group Hunan Electric Power Design Institute Co., Ltd., Changsha 410075, China; RitongLiu@163.com

[3] Department of Civil and Environmental Engineering, Tokyo Institute of Technology, Tokyo 152-8552, Japan

* Correspondence: wanggang123@csu.edu.cn; Tel.: +86-19851531805

Received: 9 December 2019; Accepted: 19 January 2020; Published: 22 January 2020

Abstract: During the construction stage of the small radius TBM (tunnel boring machine) interval, the improper control of the boring parameters and the boring posture can cause the horizontal axis deviation of the shield tunnel. In order to address this issue, the TBM segments lining structure of the small radius interval is simplified as the continuous circular curved beam based on the longitudinal equivalent continuous model and Winkler elastic foundation beam theory. The theoretical model is solved through the transfer matrix method, and its applicability is verified by comparing it with the field monitoring data. It is found that the horizontal axis deviation of the completed tunnel increases with the total jack thrust, and the lateral displacement tends to be stable when the distance between the ring and the tail is far. The horizontal axis deviation has a negative relationship with the thrust difference or path difference when the jack thrust in the outside of the shield curve is larger than that of inside the shield curve. The horizontal axis deviation has a positive relationship with the thrust difference or path difference when the jack thrust in the outside of the shield curve is smaller than that of inside the shield curve.

Keywords: small radius TBM interval; equivalent continuous model; Winkler elastic foundation beam theory; transfer matrix method; horizontal axis deviation

---

1. Introduction

Tunnel boring machine (TBM) technology has been widely used for the construction of the urban subway tunnel. The longitudinal performance research of the TBM tunnel based on the equivalent continuous model and elastic foundation theory is a hotspot. The longitudinal affects the safe operation of the subway, which has also been widely studied through field monitoring, numerical analysis and analytical solution [1–5]. The research on the tunnel longitudinal performance focuses on the longitudinal settlement and horizontal axis deviation, which are caused by the coupling effects of multi-load [6–9].

Koizumi et al. [10] and Shiba et al. [11] were the pioneers developing the analytical solution for calculating the longitudinal settlement based on longitudinal bending stiffness. Elastic foundation models such as the Winkler model [12], the Pasternak model [13], and the Kerr model [14] are the most widely used theoretical calculation models in the longitudinal deformation research of tunnels. After that, the generalized longitudinal equivalent continuous model was proposed by Zhang et al. [15]. Li et al. [16] proposed the model considering the mechanical behavior of blots. Shiba et al. [17] and Talmon and Bezuijen [18] used the continuous elastic beam to simulate the longitudinal structure model of the shield tunnel, which can analyze the longitudinal property of the shield tunnel. Yu et al. [19] derived an analytical solution for the longitudinal bending stiffness of a segmental liner, which was mainly utilized in the shield tunnels. This method is verified by the simulation results from the finite element program.

The longitudinal settlement mechanism of the tunnel can also be analyzed by the two-dimensional and three-dimensional numerical simulation methods [20–22]. Huang et al. [23] developed a finite element method for analyzing the longitudinal performance of shield tunnels considering the longitudinal variation of geotechnical parameters. Yuan et al. [9] analyzed the deformation of the shield tunnel and the surrounding soil by building a three-dimensional model of shield tunnel. A numerical model was established to simulate the earth pressure balance (EPB) in excavation processes—the simulation results were compared with those obtained by the field measurement [24,25].

Also, field monitoring is an important method to analyze the longitudinal settlement of the shield tunnel. Ocak [26] studied the relationship between shield parameters and the shield surface settlement through field surface settlement measurements. Fargnoli et al. [27] collected the measurements of tunneling-induced settlements during the construction of the new Milan underground line 5, which were back analyzed with the method of the classical Gaussian empirical expression. Then, the detailed description of the EPB tunneling performance in the transverse and longitudinal directions could be achieved.

The research on the longitudinal performance of shield tunnels is mainly focused on the longitudinal settlement of tunnels. Limited research has been conducted on the problem of horizontal axis deviation of the small radius TBM interval construction.

In the design of urban subway lines, the sharp curve of the shield tunnel with a small curvature radius is usually adopted to avoid the adverse geological conditions and surrounding existing buildings [28,29]. When the curvature radius is less than 40 times of the TBM's diameter, the construction difficulty and quality risk will be increased significantly. The damage caused by the horizontal axis deviation of the constructed tunnel is particularly prominent, which can easily affect the assembly quality of the segments. The horizontal axis deviation can result in tunnel mud leakage and local structural damage, which pose a severe threat to the safety of metro operation. Therefore, it is of great significance to study the effect of the construction load on the horizontal axis deviation of a small radius TBM interval tunnel.

The objective of this paper is to propose a solution of the tunnel horizontal axis deviation of small radius TBM interval with the transfer-matrix method, which is based on generalized longitudinal equivalent continuum model and Winkler elastic foundation beam theory. The applicability of the calculation model is verified through the monitored data in the field. The factors such as the jack thrust and shield posture affecting the tunnel horizontal axis deviation are analyzed.

2. Generalized Longitudinal Equivalent Continuous Model

2.1. Model Description

The prefabricated segment lining structure used in the TBM tunnel has a large number of longitudinal joints, which makes the deformation mechanism of the lining structure complicated. Various theoretical models for longitudinal deformation calculation have simplified the joint structures. Among these models, the longitudinal equivalent continuous model proposed by Shiba et al. [17] has the most extensive application. The tunnel cross-section is assumed as a homogeneous ring, and the tunnel longitudinal integral rigidity is also reduced when considering the joints. The tunnel is simplified as a uniform continuous beam with the equivalent stiffness on the elastic foundation.

Traditional models ignore the impacts of deformation of the tunnel cross-section and the limitation of the influential region of the longitudinal joints, which leads to the lower longitudinal effective rigidity ratio compared with the measured value. The generalized longitudinal equivalent continuous model [15] is adopted to calculate the longitudinal equivalent bending rigidity of the TBM tunnel segment lining structure in this paper. This model takes account of the influential region of the joints and lateral stiffness.

2.2. Basic Assumptions

Some assumptions are proposed in the generalized longitudinal equivalent continuous model:

(a)   The initial cross-section is uniform circular and develops into ellipse under the pressure of surrounding rock.

(b)   The cross-section before and after deformation is always plane under the longitudinal bending.

(c)   The stress distribution along the tunnel axial keeps unchanged on the neutral axis and inside the lining ring. One side of the neutral axis is under tension while the other side is under compression. The variation of the stress of lining structure along the thickness is ignored.

(d)   The bolts discretely distributed along the ring are transformed into continuously distributed bolt-ring, which can connect the annular gap. The stiffness of bolt-ring is uniformly distributed along the longitudinal length and radial thickness of the joints. The equivalent stiffness of bolt rings k_j is as follows [30].

$$k_j = \frac{nk_{ji}}{2\pi Rt} = \frac{nE_j A_j}{2\pi Rtl},$$ (1)

where k_{ji} is the elastic stiffness coefficient of a single bolt, n is the number of bolts inside the ring, R is the average of the internal and external diameter of the segment, t is the thickness of the segment, E_j is the elastic modulus of the bolt, A_j is the cross-section area of the bolt, and l is the interaction length of the bolt.

(e)   The influence coefficient of the annular gap λ is introduced, and the influence scope of the annular gap is λl. When $\lambda < 1$, the interaction length of ring direction joints within the influence scope of the annular gap is λl; when $\lambda \geq 1$, the interaction length of ring direction joints within the influence scope of the annular gap is l.

(f)   Within the influence scope of the annular gap, the lining ring undertakes all the compressive stress caused by bending, and the tensile stress is jointly undertaken by equivalent bolt-ring and segment lining ring. Outside the influence scope of the annular gap, the tensile and compressive stresses caused by bending are both undertaken by the segment lining ring.

2.3. Longitudinal Equivalent Bending Stiffness

Under the longitudinal bending, when the tensions of the bolts are all smaller than the elastic limit P_y, the whole segment is in the elastic state. Figures 1 and 2 show the longitudinal deformation and stress distribution of the segment inside and outside the influence scope of annular gaps. a and b are the long half axis and short half axis of the ellipse respectively, the ellipse is the cross-section of the segment under surrounding rock pressure, ϕ shows the position of neutral axis, c is the distance between neutral axis and y-axis, ds is the length of the micro-segmentation, $d\beta$ is the corresponding angle and x is the distance between neutral axis and the micro-segmentation, ε_t is the tension strain of the segment and ε_c is the compressive strain of the segment, M is the bending moment, E_c is the elastic modulus of tunnel reinforced concrete segment and δ is the displacement; ls is the segment length between the centerline of the two-segment rings and θ is the rotation angle.

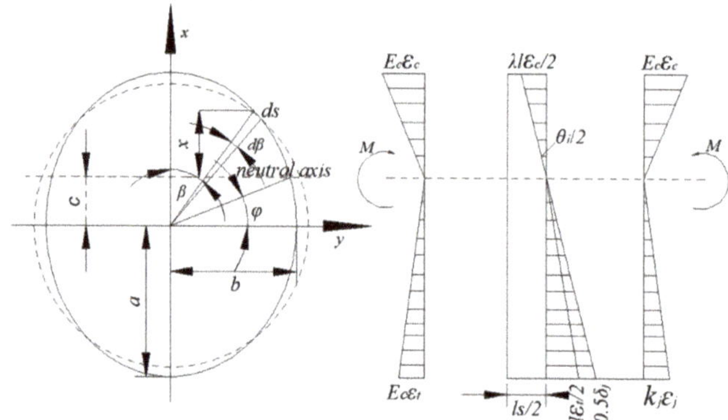

Figure 1. Diagram of stress and deformation of the lining within the influence range of annular gaps [16].

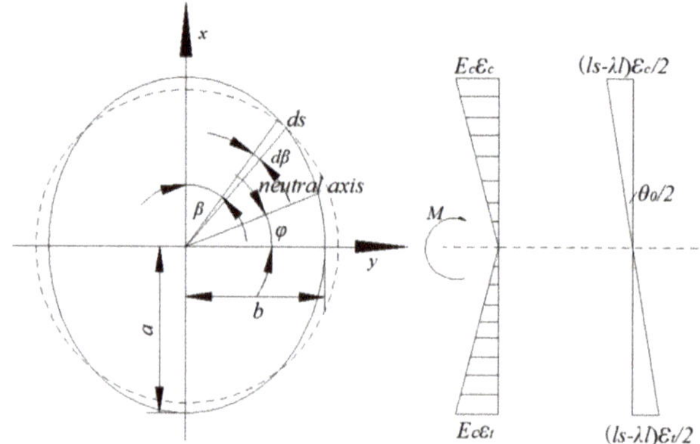

Figure 2. Diagram of stress and deformation of the lining outside the influence range of annular gaps [16].

Based on the basic assumptions, the equilibrium equation, the deformation compatibility equation, and the tunnel equivalent bending stiffness $(EI)_{eq}$ can be obtained [31].

$$(EI)_{eq} = \frac{Ml_s}{\theta} = \frac{Ml_s}{\frac{M}{\frac{E_c Z_1}{\lambda l} + \frac{E_c k_j Z_2}{\lambda E_c + \lambda k_j l}} + \frac{M(l_s - \lambda l)}{E_c I_c}} = E_c I_c \frac{l_s}{\frac{E_c I_c}{\frac{E_c Z_1}{\lambda l} + \frac{E_c k_j Z_2}{\lambda (E_c + k_j l)}} + (l_s - \lambda l)}.$$
(2)

The longitudinal stiffness effective rate η is [32],

$$\eta = \frac{(EI)_{eq}}{E_c I_c},$$
(3)

where [16],

$$Z_1 = \frac{(a+b)b^2 t\left(\frac{\pi}{2} - \varphi\right)}{2} + \frac{(a+b)^3 t \sin^2 \varphi\left(\frac{\pi}{2} - \varphi\right)}{4} + \frac{(a+b)b^2 t \sin 2\varphi}{4} - \frac{(a+b)^2 bt \sin 2\varphi}{2}$$

$$Z_2 = \frac{(a+b)b^2 t\left(\frac{\pi}{2} + \varphi\right)}{2} + \frac{(a+b)^3 t \sin^2 \varphi\left(\frac{\pi}{2} + \varphi\right)}{4} - \frac{(a+b)b^2 t \sin 2\varphi}{4} + \frac{(a+b)^2 bt \sin 2\varphi}{2}$$

I_c is the inertia moment of tunnel segment cross-section; t is the thickness of lining ring.

3. Elastic Foundation Beam Model Used in the Small Radius TBM Interval

3.1. Model Description

The elastic foundation model is the most widely used theoretical model in the study of tunnel longitudinal deformation, mainly it has three categories: (a) Winkler foundation model [12], the foundation is regarded as a series of independent springs. The soil properties are manifested by the stiffness of the spring. (b) Pasternak model [13], an incompressible shear layer, only having horizontal shear deformation, is added in the Winkler foundation model. The shearing between springs is considered in the Pasternak model. (c) Kerr model [14], the spring layer, is added based on the Pasternak model. In contrast, the parameters in the Winkler foundation model are straightforward. When the compressible strata are thin and a hard layer exists, the calculation results based on the Winkler foundation model agree well with the real present situation. Therefore, the Winkler foundation model is selected to calculate the horizontal axis deviation of a small radius TBM tunnel subjected to construction loading. The lining structure of TBM segments in the small radius interval is simplified as the continuous circular curved beam on the Winkler foundation based on the equivalent elastic foundation beam theory (Figure 3). q_r is the radial linear load on the curved beam; q_t is the tangential linear load on the curved beam. Φ is the arbitrary angle on the circular curved beam; p is the concentrated force.

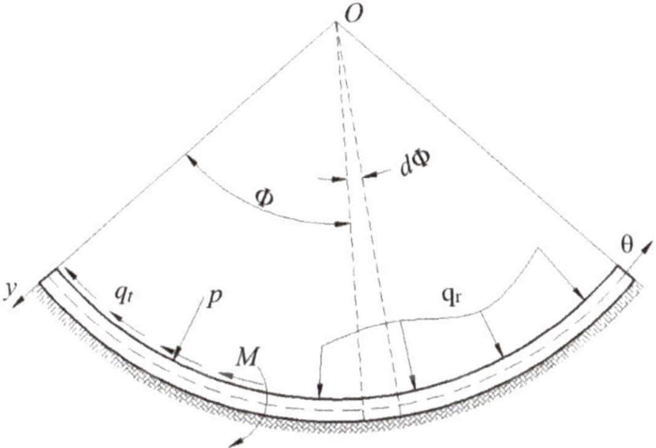

Figure 3. Schematic diagram of the circular curved beam on the Winkler elastic foundation [33].

3.2. Basic Assumptions

Some basic assumptions are:

(a)   The small radius internal segment is regarded as the Euler-Bernoulli circular curved beam on the Winkler foundation.
(b)   The material of the curved beam is the linearly elastic body and conforms to the Hook's law.
(c)   The curvature radius of the curved beam neutral axis keeps unchanged after deformation.
(d)   One end of the segment, away from shield tail and anchored in the surrounding rocks by grouting, is simplified as the fixed support.
(e)   The modified longitudinal equivalent continuous model is used to solve the longitudinal equivalent stiffness of the segment lining.

3.3. Model Solution

The uniform cross-section circular curved beam on the Winkler elastic foundation is depicted in Figure 4. The microelement body of the circular curved beam on the Winkler elastic foundation is shown in Figure 5. The equivalent equation of the microelement body, ignoring the high order, is shown below [32].

$$\frac{dQ}{dx} = p - \frac{N}{R} - q_r,$$

(4)

$$\frac{dN}{dx} = \frac{Q}{R} - q_t,$$

(5)

$$\frac{dM}{dx} = Q,$$

(6)

where Q is the shear force; N is the axial force.

For simplification, the effect of axial force on the deformation is neglected. Then, the deflection differential equation of the circular curved beam is obtained [34],

$$\left(\frac{d^2y}{dx^2} + \frac{y}{R^2}\right) + \frac{M}{EI} = 0,$$

(7)

Combining Equations from (4) to (7), we can get deflection differential equation of the circular curved beam, which ignores the tangential deformation [34],

$$\frac{d^5y}{d\Phi^5} + 2\frac{d^3y}{d\Phi^3} + \mu^2\frac{dy}{d\Phi} = \frac{R^4}{EI}\left(\frac{dq_r}{d\Phi} + q_t\right),$$

(8)

where EI is the bending stiffness of the tunnel, μ^2 is the coefficient, $\mu^2 = 1 + R^4DK/EI$, K is the coefficient of foundation bedding, D is the width of the curved beam cross-section, R is the curvature radius of the curved beam.

When there is no load on the beam span, the Equation (8) can be changed into,

$$\frac{d^5y}{d\Phi^5} + 2\frac{d^3y}{d\Phi^3} + \mu^2\frac{dy}{d\Phi} = 0.$$

(9)

This is the homogenous fifth-order differential equation with constant coefficients, whose general solution is as below [34],

$$y(\Phi) = C_1 + (C_2 ch\alpha\Phi + C_3 sh\alpha\Phi)\cos\beta\Phi + (C_4 ch\alpha\Phi + C_5 sh\alpha\Phi)sin\beta\Phi,$$

(10)

where,

$$\alpha = \sqrt{\frac{(\mu-1)}{2}}, \beta = \sqrt{\frac{(\mu+1)}{2}};$$

C_1–C_5 are undetermined coefficients, obtained by loading and boundary conditions.

With the successive derivation of Equation (10), we can obtain analytical expressions of the rotation angle θ, the bending moment M, the shear force Q, the axial force N. With the initial condition of the beam, we can get the solutions of undetermined coefficients C_1–C_5, the general solution of the defection differential equation can be converted into the below equation [34],

$$y(\Phi) = y_0 F_{y1}(\Phi) + \theta_0 F_{y2}(\Phi) + M_0 F_{y3}(\Phi) + Q_0 F_{y4}(\Phi) + N_0 F_{y5}(\Phi),$$

(11)

where

$$
\begin{cases}
F_{y1}(\Phi) = \frac{2\alpha\beta}{1+2\alpha\beta} + \frac{1}{1+2\alpha\beta}W_1(\Phi) + \frac{R^4 kD}{EI(1+2\alpha\beta)}W_2(\Phi) \\
F_{y2}(\Phi) = \frac{\alpha}{\alpha^2+\beta^2}W_2(\Phi) + \frac{\beta}{\alpha^2+\beta^2}W_3(\Phi) \\
F_{y3}(\Phi) = \frac{R^2}{(1+2\alpha\beta)EI}W_1(\Phi) - \frac{R^6 kD}{2\alpha\beta(1+2\alpha\beta)(EI)^2}W_4(\Phi) \\
F_{y4}(\Phi) = \frac{R^3}{2EIa(\alpha^2+\beta^2)}W_2(\Phi) - \frac{R^3}{2EI\beta(\alpha^2+\beta^2)}W_3(\Phi) \\
F_{y5}(\Phi) = -\frac{R^3}{2\alpha\beta EI}W_4(\Phi)
\end{cases}
$$

$$
\begin{cases}
W_1(\Phi) = ch(\alpha\Phi)\cos(\beta\Phi) \\
W_2(\Phi) = sh(\alpha\Phi)\cos(\beta\Phi) \\
W_3(\Phi) = ch(\alpha\Phi)\sin(\beta\Phi) \\
W_4(\Phi) = sh(\alpha\Phi)\sin(\beta\Phi)
\end{cases}
$$

When the arbitrary point on the curved span has the concentrated moment M_i, the concentrated force P_i, the radial distribution load q_{ri}, the tangential distribution load q_{ti}, the external load on the beam span can be regarded as the partial initial parameters. Therefore, additional deflection caused by the concentrated moment M_i is as below [34],

$$
y(\Phi)_{M_i} = M_i F_{y3}\big(\Phi - \Phi_{M_i}\big)\big(\Phi \geq \Phi_{M_i}\big), \tag{12}
$$

The additional defection caused by the concentrated force P_i when $\Phi \geq \Phi_{Pi}$ is shown below [34],

$$
y(\Phi)_{P_i} = -P_i F_{y4}\big(\Phi - \Phi_{P_i}\big)\big(\Phi \geq \Phi_{P_i}\big), \tag{13}
$$

The additional defection caused by the radial distribution load q_{ri} when $\Phi \geq \Phi_{ai}$ can be obtained [34],

$$
y(\Phi)_{q_{ti}} = -D\int_{\Phi_{ci}}^{\Phi} q_t(\delta)F_{y5}(\Phi-\delta)d\delta(\Phi \geq \Phi_{ci}), \tag{14}
$$

When $\Phi \geq \Phi_{bi}$, the integral upper limit $\Phi = \Phi_{bi}$ is selected.

The additional defection caused by the tangential distribution load q_{ti} when $\Phi \geq \Phi_{ci}$ can also be obtained [34],

$$
y(\Phi)_{q_t} = -D\int_{\Phi_c}^{\Phi} q_t(\delta)F_{y5}(\Phi-\delta)d\delta(\Phi \geq \Phi_c), \tag{15}
$$

When $\Phi > \Phi_{di}$, the integral upper limit $\Phi = \Phi_{di}$ is selected.

With the initial conditions while $\Phi = 0$ and the above external load, the deflection of the curved beam at any points can be achieved [34],

$$
\begin{aligned}
y(\Phi) = \;& y_0 F_{y1}(\Phi) + \theta_0 F_{y2}(\Phi) + M_0 F_{y3}(\Phi) + Q_0 F_{y4}(\Phi) + N_0 F_{y5}(\Phi) + \sum_{i=1}^{n_M} M_i F_{y3}\big(\Phi - \Phi_{M_i}\big) \\
& - \sum_{i=1}^{n_P} P_i F_{y4}\big(\Phi - \Phi_{P_i}\big) - \sum_{i=1}^{n_{qr}} D\int_{\Phi_{ai}}^{\Phi} q_{ri}(\delta)F_{y4}(\Phi-\delta)d\delta - \sum_{i=1}^{n_{qt}} D\int_{\Phi_{ci}}^{\Phi} q_{ti}(\delta)F_{y5}(\Phi-\delta)d\delta
\end{aligned} \tag{16}
$$

With the same procedure, we can obtain the analytical expressions of rotation angle θ, bending moment M, shear force Q, and axial force N when some loads are acting on the beam span.

The TBM tunnel segment experiences leaving the shield tail, pea gravel backfilling, and grouting in turn. There are differences in the stiffness of the medium near the segment and the deformation of segment cross-section with loading. Therefore, the bending of the limited length variable cross-section circular curved beam on the inhomogeneous Winkler foundation needs further discussion.

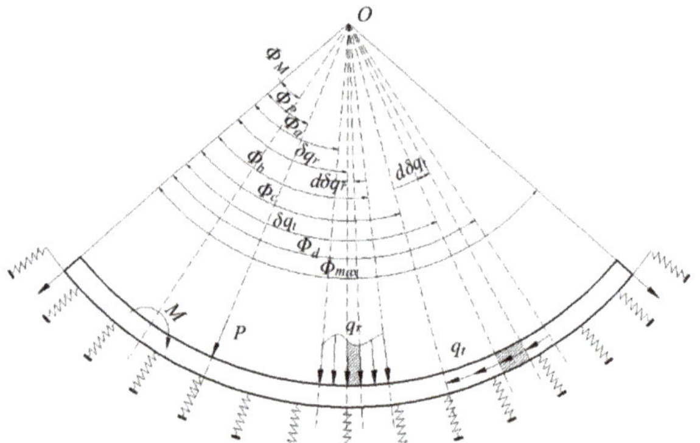

Figure 4. Uniform cross-section circular curved beam on the Winkler elastic foundation [35].

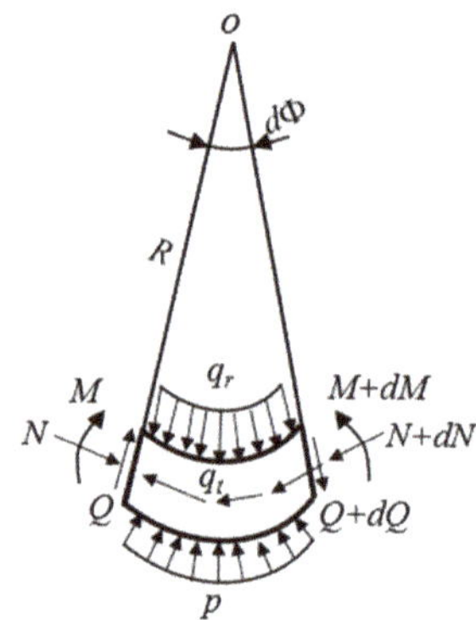

Figure 5. Diagram of force analysis on the microelement body [35].

Figure 6 shows the equal section circular curved beam subjected to the distributed load on the homogeneous Winkler foundation. The analytical expressions of the displacement and the internal force can be expressed in the following matrix form [34].

$$A(\Phi)_{5\times1} = B(\Phi)_{5\times5}A(0)_{5\times1} + C(\Phi)_{5\times1} + D(\Phi)_{5\times1},\tag{17}$$

where,

$$A(\Phi)_{5\times1} = [y(\Phi), \theta(\Phi), M(\Phi), Q(\Phi), N(\Phi)]^T;$$

$$B(\Phi)_{5\times5} = \begin{bmatrix} F_{y1}(\Phi) & F_{y2}(\Phi) & F_{y3}(\Phi) & F_{y4}(\Phi) & F_{y5}(\Phi) \\ F_{\theta1}(\Phi) & F_{\theta2}(\Phi) & F_{\theta3}(\Phi) & F_{\theta4}(\Phi) & F_{\theta5}(\Phi) \\ F_{M1}(\Phi) & F_{M2}(\Phi) & F_{M3}(\Phi) & F_{M4}(\Phi) & F_{M5}(\Phi) \\ F_{Q1}(\Phi) & F_{Q2}(\Phi) & F_{Q3}(\Phi) & F_{Q4}(\Phi) & F_{Q5}(\Phi) \\ F_{N1}(\Phi) & F_{N2}(\Phi) & F_{N3}(\Phi) & F_{N4}(\Phi) & F_{N5}(\Phi) \end{bmatrix};$$

$$A(0)_{5\times1} = [y_0, \theta_0, M_0, Q_0, N_0]^T;$$

$$C(\Phi)_{5\times1} = \left[C_y(\Phi), C_\theta(\Phi), C_M(\Phi), C_Q(\Phi), C_N(\Phi)\right]^T;$$

$$D(\Phi)_{5\times1} = \left[D_y(\Phi), D_\theta(\Phi), D_M(\Phi), D_Q(\Phi), D_N(\Phi)\right]^T;$$

The elements in the matrix $C(\Phi)_{5\times1}$ are as follows [34],

$$
\begin{cases}
C_y(\Phi) = -D\int_{\Phi_\beta}^{\Phi} q_r(\delta)F_{y4}(\Phi-\delta)d\delta \\
C_\theta(\Phi) = -D\int_{\Phi_\beta}^{\Phi} q_r(\delta)F_{\theta4}(\Phi-\delta)d\delta \\
C_M(\Phi) = -D\int_{\Phi_\beta}^{\Phi} q_r(\delta)F_{M4}(\Phi-\delta)d\delta \quad ; \\
C_Q(\Phi) = -D\int_{\Phi_\beta}^{\Phi} q_r(\delta)F_{Q4}(\Phi-\delta)d\delta \\
C_N(\Phi) = -D\int_{\Phi_a}^{\Phi} q_r(\delta)F_{N4}(\Phi-\delta)d\delta
\end{cases}
$$

The elements in the matrix $D(\Phi)_{5\times1}$ are as follows [34],

$$
\begin{cases}
D_y(\Phi) = -D\int_{\Phi_\beta}^{\Phi} q_t(\delta)F_{y5}(\Phi-\delta)d\delta \\
D_\theta(\Phi) = -D\int_{\Phi_\beta}^{\Phi} q_t(\delta)F_{\theta5}(\Phi-\delta)d\delta \\
D_M(\Phi) = -D\int_{\Phi_\beta}^{\Phi} q_t(\delta)F_{M5}(\Phi-\delta)d\delta \quad ; \\
D_Q(\Phi) = -D\int_{\Phi_\beta}^{\Phi} q_t(\delta)F_{Q4}(\Phi-\delta)d\delta \\
D_N(\Phi) = -D\int_{\Phi_a}^{\Phi} q_t(\delta)F_{N4}(\Phi-\delta)d\delta
\end{cases}
$$

The curved beam is divided into n sections into the abruptly changed sections of bending stiffness and foundation bedding coefficient, in the action points of the concentrated force and moment, and the starting action points of the distributed loading. There are concentrated moments M_i and concentrated forces P_i at the junctures $\Phi = \Phi_i$ ($i = 0, 1, 2, \dots, n$) of the adjacent curved beams. Each curved beam section has radial distributed load $q_{ri}(\Phi)$ and tangential distributed load $q_{ti}(\Phi)$.

For curved beam i, in the interval $\Phi_{i-1}{}^+ \le \Phi < \Phi_i{}^-$,

$$
\begin{aligned}
A_i(\Phi^-)_{5\times1} &= B_i(\Phi-\Phi_{i-1})_{5\times5}A_i\left(\Phi_{i-1}^+\right)_{5\times1} + C_i(\Phi-\Phi_{i-1})_{5\times1} \\
&\quad + D_i(\Phi-\Phi_{i-1})_{5\times1}(i=1,2,\dots,n-1,n)
\end{aligned}
\tag{18}
$$

where the α, β, K, EI in the factor expression should be corresponding to the α_i, β_i, K_i, and $(EI_{eq})_i$ in section i.

The interface between section $i-1$ and section i, the cross-section Φ_{i-1}, the Equation based on the deformation consistency and force equilibrium is as follows,

$$
A_i\left(\Phi_{i-1}^+\right)_{5\times1} = A_{i-1}\left(\Phi_{i-1}^-\right)_{5\times1} + E_{i-1}(i=2,3,\dots,n-1,n),
\tag{19}
$$

where,

$$
E_{i-1} = [0,0,M_{i-1},P_{i-1},0]^T (i=2,3,\dots,n-1,n);
$$

The calculation results at the end of curved beam $i-1$ can be regarded as the initial condition of the curved beam i ($2 \le i \le n$), then we can obtain,

$$
A_i(\Phi^-)_{5\times1} = \widetilde{B}(\Phi^-)_{5\times5}A_1\left(\Phi_0^+\right)_{5\times1} + \widetilde{C}(\Phi^-) + \widetilde{D}(\Phi^-)(i=2,3,\dots,n-1,n),
\tag{20}
$$

where,

$$
\widetilde{B}(\Phi) = B_i(\Phi-\Phi_{i-1})B_{i-1}(\Phi_{i-1}-\Phi_{i-2})\dots B_1(\Phi_1-\Phi_0);
$$

$$
\begin{aligned}
\widetilde{C}(\Phi) = \;& C_i(\Phi-\Phi_{i-1}) + B_i(\Phi-\Phi_{i-1})[C_{i-1}(\Phi_{i-1}-\Phi_{i-2}) + E_{i-1}] \\
& + B_i(\Phi-\Phi_{i-1})B_{i-1}(\Phi_{i-1}-\Phi_{i-2})[C_{i-2}(\Phi_{i-2}-\Phi_{i-3}) + E_{i-2}] \\
& + \dots \\
& + B_i(\Phi-\Phi_{i-1})B_{i-1}(\Phi_{i-1}-\Phi_{i-2})\dots B_3(\Phi_3-\Phi_2) \times [C_2(\Phi_2-\Phi_1) + E_2] \\
& + B_i(\Phi-\Phi_{i-1})B_{i-1}(\Phi_{i-1}-\Phi_{i-2})\dots B_3(\Phi_3-\Phi_2)B_2(\Phi_2-\Phi_1)
\end{aligned}
$$

$$\widetilde{D}(\Phi) = \; D_i(\Phi - \Phi_{i-1})$$
$$+ B_i(\Phi - \Phi_{i-1})[D_{i-1}(\Phi_{i-1} - \Phi_{i-2}) + E_{i-1}]$$
$$+ B_i(\Phi - \Phi_{i-1})B_{i-1}(\Phi_{i-1} - \Phi_{i-2})[C_{i-2}(\Phi_{i-2} - \Phi_{i-3}) + E_{i-2}]$$
$$+ \ldots$$
$$+ B_i(\Phi - \Phi_{i-1})B_{i-1}(\Phi_{i-1} - \Phi_{i-2})\ldots B_3(\Phi_3 - \Phi_2) \times [D_2(\Phi_2 - \Phi_1) + E_2]$$
$$+ B_i(\Phi - \Phi_{i-1})B_{i-1}(\Phi_{i-1} - \Phi_{i-2})\ldots B_3(\Phi_3 - \Phi_2)B_2(\Phi_2 - \Phi_1)$$

When $i = n$ and $\Phi_0{}^+ = 0$, $\Phi n^- = \Phi_{max}$, the following equation can be achieved based on Equation (20),

$$A(\Phi_{max})_{5\times1} = \widetilde{B}(\Phi_{max})_{5\times5}A(0)_{5\times1} + \widetilde{C}(\Phi_{max}) + \widetilde{D}(\Phi_{max}). \tag{21}$$

The unknown initial parameters in matrix $A(0)_{5\times1}$ can be solved with Equation (21) based on the boundary conditions of the two ends of the curved beam. Putting all the initial parameters at the initial section into the Equation (20), we can obtain the deflection, rotation angle, moment, shear force and axial force at the arbitrary angle Φ of the circular curved beam when the foundation bedding coefficient and bending stiffness stepped change.

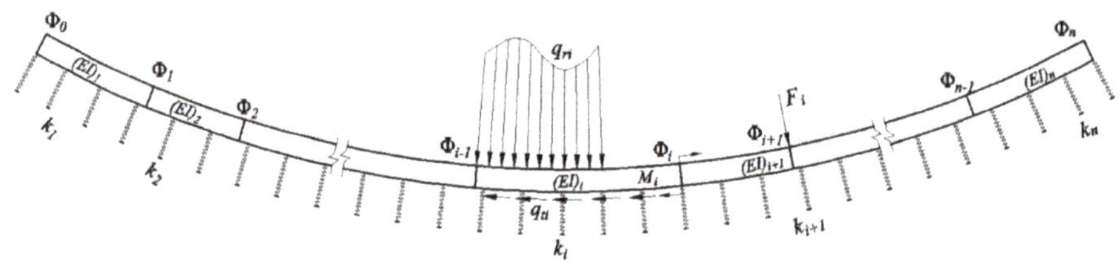

Figure 6. Sectional schematic diagram of the circular curved beam.

4. Analysis of Engineering Example

4.1. Engineering Description

The entrance/exit tunnel MRDK0+457.8~688.2 of Min-Le parking lot belongs to the Shenzhen rail transit line 6 phase II, which is a small radius TBM internal with the curvature radius R = 300 m (as illustrated in Figure 7).

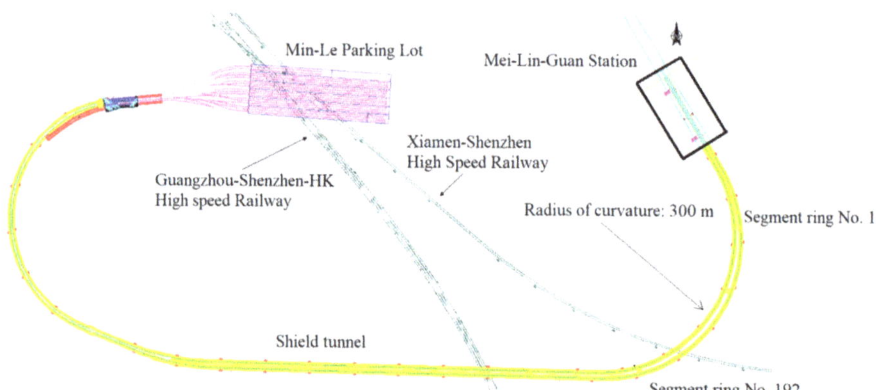

Figure 7. TBM construction interval with the small radius tunnel.

The reinforced concrete (C50) universal wedge-shaped segments with a ring width of 1.2 m are utilized for the tunnel lining structure. The outer and inner diameters are 6200 mm and 5400 mm, respectively, and the thickness is 400 mm. The rings are connected by ten longitudinal bending bolts (M24). There are a total of 192 rings in the interval. The No. 1 and 2 segments are wrapped by the shield tail brush and sealing materials, the elastic resistance coefficient of foundation at the shield tail k_1 = 100 kPa/m; The No. 3 to 5 segments have finished the backfilling of the bottom part by pea gravel, then the elastic resistance coefficient of foundation k_2 = 292 MPa/m; The No. 6 to 10 segments have already finished the backfilling of the sidewall by pea gravel, then the elastic resistance coefficient of foundation k_3 = 1701 MPa/m; The No. 11 to 96 segments are in the slightly weathered cataclasite formation, then the elastic resistance coefficient of foundation k_4 = 1993 MPa/m; The No. 97 to 120 segments are in the slighted weathered granite formation, the elastic resistance coefficient of foundation k_5 = 4770 MPa/m; The No. 121 to 192 segments are in the section where the grout is not solidified, the elastic resistance coefficient of foundation k_6 is linearly changed, and finally reaching to 10,309 MPa/m. The tangential resistance in the model is provided by the friction between the segments and the surrounding medium. The frictional coefficient between the shield tail brush and the segment is 0.3, and the frictional coefficient between the pea gravel and the segment is 0.5. The action sphere of the tangential resistance can be solved through condition $N(\Phi_{qt}) = 0$.

The influence range coefficient of the annular gap λ is set as 0.2 in this paper, and the generalized longitudinal equivalent continuous model is utilized to calculate the longitudinal equivalent bending stiffness $(EI)_{eq}$ in the tunnel. The grouting pea gravels haven't formed the ring in the wall back of No. 1 to 10 segments, and the segments haven't undertaken the surrounding rock pressure, then the bending stiffness $(EI)_{eq1}$ = 162.7 GPa·m^4; The vertical uniform pressure of No. 11 to 96 segments is 107.7 kPa, the base uniform reaction is 110.8 kPa, the horizontal uniform pressure is 32.3 kPa, and the bending stiffness $(EI)_{eq2}$ = 163.1 GPa·m^4; The vertical uniform pressure of No. 97 to 192 segments is 54.9 kPa, the base uniform reaction is 58.0 kPa, the horizontal uniform pressure is 8.2 kPa, and the bending stiffness $(EI)_{eq3}$ = 162.8 GPa·m^4.

The signs of the force and moment are stipulated for the convenience of further study. When the horizontal jack thrust on the outside of the shield curve is larger than that on the inside of the shield curve, the horizontal force couple is positive; when the path of the shield on the outside of the shield curve is larger than that on the inside of the shield curve, the jack thrust deviation angle is positive When the segment's lateral displacement points to the circle center of the circular curve, this direction is positive, and vice versa. Through the monitoring data, the average of the total auxiliary cylinder jack thrust is 5931.33 kN, the average of the jack thrust on the outside of the shield curve is larger than that on the inside of the shield curve. The average of the force couple caused by the difference of the jack thrust in the horizontal direction is 1708.20 kN·m. The average of the lateral component caused by the path difference of the auxiliary cylinder jack is 40.36 kN.

4.2. The Accumulative Value of the Segments' Lateral Displacements

The calculated lateral displacements of the ring segments and the monitored lateral displacements of the ring segments from the field are shown in Figure 8, the parameters used in this calculation are shown in Table 1. From Figure 8, we understand the following.

(a) When the segment is in the shield tail, there is a big difference between (2.267 mm) the simulated displacements and the field monitoring results. This is probably due to the simplification of the wrapping influence of the shield tail brush and sealing grease on the shield tail segments during the theoretical simulation.

(b) In the process of segments leaving the shield tail around 1 to 4 rings, the simulated accumulated lateral displacement agrees well with the lateral displacement from the field. Compared with the monitored data, the simulated results are a little underestimated. The curvature radius in this project has already reached the turning limitation of the TBM. The poorly controlled shield posture and the improper selection of assembly position can cause the squeezing action of the second ring in the shield tail from the shield shell or the tail brush. The squeezing action also causes the increase of lateral offset of some rings, which just leaving the tail and are backfilled.

The first ring leaving the tail is most influenced by this effect. From the field data, we can know that the lateral displacement increases rapidly when the ring leaves the tail.

(c)   In the process of segments leaving the shield tail around 5 to 8 rings, the simulated displacements coincide quite well with the monitored data. When the backfilling finishes, the accumulated displacements from the field tend to be stable. However, the simulated displacements still increase with low speed.

(d)   When the segments leaving the tail around nine rings, there is a big difference between theoretical results and field data. When the pea gravels form the ring, the lateral accumulated displacement declines slowly under the pressure difference of surrounding rocks and tends to be stable from the 17th ring. For the theoretical simulation, the pressure difference of surrounding rocks from both sides of the shield line is ignored as the 3D tunnel is simplified as the 2D circular curved beam in the model. The simulated accumulated displacement increases slowly from the 11th ring and tends to be stable from the 15th ring.

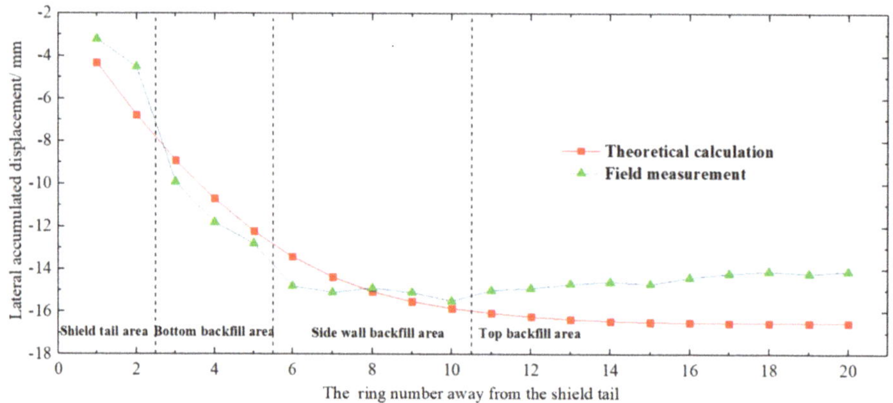

Figure 8. Comparison of horizontal accumulated displacements of the segments.

Table 1. Initial parameters for calculation model.

Total Thrust F/(kN)	Lateral Force Couple M/(kN·m)	Deviation Angle α/(°)	Monitoring Data		
			N_0/(kN)	M_0/(kN·m)	Q_0/(kN)
5931.47	−1708.20	−0.390	5931.33	−1708.20	−40.36

Finally, the simulated displacements agree well with the field data; the simulated results are more conservative. Therefore, this model is applicable in the prediction of the horizontal accumulated displacement of the shield segments in the small radius interval.

5. Single-Factor Influence Analysis of the Segments' Horizontal Displacements Based on the Model

5.1. The Effect of the Jack Total Thrust

The total jack thrust is set as N = 6000 kN, 7000 kN, 8000 kN, 9000 kN, and 10,000 kN, respectively, during the calculation in the model while other parameters are kept the same. The parameters in the theoretical model are summarized in Table 2. Then, the evolution law of the total thrust on the lateral accumulated displacements of a single ring is analyzed.

Table 2. The initial parameters under different total jack thrust.

Total Thrust F/(kN)	Lateral Force Couple M/(kN·m)	Deviation Angle α/(°)	N0/(kN)	M0/(kN·m)	Q0/(kN)
6000			5999.09		−104.71
7000			6998.93		−122.17
8000	−1000	−1	7998.78	−1000	−139.62
9000			8998.63		−157.07
10,000			9998.48		−174.52

Figure 9 shows that the lateral accumulated displacement increases with the increasing distance between the ring and the shield tail. With the increase of the leaving distance (completed rings), the lateral accumulated displacements tend to be stable. The segment results in the deformation along the axis under the jack thrust. When the outside of the segment contacts with the surrounding medium, the friction is produced to resist the jack thrust. When the frictional resistance of the first ring inside the shield tail cannot resist the jack thrust, the remaining thrust can be delivered to the next rings. The transverse component of the remaining thrust can cause a horizontal axis deviation. Therefore, with the increase of the total jack thrust, the remaining thrust acting on the segments increase, and the horizontal axis deviation also increases.

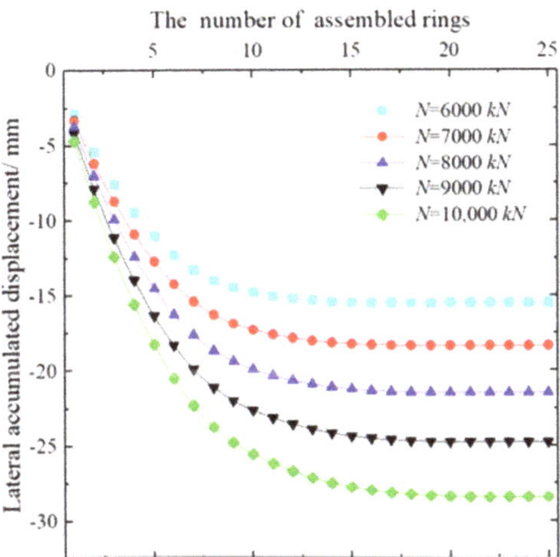

Figure 9. Lateral accumulated displacement curves of the first 20 assembled rings under different total thrusts.

Based on the stable criteria that the displacement difference of the adjacent ring is no more than 0.1 mm, the stable ring number that was leaving the tail increase from the 12th to the 17th when the total thrust increases from 6000 kN to 10,000 kN. In the stable state, the minimum lateral accumulated displacement is −15.347 mm, while the maximum lateral accumulated displacement is −28.321 mm, increasing by 84.54%.

5.2. The Effect of the Jack Thrust Misalignment

The lateral force couple M (2000 kN·m, 1000 kN·m, 0 kN·m, −1000 kN·m, and −2000 kN·m) caused by the thrust difference of the right and left jacks is selected as the variable parameters. The parameters for calculated are summarized in Table 3. The lateral accumulated displacement of a single ring is simulated based on the model.

Table 3. The initial parameters under different jack thrust misalignments.

Total Thrust F/(kN)	Lateral Force Couple M/(kN·m)	Deviation Angle α/(°)	N_0/(kN)	M_0/(kN·m)	Q_0/(kN)
6000	−2000 −1000 0 1000 2000	−1	5999.09	−2000 −1000 0 1000 2000	−104.71

Figure 10 illustrates that the lateral accumulated displacement increases with the distance of the shield tail, and tends to be stable when leaving the rail around 12 rings. The effect of the jack thrust on the shield ring will decay with the increasing distance, causing the decreasing lateral accumulated displacement. The direction of displacements is leaving the circular center; the displacement when the inside force > the outside force is larger than that when the outside force > the inside force. When the force couple is positive, the increasing force couple will prevent the lateral deviation of the shield efficiently.

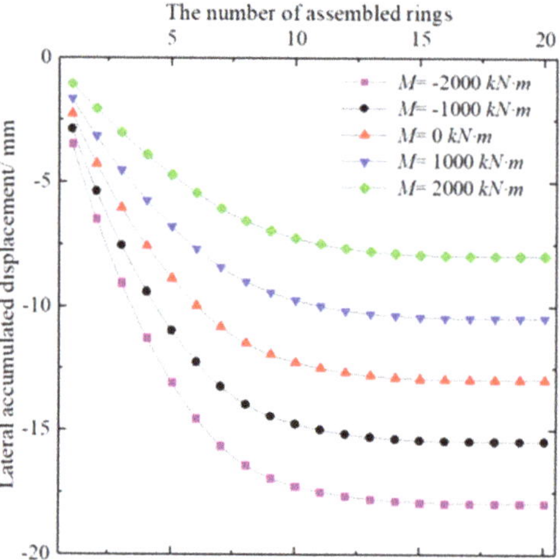

Figure 10. Lateral accumulated displacement curves of the first 20 assembled rings under different thrust misalignments.

5.3. The Effect of the Jack Thrust Deviation Angle

The deviation angle α caused by the path difference of lateral jacks is set as control variables. The lateral accumulated displacements of a single ring with different deviation angles (3°, 2°, 1°, 0°, −1°, −2°, and −3°) are simulated with the model, and the parameters used in the calculation are summarized in Table 4.

Table 4. The initial parameters under different jack thrust deviation angles.

Total Thrust F/(kN)	Lateral Force Couple M/(kN·m)	Deviation Angle α/(°)	N_0/(kN)	M_0/(kN·m)	Q_0/(kN)
6000	−1000	−3	5991.78	−1000	−314.02
		−2	5996.34		−209.40
		−1	5999.09		−104.71
		0	6000.00		0.00
		1	5999.09		104.71
		2	5996.34		209.40
		3	5991.78		314.02

As depicted in Figure 11, the lateral accumulated displacement increases with the distance between the ring and the tail and tends to be stable when the tail leaves the 12th ring. In the construction of the small radius TBM tunnel, the path of the external jack is always larger than that of the internal jack, which causes the deviation angle is negative. Therefore, the axial of the ring will have an external deviation unavoidably. The situation that the external path is larger than the internal path should be avoided in construction, especially in the process of a regripping cycle.

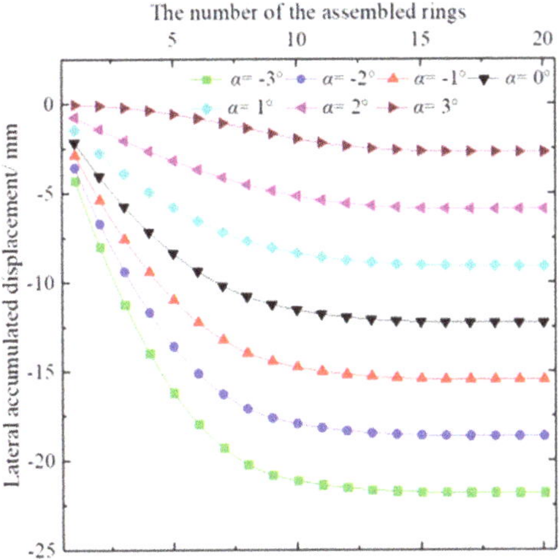

Figure 11. Lateral accumulated displacement curves of the first 20 assembled rings under different thrust deviation angles.

6. Conclusions

(a) Based on the generalized longitudinal equivalent continuous model and the elastic foundation beam theory, the horizontal axis deviation of a small radius TBM tunnel is simplified as the deflection of the circular curved beam on the Winkler foundation.

(b) The model has been solved by the transfer-matrix method, which can solve the deformation and internal force with high accuracy at any cross-section conveniently and fast. When the foundation bedding coefficient and bending stiffness stepped change, the exact solutions of the model can be achieved. When the foundation bedding coefficient and bending stiffness change continuously, the numerical solutions meeting accuracy requirements can be obtained by increasing the number of the order.

(c) In a setting of Shenzhen rail transit line 6 phase II, the calculation model of segment axial horizontal deviation for the small radius TBM tunnel construction is established. The applicability of this model is verified by the field data.

(d) The effect of the jack thrust on the lateral deviation of the tunnel segments decreases with the increase of the distance between the shield tail and the ring. The influence range depends on the total thrust and increases with the total thrust.

(e) When the lateral thrust in the external side of the curved line is larger than that in the internal side of the curved line, the horizontal axis of the segments moves to the external side, and the offset has a positive relationship with the thrust difference (or path difference). When the lateral thrust in the external side of the curved line is smaller than that in the internal side of the curved line, the horizontal axis of the segments still moves to the external side. However, the offset decreases with the increase of thrust difference (or path difference).

Author Contributions: Conceptualization, S.Q.; methodology, S.Q. and R.L.; data curation, R.L. and G.W.; formal analysis, R.L. and G.W.; validation, R.L.; investigation, P.X. and G.W.; writing—original draft preparation, G.W. and P.X.; resources, S.Q. and P.X.; writing—review and editing, G.W. and P.X.; supervision, S.Q. and G.W.; funding acquisition, S.Q. All authors have read and agreed to the published version of the manuscript.

Funding: This research was funded by China Railway No.5 Engineering Croup Co. Ltd.

Conflicts of Interest: The authors declare no conflict of interest.

References

1. Li, Q. Long-Term Settlement Mechanisms of Shield Tunnels in Shanghai Soft Clay. Ph.D. Thesis, Hong Kong University of Science and Technology, Hong Kong, China, 2013.
2. Wu, H.N.; Huang, R.Q.; Sun, W.J.; Shen, S.L.; Xu, Y.S.; Liu, Y.B.; Du, S.J. Leaking behavior of shield tunnels under the Huangpu River of Shanghai with induced hazards. *Nat. Hazard.* **2014**, *70*, 1115–1132. [CrossRef]
3. Shen, S.L.; Wu, H.N.; Cui, Y.J.; Yin, Z.Y. Long-term settlement behaviour of metro tunnels in the soft deposits of Shanghai. *Tunn. Undergr. Space Technol.* **2014**, *40*, 309–323. [CrossRef]
4. Cui, Z.D.; Tan, J. Analysis of long-term settlements of Shanghai Subway Line 1 based on the in-situ monitoring data. *Nat. Hazard.* **2015**, *75*, 465–472. [CrossRef]
5. Tan, J.; Cui, Z.D.; Yuan, L. Study on the long-term settlement of subway tunnel in soft soil area. *Mar. Georesources Geotech.* **2016**, *34*, 486–492. [CrossRef]
6. Gonzalez, C.; Sagaseta, C. Patterns of soil deformations around tunnels. Application to the extension of Madrid Metro. *Comput. Geotech.* **2001**, *28*, 445–468. [CrossRef]
7. Meguid, M.A.; Saada, O.; Nunes, M.A.; Mattar, J. Physical modeling of tunnels in soft ground: A review. *Tunn. Undergr. Space Technol.* **2008**, *23*, 185–198. [CrossRef]
8. Yang, W.; Hussein, M.F.M.; Marshall, A.M. Centrifuge and numerical modelling of ground-borne vibration from an underground tunnel. *Soil Dyn. Earthq. Eng.* **2013**, *51*, 23–34. [CrossRef]
9. Yuan, L.; Cui, Z.D.; Tan, J. Numerical simulation of longitudinal settlement of shield tunnel in the coastal city Shanghai. *Mar. Georesources Geotech.* **2017**, *35*, 365–370. [CrossRef]
10. Koizumi, J.; Murakami, H.; Saino, K. Modelling of longitudinal structure of shield tunnel. *J. Jpn. Soc. Civ. Eng.* **1988**, *394*, 79–88. (In Japanese)
11. Shiba, Y.; Kawashima, K.; Ohikata, M.; Kana, M. Evaluation of longitudinal structural stiffness of shield tunnel under earthquake loading. *J. Jpn. Soc. Civ. Eng.* **1988**, 385–394. (In Japanese) [CrossRef]
12. Winkler, E. *Die Lehre von Elastizitat und Festigkeit (on Elasticity and Fixity)*; Dominicus: Prague, Czech Republic, 1867.
13. Pasternak, P.L. *On A New Method of Analysis of An Elastic Foundation by Means of Two-Constants. Dissertation*; Gosudarstvennoe Izdatelstvo Literaturi po Stroitelstvu I Arkhitecture: Moscow, Russia, 1954. (In Russian)
14. Kerr, A.D. A study of a new foundation model. *Acta Mech.* **1965**, *1*, 135–147. [CrossRef]
15. Zhang, W.J.; Xu, X.U.; LI, X.; Shen, Y.; Zhang, M. Research on generalized longitudinal equivalent continuous model of shield tunnels. *Chin. J. Rock Mech. Eng.* **2009**, *28*, 3938–3944. (In Chinese)
16. Li, X.Y.; Liu, G.B.; Yang, X.; Han, C.L. Deformation and stress of tunnel structures based on modified longitudinal equivalent continuous model. *Chin. J. Geotech. Eng.* **2014**, *36*, 662–670. (In Chinese) [CrossRef]
17. Shiba, Y.; Kkawa Shima, K.S. Evaluation procedure for seismic stress developed in shield tunnel seismic deformation method. *J Struct. Mech. Earthq. Eng.* **1986**, *28*, 45–50. (In Japanese) [CrossRef]
18. Talmon, A.M.; Bezuijen, A. Calculation of longitudinal bending moment and shear force for Shanghai Yangtze River Tunnel: Application of lessons from Dutch research. *Tunn. Undergr. Space Technol.* **2013**, *35*, 161–171. [CrossRef]
19. Yu, H.T.; Cai, C.; Bobet, A.; Zhao, X.; Yuan, Y. Analytical solution for longitudinal bending stiffness of shield tunnels. *Tunn. Undergr. Space Technol.* **2019**, *83*, 27–34. [CrossRef]
20. Verma, A.K.; Singh, T.N. Assessment of tunnel instability-a numerical approach. *Arab. J. Geosci.* **2010**, *3*, 181–192. [CrossRef]
21. Zhao, K.; Janutolo, M.; Barla, G. A completely 3D model for the simulation of mechanized tunnel excavation. *Rock Mech. Rock Eng.* **2012**, *45*, 475–497. [CrossRef]
22. Mollon, G.; Dias, D.; Soubra, A.H. Probabilistic analyses of tunneling-induced ground movements. *Acta Geotech.* **2013**, *8*, 181–199. [CrossRef]

23. Huang, H.; Gong, W.; Khoshnevisan, S.; Juang, C.H.; Zhang, D.; Wang, L. Simplified procedure for finite element analysis of the longitudinal performance of shield tunnels considering spatial soil variability in longitudinal direction. *Comput. Geotech.* **2015**, *64*, 132–145. [CrossRef]
24. Maynar, M.M.; Rodriguez, L.M. Predicted versus measured soil movements induced by shield tunnelling in the Madrid Metro extension. *Can. Geotech. J.* **2005**, *42*, 1160–1172. [CrossRef]
25. Mirhabibi, A.; Soroush, A. Effects of building three-dimensional modeling type on twin tunneling-induced ground settlement. *Tunn. Undergr. Space Technol.* **2013**, *38*, 224–234. [CrossRef]
26. Ocak, I. Interaction of longitudinal surface settlements for twin tunnels in shallow and soft soils: The case of Istanbul Metro. *Environ. Earth Sci.* **2013**, *69*, 1673–1683. [CrossRef]
27. Fargnoli, V.; Boldini, D.; Amorosi, A. TBM tunnelling-induced settlements in coarse-grained soils: The case of the new Milan underground line 5. *Tunn. Undergr. Space Technol.* **2013**, *38*, 336–347. [CrossRef]
28. Sugimoto, M.; Sramoon, A.; Konishi, S.; Sato, Y. Simulation of shield tunneling behavior along a curved alignment in a multilayered ground. *J. Geotech. Geoenviron. Eng.* **2007**, *133*, 684–694. [CrossRef]
29. Li, S.; Gu, G.; Li, P.; Zhang, M.; Wang, G. Numerical Simulation of Mechanical Properties of the Segments for a Curved Shield Tunnel during Construction Stage: A Case Study. In Proceedings of the International Conference on Geotechnical and Earthquake Engineering 2018 (IACGE 2018) International Association of Chinese Geotechnical Engineers (IACGE), Chongqing, China, 20–21 October 2018; pp. 61–69.
30. Yang, C.S.; Mo, H.H.; Wei, L.X. Correction calculation for circumferential equivalent bolt stiffness of shield tunnel and the analysis of its influencing factors. *J. Railw. Sci. Eng.* **2017**, *14*, 1497–1504. (In Chinese) [CrossRef]
31. Geng, P.; Chen, P.L.; Zhang, J.; He, C.; Yan, Q.X. Nonlinear longitudinal equivalent bending stiffness of shield tunnel under the combined effect of axial force and bending moment. *Chin. J. Rock Mech. Eng.* **2017**, *36*, 2522–2534. (In Chinese) [CrossRef]
32. Wu, H.N.; Shen, S.L.; Yang, J.; Zhou, A.N. Soil-tunnel interaction modelling for shield tunnels considering shearing dislocation in longitudinal joints. *Tunn. Undergr. Space Technol.* **2018**, *78*, 168–177. [CrossRef]
33. Issa, M.S. Natural frequencies of continuous curved beams on Winkler-type foundation. *J. Sound VIB.* **1988**, *127*, 291–301. [CrossRef]
34. Jin, K.N. Analysis of circularly curved beam in elastical ground. *J. Civil Eng. Manag.* **1986**, *2*, 35–42.
35. Cai, W.Y.; Zhang, M.X. Longitudinal equivalent continuous model of a shield tunnel based on the ellipse form. *Mod. Tunn. Technol.* **2017**, *54*, 42–49. (In Chinese) [CrossRef]

Review

Seismic Design of Timber Buildings: Highlighted Challenges and Future Trends

Mislav Stepinac [1,*], Iztok Šušteršič [2], Igor Gavrić [2] and Vlatka Rajčić [1]

[1] Faculty of Civil Engineering, University of Zagreb, 10000 Zagreb, Croatia; vlatka.rajcic@grad.unizg.hr

[2] InnoRenew CoE, Slovenia/Faculty of Mathematics, Natural Sciences and Information Technologies, University of Primorska, InnoRenew CoE, 6310 Izola, Slovenia; iztok.sustersic@innorenew.eu (I.Š.); igor.gavric@innorenew.eu (I.G.)

* Correspondence: mislav.stepinac@grad.unizg.hr

Received: 15 December 2019; Accepted: 14 February 2020; Published: 19 February 2020

Abstract: Use of timber as a construction material has entered a period of renaissance since the development of high-performance engineered wood products, enabling larger and taller buildings to be built. In addition, due to substantial contribution of the building sector to global energy use, greenhouse gas emissions and waste production, sustainable solutions are needed, for which timber has shown a great potential as a sustainable, resilient and renewable building alternative, not only for single family homes but also for mid-rise and high-rise buildings. Both recent technological developments in timber engineering and exponentially increased use of engineered wood products and wood composites reflect in deficiency of current timber codes and standards. This paper presents an overview of some of the current challenges and emerging trends in the field of seismic design of timber buildings. Currently existing building codes and the development of new generation of European building codes are presented. Ongoing studies on a variety topics within seismic timber engineering are presented, including tall timber and hybrid buildings, composites with timber and seismic retrofitting with timber. Crucial challenges, key research needs and opportunities are addressed and critically discussed.

Keywords: seismic design; tall timber buildings; timber composites; seismic retrofitting; Eurocode 8

---

1. Introduction

In the past century, extensive demand for steel, concrete and masonry as construction materials pushed the development and significant advancement of building codes, standards and guidelines for structural systems based on these materials [1–3]. In seismically-prone areas around the globe special attention had to be paid to ensuring seismic resistance of structures as well. Seismic design of structures differs from "regular" structural design in several aspects; structural response to strong earthquakes is dynamic, nonlinear and random, whilst almost all the rest actions and responses are static, linear and deterministic. Due to globalization, seismic design of structures has recently become part of the regular structural engineering curriculum and practice, even in the areas where earthquakes are not so relevant. Past, present and future trends in analyses in seismic provisions for buildings are very well explained by Fajfar [4].

On the other hand, although serious studies on earthquakes and seismic activity began about a century ago, intense research in the field of seismic design of timber structures started only a couple of decades ago, with the advancement of engineered wood products (EWP), which enabled more complex and ambitious timber construction. Global tendency towards more sustainable, energy efficient and environmentally-friendly building solutions has further popularized timber as principle structural material.

Wood in its nature differs significantly from concrete, masonry and steel, as it is considerably lighter compared to them and it is an anisotropic natural material, while the other ones are isotropic man-made materials. These material characteristics influence significantly the overall structural and seismic performance of timber buildings. Recent technological developments and exponentially increased use of engineered wood products and wood composites reflected in deficiency of current timber norms and standards. This paper focuses on some open questions and recent developments in timber engineering regarding the use of timber in seismically active regions, on seismic design of timber structures and normative acts in Europe, and especially on the lack of information in the Eurocode 8.

Recently completed COST Action FP1402 has contributed to a better understanding and overview of broadly available scientific results and the specific information needed by the code-writers, authorities, designers and end-users in the safe, durable and efficient use of timber in structures and, consequently, increase its acceptance and use in the design of buildings. Significant progress has been made with respect to the cross-laminated timber (CLT) structures [5–9], timber–concrete composites [10–12] and understanding of the connections in timber structures [13–16]. As a result, input data for the improvement and future development of EN 1995 are given [17]. However, several topics on seismic design of timber structures still need further investigation [18].

Future trends in timber construction will require major development and research on topics of: Tall timber and hybrid buildings, new engineered wood products and connection systems related to the new technologies, modular construction with timber, composites with wood, assessment of existing timber buildings and retrofitting of historical buildings with timber (Figure 1). Due to rapid development of new timber technologies, and especially due to taller and taller timber buildings, precisely defined impact of earthquakes on these structures needs to be investigated.

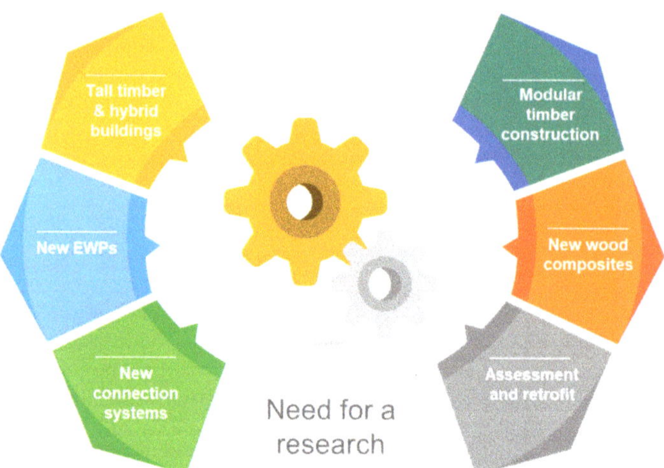

Figure 1. Future need for a research in timber engineering.

2. European Seismic Design Norms, Standards and Guidelines

Recent developments in timber engineering and exponential intense of timber in construction required the evolution of currently existing standards. Materials like cross-laminated timber (CLT) proved that timber can be an excellent material for mid- and high-rise buildings even in seismic areas [19,20]. Cross-laminated timber (CLT) is used for floors and walls and can be considered as floor diaphragms and shear walls in seismic design [21–23]. Significant development has also been achieved at the connection level. Self-tapping screws [24,25], glued-in rods [26–28] and a lot of different innovative systems [29–32] show improved behavior in seismic applications than traditional dowel-type connections. At the system level, hybrid systems with concrete [33], EWPs [34,35], steel [36,37], polymers [38] and glass [39–41] are in development and are intensely researched.

Current seismic design approaches in building codes around the world (e.g., Eurocode (EC8) [42], NBCC [43], ASCE-7 [44]) follow force-based design methods. At the moment there are only a few norms on seismic assessment of existing structures [45–48] and future development in research shall be focused in this direction as well. Nevertheless, timber as a structural material is poorly represented in the current norms (i.e., EC8 has only four pages related to seismic design of timber structures). In the current Eurocode 8—Section 8: "Specific rules for timber buildings", no information is provided for seismic design of widely used structural systems such as cross-laminated timber structures. In addition, no provisions are given regarding capacity design methods for different types of timber structural systems, which proved to be crucial in seismic design of timber buildings [49–51], as well as provisions and rules for transfer zones for continuity of shear walls along the building's height in multi-story timber buildings. The revision process of the Eurocodes began in 2015 and the final updated version is expected to be released sometime after 2020. The new proposal of timber part, prepared by Work Group 3 of CEN Technical Committee 250 (CEN/TC 250/SC 8/WG 3), is explained by Follesa et al. [18] and is based on following modifications and recommendations:

- Changes in the general definitions and design concepts,
- Update of the list of wood-based materials,
- Definition of dissipative and non-dissipative zones,
- Update of the list of timber based structural types with addition of new structural systems (modification of the description of the existing structural types including graphic presentations of structural systems)
- Modification of behavior factors values for different ductility classes,
- Introduction of capacity design rules for each structural type and of overstrength factors to be used in the design of the brittle components,
- Modification of the current equations for safety verifications
- A new provision for application of non-linear static (pushover) analysis.

Above-mentioned new provisions and concepts of timber structures seismic design will demand additional information on mechanical properties of timber connection systems such as connection ductility under cyclic loading, overstrength factors, elastic and plastic stiffness, strength degradation properties under cyclic loading, energy dissipation properties, etc. Therefore, in the near future European technical assessment documents (ETA) for timber connections shall include more information on mechanical properties of connections under cycling loading, defined in EN 12512 standard [52].

Current trend of exponential growth of new timber buildings, larger, taller and more complex projects not only requires higher volume of engineered wood products production, but also higher demand for skilled carpenters and tradespeople with proper education and training on timber construction. Thus, in addition to the updates of the current building codes, also regulation and guidelines in the area of execution and construction supervision of timber buildings shall be improved, where contracting companies shall obtain certifications as a proof of being competent to execute such buildings. Further, regulation on periodic monitoring of structural health of timber buildings, especially tall timber buildings, shall also be addressed.

3. Timber Buildings—Future Trends and Challenges in Seismic Design

In this section, current and emerging challenges in seismic-related topics in the field of timber engineering are presented, compared and critically discussed. Due to the rapid advancement in the development of engineered wood products (EWP) and structural connections, presented earlier in this paper, more and more new applications of EWPs in timber engineering and other engineering fields have emerged. These applications extend from possibilities of building taller timber buildings, to combining timber structural systems with structural systems based on other materials such as concrete and steel, forming so called hybrid timber buildings, exhibiting even higher potential for high-rise construction but also additional challenges to overcome. Further, a combination of advanced EWPs and structural connections can also serve for new composite load-bearing assemblies, such as timber–glass composite wall systems. Finally, advanced EWPs have also been proposed for seismic retrofitting of existing buildings, not only for existing buildings with timber structure, but also existing buildings with stone, masonry and concrete frame structure. Based on the current state-of-the art, key future research and development needs and trends in these selected topics are identified and presented.

3.1. Tall Timber Buildings

European strategy for a sustainable growth and sustainable society acknowledges the importance of EU research framework programs for increasing the offer of new high-quality products and services [53,54]. Sustainable construction embraces a number of aspects, such as design and management of buildings and constructed assets, choice of materials, building performance as well as interaction with urban and economic development and management. Timber as a material and as a structural system could be an ideal material for a new era of construction. Without compromising architectural requirements, the transfer of part of the on-site construction activity to off-site production, independent from weather conditions, will ensure a more continuous activity, a better quality of the finished products and an improved control of their environmental characteristics, as discussed, for example, by [55–58].

Building with timber, in recent years, has become a huge trend among the construction sector around the world. With products like CLT, the possibilities are enormous. It is obvious that the market is changing and wood as a structural material will become more relevant and more in use in the following years. It is now well understood that the concrete and steel industries are highly energy intensive and contribute to a significant portion of global carbon emissions, and wood is becoming an obvious solution to reduce it. If sustainably sourced, timber is undoubtedly one of the most environmentally-friendly materials currently available, being a natural carbon sink and truly renewable [59].

Tall timber buildings are built almost on a monthly rate and more and more are planned in the near future. Short overview of the existing tall timber building and buildings under construction are given in the Figure 2 and one erected building is shown in Figure 3.

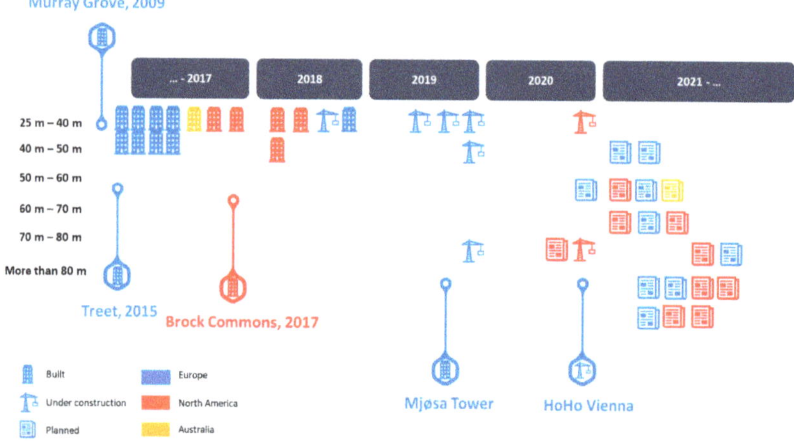

Figure 2. Tall timber buildings around the world—combined data [60–62].

Figure 3. Tall hybrid timber building Brock Commons, Vancouver, Canada.

Although, timber has a good reputation in earthquake-prone regions, in the new era of timber skyscrapers, seismic actions and detailing require additional attention in design and construction of such structures. Tendency to build taller multi-residential and non-residential buildings with more occupants will require additional precautions. In the previous chapter it was pointed out what should change and how the design of such buildings should be done [18]. Holistic approach must be appointed to a design of such structures. Namely, in addition to conventional actions such as permanent loads, imposed loads, snow and wind, there are several additional actions which can significantly affect structural behavior of timber buildings, such as thermal actions, moisture, mold, fire, etc. Therefore, special attention has to be paid to detailing in tall timber buildings (architectural, structural and seismic detailing, special acoustics details, etc.). To prevent any potential structural damage in tall timber buildings due to aforementioned actions, continuous monitoring of buildings during their lifetime could be beneficial (installation of moisture sensors in critical places, vibration monitoring, etc.). Additional potential problems which should be addressed regarding the seismic provisions of tall timber structures are:

- Acoustic insulation requirements (tendency towards physical separation of structural elements in order to prevent sound transmission [63,64]) vs. seismic design principles requirements (tendency towards connection of structural elements in order to achieve necessary strength and stiffness) should be addressed, as some studies and technical documents show the reduction in strength and stiffness of acoustically-isolated connections in timber structures [65,66]; in addition, improved acoustic performance of timber wall and floor elements can also be achieved with additional high-density layers [67,68], resulting in higher building mass and consequently higher seismic forces;
- New types of connections and assembly tools are entering the market and their performance is critical for seismic behavior of timber buildings [22,69,70];
- Higher seismic forces in taller buildings require new types of connection systems that can accommodate strength, stiffness, ductility and energy dissipation needs [7,22,71];
- In addition to seismic actions, also wind-induced vibrations shall be furtherly addressed in tall timber buildings [72–75];
- Numerical modeling of tall timber buildings—addressing problems such as determination of vibration periods and damping [76], modeling ductility in the joints [77–79], selection of appropriate seismic design assumptions and procedures [80], proper simplification assumptions in FEM models of tall timber buildings and hybrid timber buildings [81], etc.

3.2. Hybrid Systems with Timber

Combining timber with other materials, to achieve composite action, enables the strength of timber to be enhanced and its shortcomings to be strengthened or eliminated. Timber is most commonly combined with concrete and steel, but recently also with glass and polymers. Pre-stressed timber buildings [82], timber–steel hybrid systems [83,84] and timber–concrete hybrid systems [33] are already in use and are proven as earthquake-resistant structural systems. Nevertheless, due to tendency towards taller and larger timber construction, and consequently higher structural performance demand, these topics continue to be furthermore explored in terms of experimental, numerical and analytical studies [36,85,86]. In this paper, hybrid systems with structural glass will be briefly addressed. While timber–concrete and timber–steel hybrid structural systems are already widely applied to mid-rise construction in practice, the timber–glass hybrid structural system is still emerging. In this section, current state-of-the-art of this hybrid structural system is presented and key open questions and challenges for its future possible launch to the construction market are discussed.

Timber and glass composite systems are lately intensely investigated due to extremely high-aesthetic and -ecological value in addition to their cost-effectiveness and the possibility of significant load transfer [40]. Nevertheless, design models and current European standards include the usage of glass panels as the secondary elements [42], which means the positive impact of these elements when transferring transverse loads caused by the earthquake have to be ignored [39]. According to the EC8, it is necessary to calculate the primary structural elements within the allowed displacements regarding the protection of the secondary elements. If the problem is approached as defined in Eurocode 5 (EC5) [87], timber wall diaphragms shall be designed to resist both horizontal and vertical actions imposed upon them, and shall be adequately restrained to avoid overturning and sliding. Racking resistance is provided by in-plane plane stiffness of board materials, diagonal bracing or moment connections. Method A from EC5 defines that shear diaphragms with windows and doors do not contribute to stability of the structure. Method B is less restrictive and it proposes to regard panel parts from each side of the opening as separate panels. Since the openings decrease the racking resistance and significantly reduce the horizontal stiffness of precast elements, glass helps to enhance these properties. In recent years several articles on quasi-static and dynamic tests of timber–structural glass composite systems were published [39,40,88–94] (Figure 4). The researchers have concluded that timber–structural glass load-bearing systems can be used in various construction applications, depending on the required bearing or ductility levels. Žarnić et al. [95] investigated deformation capacity, lateral strength, stiffness and strength deterioration and energy dissipation capacity in order to provide data for the future development of computational models and design guidance for the new codes. Nevertheless, it is possible to devise a combined system with timber and structural glass in which each material could transfer load, and in mutual interaction of constitutive elements could be resistant to earthquake. Several extensive tests of composite systems timber–structural glass have been conducted with various types of bonding timber and glass. Bonding glass on timber proved as a good example for accomplishing high-load-bearing of composites, but deficiencies are noticed in the level of ductility along with possible problems with the durability of the structure. Further research on timber–glass composites is needed to get the structural response in seismic actions. In addition, factor q (force reduction factor) of structures constructed with hybrid laminated glass–CLT structural panels need be derived. These results may contribute to future upgrading of EC 8 where glass-based structures are not yet addressed [95]. Besides the load-bearing characteristics and development of structural design tools the special attention need to be paid to energy efficiency of developed building components [96].

Figure 4. An example of a timber–glass composite system [39,40,88–94].

3.3. Seismic Retrofitting with Timber

Seismic retrofitting can be defined as the modification of existing structures to make them more resistant to seismic activity, ground motion or soil failure due to earthquakes. Seismic retrofit strategies have been developed in the past few decades following the introduction of new seismic provisions and the availability of advanced materials. Different techniques and methods are applied for a seismic retrofitting of different structures (e.g., concrete structures [97–99], timber structures [35,100,101], historical and heritage structures [102–104] and especially masonry structures [3,105–107]).

In this chapter, retrofitting of existing structures with CLT is presented. Several researchers around the world are investigating the possibility to seismically strengthen existing structures with engineered wood products (EWPs), mostly CLT. Japanese researchers [108] proposed RC frames strengthening with several narrow CLT elements bonded onto the RC frame with epoxy resin. In their strengthening method, CLT panels are infilled in the RC frame and are acting as shear walls. An Italian group of researchers [109] is developing a novel integrated retrofit solution based on the use of CLT shear walls encased as infill in existing RC framed structures (Figure 5). The idea was to increase the overall lateral stiffness of the concrete structure and to reduce the lateral drift values. The main conclusions of preliminary experimental and numerical work are that the CLT infill allowed the RC frame to reach a lower drift value and a higher peak load with respect to common masonry infills. Numerical modeling and optimal seismic retrofit design with CLT was proposed by [110]. Bahmani et al. [111] were retrofitting four-story-soft-story timber buildings with CLT. They concluded that a retrofit, in accordance with the FEMA P-807 guidelines, using CLT panels is suitable for achieving life safety performance levels during 50% MCE level earthquakes, when retrofit of all story levels is not possible due to one or more constraints.

Figure 5. Retrofitting with CLT proposed by Sustersic [95].

Extensive research on seismic strengthening of different types of existing buildings with CLT panels was performed by Sustersic and Dujic [112]. The experimental work and numerical study were focused on typical non-earthquake-resistant older masonry and concrete structures from Southern European countries. A new outer shell is added onto existing buildings (Figure 5). It is made from CLT plates that serve as the load-bearing layer and are attached to the existing buildings with special connections. The basic idea of the proposed CLT outer envelope is that it is light (it does not contribute much to seismic forces and is easier to install), strong (timber has one of the best strength:weight ratios) and can be produced as a prefabricated system so little on-site work is necessary. Depending on the building type (concrete or masonry), the elements are anchored into buildings on floor levels, either into concrete slabs or masonry walls. The CLT is connected to the existing building with special steel brackets that offer high strength and stiffness and have a "control fuse" that enables a brittle yet predictable failure of the connection. For the unreinforced masonry walls, the most efficient mechanically-attached strengthening panel resulted in a 33% increase of idealized strength and a 166% higher idealized ultimate displacement. A substitute frame model, with concentrated plasticity for modeling URM and linear elastic shell elements combined with nonlinear springs for modeling the CLT strengthening system, could satisfactorily describe the behavior of the tested wall specimen. Finite element models of the tested RC frame, with or without masonry infill, showed that the damage in both RC frame and masonry infill is lower at equal ground acceleration when the CLT strengthening plates are installed on the structure.

The state-of-the-art research have; therefore, shown that a CLT seismic strengthening system has the potential to be used in design practice, but currently there is still a lot of "unknowns" and further research is needed. A holistic approach should be applied—seismic and energy retrofitting—as both are very important. In addition to the currently existing regulation on energy certificates of buildings, which are informing the general public, owners and potential buyers as to which condition a building is regarding the energy consumption, a similar concept shall be established for seismic evaluation of buildings—"seismic certificates". This will, on one hand raise awareness of current owners about the seismic health of the buildings where they live, and on the other hand additionally define and position a real estate worth on the market, and; therefore, influence the demand on the seismic retrofit of existing buildings in order to be competitive on the market.

4. Conclusions

Due to rapid development of high-performance engineered wood products and new timber technologies, resulting in taller and larger timber buildings with applications spanning through all building types, current building codes and standards reflect a deficiency of provisions for contemporary seismic design. State-of-the-art research in various fields of timber seismic design are presented in this paper and crucial challenges, research needs and opportunities are discussed.

The new generation of Eurocode 8—timber part—will address many topics which are not present in the current version. An updated list of timber-based structural systems with definitions of dissipative and non-dissipative zones in structures, which are needed for newly-introduced capacity design rules and overstrength factors for each type of structural system, will be included. Further, adapted q-behavior factors values for different ductility classes will be defined and a new procedure for application of non-linear static (pushover) analysis will be included.

Tall timber buildings with more than ten stories are already present in moderate- and high-seismic zones around the world. Further, due to climate and economic reasons, more and more conceptual architectural designs for taller timber buildings, including timber skyscrapers, are being proposed, which poses several additional engineering challenges to overcome. A holistic design approach including architectural, structural, durability, fire and acoustic designs as an integrated process is crucial, as all these topics are interrelated. Challenges in terms of numerical modeling of timber and hybrid structural systems, ensuring lateral stability due to wind and seismic actions, high-performing energy dissipating connections, acoustic insulation vs. seismic design philosophy, execution and building monitoring need to be addressed more in depth.

In addition to traditional timber composites with steel and concrete, recent research and developments have shown potential for timber–glass and timber–polymers composites as well. A timber–glass seismic-resistant structural system consists of a timber frame and a structural glass infill bonded together with adhesive, or based on friction contact forming a lateral-resisting wall system. The main advantage of this structural system is increasing lateral stability of buildings with high proportion of facades with glass surfaces by avoiding diagonal bracings or moment connections. Experimental and numerical studies have shown encouraging results in terms of load-bearing and stiffness, whereas the durability aspect needs further examination.

Cross-laminated timber (CLT) has proved to be a great solution for new mid- and high-rise timber buildings. Recently, CLT has also been studied for seismic and energy retrofitting of existing older masonry and concrete buildings, which do not meet current seismic design and energy efficiency criteria. In terms of seismic performance, increased strength and stiffness were observed, yet the research is still ongoing and needs additional investigation of connections between the CLT strengthening panels and the existing structure, with its application to a wider range of existing buildings.

New technologies and knowledge in timber engineering opened many new possibilities in timber application, not only for new timber buildings, but also in combination with other conventional building materials forming hybrid and composite assemblies and structural systems, and also for retrofitting of existing buildings. In this paper, in addition to an overview of some of the current challenges and emerging trends of seismic behavior of timber structures, the focus was set on three topics of advanced engineered wood products applications (tall timber buildings, composites with timber and seismic retrofitting with timber), which are representing new trends of timber engineering and push the boundaries of timber for the use in sustainable construction. All three discussed topics have shown lots of potential for their application in seismic areas, yet there are still several research challenges which need to be addressed in terms of seismic performance and seismic design.

Author Contributions: Conceptualization and development of the main idea of the paper, M.S., I.G. and I.Š.; methodology, M.S., I.G. and I.Š.; validation, all authors; formal analysis, M.S., I.G. and I.Š.; investigation, M.S., I.G. and I.Š.; resources, M.S., I.G. and I.Š.; photo credit, M.S.; literature review and manuscript writing related to Section 3.1., M.S. and I.G.; literature review and manuscript writing related to Section 3.2, M.S and V.R.; literature review and manuscript writing related to Section 3.3, I.G. and I.Š.; writing—original draft preparation, M.S. and I.G.; writing—review and editing, M.S., I.G., I.Š. and V.R.; visualization, M.S., I.G., I.Š.; supervision, I.Š. and V.R.; project administration, M.S.; funding acquisition, M.S. All authors have read and agreed to the published version of the manuscript.

Funding: This research was funded by the Unity through Knowledge Fund (UKF): Seismic behavior of multi-storey buildings, UKF Grant Agreement No. 18/19. Authors Iztok Sustersic and Igor Gavric gratefully acknowledge receiving funding from programme Horizon 2020 Framework Programme of the European Union; H2020 WIDESPREAD-2-Teaming: (#739574) and the Republic of Slovenia. Iztok Sustersic would also like to thank the Slovenian Research Agency ARRS for funding the bilateral project BI-US/19-21-014.

Conflicts of Interest: The authors declare no conflicts of interest. The funders had no role in the design of the study; in the collection, analyses, or interpretation of data; in the writing of the manuscript, or in the decision to publish the results.

References

1. Mitrovic, S.; Causevic, M. Nonlinear static seismic analysis of structures. *Građevinar* **2009**, *61*, 521–531.
2. Fischinger, M.; Kramer, M.; Isakovic, T. Seismic safety of prefabricated reinforced-concrete halls—experimental study. *Građevinar* **2009**, *61*, 1031–1038.
3. Tomaževic, M. Seismic rehabilitation of existing masonry structures. *Građevinar* **2000**, *52*, 683–693.
4. Fajfar, P. *Analysis in Seismic Provisions for Buildings: Past, Present and Future*; Springer: Cham, Switzerland, 2018; Volume 46. [CrossRef]
5. Fink, G.; Kohler, J.; Brandner, R. Application of European design principles to cross laminated timber. *Eng. Struct.* **2018**, *171*, 934–943. [CrossRef]

6. Brandner, R. Cross laminated timber (CLT) in compression perpendicular to plane: Testing, properties, design and recommendations for harmonizing design provisions for structural timber products. *Eng. Struct.* **2018**, *171*, 944–960. [CrossRef]

7. Lukacs, I.; Björnfot, A.; Tomasi, R. Strength and stiffness of cross-laminated timber (CLT) shear walls: State-of-the-art of analytical approaches. *Eng. Struct.* **2019**, *178*, 136–147. [CrossRef]

8. Nolet, V.; Casagrande, D.; Doudak, G. Multipanel CLT shearwalls: An analytical methodology to predict the elastic-plastic behaviour. *Eng. Struct.* **2019**, *179*, 640–654. [CrossRef]

9. Danielsson, H.; Serrano, E. Cross laminated timber at in-plane beam loading—Prediction of shear stresses in crossing areas. *Eng. Struct.* **2018**, *171*, 921–927. [CrossRef]

10. Schänzlin, J.; Fragiacomo, M. Analytical derivation of the effective creep coefficients for timber-concrete composite structures. *Eng. Struct.* **2018**, *172*, 432–439. [CrossRef]

11. Dias, A.M.P.G.; Kuhlmann, U.; Kudla, K.; Mönch, S.; Dias, A.M.A. Performance of dowel-type fasteners and notches for hybrid timber structures. *Eng. Struct.* **2018**, *171*, 40–46. [CrossRef]

12. Stepinac, M.; Rajčić, V.; Barbalić, J. Influence of long term load on timber-concrete composite systems. *Gradjevinar* **2015**, *67*, 235–246. [CrossRef]

13. Jockwer, R.; Dietsch, P. Review of design approaches and test results on brittle failure modes of connections loaded at an angle to the grain. *Eng. Struct.* **2018**, *171*, 362–372. [CrossRef]

14. Sandhaas, C.; Görlacher, R. Analysis of nail properties for joint design. *Eng. Struct.* **2018**, *173*, 231–240. [CrossRef]

15. Cabrero, J.M.; Yurrita, M. Performance assessment of existing models to predict brittle failure modes of steel-to-timber connections loaded parallel-to-grain with dowel-type fasteners. *Eng. Struct.* **2018**, *171*, 895–910. [CrossRef]

16. Jockwer, R.; Fink, G.; Köhler, J. Assessment of the failure behaviour and reliability of timber connections with multiple dowel-type fasteners. *Eng. Struct.* **2018**, *172*, 76–84. [CrossRef]

17. Stepinac, M.; Cabrero, J.M.; Ranasinghe, K.; Kleiber, M. Proposal for reorganization of the connections chapter of Eurocode 5. *Eng. Struct.* **2018**, *170*, 135–145. [CrossRef]

18. Follesa, M.; Fragiacomo, M.; Casagrande, D.; Tomasi, R.; Piazza, M.; Vassallo, D.; Canetti, D.; Rossi, S. The new provisions for the seismic design of timber buildings in Europe. *Eng. Struct.* **2018**, *168*, 736–747. [CrossRef]

19. Jeleč, M.; Varevac, D.; Rajčić, V. Cross-laminated timber (CLT)—A state of the art report. *Građevinar: Časopis Hrvatskog Saveza Građevinskih Inženjera* **2018**, *70*, 75–95.

20. Van De Kuilen, J.W.G.; Ceccotti, A.; Xia, Z.; He, M. Very tall wooden buildings with Cross Laminated Timber. *Procedia Eng.* **2011**, *14*, 1621–1628. [CrossRef]

21. Popovski, M.; Gavric, I. Performance of a 2-Story CLT House Subjected to Lateral Loads. *J. Struct. Eng.* **2016**, *142*, E4015006. [CrossRef]

22. Gavric, I.; Fragiacomo, M.; Ceccotti, A. Cyclic Behavior of CLT Wall Systems: Experimental Tests and Analytical Prediction Models. *J. Struct. Eng.* **2015**, *141*, 04015034. [CrossRef]

23. Pei, S.; van de Lindt, J.W.; Popovski, M.; Berman, J.W.; Dolan, J.D.; Ricles, J.; Sause, R.; Blomgren, H.; Rammer, D.R. Cross-Laminated Timber for Seismic Regions: Progress and Challenges for Research and Implementation. *J. Struct. Eng.* **2016**, *142*, E2514001. [CrossRef]

24. Hossain, A.; Popovski, M.; Tannert, T. Cross-laminated timber connections assembled with a combination of screws in withdrawal and screws in shear. *Eng. Struct.* **2018**, *168*, 1–11. [CrossRef]

25. Hossain, A.; Danzig, I.; Tannert, T. Cross-Laminated Timber Shear Connections with Double-Angled Self-Tapping Screw Assemblies. *J. Struct. Eng.* **2016**, *142*, 04016099. [CrossRef]

26. Hunger, F.; Stepinac, M.; Rajčić, V.; van de Kuilen, J.W.G. Pull-compression tests on glued-in metric thread rods parallel to grain in glulam and laminated veneer lumber of different timber species. *Eur. J. Wood Wood Prod.* **2016**, *74*, 379–391. [CrossRef]

27. Steiger, R.; Serrano, E.; Stepinac, M.; Rajčić, V.; O'Neill, C.; McPolin, D.; Widmann, R. Strengthening of timber structures with glued-in rods. *Constr. Build. Mater.* **2015**, *97*, 90–105. [CrossRef]

28. Vallée, T.; Tannert, T.; Fecht, S. Adhesively bonded connections in the context of timber engineering–A Review. *J. Adhes.* **2017**, *93*, 257–287. [CrossRef]

29. Loss, C.; Piazza, M.; Zandonini, R. Connections for steel–timber hybrid prefabricated buildings. Part II: Innovative modular structures. *Constr. Build. Mater.* **2016**, *122*, 796–808. [CrossRef]

30. Polastri, A.; Giongo, I.; Piazza, M. An Innovative Connection System for Cross-Laminated Timber Structures. *Struct. Eng. Int.* **2017**, *27*, 502–511. [CrossRef]

31. Kraler, A.; Kögl, J.; Maderebner, R.; Flach, M. Sherpa-Clt-Connector for Cross Laminated Timber Elements. In Proceedings of the World Conference on Timber Engineering, Quebec City, QC, Canada, 10–14 August 2014.

32. Pozza, L.; Scotta, R.; Trutalli, D.; Pinna, M.; Polastri, A.; Bertoni, P. Experimental and Numerical Analyses of New Massive Wooden Shear-Wall Systems. *Buildings* **2014**, *4*, 355–374. [CrossRef]

33. Dias, A.; Skinner, J.; Crews, K.; Tannert, T. Timber-concrete-composites increasing the use of timber in construction. *Eur. J. Wood Wood Prod.* **2016**, *74*, 443–451. [CrossRef]

34. Yagi, H.; Shioya, S.; Tomiyoshi, E. Innovative hybrid timber structures in Japan: Bending behaviour of T-shaped CLT-to-hybrid timber composite beam. In Proceedings of the WCTE 2016—World Conference on Timber Engineering, Vienna, Austria, 22–25 August 2016.

35. Polastri, A.; Izzi, M.; Pozza, L.; Loss, C.; Smith, I. Seismic analysis of multi-storey timber buildings braced with a CLT core and perimeter shear-walls. *Bull. Earthq. Eng.* **2019**. [CrossRef]

36. He, M.; Luo, Q.; Li, Z.; Dong, H.; Li, M. Seismic performance evaluation of timber-steel hybrid structure through large-scale shaking table tests. *Eng. Struct.* **2018**, *175*, 483–500. [CrossRef]

37. Dickof, C.; Stiemer, S.F.; Bezabeh, M.A.; Tesfamariam, S. CLT-steel hybrid system: Ductility and overstrength values based on static pushover analysis. *J. Perform. Constr. Facil.* **2014**, *28*, A4014012. [CrossRef]

38. Schober, K.U.; Harte, A.M.; Kliger, R.; Jockwer, R.; Xu, Q.; Chen, J.F. FRP reinforcement of timber structures. *Constr. Build. Mater.* **2015**, *97*, 106–118. [CrossRef]

39. Antolinc, D.; Žarnic, R.; Cepon, F.; Rajcic, V.; Stepinac, M. Laminated glass panels in combination with timber frame as a shear wall in earthquake resistant building design. In Proceedings of the Challenging Glass 3: Conference on Architectural and Structural Applications of Glass, CGC 2012, Delft, The Netherlands, 28–29 June 2012.

40. Stepinac, M.; Rajčić, V.; Žarnić, R. Timber-structural glass composite systems in earthquake environment. *Gradjevinar* **2016**, *68*. [CrossRef]

41. Antolinc, D.; Rajčić, V.; Žarnić, R. Analysis of hysteretic response of glass infilled wooden frames. *J. Civ. Eng. Manag.* **2014**, *20*, 600–608. [CrossRef]

42. European Committee for Standardization (CEN). *EN 1998-1:2004, Eurocode 8, Design of Structures for Earthquake Resistance, Part 1: General Rules, Seismic Actions and Rules for Buildings*; CEN: Brussels, The Netherlands, 2004.

43. National Research Council of Canada (NRCC). *National Building Code of Canada 2015*; Associate Committee on the National Building Code: Ottawa, ON, Canada, 2015.

44. American Society of Civil Engineers (ASCE). *ASCE 7-16, Minimum Design Loads and Associated Criteria for Buildings and Other Structures*; American Society of Civil Engineers (ASCE): Reston, VA, USA, 2017.

45. ASCE. *Seismic Evaluation and Retrofit of Existing Buildings, ASCE/SEI 41-13*; American Society of Civil Engineers: Reston, VA, USA, 2014.

46. NZSEE. *The Seismic Assessment of Existing Buildings, Technical Guidelines for Engineering Assessments, Part A: Assessment Objectives and Principles. Ministry of Business, Innovation and Employment*; Earthquake Commission, New Zealand Society for Earthquake Engineering: Wellington, New Zealand, 2017.

47. FEMA. *Next-Generation Methodology for Seismic Performance Assessment of Buildings, Prepared by ATC for FEMA, FEMA P-58*; Federal Emergency Management Agency: Washington, DC, USA, 2012.

48. CNR. *Guide for the Probabilistic Assessment of the Seismic Safety of Existing Buildings, Report CNR-DT 212/2013*; CNR—Advisory Committee on Technical Recommendations for Construction: Rome, Italy, 2014.

49. Trutalli, D.; Marchi, L.; Scotta, R.; Pozza, L. Capacity design of traditional and innovative ductile connections for earthquake-resistant CLT structures. *Bull. Earthq. Eng.* **2019**, *17*, 2115–2136. [CrossRef]

50. Gavric, I.; Fragiacomo, M.; Ceccotti, A. Capacity seismic design of X-lam wall systems based on connection mechanical properties. In Proceedings of the International Council for Research and Innovation in Building and Construction, CIB-W18/46-15-2, Working Commission W18-Timber Structures, Vancouver, BC, Canada, 26–29 August 2013.

51. Casagrande, D.; Doudak, G.; Polastri, A. A proposal for the capacity-design at wall- and building-level in light-frame and cross-laminated timber buildings. *Bull. Earthq. Eng.* **2019**, *17*, 3139–3167. [CrossRef]

52. *EN 12512 Timber Structures—Test Methods—Cyclic Testing of Joints Made with Mechanical Fasteners*; European Committee for Standardization (CEN): Brussels, Belgium, 2001.

53. United Nations. *Transforming our World: The 2030 Agenda for Sustainable Development*; United Nations: New York, NY, USA, 2015; Volume 25.

54. Council of the European Union. *A New Global Partnership for Poverty Eradication and Sustainable Development after 2015'—Council Conclusions*; Council of the European Union: Brussels, Belgium, 2015.

55. Quesada-Pineda, H.; Smith, R.; Berger, G. Drivers and Barriers of Cross-Laminated Timber (Clt) Production and Commercialization: A Case of Study of Western Europe'S Clt Industry. *Bioprod. Bus.* **2018**, *3*, 29–38.

56. Brandner, R.; Flatscher, G.; Ringhofer, A.; Schickhofer, G.; Thiel, A. Cross laminated timber (CLT): Overview and development. *Eur. J. Wood Wood Prod.* **2016**, *74*, 331–351. [CrossRef]

57. Ferdous, W.; Bai, Y.; Ngo, T.D.; Manalo, A.; Mendis, P. New advancements, challenges and opportunities of multi-storey modular buildings—A state-of-the-art review. *Eng. Struct.* **2019**, *183*, 883–893. [CrossRef]

58. Pan, W.; Gibb, A.F.; Dainty, A.R.J. Perspective of UK housebuilders on the use of offsite modern methods of construction. *Constr. Manag. Econ.* **2007**, *25*, 183–194. [CrossRef]

59. *World Green Building Council Technical Report Bringing Embodied Carbon Upfront*; World Green Building Council: London, UK, 2019.

60. Available online: https://en.wikipedia.org/wiki/List_of_tallest_wooden_buildings (accessed on 14 February 2020).

61. Kuzmanovska, I.; Gasparri, E.; Monne, D.T.; Aitchison, M. Tall timber buildings: Emerging trends and typologies. In Proceedings of the WCTE 2018—World Conf. Timber Engingeering, Seoul, South Korea, 20–23 August 2018.

62. Available online: https://usfs.maps.arcgis.com/apps/Shortlist/index.html?appid=96d170159fdd4007b19599d08592f91d (accessed on 20 January 2020).

63. Speranza, A.; Barbaresi, L.; Morandi, F. Experimental analysis of flanking transmission of different connection systems for CLT panels. In Proceedings of the WCTE 2016—World Conference on Timber Engineering, Vienna, Austria, 22–25 August 2016.

64. Soundproofing for CLT by Stora Enso. Available online: http://www.clt.info/wp-content/uploads/2015/10/Soundproofing-for-CLT-by-Stora-Enso-EN.pdf (accessed on 7 February 2020).

65. *ETA-11/0496—European Technical Approval: Three-Dimensional Nailing Plate (Angle Bracket for Timber-To-Timber or Timber-To-Concrete or Steel Connections)*; European Organization for Technical Approvals: Brussels, Belgium, 2018.

66. Titan Silent Angle Bracket for Shear Stresses with Resilient Profile—Technical Data Sheets. Available online: https://www.rothoblaas.com/products/soundproofing/resilient-profiles/titan-silent#documents (accessed on 7 February 2020).

67. Homb, A. Hybrid cross-laminated timber floors. Comparison of measurements and calculations. In Proceedings of the Proceedings of the 22nd International Congress on Acoustics, Buenos Aires, Argentina, 5–9 September 2016.

68. Homb, A.; Carter, C.G.; Rabold, A. Impact sound insulation of cross-laminated timber/massive wood floor constructions: Collection of laboratory measurements and result evaluation. *Build. Acoust.* **2017**, *24*, 35–52. [CrossRef]

69. Izzi, M.; Casagrande, D.; Bezzi, S.; Pasca, D.; Follesa, M.; Tomasi, R. Seismic behaviour of Cross-Laminated Timber structures: A state-of-the-art review. *Eng. Struct.* **2018**, *170*, 42–52. [CrossRef]

70. Niederwestberg, J.; Zhou, J.; Chui, Y.H. Comparison of theoretical and laboratory out-of-plane shear stiffness values of cross laminated timber panels. *Buildings* **2018**, *8*, 146. [CrossRef]

71. Tomasi, R.; Smith, I. Experimental characterization of monotonic and cyclic loading responses of CLT Panel-To-Foundation Angle Bracket Connections. *J. Mater. Civ. Eng.* **2015**, *27*, 04014189. [CrossRef]

72. Hu, L.; Omeranovic, A. Wind-induced vibration of tall wood buildings—Is it an issue? In Proceedings of the WCTE 2014—World Conference on Timber Engineering, Quebec City, QC, Canada, 10–14 August 2014.

73. Johansson, M.; Linderholt, A.; Jarnerö, K.; Landel, P. Tall timber buildings—A preliminary study of wind-induced vibrations of a 22-storey building. In Proceedings of the WCTE 2016—World Conference on Timber Engineering, Vienna, Austria, 22–25 August 2016.

74. Feldmann, A.; Huang, H.; Chang, W.S.; Harris, R.; Dietsch, P.; Gräfe, M.; Hein, C. Dynamic properties of tall timber structures under wind-induced vibration. In Proceedings of the WCTE 2016—World Conference on Timber Engineering, Vienna, Austria, 22–25 August 2016.

75. Connolly, T.; Loss, C.; Iqbal, A.; Tannert, T. Feasibility study of mass-timber cores for the UBC tall wood building. *Buildings* **2018**, *8*, 98. [CrossRef]

76. Andrea, P.; Luca, P. Proposal for a standardized design and modeling procedure of tall CLT buildings. *Int. J. Qual. Res.* **2016**, *10*, 607–624.

77. Sustersic, I.; Fragiacomo, M.; Dujic, B. Seismic Analysis of Cross-Laminated Multistory Timber Buildings Using Code-Prescribed Methods: Influence of Panel Size, Connection Ductility, and Schematization. *J. Struct. Eng. (U.S.)* **2016**, *142*, E4015012. [CrossRef]

78. Loss, C.; Pacchioli, S.; Polastri, A.; Casagrande, D.; Pozza, L.; Smith, I. Numerical study of alternative seismic-resisting systems for CLT buildings. *Buildings* **2018**, *8*, 162. [CrossRef]

79. Shahnewaz, M.; Alam, S.; Tannert, T. In-plane strength and stiffness of cross-laminated timber shear walls. *Buildings* **2018**, *8*, 100. [CrossRef]

80. Fragiacomo, M.; Dujic, B.; Sustersic, I. Elastic and ductile design of multi-storey crosslam massive wooden buildings under seismic actions. *Eng. Struct.* **2011**, *33*, 3043–3053. [CrossRef]

81. Vassallo, D.; Follesa, M.; Fragiacomo, M. Seismic design of a six-storey CLT building in Italy. *Eng. Struct.* **2018**, *175*, 322–338. [CrossRef]

82. Smith, T.; Pampanin, S.; Fragiacomo, M.; Buchanan, A. Design and Construction of Prestressed Timber Buildings for Seismic Areas. *NZ Timber Des. J.* **2008**, *16*, 3–10.

83. Loss, C.; Frangi, A. Experimental investigation on in-plane stiffness and strength of innovative steel-timber hybrid floor diaphragms. *Eng. Struct.* **2017**, *138*, 229–244. [CrossRef]

84. Tesfamariam, S.; Stiemer, S.F.; Dickof, C.; Bezabeh, M.A. Seismic vulnerability assessment of hybrid Steel-Timber structure: Steel moment-Resisting frames with clt infill. *J. Earthq. Eng.* **2014**, *18*, 929–944. [CrossRef]

85. Zhang, X.; Shahnewaz, M.; Tannert, T. Seismic reliability analysis of a timber steel hybrid system. *Eng. Struct.* **2018**, *167*, 629–638. [CrossRef]

86. Li, Z.; He, M.; Wang, X.; Li, M. Seismic performance assessment of steel frame infilled with prefabricated wood shear walls. *J. Constr. Steel Res.* **2018**, *140*, 62–73. [CrossRef]

87. *EN 1995-1-1:2004 Eurocode 5: Design of Timber Structures—Part 1-1: General—Common Rules and Rules for Buildings*; CEN: Brussels, Belgium, 2004.

88. Antolinc, D.; Žarni, R.; Stepinać, M.; Rajčič, V.; Krstevska, L.; Tashkov, L. Simulation of earthquake load imposed on timber-glass composite shear wall panel. In *COST Action TU0905 Mid-Term Conference on Structural Glass*; CRC Press: Poreč, Croatia, 2013.

89. Ber, B.; Premrov, M.; Sustersic, I.; Dujic, B. Innovative earthquake resistant timber-glass buildings. *Nat. Sci.* **2013**, *05*, 63–71. [CrossRef]

90. Ber, B.; Premrov, M.; Štrukelj, A.; Kuhta, M. Experimental investigations of timber-glass composite wall panels. *Constr. Build. Mater.* **2014**, *66*, 235–246. [CrossRef]

91. Ber, B.; Sustersic, I.; Dujic, B.; Jancar, J.; Premrov, M. Seismic Shaking Table Testing of Glass-Timber Buildings. In Proceedings of the WCTE 2014—Word Conference on Timber Engineering, Quebec City, QC, Canada, 10–14 August 2014; pp. 2–8.

92. Žarnić, R.; Rajčić, V. Laminated glass—timber panel as a dissipative bracing structural component for existing frame structures. In Proceedings of the 16th World Conference on Earthquake Engineering, Santiago, Chile, 9–13 January 2017.

93. Žarnić, R.; Rajčić, V. Experimental investigation of laminated glass infilled CLT frames with glued-in steel rods in joints. In Proceedings of the International Conference on Structural Health Assessment of Timber Structures, Wroclaw, Poland, 9–11 September 2015.

94. Rajčić, V.; Žarnić, R. Highly energy dissipative and ductile timber-glass hybrid element. In Proceedings of the World Conference on Timber Engineering, Vienna, Austria, 22–25 August 2016.

95. Žarnić, R.; Rajčić, V.; Kržan, M. Response of laminated glass-CLT structural components to reverse-cyclic lateral loading. *Constr. Build. Mater.* **2020**, *235*. [CrossRef]

96. Available online: https://www.grad.unizg.hr/vetrolignum (accessed on 22 January 2020).

97. Pampanin, S.; Bolognini, D.; Pavese, A. Performance-based seismic retrofit strategy for existing reinforced concrete frame systems using fiber-reinforced polymer composites. *J. Compos. Constr.* **2007**, *11*, 211–226. [CrossRef]

98. Valente, M.; Milani, G. Alternative retrofitting strategies to prevent the failure of an under-designed reinforced concrete frame. *Eng. Fail. Anal.* **2018**, *89*, 271–285. [CrossRef]

99. Kurosawa, R.; Sakata, H.; Qu, Z.; Suyama, T. Precast prestressed concrete frames for seismically retrofitting existing RC frames. *Eng. Struct.* **2019**, *184*, 345–354. [CrossRef]

100. Abdel-Aty, Y.Y.A. Proposals for seismic retrofitting of timber roofs to enhance their in-plane stiffness and diaphragm action at historical masonry buildings in Cairo. *J. Cult. Herit.* **2018**, *32*, 73–83. [CrossRef]

101. Parisi, M.A.; Piazza, M. Seismic strengthening and seismic improvement of timber structures. *Constr. Build. Mater.* **2015**, *97*, 55–66. [CrossRef]

102. Lourenço, P.B.; Ciocci, M.P.; Greco, F.; Karanikoloudis, G.; Cancino, C.; Torrealva, D.; Wong, K. Traditional techniques for the rehabilitation and protection of historic earthen structures: The seismic retrofitting project. *Int. J. Archit. Herit.* **2019**, *13*, 15–32. [CrossRef]

103. ASCE/SEI, 41-17. *Seismic Evaluation and Retrofit of Existing Buildings*; American Society Of Civil Engineers: Reston, VA, USA, 2017. [CrossRef]

104. Stepinac, M.; Rajčić, V.; Barbalić, J. Inspection and condition assessment of existing timber structures. *Gradjevinar* **2017**, *69*, 861–873.

105. Sayin, B.; Yildizlar, B.; Akcay, C.; Gunes, B. The retrofitting of historical masonry buildings with insufficient seismic resistance using conventional and non-conventional techniques. *Eng. Fail. Anal.* **2019**, *97*, 454–463. [CrossRef]

106. Michel, C.; Karbassi, A.; Lestuzzi, P. Evaluation of the seismic retrofitting of an unreinforced masonry building using numerical modeling and ambient vibration measurements. *Eng. Struct.* **2018**, *158*, 124–135. [CrossRef]

107. Bournas, D.A. Concurrent seismic and energy retrofitting of RC and masonry building envelopes using inorganic textile-based composites combined with insulation materials: A new concept. *Compos. Part B Eng.* **2018**, *148*, 166–179. [CrossRef]

108. Haba, R.; Kitamori, A.; Mori, T.; Fukuhara, T.; Kurihara, T.; Isoda, H. Development of clt panels bond-in method for seismic retrofitting of RC frame structure. *J. Struct. Constr. Eng.* **2016**, *81*, 1299–1308. [CrossRef]

109. Stazi, F.; Serpilli, M.; Maracchini, G.; Pavone, A. An experimental and numerical study on CLT panels used as infill shear walls for RC buildings retrofit. *Constr. Build. Mater.* **2019**, *211*, 605–616. [CrossRef]

110. Park, S.; van de Lindt, J.; Lee, S.-H. Optimal seismic retrofit design for residential structures using CLT panel and FEMA P-807 methodology. In Proceedings of the 2015 World Congress on Advances in Structura. Engineering Mechanics, Incheon, Korea, 25–29 August 2015.

111. Bahmani, P.; van de Lindt, J.; Iqbal, A.; Rammer, D. Mass timber rocking panel retrofit of a four-story soft-story building with full-scale shake table validation. *Buildings* **2017**, *7*, 48. [CrossRef]

112. Sustersic, I.; Dujic, B. Seismic strengthening of existing buildings with cross laminated timber panels. In Proceedings of the World Conference on Timber Engineering, Auckland, New Zealand, 15–19 July 2012; Volume 4, pp. 122–129.

Review

Methods for the Assessment of Critical Properties in Existing Masonry Structures under Seismic Loads—The ARES Project

Mislav Stepinac [1,*], Tomislav Kisicek [1], Tvrtko Renić [1], Ivan Hafner [1] and Chiara Bedon [2]

[1] Faculty of Civil Engineering, University of Zagreb, 10000 Zagreb, Croatia;
tomislav.kisicek@grad.unizg.hr (T.K.); tvrtko.renic@grad.unizg.hr (T.R.); ivan.hafner@grad.unizg.hr (I.H.)

[2] Department of Engineering and Architecture, University of Trieste, 34127 Trieste, Italy;
chiara.bedon@dia.units.it

* Correspondence: mislav.stepinac@grad.unizg.hr

Received: 12 December 2019; Accepted: 19 February 2020; Published: 25 February 2020

Abstract: Masonry structures are notoriously vulnerable to horizontal actions caused by earthquakes. Given the high seismicity of the European region, and that the European building stock comprises a lot of masonry buildings, knowledge about their structural response to seismic excitation is particularly important, but at the same time difficult to determine, due to the heterogenous nature of materials and/or constructional techniques in use. An additional issue is represented by the current methods for mechanical properties assessment, that do not provide a reliable framework for accurate structural estimations of existing buildings characterized by different typological properties. Every structure, in other words, should be separately inspected in regard to its mechanical behaviour, based on dedicated approaches able to capture potential critical issues. In this review paper, an insight on the Croatian ARES project is presented (Assessment and Rehabilitation of Existing Structures), including a state-of-the-art of the actual building stock and giving evidence of major difficulties concerning the assessment of existing structures. The most commonly used techniques and tools are compared, with a focus on their basic features and field of application. A brief overview of prevailing structural behaviours and Finite Element numerical modelling issues are also mentioned. As shown, the general tendency is to ensure "sustainable" and energy-efficient building systems. The latter, however, seem in disagreement with basic principles of structural maintenance and renovation. The aim of the ongoing ARES project, in this context, is to improve the current knowledge regarding the assessment and strengthening of structures, with a focus on a more reliable design and maintenance process for existing masonry buildings.

Keywords: structural assessment; masonry buildings; earthquakes; seismic loads; existing structures; reliability; rehabilitation; risk

---

1. Introduction

According to the literature and recent events, it is well established that masonry is one of the most commonly used materials across the world, due to its simplicity and high quality characteristics.

Even though the use of masonry for construction in earthquake-prone regions gave evidence of its intrinsic limitations (due especially to its limited tensile resistance, relevant mass and stiffness), extensive research has been carried out in the last few decades, with a focus on the material characteristics and structural behaviour, even under extreme loading conditions such as earthquake events. These efforts enabled engineers to design masonry structures on sound and safety principles, with progressively greater exactitude, economy and confidence. Accordingly, a huge number of existing buildings in the European region are composed of masonry. Given that most of the so-called "strategic" buildings of cultural significance and high historical importance are built using masonry, such a condition is the first motivation, suggesting that the assessment and rehabilitation of existing masonry structures must be conducted on a very high level.

As is known, the main goal of seismic design is to protect property—and thus life in buildings and infrastructure—in the case of earthquake events. However, an appropriate seismic design approach must necessarily develop on knowledge and feedback from existing structures. Compared to other constructional typologies and materials, past events showed that seismic loads usually cause significant damage, especially in masonry buildings, due to their large mass and stiffness. This represents an intrinsic risk for personnel, given that most people in several European countries (especially in the urban areas) work and live in masonry buildings. In addition, a huge number of masonry structures were built—over decades—before any seismic codes were developed, thus no confining elements or reinforcement members exist.

Based on all the above motivations, it is thus clear that a concise strategy for masonry buildings must be activated.

The seismic behaviour of buildings generally depends on several important factors, such as material properties, the geometry of the structure, additional non-linear effects, conceptual design and stiffness properties. The issue of seismic vulnerability assessment and rehabilitation of under-performing existing buildings is hence a complex problem [1]. "Seismic vulnerability" can be conventionally defined as a measure of the inadequacy of a given structure to resist to seismic actions [2]. In modern assessment methods, the seismic vulnerability is represented by design curves which express the physical vulnerability as a function of the intensity of the process and the degree of loss [3]. For individual cases only, some structural characteristics of the affected buildings are considered [3]. Throughout the decades, various methods have been developed to evaluate the vulnerability of buildings, and they can be divided into *empirical* and *analytical* (and thus *hybrid*) methods, and an approach based on engineering judgment by experts. Methods for vulnerability assessments mainly model damage to a discrete scale, where damage itself is commonly grouped using three to six categories [4]. However, no unified approaches on a European level are available.

The seismic vulnerability of masonry buildings is particularly difficult to assess, and notoriously requires a multitude of specialized technical skills [5]. But actually, how accurate are the methods in use for the assessment of the seismic vulnerability of existing buildings?

Unfortunately, this question generally remains unanswered, given that more extended research needs to be carried out on the topics of seismic risk and seismic vulnerability assessment. Focusing only on "visible" structural/material parameters cannot solve such an open issue. Traditional assessment methods, in most of the cases, are in fact well-known to allow the assessment of only the actual condition of a given existing structure, once its stability has already been compromised. In this paper, selected traditional assessment methods are thus discussed, pointing out some possibilities related to the use of newer technologies.

The surge of buildings of higher consequence class (and the global goals towards sustainable development) typically demands higher levels of reliability, and a more sustainable use of raw materials. That is why it the aim is the modification or extension of existing buildings rather than the demolition and substitution. There are several important aspects that have a fundamental role in the assessment of existing masonry structures, namely assessment, deterioration and damage, inspection and investigation, updating, verification, repair, rehabilitation and reinforcement and maintenance. In this paper, assessment methods of critical properties (structural and material parameters) are presented for masonry structures. A focus is set on the available methods able to provide crucial data and feedback for preventing failure mechanisms and collapses under extreme design loads.

2. The Croatian Scenario

According to the results of systematic research of Statistical Yearbooks, Croatia's national building stock consists of approx. 800,000 residential and 125,000 non-residential buildings [6]. More than 75%

of the building stock is older than 30 years, thus corresponding to a life-time requiring at least some renovation or modification of primary structural components. More than 40% of the building stock is then older than 50 years, meaning that the service life of a given structure is fully expired. In the Croatian building sector, finally, it is recognized that up to 40% of the expenses are dedicated to the rehabilitation, modification and demolition of existing structures (Figure 1a).

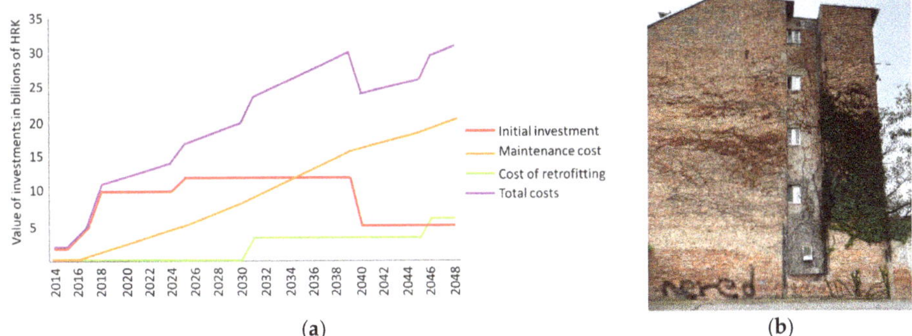

Figure 1. (**a**) Investment (values in HRK) for the renovation of Croatian national building stock (data derived from [6]); (**b**) Typical masonry building in Zagreb, with a graffiti meaning "a mess" (photo by M. Stepinac).

The main characteristic of Croatian buildings constructed in the period before the 1970s is represented by the use of traditional constructional techniques and materials, such as masonry and timber (Figure 1b). Buildings were built as full-brick masonry structures, with mostly wooden ceilings, and 30–60 cm thick walls, thus resulting in statically satisfactory structural assemblies. In the 1960s, reinforced concrete (in combination with timber and steel) started to progressively replace traditional constructional materials. In fact, most of the residential structures are still built as a combination of concrete, masonry and timber load-bearing components.

One of the most important HORIZON 2020 programme objectives is the acquisition of necessary knowledge and skills by all stakeholders in the process of energy-efficient building renovation. European Union directives clearly suggest the energy renovation of existing structures, but structural aspects are somewhat ignored and/or disregarded. At the moment, a number of existing structures are in fact under energy renovation and/or energy upgrading processes (Figure 2a). Besides such a consideration for energy performances, however, structural assessment and/or structural upgrading is mostly disregarded. According to an estimate of total investments in Europe, for the period between 2014 and 2049 (including initial investment expenditures, maintenance and replacement of worn-out equipment), around 3×10^9 Euros will be dedicated to energy processes. In such an expected scenario, it is thus clear that structural updating and retrofitting can (and must) represent an additional value for the energy renovation of buildings. This is especially the case in existing masonry structures, which, in most cases, need robust seismic strengthening interventions.

In the Croatian framework, it is in fact recognized that the majority of residential buildings older than 50 years consists of masonry structures (i.e., Figure 2b) without appropriate bonding elements to connect floors and walls [7].

 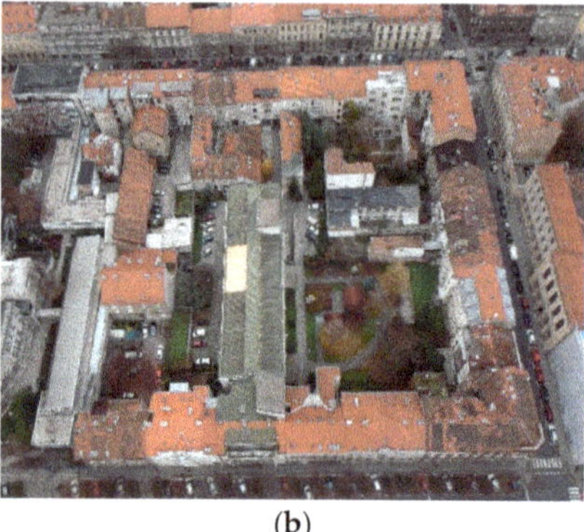

(a)   (b)

Figure 2. (a) Renovation of facades and energy upgrading of existing structures (b) Typical building blocks in Zagreb (photos by M. Stepinac).

3. Existing Masonry Structures and Maintenance Issues

The need for maintaining the built environment is supported by the global policy of the Kyoto protocol 1997 and all further World Climate Summits on existing buildings and engineering works. Sustainable development is a long-term goal of the global policy which results in modifications, substitutions or extensions of existing buildings.

In simple words, existing structures can be distinguished with regard to their value between economical (*monetary*) values and cultural (*non-material*) value. Modern existing structures commonly have a higher economical value, compared to heritage structures (with a dominating cultural value).

Accordingly, the assessment of an existing structure can be performed through different steps, with increased precision. The degree of precision thereby depends on the amount and on the quality of available information, as well as on the importance of the building being assessed. This can be reached by breaking down the assessment into different phases. The number of required phases is dependent on the remaining level of doubt, and the feasibility and simplicity of repair/strengthening, always in combination with economic considerations [8].

Advances in technology and sustainability requirements, and requirements for preservation of existing structures provoked an increased interest in scientific and professional community for assessment methods. Regarding the masonry structures, a wide variety of methods exist'; however, their frequency and scope, the decision-making approach concerning safety and the necessary interventions are far from being agreed upon. The need for an assessment of an existing structure can be based upon a multitude of reasons. The most typical are briefly explained in [8]. In situations where doubts may be raised in regard to the design assumptions, a re-assessment of the structure may be also necessary, such as [9,10]:

- extended service life;
- change in utilization;
- required increase in the level of reliability;
- lack of maintenance and inspection for an extended period;
- doubts regarding the reliability or malfunctioning of the structure (e.g., inadequate serviceability);
- exposure to accidental or unforeseen extreme loads (excessive loading, earthquake, fire, etc.);
- negative experience from other similar structures;
- availability of new knowledge and revised design codes;
- knowledge of errors in the planning or construction period.

With regard to economy and sustainability, finally, it is certainly of high interest for the building owners (as well as society) to maintain existing structures, rather than demolish and rebuild them.

4. European Standards, Norms and Guidelines

Most of the current design standards and guidelines are based on reliability-based design, as specified in ISO 2394 [11] or JCSS [12] documents. The majority of them regulate the design of new structures, such as the Eurocodes. Nevertheless, the intention is to make the Eurocodes applicable also for existing structures. Work on new technical rules for the assessment and retrofitting of existing structures is currently very intense, with special attention being given to heritage structures, aiming at elaborating the new Eurocode part for existing structures.

One of the few international guidelines for the assessment of existing structures is the ISO 13822 document [13]. Although different guidelines for the maintenance of existing structures exist in other European countries, only a few of them have issued standards for the assessment of existing structures (e.g., Switzerland [14] and the Netherlands [15]). More in detail, SIA 462 [14] standard, in combination with the SIA 469 [16] document, specify the general fundamentals for the assessment of the load-carrying capacity of existing buildings, and regulate the professional and economical maintenance of buildings according to their cultural value, respectively.

Nonetheless, the daily basis for the assessment and rehabilitation of existing structures is still based on a rather rudimentary scheme, mainly replacing and reinforcing defective members. More sophisticated procedures are needed, in addition to those available in seismic codes (such as Eurocodes). Eurocode EN 1998-3 [17] provides some suggestions for the assessment of masonry structures, but it is only informative and lacks of detailed practical suggestions. ISO 13822:2010 [13] gives general recommendations for a rough assessment of existing structures, but does not take materials into account. Similar to EN 1998-3 [17], it is mainly informative and lacks depth. Neither of these standards are very practical. The American documents ASCE 41-13 [18,19] and ASTM [20–25], in this regard, constitute a more comprehensive guide, including suggested equipment, procedures of assessment, number of tests, as well as analysis and strength calculation methods. From the perspective of an engineer, this standard is hence more practical and easily applicable.

5. Selected Assessment Methods

Numerous technical documents have been published by national and international authorities and focus on systematic and scientific methods that can be used to accurately assess the residual strength, durability and reliability of structural materials, assemblies and systems in existing buildings. Nevertheless, especially in the scope of masonry structures, all these documents need to be continuously revised, expanded and enhanced. Major issues are related to the increasing knowledge in the field of material sciences, as well as to the technological advancement in the field of the structural assessment and monitoring of structural systems, or to the practical experience and feedback derived from professional engineering activities on existing structures. The key role and relevance of research, in this context, lays in the preservation of existing building stock, restoration of objects, towards the support and development of reliable and consistent guidelines/norms.

Masonry structures are composite systems, whose main components are masonry units and mortar layers. Both can be made of various materials, with different mechanical characteristics. The geometry of masonry units (length, width, height, amount and direction of holes), the thickness of mortar and the area it covers can also strongly affect the overall mechanical behaviour. Recent research studies have shown that the inclusion of masonry infills leads to a significant increase in structural stiffness, thus affecting the overall probability of structural collapse under dynamic loads (such as those induced during an earthquake). Specifications of compressive strength, a function of parameters of brick and mortar, are often required. Assessing the variability of these properties and the uncertainty in the modelling of the masonry compressive strength is hence a topic of great importance [26].

For the assessment of existing structures, the strength classes of used materials are often unknown. Nevertheless, major advantage and benefit can derive from the fact that the reliability of structures directly depends on the actual properties of the elements in use. Assessment methods should therefore aim at identifying these properties to the highest degree of achievable certainty, in order to reduce the uncertainty with regard to the resistance of the structure as a whole. In contrast, for the design of new structures, possible uncertainties on material properties are taken into account in the partial safety factors for strength properties.

Parameters which are always measured when assessing masonry structures are thus represented by (i) the compressive strength of masonry units, (ii) the compressive strength of masonry mortar, (iii) the compressive strength of concrete infill (if any), (iv) the strength of reinforcing steel bars (if any), (v) the compressive, shear and flexural strength of masonry, (vi) the Modulus of Elasticity (MoE) and the geometry of the masonry structure (size and location of bearing walls, location and size of openings).

The most important Non Destructive Testing (NDT) methods for existing masonry structures are summarized in Table 1 and Figures 3 and 4.

Table 1. Available Non Destructive Testing (NDT) assessment methods for existing masonry structures.

NDT Method	Devices/Test	What Is Measured?	How Is It Measured?	References
Visual inspection	/	Quality of masonry (mechanical parameters, dimension, shape), mortar and wall connections	Without a device, using a base/set of rules (i.e., Masonry quality index-MQI)	Borri et al. [27]
Measurement of masonry unit hardness	Rebound hammer (Schmidt hammer)	Compressive strength of masonry units, mortars and built masonry	A predefined number of tests is conducted in both horizontal and vertical direction (with a calibration needed)	Breysse and Martínez-Fernández [28], Sýkora et al. [29]
Measurement of reinforcement location	Ground Penetrating Radar (GPR)	Location (depth) of reinforcement	The device is placed on the measured surface and moved along a linear axis (with a calibration needed), transmitting radio wave signals into a structure and detecting echoes	Agred, Klysz and Balayssac [30]
Stress wave transmission	Ultrasonic Pulse Velocity test (UPV) test/Resonant frequency test (RF)	Compressive strength of concrete or masonry	UPV-two transducers are placed on two sides of the specimen after which the time of wave travel is measured RF-a piezometric sensor is used with different attachment techniques to obtain resonant frequency	Sajid et al. [31]
Ultrasonic velocity testing	Impact hammer and accelerometer	Characterization of masonry wall homogeneity and variability	On opposite sides of the wall, an impact hammer and an accelerometer are placed. The mechanical impulse is generated by the hammer striking the material and the signal is then received by the accelerometer.	Mesquita et al. [32]
Sonic velocity testing	Impact hammer and accelerometer	Location of heterogeneities, voids or inclusions of other materials in masonry elements	On opposite sides of the wall, an impact hammer and an accelerometer are placed, after which the mechanical impulse is generated by the hammer striking the material and the signal is then received by the accelerometer	Martini et al. [33] Valluzzi et al. [34]
Surface penetrating radar	Ground Penetrating Radar (GPR)	Location (depth) of reinforcement, thickness of elements, position of voids and moisture content	The device is placed on the measured surface and moved along a linear axis (with a calibration needed) transmitting radio wave signals into a structure and detecting echoes	Martini et al. [33] Wai-Lok Lai, Dérobert and Annan [35]

Table 1. *Cont.*

NDT Method	Devices/Test	What Is Measured?	How Is It Measured?	References
Infrared thermography	Thermography cameras Visual IR thermometers	Defects in the buildings envelope, the monitoring of reinforcing steel in concrete, the detection of moisture etc.	The specimen is under thermal stimulation and its surface temperature variation is monitored during the heating or cooling phase (the presence of inhomogeneity in a material causes local temperature variations)	Meola [36]
Borescope and mortar hardness with pendulum rebound hammer	Borescope and pendulum rebound hammer	Borescope—anomalies and internal wall components, such as ties, flashing and drainage cavities Pendulum rebound hammer—mortar type and strength	The borescope is inserted into small holes drilled into mortar joints (with fiber optics and internal light source) The pendulum rebound hammer utilizes a low energy impact and the resulting rebound from the surface of a mortar joint is used to measure surface hardness	Schuller [37]
Flat-jack tests	Flat jacks	Deformability parameters in compression, compressive strength and shear strength parameters	Two cuts are made with a predefined distance between them (horizontal cuts for compression, vertical cuts for shear), after which the jack is inflated with a liquid that transmits hydrostatic pressure	Parivallal et al. [38], Simões et al. [39], Łatka and Matysek [40], Croce et al. [41]
Acoustic emission		The damage evolution in masonry, evaluation of the reliability of reinforcing techniques, analysis of residual capacity of brick masonry	A group of transducers are set to record signals, then locate the precise area of their origin by measuring the time for the sound to reach different transducers.	Invernizzi et al. [42]

Although each method mentioned in Table 1 has both advantages and disadvantages, not all of them are needed for adequate assessment. For an adequate assessment, further methods may be applied (although other combinations are possible).

Visual inspection presents a basic tool for the assessment of existing buildings. It is a significant method which should always be implemented to determine further actions. The type of masonry, building location specifics and overall state of the structure can be determined and rough estimations may be made. One of the negatives is that it requires an experienced person to determine the important parameters.

A rebound hammer is a tool that may be used to indirectly determine the compressive strength of masonry. Since this is one of the most important parameters and considering both its low cost and ease of applicability, this method seems to be a valuable tool. Although some methods may be better at determining compressive strength, they are either more expensive and complex, or not non-destructive.

Ground penetrating radar can detect the location of steel, thickness of the wall and possible voids in it. It is a versatile tool with which the geometry of a wall can be determined precisely.

Flat-jack systems can be used to assess the stresses in existing walls, strength and deformability parameters, as well as shear strength of a wall. These are all important values to obtain for a more precise assessment. One of the downsides is the fact that slits need to be cut in a wall. Either rigid or flexible flat-jacks can be used, depending on the type of wall that is being tested.

In addition to assessment, continuous Structural Health Monitoring (SHM) may assist engineers for a better understanding of the actual structural behaviour of a given assembly. Together with structural risk and reliability research and development, SHM specialists figure a comprehensive research community. Accordingly, it is generally recognized that SHM represents an important field of today's infrastructure engineering. It is a goal to enhance the benefit of SHM by the novel utilization of applied decision analysis on how to assess the value of SHM—even before it is implemented.

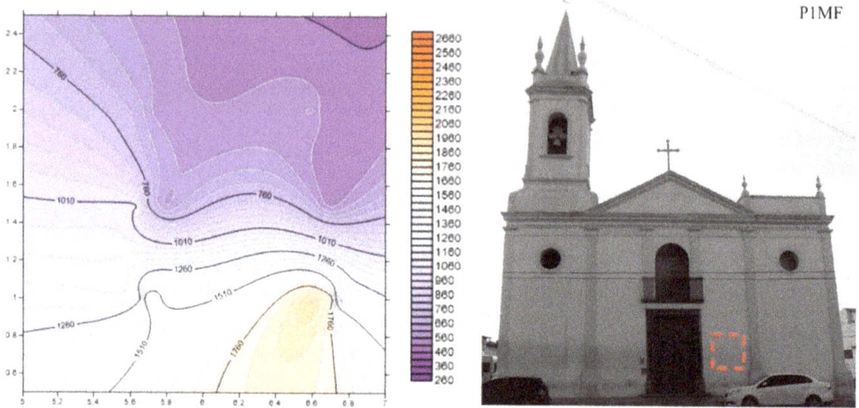

Figure 3. Example of ultrasonic velocities distribution (values in m/s) for a portion of the "P1MF" case-study building (box detail). Reproduced from [25] with permission from Elsevier® (Copyright© Agreement license n. 4726430416591, December 2019).

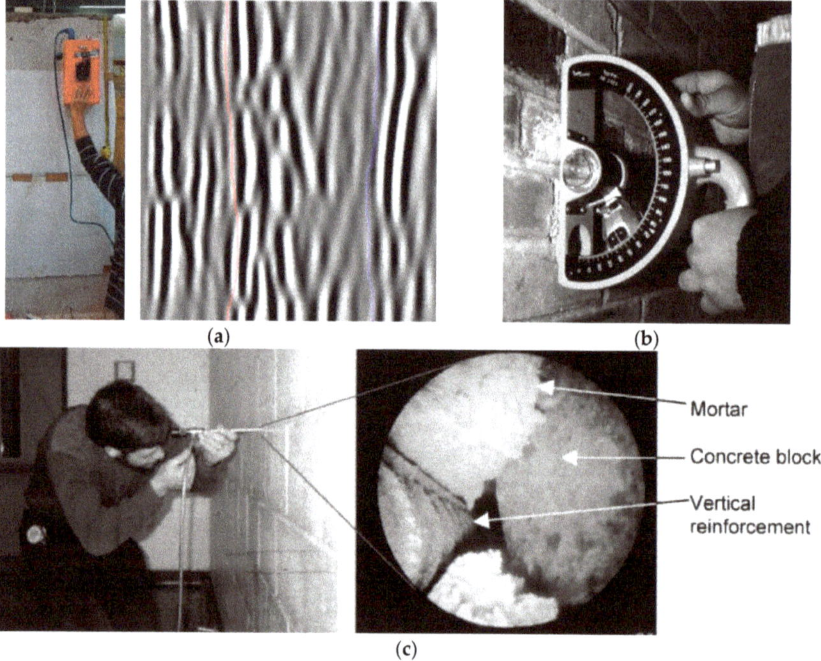

Figure 4. Example of NDT assessment of existing masonry: (**a**) GPR testing instruments and results (reproduced from [27] with permission from Elsevier® (Copyright© Agreement license n. 4726440628105, December 2019)); (**b**) pendulum rebound hammer, to evaluate masonry hardness; with (**c**) borescope investigation (figures (**b**,**c**) are reproduced from [31] with permission from John Wiley & Sons® (Copyright© Agreement license n. 4726520295075, December 2019)).

Knowing the intrinsic value of SHM techniques, the decision basis for the design, operation and life-cycle integrity management of structures can thus be improved and can finally facilitate more cost-efficient, reliable and safe strategies for maintaining and developing the built environment to the benefit of society [43]. All of the above-mentioned methods (see also Figure 5) are in fact eligible to offer support for better understanding the structural behaviour, risks and costs for the preservation of existing masonry structures.

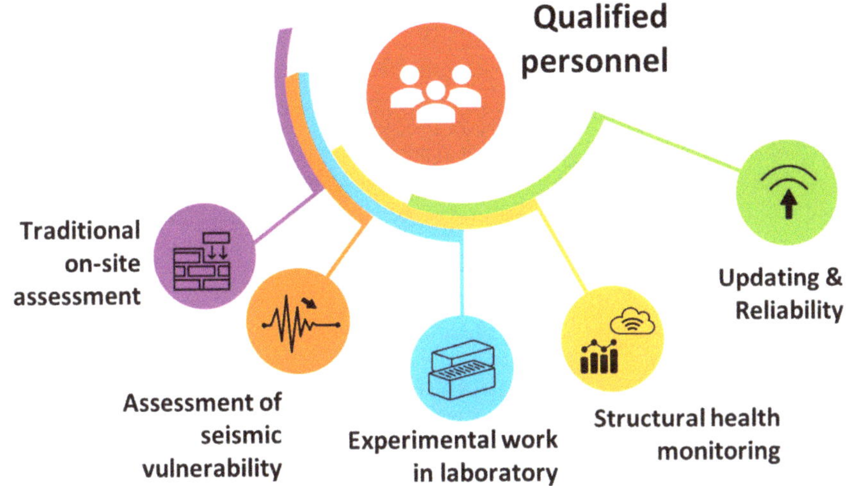

Figure 5. Graphical illustration of the typical assessment procedures for masonry structures.

Seismic Actions and Masonry Structures

In order to strengthen existing weak structures, assessment must be necessarily designed and carried out. In the past, most of the initial work on seismic assessment was based only on the visual inspection of buildings [27], NDT methods (see, for example, [32,41,44–46]), as well as the experience of engineers. The last decade, however, has seen a growth in the technological development of various tools which can greatly support the prediction of structural safety and seismic behaviour of existing structures, i.e., thermography [36], photogrammetry [47], unmanned aerial vehicles [48], etc.

The first key step is visual inspection. A structure's location, both in respect to seismic hazards [49] and with respect to micro-location (e.g., hill, valley, etc.) matters. Neighbouring structures can also influence structural behaviour. For example, different storey heights may lead to the collision of a slab with a neighbouring structure, as can be seen in Figure 6a.

While newer buildings tend to have lower stories than old buildings, this is not uncommon when a new structure is built inside a block of existing structures. Even if their storey height is the same, neighbouring structures, according to Figure 6b, may also collide.

The layout of a building's structural element defines its dynamic response, so it should be adequately measured or characterized [50]. It is not uncommon that access to the inside of a building is restricted. The time of construction can also represent another valuable piece of information, since it gives insight into the most probable material characteristics (both masonry units and mortar), used building technology and design codes. In terms of floor systems, the most commonly used solutions are timber in older constructions and reinforced concrete in newer constructions. Timber floors, as is known, are more flexible than stiff concrete slabs. This not only leads to a different distribution of seismic actions, but also means that walls are more susceptible to potential out-of-plane failure mechanisms. Concrete slabs are, in fact, recognized to ensure a box-type behaviour of walls. In structures with timber floors, horizontal steel rods are usually added to connect the walls (when they do not already exist) and overcome such an issue.

Besides the need of such a box-type behaviour and interlocking effect, the mortar and masonry quality of load-bearing elements typically degrades with time. Since different materials were used in history, it is advantageous to know the time period of construction. In addition, visual inspection of material degradation should be necessarily performed. From visual inspection, the position, direction and width of cracks can be measured and can indicate weak spots in a given structure, as well as poor soil conditions, or poor craftsmanship. In some cases, cracks may also be structurally negligible. The geometry of a wall, as well as shape, size and position of openings, can significantly influence crack formation and propagation. Cracks cause a reduction in stiffness, which in turn changes the dynamic response of a structure. To adequately model stiffness, the MoE needs to be known, which can be done with a flat-jack system or rebound hammer. Since stiffness distribution is also relevant, measurements should be done on all walls. When stiffness is correctly assessed, seismic forces can be determined. In order to check if a failure will occur, the strength of a material is needed. Compressive strength can be appraised (with careful laboratory calibration) by using the rebound hammer, which measures wall hardness. If reinforcement exists, its distribution and amount influence the crack width. Usually, reinforcement cannot be detected by visual inspection, so additional tools like Ground Penetrating Radar (GPR) are needed. Some of the reinforcement might be corroded or poorly anchored, and such an issue should be properly checked. Other important parameters of a wall can be assessed with GPR, such as the thickness of each wythe and a complete wall, or void location and size. Poor connection between different wythes may cause the falling out of a part of wall, which can potentially be dangerous for personnel. The dynamic response of a structure should be then properly checked using a Finite Element (FE) model. In the latter, however, the previously measured material parameters (or appraised on-site) should be carefully calibrated.

In addition to previously mentioned data, an important role is also that of soil, whose stiffness and strength should be carefully taken into account. The variation of soil conditions under the building should also be considered. Based on a structural model, stresses and strains may be calculated in each point of a wall. The vulnerability of a structure under different earthquake records may also be calculated and, in combination with hazard risk, can be defined. Based on risk, expected cost can be estimated. When the expected risk is unacceptable, the structure should be properly strengthened. This can be done by adding steel reinforcement or some sort of non-metallic material with high tensile resistance, such as Fiber-Reinforced Polymers (FRP) or Textile Reinforced Mortar (TRM) solutions. Based on the expected critical failure mechanisms, different configurations of reinforcement may be adopted. Strengthening can also be effective if maximum deformations either cause the failure of non-load-bearing elements or stability problems. Since stiffness changes with adding reinforcement, however, additional changes in the reference models should be necessarily implemented.

(a)   (b)

Figure 6. (**a**) Reconstruction works in urban areas—building of new facilities which directly influence the seismic safety of existing structures. (**b**) Typical street in the old city of Zagreb, where the majority of buildings are made of masonry (photos by M. Stepinac).

6. Open Challenges in the Framework of the ARES Project

The preservation of the building stock tackles societal challenges, environmental issues, and resource efficiency and represents the complex process of engineering and technical work.

The broad range of available methods shows the extensive experience in the measurement techniques gained in recent years. However, this experience, so far, has not led to a considerable advance in the quality of assessment with regard to the evaluation of the reliability of structures. Most methods only allow for the determination of the level of degradation and the localization of damages or the determination of localized material properties that do not represent the overall properties and load-carrying capacity of the entire structure. So far, it is often not possible to relate and interpret these measurements with regard to strength and stiffness properties required for the design and evaluation of reliability of the structure with sufficient certainty. Advanced methods for the assessment and probabilistic evaluation of existing structures make use of updated information and are already successfully applied in different other fields [51] but are still scarce in masonry engineering [29,52–56].

At the moment, there is a huge knowledge gap in the assessment methods and design checks of existing structures. Without appropriate guidelines, it is difficult to approach the problem, and this often leads to the wrong interpretation of collected data, and, thus, wrong decision scenarios for the reinforcement and rehabilitation of existing structures. The development of investigation techniques for the updating of material properties will help in reducing the uncertainty associated with the prediction of the structural behaviour of existing structures.

The ARES project, in this context, currently tries to fill the mentioned gaps. The project is financially supported by the Croatian Science Foundation and aims at delivering the basis for advanced assessment and design of existing structures, allowing a more economic design and a more accurate analysis of the consequences of failure. The ARES project aims to improve the way assessment is actually carried out. Standard methods will be compared, and necessary procedures will be determined to simplify assessment for practical use. General guidelines will also be developed, and safety factors will be reviewed.

The research project will provide knowledge about the building stock in Croatia, assessment of existing masonry structures and prediction of material properties from NDT. In addition, one of the current project goals is to evaluate updating methods of properties for the application of the assessment and verification methods for structures, including consideration of time-dependent behaviour and the influence of environmental conditions.

Furthermore, it is expected that in the current revision of the Eurocodes the consideration of existing structures will be more prominent, but, so far, no adequate international rules for the assessment, reuse and rehabilitation of existing structures exist.

In the most developed societies, as they progress, the feeling grows that it is necessary to maintain the existing building stock to preserve cultural identities. Preserving old buildings also benefits businesses and the local economy of societies. Preservation reduces waste, demolition energy use and new construction. Adaptive re-use concepts, renovations for less energy use, maintenance and type of use also affect building sustainability.

The following main challenges for future research and development in the field of assessment of existing structures can be identified:

- efficient determination of properties of structures
- precise prediction of material properties
- reliable prediction of the structural performance
- accounting for updated information in the assessment process
- optimization of verification and design procedures
- quantification of the impact of loading history and load duration on the load-carrying capacity in the remaining service life
- development of low invasive measures for intervention and rehabilitation

- enhancement of the communication with decision-makers.

Solving the above-mentioned issues can be of valuable use and precious help for consultants and designers, as well as for researchers and scholars dealing with the assessment of masonry structures.

7. Conclusions

European building stock comprises a large number of masonry structures which are vulnerable to seismic excitation. To reduce vulnerability, they usually need to be strengthened with either steel or non-metallic reinforcement. Configuration and the amount or reinforcement need to be determined after the assessment is done. Since sustainability is becoming a prevailing issue, the general aim seems to be shifting from building new structures to the maintenance of existing ones. Thus, the evaluation of structural behaviour and strengthening techniques is becoming more important than it was before. Many questions regarding seismic vulnerability assessment of masonry structures remain unanswered, although many tools and techniques for estimation of material properties exist. Most of the historic structures and a lot of residential buildings are built with masonry, so both economic and cultural aspects are at risk. In many cases, access to the interior of a structure is limited. At the same time, energy efficiency is currently being improved in a lot of structures without any seismic or even structural considerations. Some simple solutions may lead to a large improvement in seismic behaviour, such as adding horizontal steel rods to ensure a box-type behaviour of a structure.

The majority of existing design standards regulate the design of new structures but provide poor information for their maintenance and repair. Although some guidelines and technical documents exist, they need to be improved and adapted to different solutions. Further research on material properties (and also its variability), risk assessment and modelling is, hence, necessary. Although there is a lot of experience regarding existing structures, no significant advancement in the quality of assessment can actually be perceived. The interpretation of measurements and their relation to calculation parameters still represents a weak aspect of the overall process. Currently, there is a huge knowledge gap, especially in the assessment methods and design checks for existing structures, in every aspect of the assessment process.

Firstly, it is important to clearly define which parameters must be tested, and which may be calculated (i.e., is it necessary to measure mortar characteristics or are they similar for the Croatian building stock? Is it necessary to measure shear strength or will the well-known expressions approximate the values appropriately?). In this regard, the plan is to make case studies on existing structures to develop the exact procedure.

Secondly, the focus should be spent on how to properly model such a series of structures, considering economy. This is in contrast with the current practice for existing structures, which is usually either overly complex or overly simplified. The plan is thus to model structures with different input parameters, and afterwards try to assess them before testing. The values will be compared to the ones provided by tests and models will be calibrated. The procedure will be repeated until a stable system (i.e., where no further calibration is needed) will be obtained.

Thirdly, gaps exist in the way strengthening is ensured (which safety factors can be expected for a specific strengthening technique, masonry and standard practice characteristic of Croatia? How does strengthening influence the parameters for modelling? etc.).

The ARES project, in this context, tries to fill these gaps, allowing for a more economic design and a more accurate analysis of the consequences of failure. In order to achieve sustainability, SHM should be implemented to ensure continuous tracking of structural behaviour and provide owners with information important for maintenance.

Author Contributions: Conceptualization, M.S. and C.B.; methodology, M.S., C.B. and T.K.; validation, M.S. and T.K.; formal analysis, M.S., T.R. and I.H.; investigation, M.S., T.R. and I.H.; resources, M.S., C.B., T.R. and I.H.; photo credit, M.S.; writing—original draft preparation, M.S. and T.R.; writing—review and editing, M.S., C.B.; visualization, M.S. and I.H.; supervision, M.S. and T.K.; project administration, M.S.; funding acquisition, M.S. All authors have read and agreed to the published version of the manuscript.

Funding: This research was funded by Croatian Science Foundation, grant number UIP-2019-04-3749.

References

1. Negro, P.; Mola, E. A Performance Based Approach for the Seismic Assessment and Rehabilitation of Existing RC Buildings. *Bull. Earthq. Eng.* **2017**. [CrossRef]
2. Barbieri, G.; Biolzi, L.; Bocciarelli, M.; Fregonese, L.; Frigeri, A. Assessing the Seismic Vulnerability of a Historical Building. *Eng. Struct.* **2013**, *57*, 523–535. [CrossRef]
3. Papathoma-Köhle, M. Vulnerability Curves vs. Vulnerability Indicators: Application of an Indicator-Based Methodology for Debris-Flow Hazards. *Nat. Hazards Earth Syst. Sci.* **2016**. [CrossRef]
4. Goretti, A.; Di Pasquale, G. An Overview of Post-Earthquake Damage Assessment in Italy. In *EERI Invitational Workshop an Action Plan to Develop Earthquake Damage and Loss Data Protocols*; Earthquake Engineering Research Institute: Oakland, CA, USA, 2002.
5. Lourenco, P.; Karanikoloudis, G. Seismic Behavior and Assessment of Masonry Heritage Structures. Needs in Engineering Judgement and Education. *RILEM Tech. Lett.* **2019**. [CrossRef]
6. Republic of Croatia Ministry of Construction and Physical Planning. *Proposal of the Long-Term Strategy for Mobilising Investment in the Renovation of the National Building Stock of the Republic of Croatia*; National Renovation Report; Ministry of Construction and Physical Planning: Zagreb, Croatia, 2014.
7. Sigmund, Z.; Radujkovic, M.; Lazarevic, D. Decision Support Model for Seismic Strengthening Technology Selection of Masonry Buildings. *Teh. Vjesn. Tech. Gaz.* **2016**, *23*, 791–800. [CrossRef]
8. Dietsch, P.; Kreuzinger, H. Guideline on the Assessment of Timber Structures: Summary. *Eng. Struct.* **2011**. [CrossRef]
9. Steiger, R.; Kohler, J. Development of New Swiss Standards for the Reassessment of Existing Load Bearing Structures. In *Proceedings of the 41th Meeting, International Council for Research and Innovation in Building and Construction, Working Commission W18 – Timber Structures, CIB-W18 Meeting 41, Paper No. 41-102-2*; St. Andrews-by-the-Sea, Canada; University of Karlsruhe: Karlsruhe, Germany, 2008.
10. Diamantidis, D. Reiability assessment of existing structures. *Eng. Struct.* **1987**, *9*, 177–182. [CrossRef]
11. *ISO 2394:2015: General Principles on Reliability for Structures*; ISO: Geneva, Switzerland, 2015. [CrossRef]
12. JCSS. *Probabilistic Model Code*; Joint Committee on Structural Safety (JCSS): Zurich, Swotzerland, 2001; ISBN 978-3-909386-79-6.
13. ISO 13822. *Bases for Design of Structures—Assessment of Existing Structures*; ISO: Geneva, Switzerland, 2010.
14. SIA. *SIA-Richtlinie 462: Beurteilung Der Tragsicherheit Bestehender Bauwerke (Guideline SIA 462: Assessment of the Structural Safety of Existing Buildings)*; SIA: Zurich, Switzerland, 1994.
15. *NEN 8700: Assessment of Existing Structures in Case of Reconstruction and Disapproval—Basic Rules*; NEN: Delft, The Netherlands, 2011.
16. *SIA-Norm 469: Erhaltung von Bauwerken (Standard SIA 469: Maintenance of Buildings)*; SIA: Zurich, Switzerland, 1997.
17. *EN 1998-3—Eurocode 8: Design of Structures for Earthquake Resistance—Part 3: Assessment and Retrofitting of Buildings*; CEN: Brussels, Belgium, 2004.
18. Derakhshan, H. Proposed Update to Masonry Provisions of ASCE/SEI 41: Seismic Evaluation and Retrofit of Existing Buildings. In Proceedings of the 15th World Conference Earthquake Engineering, Lisbon, Portugal, 24–28 September 2012.
19. Pekelnicky, R.; Poland, C. ASCE 41-13: Seismic Evaluation and Retrofit Rehabilitation of Existing Buildings. In *SEAOC 2012 Convention Proceedings, Santa Fe, New Mexico*; Structural Engineers Association of California (SEAOC): Sacramento, CA, USA, 2012; 12p.
20. ASTM C1197-14a. *Standard Test Method for In Situ Measurement of Masonry Deformability Properties Using the Flatjack Method*; ASTM International: West Conshohocken, PA, USA, 2014.
21. ASTM C1587/C1587M-15. *Standard Practice for Preparation of Field Removed Manufactured Masonry Units and Masonry Specimens for Testing*; ASTM International: West Conshohocken, PA, USA, 2015.

22. ASTM C1531-16. *Standard Test Methods for In Situ Measurement of Masonry Mortar Joint Shear Strength Index*; ASTM International: West Conshohocken, PA, USA, 2016.
23. ASTM C1532/C1532M-19a. *Standard Practice for Selection, Removal, and Shipment of Manufactured Masonry Units and Masonry Specimens from Existing Construction*; ASTM International: West Conshohocken, PA, USA, 2019.
24. ASTM C1196-14a. *Standard Test Method for in situ Compressive Stress within Solid Unit Masonry Estimated Using Flatjack Measurements*; ASTM International: West Conshohocken, PA, USA, 2014.
25. ASTM C1019-19. *Standard Test Method for Sampling and Testing Grout for Masonry*; ASTM International: West Conshohocken, PA, USA, 2019.
26. Dymiotis, C.; Gutlederer, B.M. Allowing for Uncertainties in the Modelling of Masonry Compressive Strength. *Constr. Build. Mater.* **2002**, *16*, 443–452. [CrossRef]
27. Borri, A.; Corradi, M.; De Maria, A.; Sisti, R. Calibration of a Visual Method for the Analysis of the Mechanical Properties of Historic Masonry. *Procedia Struct. Integr.* **2018**, *11*, 418–427. [CrossRef]
28. Breysse, D.; Martínez-Fernández, J.L. Assessing Concrete Strength with Rebound Hammer: Review of Key Issues and Ideas for More Reliable Conclusions. *Mater. Struct. Constr.* **2014**, *47*, 1589–1604. [CrossRef]
29. Sýkora, M.; Diamantidis, D.; Holický, M.; Marková, J.; Rózsás, Á. Assessment of Compressive Strength of Historic Masonry Using Non-Destructive and Destructive Techniques. *Constr. Build. Mater.* **2018**, *193*, 196–210. [CrossRef]
30. Agred, K.; Klysz, G.; Balayssac, J.P. Location of Reinforcement and Moisture Assessment in Reinforced Concrete with a Double Receiver GPR Antenna. *Constr. Build. Mater.* **2018**, *188*, 1119–1127. [CrossRef]
31. Sajid, S.H.; Ali, S.M.; Carino, N.J.; Saeed, S.; Sajid, H.U.; Chouinard, L. Strength Estimation of Concrete Masonry Units Using Stress-Wave Methods. *Constr. Build. Mater.* **2018**, *163*, 518–528. [CrossRef]
32. Mesquita, E.; Martini, R.; Alves, A.; Antunes, P.; Varum, H. Non-Destructive Characterization of Ancient Clay Brick Walls by Indirect Ultrasonic Measurements. *J. Build. Eng.* **2018**, *19*, 172–180. [CrossRef]
33. Martini, R.; Carvalho, J.; Barraca, N.; Arêde, A.; Varum, H. Advances on the Use of Non-Destructive Techniques for Mechanical Characterization of Stone Masonry: GPR and Sonic Tests. *Procedia Struct. Integr.* **2017**, *5*, 1108–1115. [CrossRef]
34. Valluzzi, M.R.; Cescatti, E.; Cardani, G.; Cantini, L.; Zanzi, L.; Colla, C.; Casarin, F. Calibration of Sonic Pulse Velocity Tests for Detection of Variable Conditions in Masonry Walls. *Constr. Build. Mater.* **2018**, *192*, 272–286. [CrossRef]
35. Wai-Lok Lai, W.; Dérobert, X.; Annan, P. A Review of Ground Penetrating Radar Application in Civil Engineering: A 30-Year Journey from Locating and Testing to Imaging and Diagnosis. *NDT E Int.* **2018**, *96*, 58–78. [CrossRef]
36. Meola, C. Infrared Thermography of Masonry Structures. *Infrared Phys. Technol.* **2007**, *49*, 228–233. [CrossRef]
37. Schuller, M.P. Nondestructive Testing and Damage Assessment of Masonry Structures. *Prog. Struct. Eng. Mater.* **2003**, *5*, 239–251. [CrossRef]
38. Parivallal, S.; Kesavan, K.; Ravisankar, K.; Sundram, B.A.; Ahmed, A.K.F. Evaluation of In-Situ Stress in Masonry Structures by Flat Jack Technique. *Natl. Semin. Exhib. Non-Destr. Eval.* **2011**, *2011*, 8–13.
39. Simões, A.; Gago, A.; Bento, R.; Lopes, M. Flat-Jack Tests on Old Masonry Buildings. In Proceedings of the 15th International Conference Experimental Mechanics, Porto, Portugal, 22–27 July 2012; Volume 1, p. 3056.
40. Łatka, D.; Matysek, P. The Estimation of Compressive Stress Level in Brick Masonry Using the Flat-Jack Method. *Procedia Eng.* **2017**, *193*, 266–272. [CrossRef]
41. Croce, P.; Beconcini, M.L.; Formichi, P.; Cioni, P.; Landi, F.; Mochi, C.; De Lellis, F.; Mariotti, E.; Serra, I. Shear Modulus of Masonry Walls: A Critical Review. *Procedia Struct. Integr.* **2018**, *11*, 339–346. [CrossRef]
42. Invernizzi, S.; Lacidogna, G.; Lozano-Ramírez, N.E.; Carpinteri, A. Structural Monitoring and Assessment of an Ancient Masonry Tower. *Eng. Fract. Mech.* **2018**. [CrossRef]
43. Stepinac, M.; Rajčić, V.; Honfi, D. Condition Assessment of Timber Structures—Quantifying the Value of Information. In *IABSE Symposium Nantes, 2018 Tomorrow's Megastructures International Association for Bridge and Structural Engineering*; IABSE: Zurich, Switzerland, 2018.
44. Ramos, L.F.; De Roeck, G.; Lourenço, P.B.; Campos-Costa, A. Damage Identification on Arched Masonry Structures Using Ambient and Random Impact Vibrations. *Eng. Struct.* **2010**. [CrossRef]
45. Mendes, N.; Lourenco, P.B. Seismic Assessment of Masonry Gaioleiro Buildings in Lisbon, Portugal. *J. Earthq. Eng.* **2010**. [CrossRef]

46. Grandić, I.Š.; Grandić, D. Estimation of Damage Severity Using Sparse Static Measurement. *J. Civ. Eng. Manag.* **2017**, *23*, 213–221. [CrossRef]

47. Mader, D.; Blaskow, R.; Westfeld, P.; Weller, C. Potential of UAV-Based Laser Scanner and Multispectral Camera Data in Building Inspection. *Int. Arch. Photogramm. Remote Sens. Spat. Inf. Sci.* **2016**, *41*. [CrossRef]

48. Ellenberg, A.; Kontsos, A.; Bartoli, I.; Pradhan, A. Masonry Crack Detection Application of an Unmanned Aerial Vehicle. In Proceedings of the 2014 International Conference on Computing in Civil and Building Engineering, Orlando, FL, USA, 23–25 June 2014. [CrossRef]

49. Atalić, J.; Šavor Novak, M.; Uroš, M. Seismic Risk for Croatia: Overview of Research Activities and Present Assessments with Guidelines for the Future. *Građevinar* **2019**, *71*, 923–947.

50. Beeson, S.; Kubin, J.; Unav, A.I. Potresna Osjetljivost Povijesnih Zidanih Konstrukcija Nepravilne Geometrije. *Gradjevinar* **2015**, *67*, 151–158. [CrossRef]

51. Rücker, P.W.; Hille, D.F.; Rohrmann, D.R. F08a Guideline for the Assessment of Existing Structures. *SAMCO Final Rep.* **2006**, *48*.

52. Holicky, M. Probabilistic Model for Masonry Strength. *Eng. Mech.* **2010**, *17*, 61–70.

53. Schueremans, L. Reliability Analysis in Structural Masonry Engineering 2. Local Probability of Failure of Masonry Shear Panels. *Civ. Eng.* **1995**, *3*, 553–568.

54. Vailati, M.; Monti, G.; Khazna, M.J.; Napoli, A.; Realfonzo, R. Probabilistic Assessment of Masonry Building Clusters. In Proceedings of the 15th World Conference Earthquake Engineering (WCEE), Lisbon, Portugal, 24–28 September 2012.

55. Asteris, P.G.; Moropoulou, A.; Skentou, A.D.; Apostolopoulou, M.; Mohebkhah, A.; Cavaleri, L.; Rodrigues, H.; Varum, H. Stochastic Vulnerability Assessment of Masonry Structures: Concepts, Modeling and Restoration Aspects. *Appl. Sci.* **2019**, *9*, 243. [CrossRef]

56. Rota, M.; Penna, A.; Magenes, G. A Framework for the Seismic Assessment of Existing Masonry Buildings Accounting for Different Sources of Uncertainty. *Earthq. Eng. Struct. Dyn.* **2014**, *43*, 1045–1066. [CrossRef]

Article

Numerical Evaluation of Dynamic Responses of Steel Frame Structures with Different Types of Haunch Connection Under Blast Load

Mustafasanie M. Yussof [1,*], Jordan Halomoan Silalahi [2,*], Mohd Khairul Kamarudin [3], Pei-Shan Chen [4] and Gerard A. R. Parke [5]

[1] School of Civil Engineering, Universiti Sains Malaysia, Nibong Tebal 14300, Malaysia
[2] Damit Worley Parsons Engineering Sdn. Bhd, Kuala Belait KA1131, Brunei
[3] Faculty of Civil Engineering, Universiti Teknologi MARA, Shah Alam 40450, Selangor, Malaysia; mkhairul3965@uitm.edu.my
[4] Department of Civil and Architectural Engineering, Kyushu Institute of Technology, Fukuoka 804-8550, Japan; chen@civil.kyutech.ac.jp
[5] Department of Civil and Environmental Engineering, University of Surrey, Guildford GU2 7XH, UK; g.parke@surrey.ac.uk
* Correspondence: cemustafa@usm.my (M.M.Y.); Jordan.silalahi@worleyparsons.com (J.H.S.)

Received: 16 December 2019; Accepted: 12 February 2020; Published: 6 March 2020

Abstract: This research is aimed at investigating the dynamic behaviour of, and to analyse the dynamic response and dynamic performance of steel frames strengthened with welded haunches subjected to a typical hydrocarbon blast loading. The structural dynamic analysis was carried out incorporating the selected blast load, the validated 3D model of the structures with different welded haunch configurations, steel dynamic material properties, and non-linear dynamic analysis of multiple degree of freedom (MDOF) structural systems. The dynamic responses and effectiveness of the reinforced connections were examined using ABAQUS finite element software. Results showed that the presence of the welded haunch reinforcement decreased the maximum frame ductility ratio. Based on the evaluation of the results, the haunch reinforcements strengthened the selected steel frame and improved the dynamic performance compared to the frame with unreinforced connections under blast loading, and the biggest haunch configuration is the "best" type.

Keywords: blast loading; welded haunch connection; steel frame structures; non-linear dynamic analysis; ABAQUS; multiple degree of freedom (MDOF); frame ductility ratio

---

1. Introduction

The record of oil and gas industry accidental events shows that the historical offshore disasters have generally caused very significant losses in term of human lives, economy, and environmental pollution. Steel frames, as shown in Figure 1, are typical structures used in the offshore industry. The main functions of these frames is to support mechanical equipment, electrical and instrument cables, and hydrocarbon pipelines. Its function as a pipe rack is one of the reasons for selecting the frame because a ruptured pipeline which contains hydrocarbon liquid or gases may escalate the damage in a blast accident by intensifying the fire from the explosion. Therefore, this frame type was considered in our research.

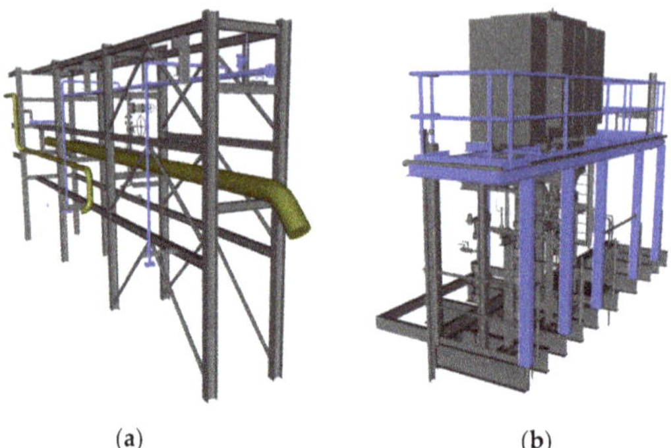

(a)   (b)

Figure 1. Typical steel frames used in offshore structures. (**a**) Pipe rack; (**b**) Mechanical Equipment Support.

During the period from 1993–2013, the worldwide petroleum and chemical insurance market reported that about 1100 insurance claims on major incidents were made [1]. Among these accidents, hydrocarbon explosions are the most hazardous events. Although a hydrocarbon explosion is a very rare accidental event, totally eliminating such incidents is difficult. Therefore, efforts should be made to keep the risk as low as possible and reduce the amount of structural damage if an explosion incident occurs. One of the approaches to minimize the risk to people and the facility is to utilize a blast resistant design or structural strengthening. The frames must have adequate stiffness, strength, and ductility capacity to resist blast loading. The connections of the structural component need to provide a significant contribution to maintain the integrity of the structural system. In many cases, connection strengthening can be implemented to enhance the structural dynamic performance.

In recent years, a few studies [2–7] were conducted on the investigation of the steel connection performance subjected to blast loading. Several steel joint reinforcement methods which strengthen and provide a better performance were proposed. An additional plate attached to beam flanges can improve the connection performance under blast loading [3]. Blast events are dynamic phenomena which are similar to earthquakes. Both blast resistant structure design and seismic design involve the ductile behaviour of the structures. Under seismic loading, the performance of the welded haunch connection reinforcements was studied by many researchers [8–11]. In seismic design, the welded triangular haunch, as shown in Figure 2, is one of the steel connection reinforcement methods that has been effectively implemented for enhancing steel moment connection performance as recommended by several seismic design guidelines [12–14]. Since seismic and blast loading characteristics are different, design for seismic load does not guarantee adequate performance for blast loading. Therefore, seismic design recommendations on connection reinforcement, and using welded haunches, cannot be directly implemented in blast design.

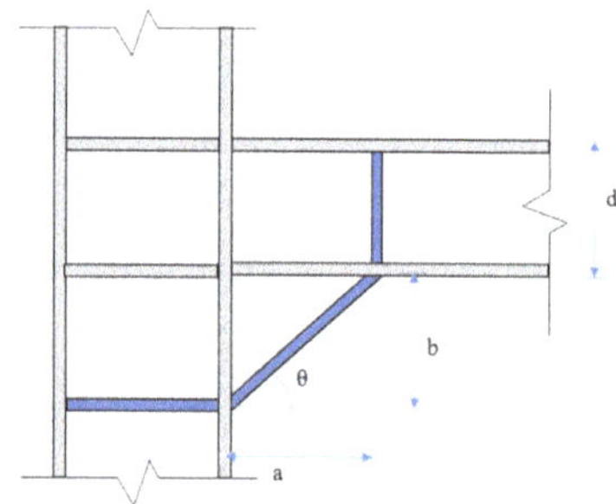

Figure 2. Typical sketch of a haunch connection configuration.

This study investigates the dynamic behaviour of a simple single-bay two-storey steel frame structure subjected to blast loading caused by hydrocarbon explosions. To achieve these objectives, a typical hydrocarbon explosion dynamic pressure was considered and the selected structural configuration including the connection details were modelled by numerical simulation using the finite element analysis (FEA) computer program ABAQUS/Explicit. To facilitate the model validation, a single-bay two-storey steel frame configuration from Chan and Chui [15] was adopted and developed as a 3D model. Before modifying the model for further analysis, a validation exercise was performed by comparing the numerical simulation responses with the responses as reported by Chan and Chui [15]. The model linear response was also compared with the theoretical approach/method computed using MATLAB Toolbox.

2. Non-Linear Finite Element Analysis

An explosion event typically happens in a very short duration of less than 1 s which indicates that the loading is time-dependent and requires a dynamic approach in the structural analysis. To solve the dynamic problem, a few techniques for structural dynamic analysis for blast resistant design have been developed; namely, the conventional equivalent static load or an elastic approach, a single degree of freedom (SDOF), and non-linear FEA. Offshore steel structures or petrochemical facility structures are generally constructed as a structural module with complex configurations and highly occupied by equipment and pipelines. This complexity leads to the classification of the structure as a multiple degree of freedom (MDOF) system and requires the non-linear MDOF analysis to provide the most comprehensive approach in the structural response computation.

The basic concepts and problem-solving techniques for MDOF systems can be found in many references [16–19]. A simple structure consisting of a MDOF system with the general dynamic equation of motion given in Equation (1).

$$M\ddot{u}(t) + C\dot{u}(t) + F_{int}(t) = F_{ext}(t) \tag{1}$$

M is the mass matrix, C is the damping matrix, u, $\dot{u}$, $\ddot{u}$ are the displacement, velocity, and acceleration vectors, respectively, F_{int} is the internal forces vector, and F_{ext} is the external forces vector. For dynamic events like the explosion case with a very high loading rate within short duration, the calculation requires small time increments to obtain a high-resolution for accurate solutions. Additionally, this technique is stable for small-time increment steps and requires relatively small computational cost per increment [20]. Therefore, the explicit method is efficient and suitable to be used in transient dynamic cases such as blasts, explosions, and impacts [21].

2.1. Finite Element Modeling

2.1.1. Geometry

Pipe rack framings are usually symmetric with uniformly distributed loadings. The braces support any lateral loadings in the longitudinal direction and restrain lateral movements whereby relatively small deflections will be experienced in this direction. This is a reasonable assumption to simplify the model by considering only a single-bay two-storey moment frame in the analysis.

To provide sufficient space for an access walkway and escape route, a minimum clear space of 1.0m width and 2.1 m height must be included within the structure [22]. Therefore, a column distance of 4 m that represents a typical offshore steel pipe racks framing configuration is considered sufficient. Regarding the requirement for vertical space between beams, the first author found no restrictions as long as the space for pipelines is sufficient in accordance with the piping engineers design, and the structural integrity can fulfil the design criteria. Therefore, a single-bay two-storey frame [15] was adopted and modified. The selected frame configuration is shown in Figure 3. This selected configuration facilitates the final model development after conducting the validation.

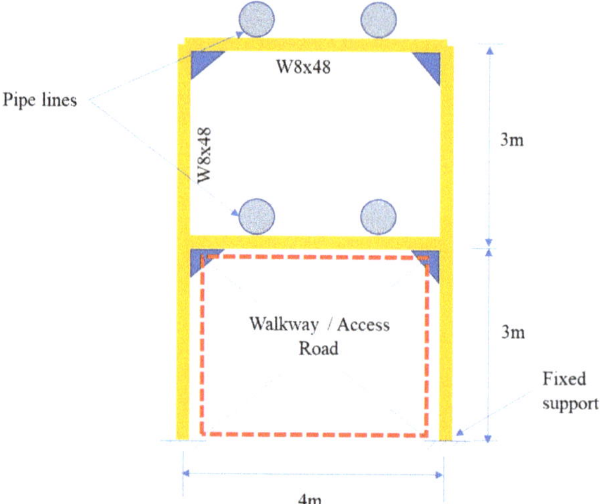

Figure 3. The steel frame model description.

The haunch length, a, and angle, θ, were dimensioned in accordance with the provision given in Gross et al. [13]. The haunch is usually fabricated by cutting a structural beam sections or plates with standard thickness. In this research, plates with a standard thickness of 10 mm were considered for the haunch and stiffener plates. The thickness was selected because it is the maximum standard plate thickness that is less than the beam and column web/flange thicknesses. The haunch flanges are attached to the beam and column flanges by groove welding, and the webs are then fillet welded to the beam and column flanges as described in Figure 4. The haunches are placed only on the lower side of the beams to avoid obstruction for the pipelines on the top of the beams.

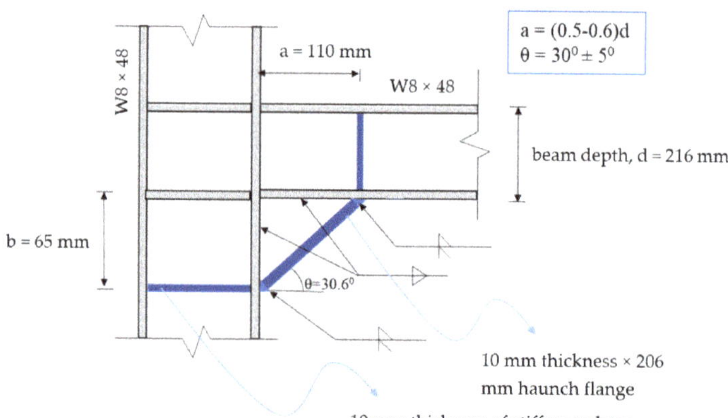

Figure 4. The welded triangular haunch configuration.

Varying details of beam-column connections were considered in this study. In addition to the joint type based on the configuration shown in Figure 4, a connection without reinforcement and a connection with a haunch slope angle of 45° were also considered in the analyses, as presented in Figure 5. The latter is a typical connection that is usually used in many offshore frame structures. The FE model of the steel frame was developed using ABAQUS/Explicit. The model geometry was modelled in accordance with the single-bay two-storey frame configuration, as depicted in Figure 3. The cross section of W8 × 48 was modelled and assigned to both the beams and columns.

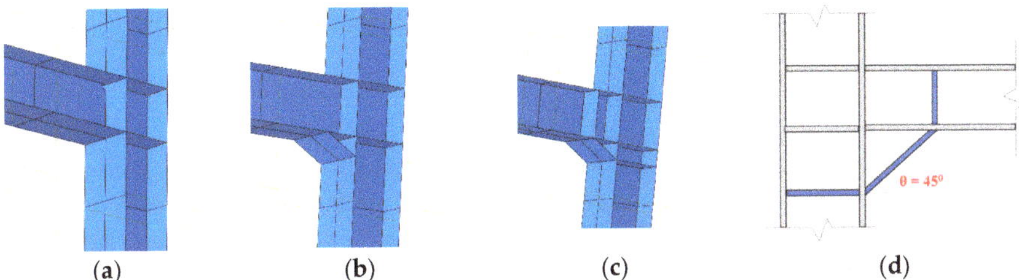

(**a**)   (**b**)   (**c**)   (**d**)

Figure 5. Different details of joint configurations with reinforcement. (**a**) Case 1—Beam-column connection without haunch; (**b**) Case 2—Haunch_01 Beam-column connection with haunch by Chopra [16] but no web stiffener plate on the beam and no continuity plates for haunch flange; (**c**) Case 3—Haunch_02 Beam-column connection with haunch by Chopra [16]. Web stiffener plates and haunch continuity plates were provided on the beam and column, respectively; and (**d**) Case 4—Haunch_03 Beam-column connection with haunch but the haunch configuration is modified by providing the slope angle of 45°.

2.1.2. Constraint and Boundary Conditions

The beams-columns, plates, and haunches were connected using the tie constraint that connects two surfaces regardless of the mesh size of each surface. The tie constraint is equivalent to a welded joint that prevents penetration, separating or sliding in the interaction between the modelled surface relative to another surface [20]. The application of this constraint in the model is illustrated in Figure 6. The column bases were considered as fixed supports which were created by restraining all the degrees of freedom of the nodal points concerned in the model boundary conditions.

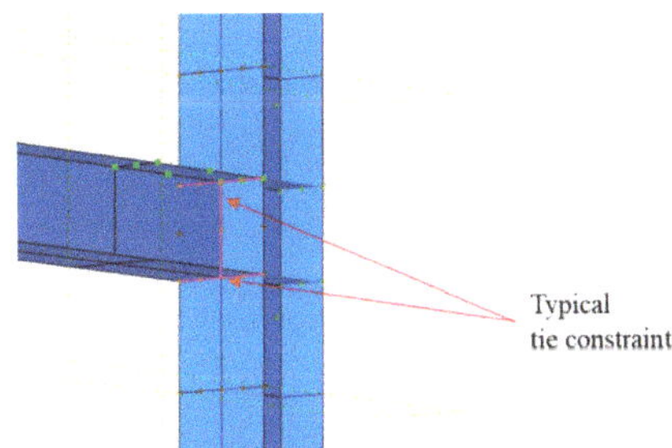

Figure 6. Typical tie constraint at beam-column connection.

2.1.3. Material Properties

For large deformation cases, the cross-sectional area undergoes a significant reduction and therefore true stress-strain indicating true material deformation must be considered. The following Equations (2) and (3) are the nominal stress-strain and true stress-strain relationships [23].

$$A = \pi r^2 \sigma_{true} = \sigma_{eng}\left(1 + \varepsilon_{eng}\right) \tag{2}$$

$$\varepsilon_{true} = ln\left(1 + \varepsilon_{eng}\right) \tag{3}$$

where σ_{true} is the true stress, σ_{eng} is the engineering stress, ε_{true} is the true strain and ε_{eng} is the engineering strain. It is recommended that the material properties for nonlinear analysis is based on the actual test results. However, in many cases this data is unavailable. In the absence of actual tensile test results, Det Norske Veritas (DNV) [23] established idealized material curves according to the European Standards. The elasto-plastic material properties for steel grade S235 with isotropic hardening was selected for the non-linear properties of steel [23]. ABAQUS defines the rate-dependent behaviour in term of plastic strain rate $\bar{\varepsilon}^{pl}$ expressed as:

$$\bar{\varepsilon}^{pl} = D(R-1)^q \tag{4}$$

where R is the ratio of the dynamic yield stress to the static yield stress, D and q are material constants. To match the Cowper-Symonds constitutive equation, the material constants for structural steel in Equation (4) were specified as $D = 40.4 \text{ s}^{-1}$ and $q = 5$ [24] which were included in the power law rate dependence definition input.

Typically, Gurson's porous model is considered to model material with relative density greater than 0.9. Amadio et al. [25] considered Gurson's porous model in the material attributes to overcome the limitation of the Von Mises constitutive law. However, Gurson's damage model on material plasticity is not considered in this study. The study is more concentrated on the structural performance based on global response without investigation of the material constitutional law, sensitivity of the local effect in the beam-column connections, and detail failure mechanism.

2.1.4. Element Type and Mesh

The required numerical results obtain from the FE analysis are depending on the selection of element type for the FE model. The accuracy of FE results is relatively influenced by the type of elements defined onto the FE model [26]. Solid element is significantly useful to obtain numerical stress components of the FE model. However, in this study, only the displacement component of the FE model is required to estimate the dynamic behaviour of the structure considered under blast load. Both solid and shell elements can significantly produce the same results of the displacement component of the element under static and even extreme load such as blast load [26,27]. All structural components were modelled using deformable four-node doubly curved with reduced integration S4R shell element. The S4R element is suitable for large-rotation problems because it includes finite membrane strains and arbitrarily large rotations [20]. A fine mesh size of 25 mm was used in the connection region and a coarse mesh size of 100 mm for the regions other than the connection area were selected as shown in Figure 7.

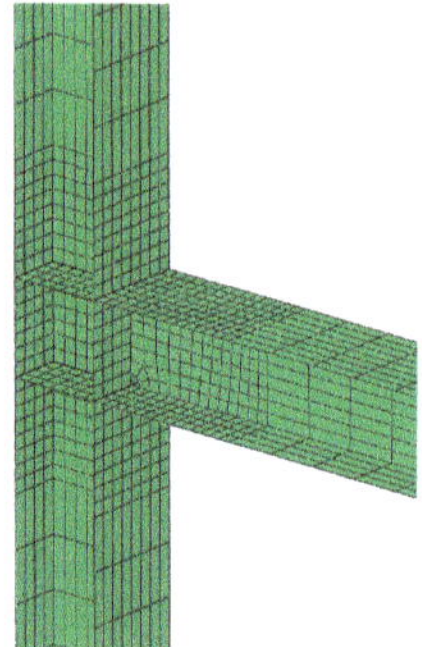

Figure 7. Finer mesh size at used in the connection region.

2.1.5. Loadings

All loadings associated with in-place actions that contribute to the dynamic masses must be included in the dynamic analysis [28]. The dead loads not contributing to the overall structural stiffness were incorporated as inertia masses input because these masses, including the weight of all pipes, contribute to the inertia mass in the dynamic behaviour of the overall structure. The inertia mass of pipes was determined by assuming all pipe size as 400 mm diameter × 12.7 mm thickness with a span length of 5 m. The weight of each pipe was 1.2 tonnes. At the same time, the miscellaneous dead loads such as smaller pipes, cable tray, electrical lines, grating, and handrails were included in the model. A total inertia mass of 10.2 tonne (100 kN) at each pipe location were assumed and included in the model. The total mass based on the assumed inertia masses gives a conservative scenario, besides maintaining the same total mass as in the original steel frame presented by Chan and Chui [15]. The weight of the pipes on the rack was modelled by applying mass inertia at four prescribed locations as shown in Figure 8.

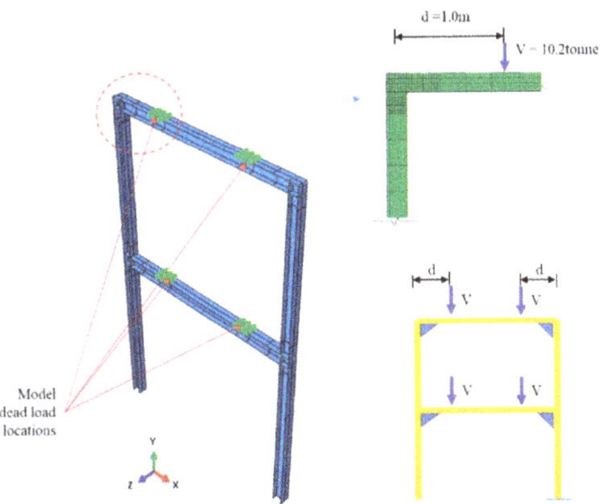

Figure 8. Pipe inertia masses and locations.

During the gas explosion process, the atmospheric pressure increases dramatically to a maximum pressure with the propagation of the blast wave, and then it slowly decreases to a negative value with respect to the standard atmospheric pressure. The negative phase pressure occurs because the shockwave forces the air to move as it spreads outward from the explosion centre and creates a lack of air behind, causing a partial vacuum or negative pressure phase. This negative phase can be ignored in structural design [29]. However, the negative phase of the blast pressure is important in order to accurately predict the responses of blast loading on structures [30], and it is unconservative if the negative phase is ignored [31]. The original pressure-time history output of blast simulation is not practicable to be used in structural dynamic analysis and needs to be idealized [32]. Mohamed et al. reference [30] summarised typical nominal blast overpressures for offshore structure according to industrial standard guidelines. The durations of these loads were reported to be between 50 and 200 msec and considered to be close to typical offshore structure natural periods (between 300 to 1100 msec). Yasseri et al. [33] also proposed an overpressure load of 0.9 bar and 2.25 bar as the lower and upper level events for hydrocarbon explosions, respectively.

During an explosion event, the following three fundamental consequences can occur; namely, the blast overpressures, dynamic pressures (drag loads) and projectiles, missiles, and shrapnel. Among these consequences, the explosion overpressures are generally considered to be the most critical measurement. In the absence of project-specific data, it has been suggested in a Chevron Engineering Standard that a load equal to one-third of the positive phase load can be considered for the negative phase pressure. The negative pressures are usually within the range 10–30% of the maximum pressure [34].

The explosion loads on open frame structures, structural components, equipment items and pipework are usually caused by dynamic pressure loads [35] containing drag loads, inertia loads, and a pressure difference load. The first two loadings are similar to the fluid force terms in the Morrison equations (refer Equation (5)) [36].

$$F_D = \frac{\pi}{4} C_m \rho D^2 \frac{\partial U}{\partial t} + \frac{1}{2} \rho v_{gas}^2 C_D A \qquad (5)$$

F_D is the drag force, C_m and C_D are the inertial and drag coefficients. For small objects and typical gas velocities, the contribution of the inertia load is less than 1% of the force hence it may be neglected [37]. Therefore, the Morrison's equation can be reduced and the magnitude of the drag forces on the steelwork elements can be calculated using Equation (6).

$$F_D = \frac{1}{2} \rho v_{gas}^2 C_D A \qquad (6)$$

where C_D is the drag coefficient of the object which is dependent on the shape of the structure (projected area), ρ is density of gas (kg/m^3), v_{gas} is velocity of the unburned gas mixture (m/s) and A is structures projected area (m^2). Since the prediction of gas density and velocity in an explosion event are very difficult, Mohamed et al. Reference [30] reported that the dynamic pressure (drag load) calculation in Equation (6) can be simplified to the following empirical expression given by Equation (6).

$$q_D(t) = C_D \times p(t) \times OD \qquad (7)$$

where $q_D(t)$ is the line load on a pipe function with respect to time, $p(t)$ is maximum overpressure time-history and OD is outer pipe diameter. According to Equation (7), the dynamic pressure can be calculated by using the maximum overpressure values. The maximum dynamic pressure P_{max} of 2.5 bar was selected based on Mohamed et al. [30] whereas the peak negative pressure P_{min} of −0.83 bar was taken as one-third of the maximum dynamic pressure as suggested by in the Chevron Engineering Standard and Hansen et al. [34]. The pressure duration of 0.136 s for pressures P_{max} of 2.5 bar and the duration of 0.24 s for P_{min} of −0.83 bar were determined using the curve in Figure 9.

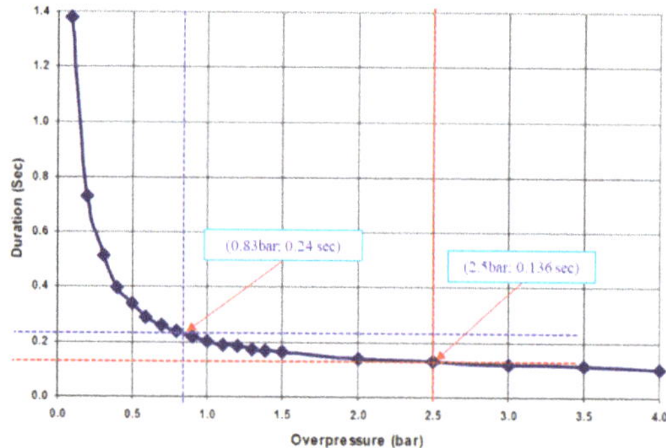

Figure 9. Pressure durations relationship based on API RP 2FB [24].

The dynamic pressure wave profile can be developed using the pressures and durations information. The idealized triangular waveform shown in Figure 10 was considered to describe the dynamic pressure history that was applied in the FE models.

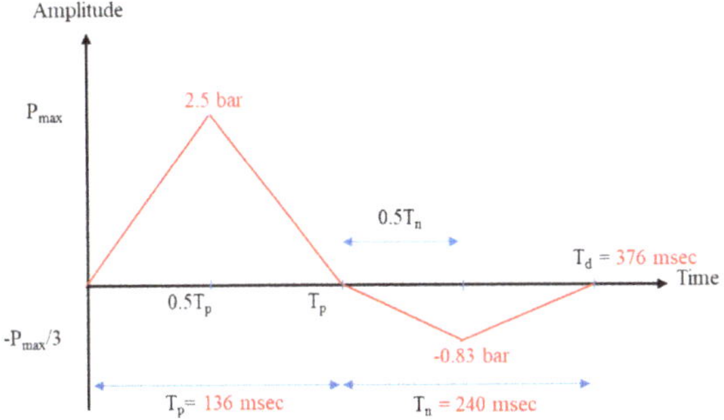

Figure 10. Dynamic pressure history.

In the analyses, a uniform distributed pressure of 2.5 bar was applied on the column flanges. To simplify the problem, the drag loads on the unmodeled pipes and non-structural items on the frame were ignored. Since the blast direction and the beam longitudinal direction are parallel, the beams are to be minimally affected by the lateral pressures, and thus no blast loads were applied to the beams.

In a blast event, the structures might buckle and undergo large geometry changes, hence the necessity to also consider geometric non-linearity [20]. In the initial step, initial conditions, boundary conditions, and predefined fields that are applicable with the analysis can be specified. In addition to the initial step, three steps were specified to simulate the blast analysis, namely; self-weight (quasi-static analysis with structural weight and dead load), blast (dynamic analysis using dynamic pressure history with self-weight propagation) and post-blast (only dead load is set active, no propagation of blast load from previous analysis step). In each analysis step, the incrementation settings were set to automatic incrementation so that ABAQUS automatically adjusts the time increment size depending on the numerical stability limit [20].

Before applying dynamic pressure, the initial condition of the structure was in a static condition under self-weight consisting of structural dead load and piping masses. Unfortunately, ABAQUS cannot directly combine static analysis with dynamic explicit analysis in the same model. When dynamic explicit analysis is being used, all the following analysis in the next steps must also be in the dynamic explicit mode. Therefore, the self-weight member forces also need to be analysed using dynamic explicit analysis. When the self-weight is applied instantaneously in the dynamic analysis, oscillatory vertical reaction force occur as shown in Figure 11. This oscillation does not represent the correct static condition in which the structure should be before the blast is initiated. This oscillation will create incorrect stress variations during and after the actual blast dynamic analysis. To overcome this problem, the ramp loading was implemented in the self-weight dynamic analysis whereby the load was applied slowly to avoid the dynamic effect. The basis of this method is related to the concept of the dynamic case of "step force with finite rise time" [16–18] in which if the rising time of the self-weight dynamic loading is relatively long and greater than three times the structural natural period, the dynamic response is quasi-static or like a static force. Therefore, a rise time of 2.15 s (=3 × 0.717 sec) was used in the self-weight dynamic step as shown in Figure 12. The result of ramp loading implementation in the self-weight step is presented in Figure 13 whereby no oscillation has occurred in the self-weight dynamic step.

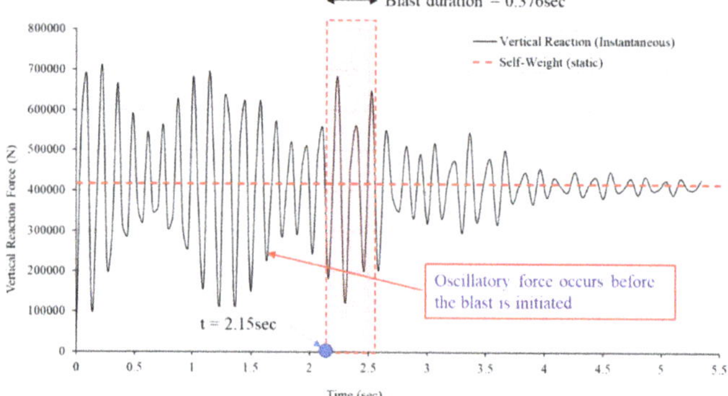

Figure 11. Vertical reaction force for instantaneously applied self-weight.

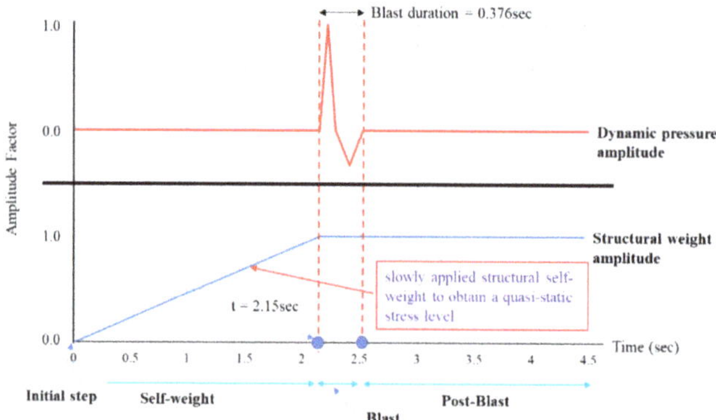

Figure 12. Dynamic analysis steps and durations.

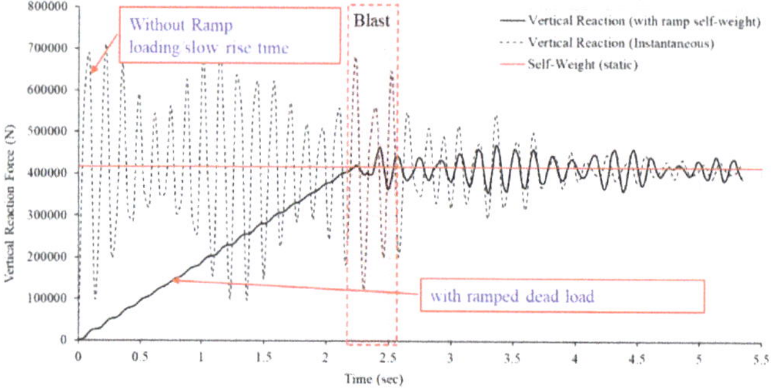

Figure 13. Vertical reaction force after ramp loading implementation in the self-weight step.

In the "blast" analysis step, the dynamic pressure was introduced. The duration of the blast analysis step was specified in line with the total duration of the blast amplitude definition as 0.376 sec. During this analysis step, the self-weight propagating from the previous self-weight analysis step was included in the analysis. After the blast analysis step, the last analysis step, called the post blast, was created with the purpose of studying the post-blast response of the structure. No load was applied in this step except the self-weight propagation from the previous step. Since the transient dynamic response is the primary interest of the study, a step duration of 2 s was considered in the post-blast analysis. In the explosion dynamic analysis, the maximum dynamic response usually occurred within the transient response duration. Therefore, the duration of 2 s is considered adequate to describe the steady state response of the structure. Moreover, a longer duration needs to be avoided to minimize the total computation duration.

2.2. Structural Assessment

2.2.1. Ductility Ratio of the Steel Frame

The level of structural performance can be measured by quantifying its stresses, strains, displacements, ductility, functionality, and similar parameters. Like seismic design, blast resistant structural design also considers inelastic response under dynamic load. In terms of the structural response, the impulsive blast load yields higher response frequency than the response due to seismic load [38] and the inter-story drifts due to the blast loading on a building are generally higher than the drifts due to earthquakes [39]. In the case of blast loading, the design is based on the member's plastic behaviour and the structural component is governed by its ductility and rotation [40]. The ductility is defined as the structural capability of plastic deformation without fracturing. The maximum plastic displacement of a structure can be measured by the ductility ratio (μ) relative to first yield [41] as given in Equation (8).

$$\mu = \frac{\delta_u}{\delta_y} \tag{8}$$

where δ_u is the maximum plastic displacement, and δ_y is the displacement at first yield. The displacement at first yield is not obvious for most material. It can be determined by using graphical methods with several alternative definitions [41]. The method based on initial yielding, also called the Tangent Method, is usually implemented to approximate the yield point which is defined as the intersection between the tangential lines of elastic and plastic behaviour [42].

In design practice, ductility ratio values and rotation criteria are used for designing structure particularly topside offshore structures. The ductility ratio and the plastic strain are related, and the failure of the structure is assumed to occur when the plastic strain approaches the material failure strain [43]. The allowable range of the structural member's plastic deformations is measured based on the calculated ductility ratio, whereas the limit on the support rotation makes sure that tension membrane action, that may develop in a member, will be in a safe range where no connection failures occur.

2.2.2. Evaluation Criteria

In industrial practice, various design codes and guidelines [24,31,40,44–46] are used for blast design evaluation criteria. Specifically, for a petroleum facility, API RP 2FB [24] defines two important levels of assessment namely ductility level blast (DLB) and the strength level blast (SLB). These levels are associated with the explosion loading risks by analogy with earthquake assessment. A low-probability high-consequence event representing the extreme design event may be assessed using DLB. While a higher-probability lower-consequence event, such an explosion case with low overpressure, can be assessed using SLB.

Louca et al. [6] summarised the ductility values that were used in the offshore structure design practice. By correlating the ductility ratio and damage levels, Yasseri [47] proposed the ductility ratio criteria. Several design guidelines in the industry present the response criteria [40,46].

3. Results

3.1. Model Validations Results

A single-bay two-storey steel framing configuration as presented in Figure 14 was chosen to develop the model. The selected steel frame was considered appropriate for the object in the study because a simple and small structure does not require excessive computation resources. Chan and

Chui [15] investigated the transient dynamic response of the inelastic steel frames with nonlinear connections. Geometric imperfections and residual stress with a maximum magnitude of 50% of yield stress were included in their study.

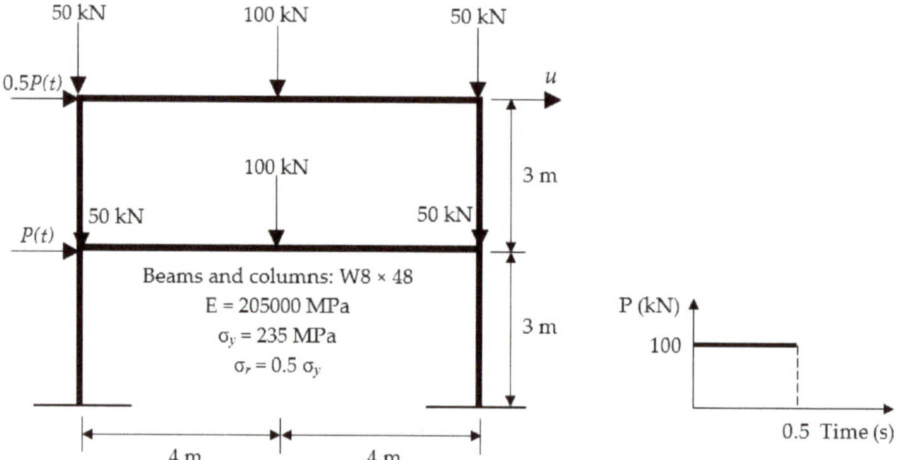

Figure 14. A two-storey steel frame under dynamic loading [15].

The model of the selected steel frame was developed in ABAQUS. Before starting the simulations for the research, two different analyses namely elastic and elastic-perfectly-plastic analyses were used for validating the model by comparing the dynamic responses from numerical analyses against the dynamic responses presented by Bathe [19]. To differentiate the analysis, two different material properties were created representing the elastic and elastic-perfectly plastic with a yield stress of 235 MPa. The elastic and elastoplastic analyses were carried out. To verify the dynamic characteristic of the frame using theoretical calculations, elastic dynamic analysis was also carried out using CALFEM which is a MATLAB computational toolbox for teaching FEM developed at Lund University [48]. The structural lateral displacement responses are presented in Figures 15 and 16.

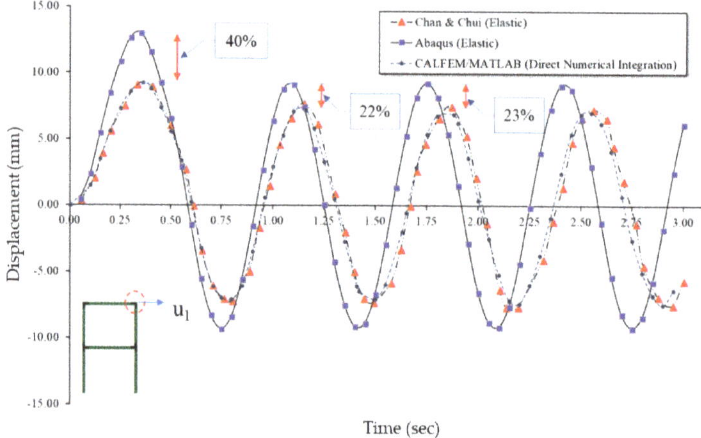

Figure 15. Elastic analysis dynamic response ABAQUS, Chan and Chui [15] and CALFEM/MATLAB.

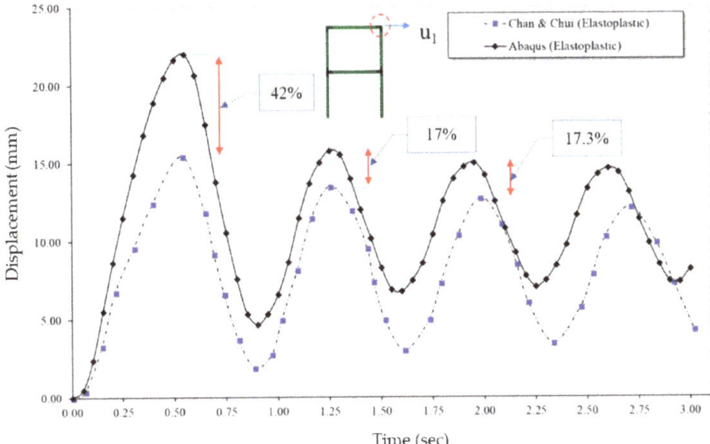

Figure 16. Elastoplastic analysis dynamic response ABAQUS and Chan and Chui [15].

Since the response based on CALFEM calculation matches the elastic response presented by Chan and Chui [15], their dynamic characteristics are considerably identical. This implies that the dynamic characteristic based on CALFEM calculations represents the similar dynamic characteristic of the frame presented by Chan and Chui [15]. According to the dynamic response in Figures 15 and 16, the maximum response from the ABAQUS analysis is 42% higher than the maximum response given by Chan and Chui [15]. In addition to this, the displacement response wave period of the ABAQUS results are shorter than the results from Chan and Chui [15].

To investigate these discrepancies, the natural frequencies that represent the dynamic characteristic of the frames were calculated using ABAQUS and CALFEM. As shown in Figure 17, the structure natural frequency of 1.4978 Hz extracted from ABAQUS is 5.4% higher than the CALFEM result (1.4214 Hz). Tedesco et al. [18] showed that the effect of mass and load distribution on the dynamic response of a structural system in which a structural system with distributed mass is stiffer than a system where the entire mass is concentrated at midspan points. Unlike beam finite element formulation, the structural masses in the ABAQUS model is distributed throughout the structure and not concentrated at points. This implies that the ABAQUS model should be stiffer than the CALFEM calculation, hence it will have a higher natural frequency.

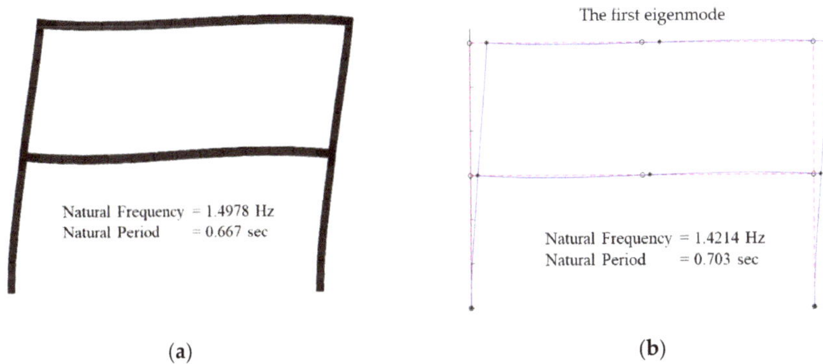

Figure 17. Structural natural frequency from (**a**) ABAQUS and (**b**) CALFEM/MATLAB.

The effect of the nonlinearity of the system was also investigated. As shown in Figure 18, the Von Mises stress reaches the yield stress value. This situation is considered overstress, in which the validated model undergoes yielding and buckling. In this case, analysis using ABAQUS does have capability to consider inelastic moment redistribution [49] and P-delta effects by activating the geometric non-linearity option in the dynamic analysis. These features allow the analysis using ABAQUS to capture the more realistic dynamic behaviour of the structure.

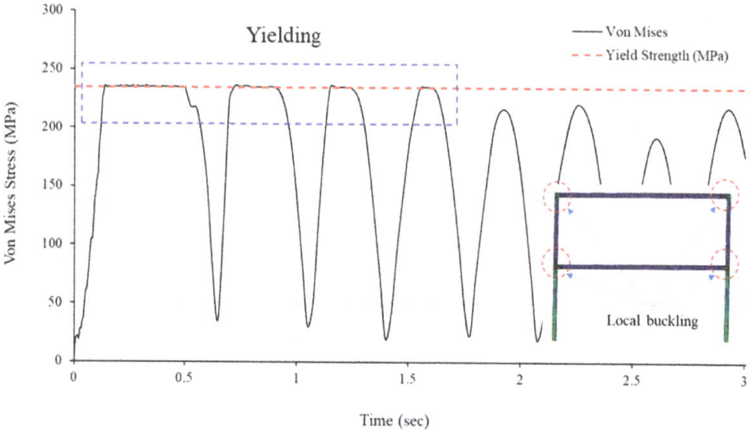

Figure 18. Yielding and buckling in the validated model.

The preceding section explained possible factors affecting the higher dynamic responses obtained from ABAQUS compared to the dynamic response presented by Chan and Chui [15].

3.2. Eigenmodes and Eigenvalues

The natural period of the structure was obtained by eigenvalue analysis. To perform this, the analysis model was copied and modified by removing the blast loading and maintaining the structural and piping masses. The dynamic analysis step was replaced by linear perturbation frequency analysis type. Ten values of the eigenvalues were requested for the analysis output. The mode shape which is parallel with the direction of blast loading was selected as the representation of the dynamic characteristics of the structure. Therefore, the structural natural period associated with this mode shape was used in the step duration of self-weight dynamic analysis. By relating the structural natural period to the dynamic loading duration, the structural dynamic response was classified into impulsive $t_d/T_n < 0.3$, dynamic $0.3 < t_d/T_n < 3$, or quasi-static $3 < t_d/T_n$ categories [50]. Where t_d is the dynamic load duration and T_n is the structural natural period. The structural dynamic characteristic was identified by undertaking the eigenvalue analysis. Four mode shapes considered to to be the most important shapes are presented in Figure 19. The first mode as the highest eigenvalue is usually considered as the critical case. However, mode shapes 1 and 2 in this study were considered not to be the critical cases because the columns, bracings, and lateral beams provide lateral restrains in the longitudinal direction of the actual structure. The mode shape-3 that deflects in the same direction as the blast excitation loading was selected to represent the dynamic characteristic for the structure. Subsequently, the eigenvalue based on mode shape-3 was used in all calculations that are relevant with the structural dynamic characteristics.

By considering the dynamic load duration of $t_d = 0.376$ s and the structural natural period of 0.717 s, the ratio of the dynamic loading duration to the natural period is determined as $t_d/T_n = 0.52$. The ratio is greater than 0.3 and less than 3, hence falls into the dynamic category. Therefore, dynamic analysis was solved using numerical integration of the dynamic equations of equilibrium that is already implemented in the ABAQUS numerical computational technique.

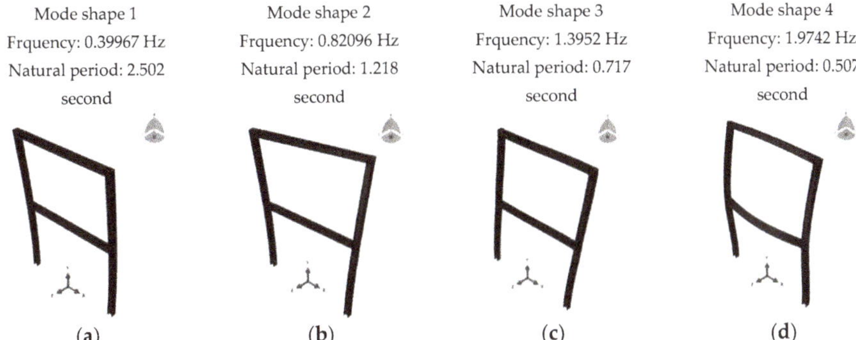

Figure 19. Structural eigen modes and eigen periods. (**a**) Mode shape 1; (**b**) Mode shape 2; (**c**) Mode shape 3; and (**d**) Mode shape 4.

3.3. Lateral Displacement Response and Ductility Ratio

Structural ductility was calculated using Equation 8 and the value of structural displacement at first yield was determined using initial yield or the tangent method [41]. A new model was developed by modifying the analysis model. The static pushover analysis was performed with only lateral loading applied at the top corner of the frame considered.

The force-displacement resulting from the analysis is presented in Figure 20. By using the curve, the structural displacement at first yield was determined by estimating the yield point location which is defined as the intersection between the tangential lines of elastic and plastic behaviour, where the displacement at first yield is 71 mm. Subsequently, the structural ductility ratio was calculated as the ratio of peak displacement under blast load to the displacement at first yield.

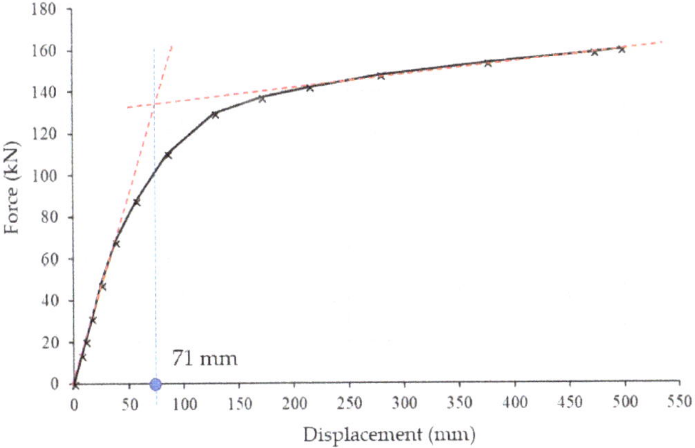

Figure 20. Structural displacement at first yield approximation.

Figure 21 describes the general displacement direction of the frame swaying horizontally in the same direction as the blast excitation load. Two different points depicted in the figure that represents structural height and beam elevations were selected as the node locations to extract the global lateral displacement responses. The lateral displacement at the top right corner (Point A) was considered as a good measurement point for the frame system deformation, whereby the maximum lateral displacements at Point B were used in structural ductility ratio calculation.

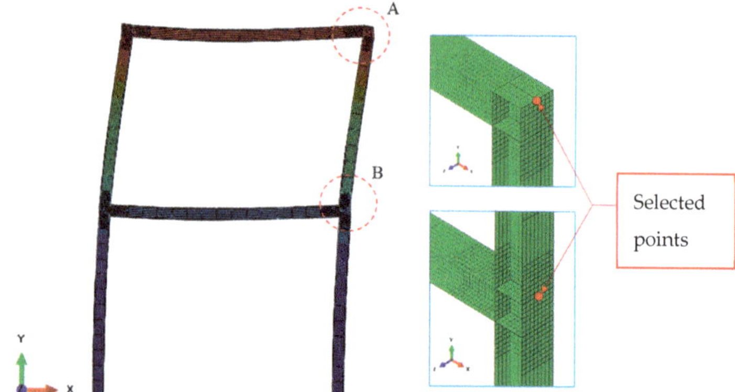

Figure 21. Selected points for structural displacement location.

The sway histories for these two observation points are presented in Figure 22a,b for all frame analysis with different beam-column connection. The maximum sway responses at lower and upper beam elevations, points A and B, were observed during the blast duration. The maximum responses of the frame without reinforcement are higher than the frames that are reinforced with haunches.

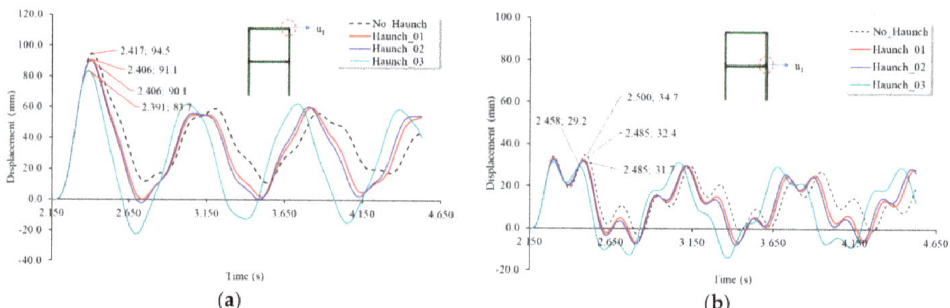

Figure 22. (**a**) Lateral displacement response history at top right corner of the frame; (**b**) Lateral displacement response history at right end of the lower beam.

To measure the frames dynamic performance based on the sway responses, the extracted maximum sway displacements at the top elevation for each frame during blast and post blast are presented in Table 1 and Figure 23. Subsequently these maximum sway responses were used to calculate the ductility ratios as tabulated in Table 1. The maximum lateral displacement of 94.6 mm occurred in the frame without haunch reinforcement. Since the ductility ratio is proportional with the maximum displacement, the maximum ductility ratio of 1.33 also occurred in the same frame without haunch. The calculated results as presented in Table 1 show that all maximum sway responses and maximum ductility ratios are less than the maximum criteria recommended [40,46]. The allowable deflection and ductility ratio of frame structures are 1.5 and 240 mm (Height/25), respectively [40,46].

Table 1. Summary of maximum displacement at all beam elevations for connection types of all cases considered in this study.

Case	Location	Maximum Displacement, u (mm)		Ductility μ [1]	
		Blast	**Post-Blast**	**Blast**	**Post-Blast**
No haunch	Upper beam	94.6	61.4	1.33	0.86
	Lower beam	34.9	29.6	-	-
Haunch_01	Upper beam	91.1	60.0	1.28	0.84
	Lower beam	33.1	29.6	-	-
Haunch_02	Upper beam	90.2	60.6	1.27	0.85
	Lower beam	32.7	30.0	-	-
Haunch_03	Upper beam	83.7	62.4	1.18	0.88
	Lower beam	31.4	31.7	-	-

[1] From Equation (8).

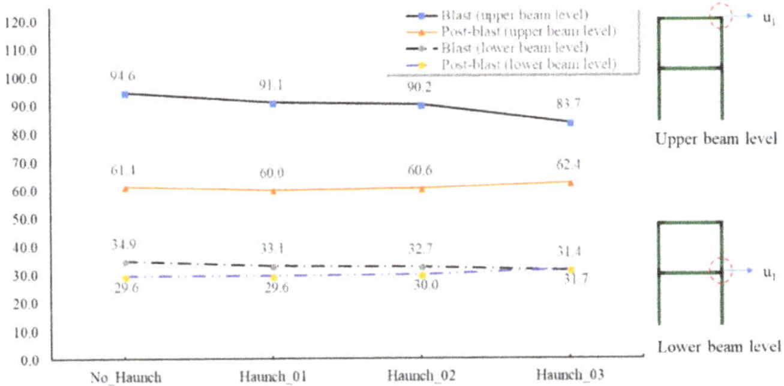

Figure 23. Maximum lateral displacement response.

To evaluate the relative differences between frames performance using ductility criteria, the ductility ratios are presented in Figure 24. The trends show that the frames with haunches performed better compared to the frame without haunches. Qualitatively, the Haunch_03 impact on the frame sway response is the highest among all the haunch types. The Haunch_03 connection reinforcement reduced the ductility ratio by 12% against the frame without haunches, while in the case of the Haunch_02 the reduction is 5% and for the Haunch_01 is 4%. The results demonstrated that the Haunch_03 provides a enhanced dynamic performance. It was also observed that the plate stiffeners on beam-column joints (Haunch_02) provided slightly better ductility ratio than the case of the haunch without the plate stiffeners (Haunch_01). In general, the biggest haunch configuration provided the best frame dynamic performance. This finding is relevant to the beam-to-column connections flexibility whereby connection strengthening increases the stiffness of the frame.

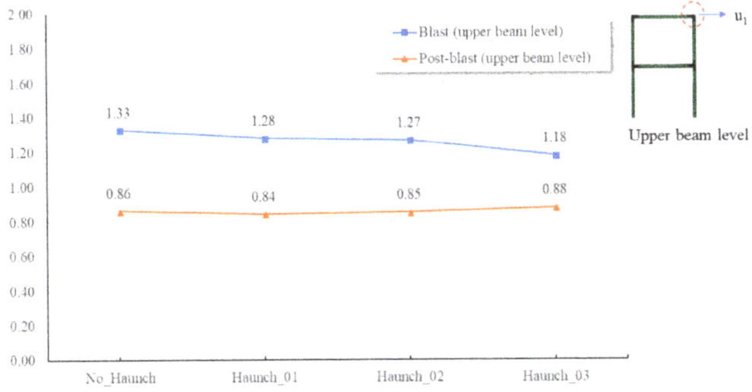

Figure 24. Frames ductility.

4. Discussion

The results of this research are limited and cannot be extrapolated to cover the structural dynamic performance of steel frames with connection haunch reinforcement because this research considered only four cases of frames with different joint haunch configurations. To further understand the dynamic performance aspects of typical steel frame subjected to blast loading, the following points are recommended to be undertaken:

- In this research, the blast loadings were applied on the columns of a typical steel frame structure, whereas in actual structures, the blast pressure are also applied onto unmodeled items such as pipelines and equipment located on the structure. Further study is recommended to investigate the effect of blast loading on the pipelines and additional equipment attached to the structures, both with and without haunch reinforcement.
- The Gurson porous model has an important local effect in the beam-column connection. Therefore, further sensitivity study comparing the effects of using material constitutive law of Von Mises versus Gurson model on material plasticity is suggested to investigate the stresses evolution in joint area. In addition to the material model aspect, a solid element is to be included in the study as element selection plays major impact in the analysis result as shown in the study
- Generally, the analysis results demonstrated an enhanced performance when the haunches with a size greater than the size recommended by AISC [14] was used. To investigate the optimum haunch size, it is recommended to carry out parametric study on haunch sizes and stiffener plate thicknesses. A parametric study on the beam and column sizes are also suggested to understand the effect of a combination of haunches and frame configurations.
- In the absence of actual blast loading data, it is also recommended that a parametric study on pressure time histories is carried out.

5. Conclusions

The structural eigenvalue was extracted and compared to the dynamic loading duration in order to characterize the dynamic response. The obtained ratio of 0.52 that is greater than 0.3 and less than 3 has classified the dynamic response under the influence of dynamic category. Therefore, the dynamic analysis was solved using numerical integration of dynamic equations of equilibrium that is already implemented in ABAQUS. The structural maximum ductility ratio achieved using haunches was 1.33 and less than the allowable criteria of 1.5. Generally, the structural ductility ratios decreased due to the presence of the haunch reinforcement. The ductility ratio of the frame with Haunch_03 reinforcement was reduced by 12% compared with the frame without haunches, while in the case of the Haunch_02 the reduction was 5% and for Haunch_01 was 4%. According to the evaluation results, the haunch reinforcements have strengthened the selected steel frame and improved dynamic performance compared to the frame with unreinforced connections under blast loading. The Haunch_03 as the biggest reinforcement configuration performed better compared to other connection configurations.

Author Contributions: Conceptualization, supervision, project administration, visualization, M.M.Y; methodology, software, validation, resources, data curation, writing—original draft preparation, M.Y. and J.S; formal analysis, investigation, J.H.S.; supervision, project administration M.M.Y.; writing—review and editing, M.M.Y., M.K.K., P.-S.C. and G.A.R.P. All authors have read and agreed to the published version of the manuscript.

Funding: This research was supported by the Universiti Sains Malaysia bridging grant (304/PAWAM/6316571).

References

1. Nolan, D.P. *Handbook of Fire and Explosion Protection Engineering Principles: In For Oil, Gas, Chemical and Related Facilities*, 3rd ed.; Elsevier: Amsterdam, The Netherlands, 2014.

2. Krauthammer, T. Blast-resistant structural concrete and steel connections. *Int. J. Impact Eng.* **1999**, *22*, 887–910. [CrossRef]

3. Sabuwala, T.; Linzell, D.; Krauthammer, T. Finite element analysis of steel beam to column connections subjected to blast loads. *Int. J. Impact Eng.* **2005**, *31*, 861–876. [CrossRef]

4. Krauthammer, T.; Yim, H.C.; Astarlioglu, S.; Starr, C.; Lim, J. Blast-Induced Response of Moment Connections. In *Structural Engineering Research Frontiers*; Wallace, J.W., Ed.; American Society of Civil Engineers: Reston, VA, USA, 2007; pp. 1–15.

5. Yim, H.C.; Krauthammer, T. Mechanical properties of single-plate shear connections under monotonic, cyclic, and blast loads. *Eng. Struct.* **2012**, *37*, 24–35. [CrossRef]

6. Louca, L.A.; Mohamed Ali, R.M. Improving the Ductile Behaviour of Offshore Topside Structures Under Extreme Loads. *Eng. Struct.* **2008**, *30*, 506–521. [CrossRef]

7. Urgessa, G.S.; Arciszewski, T. Blast response comparison of multiple steel frame connections. *Finite Elem. Anal. Des.* **2011**, *47*, 668–675. [CrossRef]

8. Yu, Q.S.K.; Uang, C.-M.; Gross, J. Seismic Rehabilitation Design of Steel Moment Connection with Welded Haunch. *J. Struct. Eng.* **2000**, *126*, 69–78. [CrossRef]

9. Valente, M. Welded Triangular Haunch for the Seismic Improvement of Steel Beam-to-column Connections. In *Design, Fabrication and Economy of Welded Structures*; Elsevier: Amsterdam, The Netherlands, 2008; pp. 245–252.

10. Jiang, L.Y.; Liu, G.J.; Sun, X.D. Mechanic Behavior Analysis of Connection with Haunch under Single and Cyclic Load. *Appl. Mech. Mater.* **2011**, *105–107*, 848–852. [CrossRef]

11. Liu, Y.J.; Wang, Y. The Study on Mechanical Property for Welded Haunch-Reinforced Seismic Connections of Steel Frame. *Appl. Mech. Mater.* **2014**, *501–504*, 485–494. [CrossRef]

12. FEMA-351. *Recommended Seismic Evaluation and Upgrade Criteria for Existing Welded Steel Moment-Frame Buildings*; Federal Emergency Management Agency: Washington, DC, USA, 2000.

13. Gross, J.L.; Engelhardt, M.D.; Uang, C.-M.; Kasai, K.; Iwankiw, N. *Steel Design Guide Series 12: Modification of Existing Welded Steel Moment Frame Connections for Seismic Resistance*; American Institute of Steel Construction: Chicago, IL, USA, 1999.

14. AISC. *Seismic Design Manual*, 2nd ed.; American Institute of Steel Construction: Chicago, IL, USA, 2012.

15. Chan, S.L.; Chui, P.P.T. *Nonlinear Static and Cyclic Analysis of Steel Frames with Semi-Rigid Connections*; Elsevier: Amsterdam, The Netherlands, 2000.

16. Chopra, A.K. *Dynamics of Structures: Theory and Applications to Earthquake Engineering*, 4th ed.; Prentice Hall: Upper Saddle River, NJ, USA, 2012.

17. Craig, R.R.; Kurdila, A. *Fundamentals of Structural Dynamics*, 2nd ed.; John Wiley & Sons: Hoboken, NJ, USA, 2006.

18. Tedesco, J.W.; McDougal, W.G.; Ross, C.A. *Structural Dynamics: Theory and Applications*; Addison Wesley Longman: Menlo Park Calif, CA, USA, 1999.

19. Bathe, K.J. *Finite Element Procedures*; Prentice Hall, Pearson Education, Inc.: Upper Saddle River, NJ, USA, 2006.

20. ABAQUS 6.14. Documentation: Getting Started with Abaqus-Interactive Edition. 2014. Available online: http://130.149.89.49:2080/v6.14/pdf_books/GET_STARTED.pdf (accessed on 16 December 2019).

21. Kang, K.-Y.; Choi, K.-H.; Choi, J.W.; Ryu, Y.H.; Lee, J.-M. Explosion Induced Dynamic Responses of Blast Wall on FPSO Topside: Blast Loading Application Methods. *Int. J. Nav. Archit. Ocean. Eng.* **2017**, *9*, 135–148. [CrossRef]

22. S-DP-001. *Desing Principles Technical Safety*; Norwegian Technology Standards Institution: Oslo, Norway, 1994.

23. DNV. *DNV-RP-C208: Determination of Structural Capacity by Non-linear FE Analysis Methods*; Det Norske Veritas; DNV: Oslo, Norway, 2013. Available online: https://pdfs.semanticscholar.org/9051/79af16102318a837f9f2315f2b1064c777ee.pdf (accessed on 16 December 2019).

24. API RP 2FB. *Recommended Practice for the Design of Offshore Facilities Against Fire and Blast Loading*; API: Washington, DC, USA, 2006.

25. Amadio, C.; Bedon, C.; Fasan, M.; Pecce, M.R. Refined numerical modelling for the structural assessment of steel-concrete composite beam-to-column joints under seismic loads. *Eng. Struct.* **2017**, *138*, 394–409. [CrossRef]

26. Peeters, M.; Santo, G.; Degroote, J.; Van Paepegem, W. Comparison of shell and solid finite element models for the static certification tests of a 43 m wind turbine blade. *Energies* **2018**, *11*, 1346. [CrossRef]

27. Momeni, M.; Hadianfard, M.A.; Bedon, C.; Baghlani, A. Numerical damage evaluation assessment of blast loaded steel columns with similar section properties. In *Structures*; Elsevier: Amsterdam, The Netherlands, 2019.

28. ISO 19901-3. *Petroleum and Natural Gas Industries-Specific Requirements for Offshore Structures Part 3: Topsides Structure*; BSI: London, UK, 2010.

29. Biggs, J.M. *Introduction to Structural Dynamics*; McGraw-Hill: New York, NY, USA, 1964.

30. Mohamed Ali, R.M.; Louca, L.A. Performance based design of blast resistant offshore topsides, Part I: Philosophy. *J. Constr. Steel Res.* **2008**, *64*, 1030–1045. [CrossRef]

31. Gilsanz, R.; Hamburger, R.; Barker, D.; Smith, J.L.; Rahimian, A. *Steel Design Guide 26: Design of Blast Resistant Structures*; American Institute of Steel Construction: Chicago, IL, USA, 2013.

32. Burgan, B.A.; Hamdan, F.H.; Burgan, B.A.; Hamdan, F.H. Response of topside structures to fires and explosions: Design considerations. In Proceedings of the Offshore Technology Conference, Houston, TX, USA, 6–9 May 2002.

33. Yasseri, S.F. Performance Based Blast Resistant Design. In *FABIG Newsletter Sept 2002*; FABIG Newsletter: Ascot, UK, 2002.

34. Hansen, O.R.; Kjellander, M.T.; Pappas, J.A. Explosion Loading on Equipment from CFD Simulations. *J. Loss Prev. Proces. Ind.* **2016**, *44*, 601–613. [CrossRef]

35. Walker, S.; Bleach, R.; Carney, S.; Fairlie, G.; Louca, L.A. New Guidance on the Design of Offshore Structures to Resist the Explosion Hazard. In Proceedings of the ASME 2003 22nd International Conference on Offshore Mechanics and Arctic Engineering, Cancun, Mexico, 8–13 June 2003; pp. 111–117.

36. Zhang, D.; Paterson, E.G. A study of wave forces on an offshore platform by direct CFD and Morison equation. *E3S Web Conf.* **2015**, *5*, 04002. [CrossRef]

37. Corr, R.B.; Tam, V.H.Y. Gas Explosion Generated Drag Loads in Offshore Installations. *J. Loss Prev. Process. Ind.* **1998**, *11*, 43–48. [CrossRef]

38. Dusenberry, D.O. *Handbook for Blast-Resistant Design of Buildings*; John Wiley & Sons: Hoboken, NJ, USA, 2010.

39. ASCE. *Design of Blast-Resistant Buildings in Petrochemical Facilities*, 2nd ed.; American Society of Civil Engineers: Reston, VA, USA, 2010.

40. Nourzadeh, D.; Humar, J.; Braimah, A. Comparison of Response of Building Structures to Blast Loading and Seismic Excitations. *Procedia Eng.* **2017**, *210*, 320–325. [CrossRef]

41. Yasseri, S.F. *An Approximate Method for Blast Resistant Design*; FABIG Newsletter: Ascot, UK, 2002.

42. Mohamed Ali, R.M.; Louca, L.A. Performance-based design of blast resistant offshore topsides, Part II: Modelling and design. *J. Constr. Steel Res.* **2008**, *64*, 1046–1058. [CrossRef]

43. Goel, M.D.; Matsagar, V.A. Blast-Resistant Design of Structures. *Pract. Period. Struct. Des. Constr.* **2014**, *19*, 4014007. [CrossRef]

44. DNV. Design against Accidental Loads. In *DNV-RP-C204*; Det Norske Veritas; DNV: Oslo, Norway, 2010. Available online: https://rules.dnvgl.com/docs/pdf/DNV/codes/docs/2010-10/RP-C204.pdf (accessed on 16 December 2019).

45. Bowerman, H.; Owen, G.W.; Rumley, J.H.; Tolloczko, J.A. *Interim Guidance Notes for the Design and Protection of Topside Structures Against Explosion and Fire*; SCI-P-112/299UK; Steel Construction Institute: Ascot, UK, 1992.

46. Unified Facilities Criteria (UFC) 3-340-02. Structures to Resist the Effects of Accidental Explosions. In *US Department of the Army, Navy and Air Force Technical Manual*; U.S. Department of Defense: Washington, DC, USA, 2008.

47. Yasseri, S.F. Iso-Damage Diagrams for Blast Resistant Design. In *FABIG Newsletter April 2005*; FABIG Newsletter: Ascot, UK, 2005.

48. Austrell, P.E.; Dahlblom, O.; Lindemann, J.; Olsson, A.; Olsson, K.G.; Persson, K.; Wernberg, P.A. *CALFEM: A Finite Element Toolbox Version 3.4*; Division of Structural Mechanics, LTH: Lund, Sweden, 2004.

49. Kim, S.E.; Lee, D.H. Second-Order Distributed Plasticity Analysis of Space Steel Frames. *Eng. Struct.* **2002**, *24*, 735–744. [CrossRef]

50. N-004. Design of Steel Structures, No. N-004, Rev. 2. In *NORSOK Standard*; McGraw-Hill: New York, NY, USA, 2004.

MDPI

St. Alban-Anlage 66

4052 Basel

Switzerland

Tel. +41 61 683 77 34

Fax +41 61 302 89 18

www.mdpi.com

Applied Sciences Editorial Office

E-mail: applsci@mdpi.com

www.mdpi.com/journal/applsci